T0215776

Six Sigma for Organizational Excellence

K. Muralidharan

Six Sigma for Organizational Excellence

A Statistical Approach

 Springer

K. Muralidharan
Department of Statistics
Faculty of Science
M. S. University of Baroda
Vadodara, Gujarat
India

ISBN 978-81-322-3434-0   ISBN 978-81-322-2325-2   (eBook)
DOI 10.1007/978-81-322-2325-2

Springer New Delhi Heidelberg New York Dordrecht London
© Springer India 2015
Softcover reprint of the hardcover 1st edition 2015

Printed on acid-free paper

Springer (India) Pvt. Ltd. is part of Springer Science+Business Media (www.springer.com)

Preface

Organizations run successful business only when they provide satisfaction to consumers. Competitiveness in quality is not only central to profitability, but also crucial to business survival. Consumer should not be required to make a choice between price and quality of products. Manufacturing and service organizations exist if they learn how to manage quality. In today's tough and challenging business environment, the development and implementation of a comprehensive quality policy is not merely desirable, it is essential. *Six Sigma* is a business process improvement tool to achieve customer satisfaction through a systematic problem-solving approach. It is uniquely driven by close understanding of customer requirements and reinventing business processes. It facilitates people excellence as well as technical excellence in terms of creativity, collaboration, communication, dedication and above all increases the accountability of what one does in an organization.

The Six Sigma philosophy works under a five-phase improvement cycle, called DMAIC, where D for define, M for measure, A for analysis, I for improvement and C for control. It can apply to both process improvement and product improvement or even design redesign efforts. A Six Sigma initiative includes enterprise resource planning (ERP), e-commerce and services, lean manufacturing, customer relationship management systems, strategic business partnerships, knowledge management, activity-based management, just-in-time inventory and globalization. Organizational excellence is a result of continuous improvement, which can attain only through systematic reduction of defects and variations in the process activities. Here comes the importance of Six Sigma methodology.

Who Will Read?

This book serves three main purposes: first, an academic book for students and teachers; second, this book can be used as a reference material for engineers and managers working as Six Sigma professionals and Black Belt trainers; third, this book could be a user manual for practitioners and project consultants. The emphasis is laid on understanding and applying the concepts of quality through project management and technical analysis by using statistical methods. The contents are prepared in a ready-to-use form with continuity established for each phases of Six Sigma project. This will help practitioners to implement the Six Sigma projects without any hurdles. Three most important aspects of Six Sigma project—Sigma estimation, sample size calculation and Sigma-level estimation—are separately treated in different chapters. Necessary tables, graphs, descriptions and checklists are provided to ease the referencing of tools and techniques. The concepts are critically assessed, reasoned and explained to enable their uses in managerial decision making. The objectives of each chapter and its continuity with subsequent chapters are also clearly established for a smooth reading. Charts and plots, a number of worked-out examples, case studies and necessary tables are provided for better understanding of the concepts.

Students of undergraduate, postgraduate and research students can make optimum use of the integrated concepts of quality engineering and management tools of statistics. The science of Six Sigma project management, integrated through engineering concepts, is explained through statistical tools and that is the uniqueness of this book. The book could also serve as a concise book for Six Sigma Green Belt, Black Belt and Master Black Belt training.

Inspiration

The content is based on the author's own teaching experience, lecture notes, research publications, private communications, book references, article citations and training and consulting materials. The content is highly inspired with some available books: The Six Sigma Way by Peter S. Pande et al. (2003), Juran's Quality Planning and Analysis for Enterprise Quality by Frank M. Gryna et al. (2008), Lean Six Sigma Statistics by Alastair Muir (2005), The Certified Six Sigma Black Belt Handbook by T.M. Kubiak and Donald W. Benbow (2009), Introduction to Statistical Quality Control by Douglas C. Montgomery (2009), and Statistical Process Control by John S. Oakland (2012). The author has consulted a number of other books on Six Sigma, lean, management, engineering and general statistical books for integrating the things, as required by Six Sigma professionals.

About the book

This book integrates three main disciplines: Science, Engineering and Management. The author has tried to maintain a balance of these three disciplines from a practitioner's point of view. Chapter 1 is like an introduction to various Six Sigma concepts practiced by professionals and organizations, globally. Various perceptions and their implementation styles are critically examined in this chapter. Chapter 2 details the importance of Six Sigma project management concepts. The necessity of model-based projects is statistically emphasized. Apart from this, the importance of quantitative project management and its risk assessment and critical evaluation are also discussed in this chapter. The importance of process-based projects and models are included in Chap. 3. This is followed by the understanding of the process variation, which is an essential part of a Six Sigma project. The sources of identifying variation and possible identification of variations are discussed in Chap. 4. Since Sigma is being considered as a measure of variation, the estimation of Sigma is a vital issue in a statistical study. This is considered in detail in Chap. 5. A number of methods of estimation of Sigma is considered in the chapter. Chapter 6 details one of the important issues of project management, and that is the sample size determination. From a practitioner's point of view, many simple and easy-to-implement methods of sample size calculations are presented in the chapter. Chapters 7–11 discuss the Six Sigma philosophy, systematically, detailing with the necessary tools and techniques of statistics to execute a project. The management quality improvement topics covered in these phases are SIPOC, voice of customer, value stream mapping, brainstorming, root-cause analysis, failure-mode effect analysis, seven quality tools, Kaizen, 5S, designed for Six Sigma, quality function deployment, understanding of defect per unit (DPU) and defect per million opportunities (DPMO), Sigma-level estimation, cost of poor quality, etc.

Statistical topics include descriptive statistics, the basic notions on probability and probability distributions, point and interval estimation of parameters, parametric and nonparametric testing, correlation and regression techniques, design of experiments including factorial experiments, control charts, etc. A care has been taken to show every method in a simple and practical way without involving any rigorous theoretical steps involved. The SQC/SPC part of the project management is given the maximum emphasis, as they naturally become the essential tools to improve and control the phase of the philosophy. Most of the tools and techniques are explained through numerical and illustrative examples. The performance of a Six Sigma project is evaluated through its Sigma level. The understanding of DPU/DPMO and short-term and long-term variability of a process, etc., are the requirements for the Sigma-level estimation. This is carried out in Chap. 12. One can also refer to this chapter in the beginning of the actual project execution for setting up the goal and can be used to baseline the performance evaluation. This chapter will be handy for those involved with project evaluation and target setting at any time during the project.

The methods for continuous improvement is presented in Chap. 13, where the author has discussed various quality improvement programs offered by Deming, Juran, Feigenbaum, Crosby, Ishikawa, Taguchi, etc. Chapter 14 offers the importance of Six Sigma marketing, which is a growing area of research in Six Sigma philosophy. A Six Sigma marketing is a fact-based data-driven disciplined approach to growing market share by providing targeted product/markets with superior value. Various issues associated with Six Sigma marketing, like strategic, tactical, and operational processes of marketing, are discussed in the chapter. The chapter on Green Six Sigma emphasizes the importance of Six Sigma projects from a sustainable business practices. Green Six Sigma is nothing but the qualitative and quantitative assessment of the direct and eventual environmental effects of all processes and products of an organization. The activities involve the systematic usage of infrastructure and manpower, optimum use of technology and accountability of sustainable business practices. The benefits of Green Six Sigma are also detailed in the chapter.

The pros and cons of Six Sigma are presented in Chap. 16. A detailed discussion on advantages and disadvantage; various limitations, dos and don'ts of Six Sigma are also discussed in the chapter. The concern about the future of Six Sigma is also given at the end of the chapter. Chapter 17 is allotted to the discussion of case studies.

Apart from this, a separate session on "Relevance for Managers" is also added at the end of each chapter to increase the usefulness of each tools and methods. The citations and references for each chapter are given at the end of each chapter. Although Microsoft® Excel®, Minitab® and R® software have been used in the book for preparing charts and plots, this is not a prerequisite for using this book.

Vadodara, Gujarat, India   K. Muralidharan

Acknowledgments

This book has come to fruition because of the generosity shown by many professional colleagues, friends and well-wishers. I express my deep sense of gratitude and regards to all of them. I sincerely acknowledge Prof. Subha Chakraborthy (University of Alabama), Prof. W.H. Woodall (Virginia Tech University), Prof. D.C. Montgomery (Arizona State University) and Prof. Jiju Antony (Heriot-Watt University) for their helpful contributions to the content of the book. I sincerely thank the head, Department of Statistics, University of Pune, for permitting me to use the statistical tables. In putting together the professional perspectives, I place on record many university department heads, colleagues and librarians for their assistance in materials and academic collaborations.

My colleagues at the Maharajah Sayajirao University of Baroda have been very supportive and cooperative in rendering their assistance towards this project. Two of my colleagues Dr. (Mrs.) Khimiya Tinani and Dr. (Mrs.) Rupal Shah need special mention as they have done an excellent job of verifying and correcting the problems and their solutions in the book. I sincerely acknowledge Prof. Venkateswarlu and Prof. J.P. Parikh at the Department of English, Maharajah Sayajirao University of Baroda, for their assistance towards reading the material for language, grammar and syntax. I also place on record Dr. Aarti Mujumdar and Dr. Milan Sagar of Cambridge Education for providing their professional editorial service.

I also thank Prof. B.K. Kale (University of Pune) and Prof. Ashok Shanubhogue (Sardar Patel University), as they remain as a constant inspiration to me to take up the task and assignments throughout my professional career. Thanks are also due to my industry friends Mr. Rakesh Singh (from Trendz Process Consulting, Hyderabad), Mr. Rajesh Nambiar (Panacea Software, Baroda) and Mr. Vimal Vyas (Siemens, Vadodara) for enlightening me on the importance of Six Sigma concepts in industries and manufacturing industries.

I thank my wife Lathika and sons Vivek and Varun for their patience and silence, rendered throughout the writing of the book.

If the readers find that proper acknowledgement has not made to any particular individuals, authors, references, citations, etc., kindly treat this it is not intentional and request that you bring the same to my notice for future mention.

Contents

About the Author

K. Muralidharan is professor and head of the Department of Statistics, Faculty of Science, Maharajah Sayajirao University of Baroda, Vadodara. As well as director of the Population Research Centre, Maharajah Sayajirao University, Professor Muralidharan is an adjunct faculty at IIT Gandhinagar. He did post-doctoral fellowship from the Institute of Statistical Science, Academia Sinica, Taiwan. He is an internationally certified Six Sigma Master Black Belt from Indian Statistical Institute, Bangalore.

Abbreviations

AIAG	Automotive Industry Action Group
AIC	Akaike information criterion
AMA	American Marketing Association
ANOVA	Analysis of variance
AP	Action Plan
ASQ	American Society for Quality
BB	Black belts
BIC	Bayesian information criterion
BIBD	Balanced incomplete block design
BIS	Bureau of Indian Standards
BSC	Balanced score card
CA	Comparative Analysis
C&E	Cause and effect
CED	Cause and effect diagram
CI	Confidence interval
CL	Control limit
CLT	Central limit theorem
CMM	Capability maturity model
CMMI	Capability maturity model integration
COPIS	Customer-Output-Process-Input-Supplier
COQ	Cost of quality
COPQ	Cost of poor quality
COV	Covariance
CPM	Critical path method
CRD	Completely randomized design
CRT	Current reality tree
CSF	Critical success factors
CTC	Critical to cost
CTD	Critical to delivery
CTQ	Critical to quality
CTP	Critical to process
CTS	Critical to safety

CUSUM	Cumulative sum
CV	Coefficient of variation
CVM	Customer value mapping
DER	Debt-equity ratio
DFSS	Designed for Six Sigma
DIDOV	Define-Identify-Design-Optimize-Verify
DMAIC	Define-Measure-Analyze-Improve-Control
DMADV	Define-Measure-Analyze-Design-Validate
DMEDI	Define-Measure-Explore-Develop-Implement
DOE	Design of experiments
DPO	Defects per opportunity
DPMO	Defects per million opportunities
DPU	Defect per unit
EDA	Exploratory data analysis
ELEV	Empirical limited expected value
EMAS	Eco-Management and Audit Scheme
EMS	Environmental management standards
ERP	Enterprise Resource Planning
ESS	Error sum of squares
EWMA	Exponentially weighted moving average
FCF	Free cash flow
FMEA	Failure mode effect analysis
FTA	Fault tree analysis
GB	Green belts
GOFT	Goodness-of-fit-test
GRPI	Goals-roles and responsibilities-processes-procedures
GSS	Green Six Sigma
ID	Interrelationship diagram
IDEA	Identify-Define-Evaluate-Activate
IPO	Input-Process-Output
IQR	Inter-quartile range
ISO	International Organization for Standardization
IT	Information technology
ITES	Information technology enabled services
JUSE	Union of Japanese Scientists and Engineers
KBD	Key business drivers
KMI	Key marketing indicators
KPI	Key performance indicators
KPIV	Key process input variables
KPOV	Key process output variables
LCL	Lower control limit
LER	Loss elimination ratio
LMAD	Launch-Manage-Adapt-Discontinue
LSD	Latin square design
LSL	Lower specification limit

LSS	Lean Six Sigma
MAD	Median absolute deviation
MBB	Master black belts
MBO	Management by objectives
MD	Mean deviation
MIS	Management information system
MRL	Mean residual life
MSA	Measurement system analysis
MSE	Mean square error
MTBF	Mean time between failures
MVLUE	Minimum Variance Linear Unbiased Estimate
MVUE	Minimum variance unbiased estimate
NVA	Non value added
PERT	Project (or program) evaluation and review technique
PBIBD	Partially balanced incomplete block design
PDCA	Plan–Do–Check–Act
PM	Project management
PPM	Parts per million
PVM	Product value mapping
QC	Quality control
QCI	Quality council of India
QE	Quality engineering
QFD	Quality function deployment
QM	Quality management
QPM	Quantitative process management
QD	Quartile deviation
RBD	Randomized block design
RCA	Root cause analysis
ROI	Return of investment
R&R	Repeatability and reproducibility
RPN	Risk priority number
SC ED	Stratification Cause and Effect Diagram
SD	Standard deviation
SIC	Schartz information criterion
SIPOC	Supplier-Input-Process-Output-Customer
SMART	Specific-Measurable-Achievable-Relevant-Timely
SOP	Standard operating procedures
SOW	Statement of work
SPC	Statistical process control
SQC	Statistical quality control
SS	Six Sigma
SSM	Six Sigma Marketing
SWOT	Strength-Weakness-Opportunities-Threats
TLA	Time Line Analysis
TOP	Total opportunity

TPM	Total productivity maintenance
TQC	Total quality control
TQM	Total quality management
TRIZ	Theory of inventive problem solving
UCL	Upper control limit
UMVUE	Uniformly minimum variance unbiased estimate
UAPL	Understand-Analyze-Plan-Launch
UMC	Unit manufacturing cost
USL	Upper specification limit
VA	Value added
VAM	Value analysis mapping
VSM	Value stream mapping
VIF	Variance inflation factor
VOC	Voice of customer
VOP	Voice of process
WIP	Work in progress
WIQ	Work in queue
WTA	Willingness to accept
WTP	Willingness to pay

Chapter 1
Six Sigma Concepts

1.1 Introduction

Six Sigma is a disciplined, project-oriented, data-driven approach and a methodology for eliminating defects in a process—from manufacturing to transactional and from product to service. It is a management philosophy attempting to improve effectiveness and efficiency. It is uniquely driven by close understanding of customer requirements and reinventing business processes. The approach relies heavily on advanced statistical methods that complement the process and product knowledge to reduce variation in processes [6, 8, 12, 14]. The concept, if implemented properly, helps to reduce the defects significantly and exceptionally raises the level of performance. It is a cornerstone of strategic planning to provide first-class service and products to customers. More detailed reviews and extensive Six Sigma bibliographies were provided by Brady and Allen [3] and Nonthaleerak and Hendry [25]. For a recent overview of Six Sigma, one may refer to the article by Montgomery and Woodall [20].

The main goal of Six Sigma is to identify, isolate, and eliminate variation or defects in a manufacturing or transaction process. Six Sigma is a business process tool to achieve customer satisfaction by reducing variations. To make a product defect free, it is essential to minimize the variation of a process. At its core, Six Sigma revolves around few key concepts like the following:

- *Defect*: Failure to deliver what the customer wants
- *Variation*: What the customer sees and feels
- *Critical to Quality*: Those attributes most important to the customer
- *Stable Operations*: Ensuring consistent, predictable processes to improve what the customer sees and feels
- *Process Capability*: What your process can deliver
- *Design for Six Sigma*: Designing to meet customer needs and process capability

© Springer India 2015
K. Muralidharan, *Six Sigma for Organizational Excellence*,
DOI 10.1007/978-81-322-2325-2_1

Six Sigma as a quality improvement tool advocates the practice of measuring variability of a process, which may then be controlled by continuous improvement. In fact, the application of Six Sigma begins with translating a practical problem into a statistical one. An optimal solution is found using appropriate statistical tools and then implemented as a practical solution to real-life situation [12].

At General Electric, Six Sigma is a measurement. "The Six Sigma quality initiative, very briefly, means going from approximately 35,000 defects per million operations, which is average for most companies, including GE, to fewer than four defects per million in every element in every process that this company engages in every day" (General Electric Company) [13].

The other perceptions of Six Sigma concept as proposed by various authors are as follows:

"Six Sigma is a quality initiative that employs statistical measurements to achieve 3.4 defective parts per million—the virtual elimination of errors" [23].
"Six Sigma is a comprehensive, statistics-based methodology that aims to achieve nothing less than perfection in every single company process and product" [27].
"Six Sigma alters the paradigm from fixing defective products to correcting the process so that perfect products are made" [17].
"A Six Sigma initiative is designed to change the culture in an organization by way of breakthrough improvement in all aspects of the business" [5].
"Six Sigma is a program that combines the most effective statistical and non-statistical methods to make overall business" [28].
"Six Sigma is a highly disciplined process that helps us focus on developing and delivering near-perfect products and services. The central idea behind Six Sigma is that you can measure how many defects you have in a process, you can systematically figure out how to eliminate them and get as close to "zero defects" as possible. Six Sigma has changed the DNA of GE—it is the way we work—in everything we do in every product we design" (General Electric at www.ge.com).
"Six Sigma is a business improvement approach that seeks to find and eliminate causes of mistakes or defects in business processes by focusing on process outputs that are of critical importance to customers" [30].
"Six Sigma is a useful management philosophy and problem-solving methodology but it is not a comprehensive management system".
"Six Sigma is a methodology with accompanying highly structured processes using efficient statistical approaches for acquiring, assessing, and applying the customer, competitor, enterprise, and market intelligence to produce superior product, process and enterprise innovations and designs with the goal of creating a sustainable competitive advantage" [19].

To this list of definitions, I would like to add a practical definition underlying the necessity of Six Sigma process improvement.

"A Six Sigma initiative is a customer focused problem-solving approach with reactive and proactive improvements of a process leading to sustainable business practices. The sustainable business practices include innovation, improvement, competition, environmental compliance, customer satisfaction, and growth of the organization."

The above definition entails organization to undergo structured problem-solving approach through proper data collection and deliver the expected customer satisfaction. The growth of the organization may be valued in terms of its financial gain, stakeholder confidence, employee retention, productivity, and resource utilization. This definition also warrants the importance and necessity of dedicated people who can improve a process with zero variation and sustain the improvements for a long period of time ensuring the success of a Six Sigma initiative.

Technically speaking, Six Sigma is described as a data-driven approach to reduce defects in a process or cut costs in a process or product, as measured by "six standard deviations" between the mean and the nearest specification limits. "Sigma" (or σ) is the Greek letter used to describe variability or standard deviation, such as defects per units. Figure 1.1 shows a normal distribution of a population, with its mean (μ) in the center and a data point on the curve indicating one standard deviation (1 σ) to the right of the mean.

How well a desired outcome (or target) has been reached can be described by its average or mean (μ), which is nothing but the sum of all data points divided by the number of data points. The standard deviation (σ) describes how much variation actually exists within a data set, which is calculated as the square root of the variation from the mean. A detailed discussion on the measures of standard deviation and the estimation procedures are given in Chap. 5.

If a process is described as within "Six Sigma," the term quantitatively means that the process produces fewer than 3.4 defects per million opportunities (DPMO). That represents an error rate of 0.0003 %; conversely, that is a defect-free rate of 99.9997 %. Note that the sigma measure compares your performance to customer requirements (defined as target), and the requirements vary with the type of industry or business.

Fig. 1.1 A normal distribution

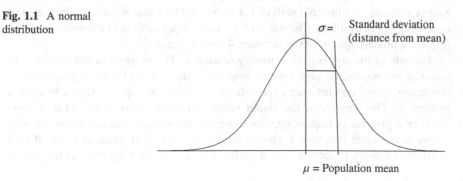

$\sigma =$ Standard deviation (distance from mean)

$\mu =$ Population mean

Table 1.1 DPMO for different sigma level

Off-centering	3 σ	3.5 σ	4 σ	4.5 σ	5 σ	5.5 σ	6 σ
0	2700	465	63	6.8	0.57	0.038	0.002
0.25	3557	665	99	11.7	1.09	0.0805	0.0047
0.5	6442	1382	236	32	3.4	0.29	0.019
0.75	12,313	2990	578	88.5	11	1.02	0.1
1	22,782	6213	1350	233	32	3.4	0.29
1.25	40,070	12,225	2980	577	88.4	10.7	1
1.5	66,811	22,750	6210	1350	233	32	3.4
1.75	105,651	40,059	12,224	2980	577	88.4	11
2	158,656	66,807	22,750	6210	1350	233	32

The sigma value of a process describes the quality level of that process. A quality level of K sigma exists in a process when the half tolerance of the measured product characteristic is equal to K times the standard deviation of the process [14]:

$$K \times \text{process standard deviation} = \text{half tolerance of specification}$$

However, this definition alone does not account for the centering of a process. A process is centered when $X = T$, where X is the process average or mean and T is the target value, which is typically the midpoint between the customer's upper specification limit (USL) and the lower specification limit (LSL). A process is off-centered when the process average, X, does not equal the target value T. The off-centering of a process is measured in standard deviations or sigma. Table 1.1 presents the DPMO corresponding to different Sigma levels under various off-centering of the process.

Note that the value or number of defects of a process is a function of the sigma value (quality level) of the process. The true value of the quality level of a process is the number of defects that occur when the process is centered; that is, when the off-centering value is 0 sigma. In the case of Six Sigma, there are 0.002 defects per million or 2 defects per billion. On the other hand, "Motorola's concept of 6 sigma allows a shift in the mean of 1.5 sigma" [9]. Therefore, Motorola's value of Six Sigma assumes an allowable shift of 1.5 sigma and thus also shows a defect rate not exceeding 3.4 per million. The value of 3.4 defects per million in a centered process implies a process quality level between 4 and 5 sigma.

The role of the sigma shift is mainly academic. The purpose of Six Sigma is to generate organizational performance improvement. It is up to the organization to determine, based on customer expectations, what the appropriate sigma level of a process is. The purpose of the sigma value is a comparative figure to determine whether a process is improving, deteriorating, stagnant, or non-competitive with others in the same business. It should not be the goal of all processes. Geoff [10] justifies that every process does not perform as well in the long term as they do in

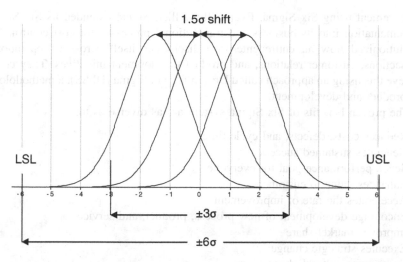

Fig. 1.2 A Six Sigma process with 1.5 σ shift

the short term. As a result, the number of sigma's that will fit between the process mean and the nearest specification limit may well drop over a period of time. To account for this real-life increase in process variation over time, an empirically based 1.5 sigma shift is introduced into the calculation. According to this idea, a process that fits 6 sigma between the process mean and the nearest specification limit in a short-term study will in the long term fit only 4.5 sigma—either because the process mean will move over time, or because the long-term standard deviation of the process will be greater than that observed in the short term, or both. This is explained in Fig. 1.2.

Hence, the widely accepted definition of a Six Sigma process is a process that produces 3.4 defective parts per million opportunities (DPMO). This is based on the fact that a process that is normally distributed will have 3.4 parts per million beyond a point that is 4.5 standard deviations above or below the mean. So the 3.4 DPMO of a Six Sigma process in fact corresponds to 4.5 sigma, namely 6 sigma minus the 1.5 σ shift introduced to account for long-term variation. This allows for the fact that special causes may result in deterioration in process performance over time and is designed to prevent underestimation of the defect levels likely to be encountered in real-life operation.

A successful implementation of a Six Sigma approach requires companies to consider changes in methodologies across the enterprise, introducing new linkages and communications. Three benchmark examples of how Six Sigma permeates a corporate philosophy and becomes a business initiative can be found by studying Motorola, Allied Signal, and General Electric (GE). Motorola created Six Sigma as a rallying point to change the corporate culture to compete better in the Asia-Pacific telecommunications market. Motorola's main focus was on manufacturing and defect reduction. Allied Signal rebuilt its business with bottom-line cost improvement using Six Sigma. Eventually, Allied Signal extended its Six Sigma implementation into its business and transactional processes for cost control. GE revolutionized how an entire enterprise disciplines itself across its operations, transactions, customer relations, and product development initiatives. They could achieve this using an approach called Design for Six Sigma (DFSS), a methodology for product and development.

The proven benefits of Six Sigma system are as diverse as to:

- Reduces costs, defects, and cycle time
- Generates sustained success
- Sets a performance goal for everyone
- Enhances value to customers
- Accelerates the rate of improvement
- Encourage development of new process, product, and service
- Improves market share
- Executes strategic change
- Encourages cultural change
- Promotes learning and cross-pollination.

For a service industry, it helps to impart best services and skills. It facilitates people excellence as well as technical excellence in terms of creativity, collaboration, communication, dedication, and above all increases the accountability of what one does in an organization [14, 22, 26].

1.2 Six Sigma Methodology

The Six Sigma philosophy works under a structured problem-solving approach. The problems generally concerned with eliminating variability, defects, and waste in a product or process all undermine customer satisfaction. The working philosophy is generally called define–measure–analyze–improve–control (DMAIC), where

- *Define*—Define the problem or project goals that need to be addressed
- *Measure*—Measure the problem and process from which it was produced
- *Analyze*—Analyze data and process to identify the root causes of defects and opportunities
- *Improve*—Improve the process by finding solutions to fix, diminish, and prevent future problems
- *Control*—Implement, control, and sustain the improvements and solutions to keep the process on the new course

The DMAIC philosophy applies to both process improvement and product improvement or even design–redesign efforts. The key elements in a DMAIC project are team discipline, structured use of metrics and tools, and execution of a well-designed project plan that has clear goals and objectives. Lean Six Sigma (LSS) modifies the DMAIC approach by emphasizing speed. Lean focuses on streamlining a process by identifying and removing non-value-added steps. A leaned production process eliminates waste. Target metrics include zero wait time, zero inventory, scheduling using customer pull, cutting batch sizes to improve flow, line balancing, and reducing overall process time. Lean Six Sigma's goal is to produce quality products that meet customer requirements as efficiently and effectively as possible.

If a process cannot be improved as it is currently designed, another well-known Six Sigma problem-solving approach can be applied. The DMADV process is used to fundamentally redesign such a process. This five step approach is as follows:

- *Define*—Define the problem and/or new requirements
- *Measure*—Measure the process and gather the data that are associated with the problem or in comparison with the new requirements
- *Analyze*—Analyze the data to identify a cause-and-effect relationship between key variables
- *Design*—Design a new process so that the problem is eliminated or new requirements are met
- *Validate*—Validate the new process to be capable of meeting the new process requirements

A second redesign approach has been developed to incorporate elements from a Lean Six Sigma approach—the DMEDI process. This is similar to DMADV and adds tools from the Lean methodology to ensure efficiency or speed. The steps are as follows:

- *Define*—Define the problem and/or new requirements
- *Measure*—Measure the process and gather the data that are associated with the problem or in comparison with the new requirements
- *Explore*—Explore the data to identify a cause-and-effect relationship between key variables
- *Develop*—Develop a new process so that the problem is eliminated or new requirements are met
- *Implement*—Execute the new process under a control plan

Another variant of Six Sigma improvement cycles is designed for Six Sigma (DFSS) approach. DFSS ensures that the new product or service meets stakeholder's needs, provides a high level of performance, and is robust to process variations. While DMAIC is a reactive approach fixing past problems, DFSS is used in a proactive way to avoid problems in the first place. DFSS approach enables launching new products and services on time and on budget and target gaining incremental revenues sooner, achieving greater market share, and ensuring that the company generates differentiated products and services that target customer and stakeholder needs. DMADV is one of the important methodologies used in DFSS. This methodology works with the following five steps:

- *Define*—Identify the Customer and project
- *Identify*—Define what the customers want, or what they do not want

- *Design*—Design a process that will meet customer's needs
- *Optimize*—Determine process capability and optimize design
- *Verify*—Test, verify, and validate design

The concept of Six Sigma at GE deals with measuring and improving how close the company comes in delivering on what it planned to do. Six Sigma provides a way for improving processes so that the company can more efficiently and predictably produce world-class products and services. Specifically, the DMAIC according to Paul [27] is as follows:

- *Define*—Who are the customers and what are their priorities?
- *Measure*—How is the process measured and how is it performing?
- *Analyze*—What are the most important causes of defects?
- *Improve*—How do we remove the causes of the defects?
- *Control*—How can we maintain the improvements?

1.3 Six Sigma Tools

Most of the decision-making tools for Six Sigma are borrowed from the subject statistics. These tools along with the management inputs are systematically used in the organizational process to get the maximum yield. Some of the tools may be integrated in two or more phases to establish the initial performance of the process and some of them are exclusively used for confirmatory analysis and continuous improvement. The DMAIC phasewise tools are presented in Fig. 1.3. It is expected that a basic- to advance-level knowledge is essential to carry out proper analysis and interpretation.

Apart from the above tools, some of the tools used in DMADV that are not covered in DMAIC are as follows: Multi-Generational Project Plans (MGPP), Analytical Hierarchy Process (AHP), Kano Analysis, KANSEI Engineering, TRIZ (Theory of inventive problem solving), Pugh Analysis, Taguchi Optimization, and Capability Maturity Model Integration (CMMI) and Simulation methods etc.

Six Sigma is being utilized in almost every industry these days, like manufacturing industries, automotive and aerospace industries, IT/ITES development industries, process industries, pharmaceutical industries, etc. As discussed earlier, Six Sigma is as much about people's excellence as it is about technical excellence. Creativity, collaboration, and dedication are infinitely more powerful than a corps of super-statisticians. Thus, a Six Sigma process can inspire and motivate better ideas and performance from people and create synergy between individual talents and technical prowess. Very often, the significant financial savings from Six Sigma may exceed in value by the intangible benefits. In fact, the changes in attitude and enthusiasm that come from improved processes and better informed people are often easier to observe, and more emotionally regarding than dollar savings.

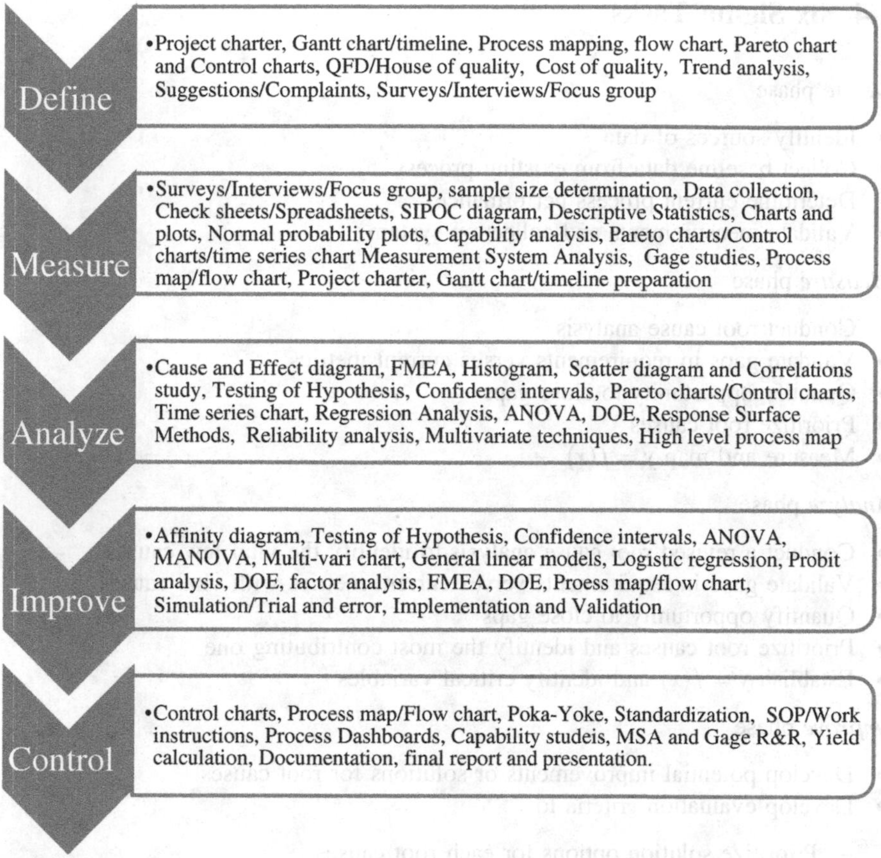

Define
- Project charter, Gantt chart/timeline, Process mapping, flow chart, Pareto chart and Control charts, QFD/House of quality, Cost of quality, Trend analysis, Suggestions/Complaints, Surveys/Interviews/Focus group

Measure
- Surveys/Interviews/Focus group, sample size determination, Data collection, Check sheets/Spreadsheets, SIPOC diagram, Descriptive Statistics, Charts and plots, Normal probability plots, Capability analysis, Pareto charts/Control charts/time series chart Measurement System Analysis, Gage studies, Process map/flow chart, Project charter, Gantt chart/timeline preparation

Analyze
- Cause and Effect diagram, FMEA, Histogram, Scatter diagram and Correlations study, Testing of Hypothesis, Confidence intervals, Pareto charts/Control charts, Time series chart, Regression Analysis, ANOVA, DOE, Response Surface Methods, Reliability analysis, Multivariate techniques, High level process map

Improve
- Affinity diagram, Testing of Hypothesis, Confidence intervals, ANOVA, MANOVA, Multi-vari chart, General linear models, Logistic regression, Probit analysis, DOE, factor analysis, FMEA, DOE, Proess map/flow chart, Simulation/Trial and error, Implementation and Validation

Control
- Control charts, Process map/Flow chart, Poka-Yoke, Standardization, SOP/Work instructions, Process Dashboards, Capability studeis, MSA and Gage R&R, Yield calculation, Documentation, final report and presentation.

Fig. 1.3 DMAIC phasewise tools

A Six Sigma initiative includes Enterprise Resource Planning (ERP), e-Commerce and Services, Lean Manufacturing, Customer Relationship Management (CRM) Systems, Strategic business Partnerships, Knowledge Management, Activity-based Management, Just-in-time inventory and Globalization. For Six Sigma measures to be applied effectively across the organization, strict guidelines need to be followed. Otherwise, the measurement system and identification of critical-to-quality (CTQ) characteristics will be redundant, ultimately leading to a total chaos in the organization.

In the next two subsections, we give the Six Sigma tasks and deliverables for each phase in a concise form.

1.4 Six Sigma Tasks

Define phase

- Identify sources of data
- Collect baseline data from existing process
- Determine current process performance
- Validate measurements and collection system

Measure phase

- Conduct root cause analysis
- Validate gaps in requirements versus current metrics
- Quantify opportunity to close gaps
- Prioritize root causes
- Measure and map $y = f(x)$

Analyze phase

- Conduct a revised root cause analysis to identify the vital few causes
- Validate gaps in requirements versus current metrics with vital causes
- Quantify opportunity to close gaps
- Prioritize root causes and identify the most contributing one
- Establish $y = f(x)$ and identify critical variables

Improve phase

- Develop potential improvements or solutions for root causes
- Develop evaluation criteria to

  - Prioritize solution options for each root causes
  - Examine solutions with a short-term and long-term approach.
  - Weigh the costs and benefits of "quick-hit" versus more difficult solution options

- Select and implement the improved process and metric
- Measure results
- Evaluate whether improvements meet targets
- Evaluate for risk

Control phase

- Document a new or improved process and measurements
- Validate collection systems and the repeatability and reproducibility of metrics in an operational environment
- Define the control plan and its supporting plans like the following:

  - Communications plan of the improvements and operational changes to the customers and stakeholders
  - Implementation plan

- Risk management plan
- Cost/benefit plan
- Change management plan

- Train the operational stakeholders
- Establish the tracking procedures in an operational environment:

- Monitor implementation
- Validate and stabilize performance gains
- Jointly audit the results and confirm final financials.

1.5 Six Sigma Deliverables

Define phase

- Preparation of project charter and statement of work (SOW), which includes

- Process and problem
- Scope and boundaries
- Team, customers, and critical concerns
- Improvement goals and cost of poor quality (CoPQ)

- Gantt chart/timeline preparation
- Process map/flowchart
- Stepwise documentation and follow-up steps
- Exit Review

Measure phase

- Initial Sigma level and cost calculations
- Process capability analysis
- Measurement System Analysis and Gage R&R
- Gantt chart/timeline
- SIPOC diagram/high level process map
- Stepwise documentation and follow-up steps
- Exit Review

Analyze Phase

- Identify root causes using cause-and-effect diagram, and through statistical analysis
- Validate root causes
- Stepwise documentation and follow-up steps
- Exit Review

Improve Phase

- Select root causes and countermeasures
- Improve implementation plan
- Validate solutions or suggest improvement using statistical analysis
- Stepwise documentation and follow-up steps
- Exit Review

Control Phase

- Prepare control plan for

  - Tolerances, controls, and measures
  - Charts and monitor
  - Standard operating procedures (SOP)

- Design a response plan, design re-design of control plans
- Validated in-control process and benefits for

  - process capability
  - MSA and Gage R&R
  - documentation

- Stepwise documentation and follow-up steps
- Exit review

1.6 Lean Six Sigma

Lean Six Sigma is a process improvement program that combines two ideas: *Lean*—
a collection of techniques for reducing the time needed to provide products or
services—and Six Sigma—a collection of techniques for improving the quality by
combining the two. Lean Six Sigma is a proven business management strategy that
helps organizations operate more efficiently. As we know, a business process is a
series of interconnected sub-processes. The tools and focus of Six Sigma is therefore
to fix processes. Lean concentrates on the interconnection between the processes.
The central theme of Six Sigma is defect reduction, where the root causes of the
defects are examined and improvement efforts are focused on those causes.

In Lean Six Sigma (sometimes referred to Lean or Lean manufacturing), the
emphasis is on reducing costs by eliminating product and process waste [29]. It also
focuses on eliminating non-value-added activities such as producing defective
product, excess inventory charges due to work-in-process and finished goods
inventory, excess internal and external transportation of product, excessive
inspection, and idle time of equipment or workers due to poor balance of work steps
in a sequential process. The goal of Lean manufacturing has long been one of the

goals of industrial engineering [11]. According to Muir [21], the two philosophies can be summed up as: "Reduce the time it takes to deliver a defect-free product or service to the customer." It is not the question of whether the customer wants it right or quickly, but both.

The operating philosophy of Lean is to eliminate waste through continuous improvement. This is achieved through the following:

- defining value from the client's perspective
- identifying the value stream
- only making what the client pulls
- keeping the flow moving continuously and
- always improving the process

As seen earlier, the Six Sigma process improvement works with the philosophy of DMAIC techniques. In Lean Six Sigma, it develops with the identification of the problems and ends with retaining the benefits of the program. Thus, the steps followed in Lean Six Sigma are R–DMAIC–S, where the additional R stands for recognize and S stands for sustain. It is important to identify the significant gaps in a business problem by articulating the business strategy. This will decide what the management wants to deliver. Once a Six Sigma project completes through the DMAIC procedure, the management will be able to identify the tangible benefits coming out of the project and its immediate impact can be assessed for retaining similar such projects.

In order to achieve the required target, a process flow is necessary to be established. A perfect process has continuous flow, as products, services, and knowledge are transformed continuously without delay from step to step. A flow is created by eliminating queues and stops and improving process flexibility and reliability. The other aspects of establishing the continuous flow is by identifying the value-added and non-valued activities in a process. Any activities that add no value to the client are by definition "waste." The mapping of all activities and steps according to their time of occurrence required bringing a product, service or capability to the client is called Value Stream Mapping (VSM).

Implementing Lean will address waste and its root causes. Waste points us to problems within the system. Lean also focuses on efficient use of equipment and people and minimizes issues by standardizing work. A Lean cost model says that decreased cost always leads to increased profit. If the focus of Six Sigma is to reduce variation in a process, then the focus of Lean Six Sigma is to reduce waste. Hence, the tools for both philosophies differ depending on the business issue to be resolved.

Interestingly, both methodologies follow DMAIC problem-solving approach, Lean/Kaizen events follow a faster improvement cycle than Six Sigma, and Lean pursues constant and continuous improvement until all non-value-added activities are eliminated.

Six Sigma characteristics

- In-depth root cause analysis and solutions
- Builds highly trained and skilled staff
- Used for solving more complex, larger issues
- Strong, positive results take longer time to achieve
- Robust infrastructure

Lean Six Sigma characteristics

- Speed and flexibility
- Involves all employees
- Positive results in short time frame
- Focused on smaller-scale projects
- Less scientific: often trial and error

According to many business analysts and quality improvement experts, Lean Six Sigma is the most popular business performance methodology in the history of corporate development of quality of products and services, substantially contributing to increased customer satisfaction. GE under Jack Welch was a leading proponent of Lean Six Sigma globally. Genpact (at that time known as GE Capital International Services) was the first service provider in the world to apply this methodology at scale for business processes, making this practice a tremendous success. Lean Six Sigma permeates what we do and is highly visible in our operations, people, processes, and leadership direction (http://www.genpact.com/home/about-us/lean-six-sigma-dna).

1.7 Six Sigma: The Belt Systems

A Six Sigma project is executed through a coordinated effort with highly responsible personnel of organizations. The success of any Six Sigma project also depends on these personals. The system adopts different belt training to facilitate this. The main roles associated with a specific belt of a Six Sigma projects can be summarized as follows:

> Green Belt (GB): A Six Sigma role associated with an individual who retains his or her regular position within the firm but is trained in the tools, methods, and skills necessary to conduct Six Sigma improvement projects either individually or as part of larger teams. GB training is typically for one or two weeks, and they generally assist on major project teams or lead teams engaged in smaller, more highly specific projects.
>
> Black Belts (BB): A Six Sigma role associated with an individual who is typically assigned full time to train and mentor Green Belts as well as lead improvement projects using specified methodologies such as define, measure, analyze, improve, and control (DMAIC); define, measure, analyze, design, and

verify (DMADV); and desiged for Sigma (DFSS) tools. BBs typically have a minimum of four weeks of specialized training, sometimes spread over a four-month period and usually combined with concurrent work on a Six Sigma project. They lead teams that are focused on projects with both quality and business impact for the organization. In most organizations, BBs train GBs and work on other functions such as new project identification.

Master Black Belt (MBB): A Six Sigma role associated with an individual typically assigned full time to train and mentor Black Belts as well as lead the strategy to ensure improvement projects chartered are the right strategic projects for the organization. MBBs are usually the authorizing body to certify Green Belts and Black Belts within an organization. They often write and develop training materials, are heavily involved in project definition and selection, and work closely with business leaders called Champions. The BBs and MBBs have specialized training and education on statistical methods and other quality and process improvement tools that equip them to function as team leaders, facilitators, and technical problem solvers. These days, many national and international organizations offer education in Six Sigma quality training and offer belt certifications. The American Society for Quality (ASQ) maintains a Six Sigma Black Belt body of knowledge on their Web site (see http://www.asq.org/certification/six-sigma/bok.html).

The role of Champions is to ensure that the right projects are being identified and worked on, that teams are making good progress, and that the resources required for successful project completion are in place. Champions are project sponsors. Champions are generally guided by project managers in the organization. The appointments of higher officials are altogether the organization's own matter of policy (for more discussion on these aspects, see Chap. 7). The flow of involvement and responsibilities described above is the essence of how Six Sigma has been implemented to date and how implementation is changing the organization in terms of its growth and financial gains. A current trend consistent with administration of quality and certain management functions is to push responsibility to lower levels within organizations. How this applies to implementation of Six Sigma is through greater responsibility for problem or opportunity identification, data collection, and analysis, and corrective action is being levied on Green Belts. In order to facilitate this, many organizations initiate Yellow Belts and even White Belts training for beginners to be ready for larger roles in Six Sigma projects and activities. They are generally trained for technical aspects of management decision-making opportunities.

Since Six Sigma focuses on reducing defects as a top priority for quality improvements [16], it is important here to note that often the large savings obtained from Six Sigma efforts are savings from reducing the costs of poor quality—obtained by "extracting gold in the mine," as Juran had already said 50 years ago. Further, the focus on processes and on eliminating variation has certainly increased knowledge about variation, an important part of Deming's Profound Knowledge system [7]. In recent years, also a focus on reducing lead times is also emphasized as a side effect of process improvement.

It is with these requirements, new variants of Six Sigma concepts have appeared. Some of them are FIT SIGMA [2], Ultimate Six Sigma, and Strategic Six Sigma. Furthermore, it is fundamentally important to understand what the customer wants and needs and to use this information to guide R&D efforts on existing products or design of new ones [18]. While an increasing number of organizations are engaging in DFSS, it must be stressed that DFSS is hard work, requiring a relatively imposing amount of expertise—and it is still relatively new—so that there is probably "more talk than work" done with respect to DFSS application.

Hoerl [15] states that "perhaps the most critical question about the future of Six Sigma is when it will begin to wind down and perhaps morph into something else." While Six Sigma will evolve over time, as in the case of TQM has and will, there are some core strengths of Six Sigma that will be maintained so whatever "the next big thing" is, it will look at least vaguely familiar to Six Sigma. Some of these core strengths are the use of infrastructure to supply the core people, money and other resources, freeing top talent to work on new initiatives, and, of course, reliance on senior leadership commitment.

According to Woodall [31], and Montgomery and Woodall [20], the healthcare service is one of the potential area, where Six Sigma can play significant role in improving the services and delivery. For continued success of Six Sigma, it is necessary to incorporate a broader array of statistical methods, particularly those that are more appropriate for the increasing amount of data available in applications. Nair et al. [24], for example, pointed out the need for methods that allow the effective analysis of functional data and spatiotemporal data. The analysis of functional data in process monitoring was recently reviewed by Woodall et al. [32]. There are an increasing number of applications where somewhat more sophisticated statistical methods, such as these, are required in Six Sigma applications.

1.8 Relevance for Managers

Six Sigma can take on various definitions across a broad spectrum depending on the level of focus and implementation. A management can perceive it either as a philosophy, a set of tools, a methodology, or as a set of metrics; Six Sigma is all of this. This chapter details all these points in an exhaustive manner highlighting the importance of each from a project management point of view. The essence of Six Sigma project is the necessity of defect reduction to the minimization of variation and then quality improvement. Therefore, the tools, tasks, and deliverables essential for the implementation of Six Sigma project are clearly suggested for better management and business excellence.

Exercises

1.1. What is Six Sigma? What are the objectives of Six Sigma?
1.2. Explain various perceptions of Six Sigma concept.
1.3. State various tasks and deliverables associated with each phase of DMAIC.
1.4. What is defects per million opportunities? How it helps to find the sigma level of a process?
1.5. Distinguish between a centered and off-centered process? What adjustment is done to find the sigma level of such a process?
1.6. What are the benefits of Six Sigma quality philosophy?
1.7. Explain DMAIC philosophy by detailing various statistical tools applied in each phase.
1.8. Distinguish between Lean and Six Sigma. How does the operating philosophy of both differ?
1.9. What is design for Six Sigma? Explain, how the technique is applied for organizational excellence?
1.10. Explain various tasks associated with each phases of DMAIC technique.
1.11. Explain various deliverables associated with each phases of DMAIC technique.
1.12. What is the significance of Lean or Lean manufacturing? State important features of Lean philosophy.
1.13. What is value stream mapping? How does it help to speed up a Lean process?
1.14. What are the management roles of a Six Sigma Black Belt, Six Sigma Master Black Belt and Champion?

References

1. Basu, R.: Six sigma to operational excellence: role of tools and techniques. Int. J. Six Sigma Competitive Advantage 1(1), 44–64 (2004)
2. Basu, R., Wright, J.N.: Quality Beyond Six Sigma. Butterworth-Heinemann, Oxford (2003)
3. Brady, J.E., Allen, T.T.: Six sigma literature: a review and agenda for future research. Qual. Reliab. Eng. Int. 22(3), 335–367 (2006)
4. Breyfogle III, F.W.: Implementing six sigma-smarter solutions using statistical methods. Wiley, New York (1999)
5. Breyfogle, F., Cipello, J., Meadows, B.: Managing Six Sigma. Wiley Inter science, New York (2001)
6. Coronado, R.B., Antony, J.: Critical success factors for the successful implementation of six sigma projects in organization. TQM Mag. 14(2), 92–99 (2002)
7. Deming, W.E.: The New Economics for Industry, Government and Education. MIT Center for Advanced Engineering Study, Massachusetts (1993)
8. Eckes, G.: The Six Sigma Revolution. Wiley, New York (2000)
9. Evans, J.: Letters, Quality Progress (1993)
10. Geoff, T.: Six Sigma: SPC and TQM in Manufacturing and Services. Gower Publishing Ltd., Aldershot (2001)

11. Gryna, F.M., Chua, R.C.H., Defeo, J.: Juran's Quality Planning and Analysis for Enterprise Quality. Tata McGraw-Hill, New Delhi (2007)
12. Harry, M., Schroeder, R.: Six sigma: the breakthrough management strategy revolutionizing the world's top corporations. Doubleday Currency, New York (2000)
13. Hendericks, C., Kelbaugh, R.: Implementing six sigma at GE. J. Qual. Participation (1998)
14. Henderson, K.M., Evans, J.R.: Successful implementation of six sigma: benchmarking general electric company. Benchmarking Int. J. 7(4), 260–281 (2000)
15. Hoerl, R.: One perspective of the future of six sigma. Int. J. Six Sigma Competitive Advantage 1(1), 112–119 (2004)
16. Hong, G.Y., Goh, T.N.: A comparison of six sigma and GQM approaches in software development. Int. J. Six Sigma Competitive Advantage 1(1), 65–75 (2004)
17. Kane, L.: The quest for six sigma. In: Hydrocarbom Processing (International Edn), vol. 77, no. 2, 1998
18. Klefsjö, B., Wiklund, H., Edgeman, R.: Six sigma seen as a methodology for total quality management. Measuring Bus. Excellence 5(1), 31–35 (2001)
19. Klefsjö, B., Bergquist, B., Edgeman, R.L.: Six sigma and total quality management: different day, same soup? Int. J. Six Sigma Competitive Advantage 1–17 (2008)
20. Montgomery, D.C., Woodall, W.H.: An overview of Six Sigma. Int. Stat. Rev. 76(3), 329–346 (2008)
21. Muir, A.: Lean Six Sigma way. McGraw Hill, New York (2006)
22. Muralidharan, K.: Data mining: a subject to explore for management. Qual. Council Forum India 45, 1–2 (2010)
23. Murphy, T.: Close enough to perfect. Ward's Auto World 34(8) (1998)
24. Nair, V.N., Hansen, M., Shi, J.: Statistics in advanced manufacturing. J. Am. Statist. Assoc. 95 (451), 1002–1005 (2000)
25. Nonthaleerak, P., Hendry, L.C.: Six sigma: literature review and key future research areas. Int. J. Six Sigma Competitive Advantage 2(2), 105–161 (2006)
26. Pande, P.S., Newuman, R.P., Cavanagh, R.R.: The Six Sigma Way. Tata McGraw-Hill, New Delhi (2003)
27. Paul, L.: Practice Makes Perfect, CIO Enterprise, vol. 12 no. 7, Section 2, 15 Jan 1999
28. Pearson, T.A.: Measure for six sigma success. In: Quality Progress, vol. 34, pp. 35–40, Feb 2001
29. Shuker, T.J.: The leap to lean. In: Annual Quality Congress Proceedings, ASQ, Milwaukee, pp. 105–112 (2000)
30. Snee, R.: Why should statisticians pay attention to six sigma?: an examination for their role in the six sigma methodology. Qual. Progr. 32(9), 100–103 (1999)
31. Woodall, W.H.: Use of control charts in health-carer and public-health surveillance (with discussion). J. Qual. Technol. 38(2), 89–104 (2006)
32. Woodall, W.H., Spitzner, D.J., Montgomery, D.C., Gupta, S.: Using control charts to monitor product quality profiles. J. Qual. Technol. 36(3), 309–320 (2004)

Chapter 2
Six Sigma Project Management

2.1 Project Management

A *project* is a temporary endeavor to achieve some specific objectives in a defined time [10]. A project may vary in size and duration, involving a small group of people or large numbers in different parts of the organization. It is usually unique in content and unlikely to be repeated again in exactly the same way. *Project management* is a dynamic process that utilizes the appropriate resources of the organization in a controlled and structured manner to achieve some clearly defined objectives, identifies as strategic needs conducted within a defined set of constraints. A project involves many processes and each such process progresses with specific objectives (see Chap. 3 for details). A project can be any of the following types:

- *Personal projects*: Preparations for writing a thesis, books, dissertations; student projects; any family functions; conducting an examination; conducting a live show, arranging a tour program, etc.
- *Local projects*: Organization of a public program, organization of a conference or a seminar program, any voluntary projects executed by NGOs and private organizations, etc.
- *Organizational projects*: Construction of buildings or a highway, planning and launching a new product, setting up an automobile plant, establishing a new office, investigating cause and effect of a product's defects, brainstorming a session, organizing an audit check, etc.
- *Projects of national importance*: Launching a vaccination drive, launching a new satellite, introducing a literacy campaign/poverty removal, preparation of annual budget, construction of metro rail/road transports, conducting a national level sporting event, etc.
- *Projects of global importance*: Organizing peace missions, space explorations, environmental sustainability drives, conducting an international level sporting event, etc.

© Springer India 2015
K. Muralidharan, *Six Sigma for Organizational Excellence*,
DOI 10.1007/978-81-322-2325-2_2

A project is identified through a project identification process. This is the stage where new opportunities and threats emerging in the environment are investigated and suitable proposals that can be adopted by the organization are generated. This is done through the generation of new ideas by the company's think tank. *Brainstorming* is a very effective technique for doing this exercise in a group. Brainstorming may be structured in which each member of the group is asked for his/her idea in a sequential manner. These ideas are then scrutinized for their viability of execution and implementation thereafter.

Project characterization

- Every project involves various processes and these processes are characterized by their inputs and outputs
- The output of the business processes depends on the strategic planning, customer surveys, competitive analysis and benchmarking of the process, etc.
- For the success in projects, it is essential for everyone involved to commit to using a common set of processes and procedures
- This makes the sharing of information considerably easier, particularly when working across different sites, organizations, and countries.

2.2 SWOT Analysis

After having identified the objectives to be achieved through a project, it is generally worthwhile to conduct a strength–weakness–opportunities–threats (SWOT) analysis so that the organization's strengths and weaknesses are highlighted and the opportunities and threats emerging from the environment are viewed in an objective manner. The purpose of this analysis is to be able to generate ideas exploring the emerging opportunities, guarding against the threats while keeping the organization's strengths and weaknesses in mind. The participants of brainstorming may also be made a part of these exercises, so that they become aware of the requirements and limitations of the system they are dealing with. The extent of conformance of the various proposed solutions with the SWOT profile could also be used to evaluate the various ideas after they are proposed during brainstorming. Some of the factors that ought to be considered while doing a SWOT analysis are as follows:

Strengths

- Experience and expertise
- Financial position
- Capital raising capability
- Industrial contacts
- Foreign collaborations

Weaknesses

- Lack of experience
- Lack of trained personnel
- Inability to cope with newer technologies
- Inability to raise huge investments
- Inability to forecast market trends

Opportunities

- Emerging technologies
- New products with new markets
- New processes with better features
- Special financing schemes
- Government and other incentives

Threats

- Competitors
- Poor state of the economy
- Outdated process and technology
- Unprofessional management skills
- New products and services.

Understanding the key customer, market, and operational conditions is input to setting strategic direction. Identifying these components through a SWOT analysis also helps identify the "gaps" in the projects (see also Thompson and Strickland [9]).

2.3 Project Phases

We know that every project involves various processes, and these processes are characterized by their inputs and output. In the previous chapter, we have discussed various components and characteristics of a process. The output of the business processes depends on the strategic planning, customer surveys, competitive analysis, benchmarking of the process, etc. For success in projects, it is essential for everyone involved to commit to using a common set of processes and procedures. This makes the sharing of information considerably easier, particularly when working across different sites, organizations, and countries. The process can be broken down into a number of definable phases with decision gates between each of the phases:

- Project conception
- Project definition
- Project planning

- Project launch and execution
- Project closure
- Post-project evaluation.

These activities are often referred to as "key strategies" as they may comprise several actual tasks carried out by more than one person. It is also expected that each phase is carried out sequentially to generate useful data for decision making. Although each phase is treated as discrete with specific work to be completed, this does not signify they are "one-off" activities. In reality, the phases are often revisited during a project. Once a project is initiated, the need to reiterate some or all of the work done in the definition or planning phases is always a possibility as the project moves ahead in the execution phase.

For many organizations, project management is a way of management for change. The idea for project will be created from the knowledge and experience of the organization, customer requirements, and market trend. The climate of project execution depends on the organizational culture, organizational structure, and business strategy. It is essential for senior managers to create and work continuously to sustain the climate for success. Failure to do this will be a disaster for the organization and the people associated with the projects. Collaborative working across the whole structure is a key to project success, as is recognition that assigning an individual to a project team is a dedicated assignment of the whole project.

Projects are an integral component of Six Sigma. Selecting, managing, and completing projects successfully are critical in deploying any systematic business improvement effort, not just Six Sigma. A project should represent a potential breakthrough in the sense that it could result in a major improvement in the product or service. Project impact should be evaluated in terms of its financial benefit to the business, as measured and evaluated by the finance or accounting unit. Obviously, projects with high-potential impact are the most desirable. This financial systems' integration is standard practice in Six Sigma and should be a part of any DMAIC project, even if the organization as a whole is not currently using Six Sigma [6].

2.4 Alignment with the Business Strategy

Projects and programs are selected only if they support achieving the business strategy and contribute to business growth. A carefully constructed business case is an essential document supporting the decision. Some of the important strategic inputs are as follows:

- Forward planning
- Resource management
- Financial management
- Portfolio management.

In order to start any new project, it is essential to know the commitments and liabilities of the organization in prior. This will decide the future of all ongoing activities and its successes. The selection of projects requires the organization to plan ahead using adequate intelligence gathered from the marketplace and customers. This will help to address the critical areas of potential business growth and lost opportunities. Adequate funding must be available to satisfy the budget of all active projects. Otherwise, the chances of derailment of the projects become certain. It is also important to maintain a visible, authorized list of active projects and those waiting to start, to inform every one of the priorities and relative importance of those on the list. Timescale and completion targets need to be agreed to meet the business and/or customer needs and plan the effective deployment of resources.

2.5 Project Stakeholders

The relationship in a project environment can only lead to success when there is a clear definition of ownership at each level in the organization with clearly defined roles and responsibilities. This avoids confusion and clarifies where authority exists to make decisions and avoid unnecessary slippage and delays in projects. According to Young [10], the people associated with a project can be:

- Someone who needs the benefits—the company senior management
- Someone who commits to provide the resources—the line managers
- Someone who is accountable for achieving the benefits—the sponsor
- Someone who is accountable for the project work—the project manager
- Someone who is responsible for the project work—the project team
- Someone who wants to use, influence or is affected by the outcomes—customer, the stakeholders.

Together this, whole group creates an infrastructure that is overlaid on the functional hierarchy, and their behavior collectively can determine the degree of success that is achievable with all projects. The influence of organizational culture has a significant impact on climate of a project. The behavior is strongly influenced by the perceptions people have of the internal climate. Some of the other obvious cultural influences are as follows:

- morale
- mutual trust, support and respect for decisions, openness, and integrity
- risk taking and optimism—recognition of risks and sharing in success
- freedom of action—through accountability, pride, and participation in decision making
- commitment—a sense of belonging, avoiding confusion with clear responsibilities
- collaboration—shared benefits, teamwork and mutual assistance, minimizing stress
- training—opportunities to learn both on and off the job.

Paying specific attention to these influences is important for any ongoing projects. It is not enough to blame the management if the climate is going wrong. Perceptions of the climate are always stronger in the staff than among the management. This, in no way, means that the staff is better than the management, but for a cohesive atmosphere, it is better to ensure each influence is given adequate attention.

2.6 Managing the Stakeholders

There are two groups of stakeholders for any publicly traded company [7]. They are company's shareholders or project sponsor and customers. A successful business is one that understands and meets the needs of both the groups. The first group of customers is the company's shareholders. They are mostly focused on the financial aspects of the business including the market share, growth, and profit. The second group is the customers paying for your business services or products. Customers make decisions based on value and quality. All the stakeholders have an open and a closed or hidden agenda about what they expect from the project. These expectations are finalized before scoping the project. This task may not be easy as there will be lots of pressures and influences coming from all the quarters of the project. The project team should have the fortitude to resist such pressures for the larger interest of the company.

Identifying stakeholders is not just part of the project start-up. As many appear later, the leader must review the list at regular intervals. The relative importance of each stakeholder changes with time and through the stages of the project. It is a serious risk to fail to cooperate with or recognize a stakeholder. Set the ground rules at the outset to control the poor stakeholders. The stakeholders are inside and outside the organization. It is a good idea to interact with the sponsors and customers to get involved with the identification of the stakeholders, since some stakeholders impact both. They can come from departments such as finance, sales and marketing, development, production, strategic, and production. They can be consultants, contractors, suppliers, government agencies, public representatives, and supply chain partners.

The involvement of the sponsor throughout the project is an essential success factor for any project as this individual has the authority and power to make decisions about money and the resources—the people that you need to get the work done. The sponsor cannot be effective if the individual has no authority in the organization. The sponsor is accountable for the project and therefore is the appointed guardian of the project on behalf of the organization. An effective sponsor can provide a significant amount of support through:

- Responding rapidly to issues requiring senior management decisions
- Sustaining the agreed priority of the project in the organization
- Ensuring the project stays focused on the organizations strategic needs

- Building a working relationship with the customer
- Influencing the peer-group to provide cross-organization resources and services on time for the project
- Demonstrating concern for success by visible leadership
- Influencing other stakeholders in the approval and sign-off of the phases of the project.

It is the responsibility of the team members to establish a good working relationship with the sponsor to benefit from the above supports. This can be achieved through frequent meeting and discussions held for shorter time in the organization.

2.7 A Six Sigma Project

The main objective of a Six Sigma project is to improve quality of the process by reducing variation. There are two approaches for selecting the right project: a model generating *quantitative* data and a model only generating *qualitative* data. A Six Sigma project is considered to be a quantitatively managed project. Some of the important issues need special attention while selecting a Six Sigma project is:

- Will the project maximize profits?
- Will the project maintain the market share?
- Will the project consolidate the market position?
- Will the project open up new markets?
- Will the project maximize profits?
- Will the project maximize utilization of existing resources?
- Will the project boost company's image?
- Will the project increase risk faced by the company?
- Is the project scope within the company's current skills and experience?

Most of the Six Sigma projects are the result of customer needs and expectations, and the projects are selected for their potential impact on business. The value opportunity of projects must be clearly identified and projects must be well aligned with corporate business objectives at all levels. At the highest level, the stockholders, top executives, members of the board of directors, and business analysts who guide investors typically are interested in return on equity, return on invested capital, stock price, dividends, earnings, products and patents, and development of future business leaders. At the business unit or operations level, managers and executives are interested in factory metrics such as yield, cycle time and throughput, cost reduction, safety of employees and customers, efficient use of assets, new product introduction, sales and marketing effectiveness, development of people, and supply chain performance (cost, quality, service). Aligning projects with both business unit goals and corporate-level metrics helps ensure that the best projects are considered for selection. The DMAIC approach is an extremely effective framework for meeting these requirements.

The first types of projects that companies usually undertake are designed to demonstrate the potential success of an overall improvement effort. The projects often focus on the areas of the business that are full of opportunities, but they also tend to be driven by current problems. Issues that are identified by customers or from customer satisfaction (or dissatisfaction) feedback, such as analysis of field failures and customer returns, sometimes are the source of these projects. Hence, project selection is probably the most important part of any business improvement process. Projects should be able to be completed within a reasonable time frame and should have real impact on key business metrics. This means that a lot of thought must go into defining the organization's key business processes, understanding their relationships, and developing appropriate performance measures [6, 8].

The results of any project, (especially, a Six Sigma project) depend on a reliable and quality data. Any data collected contain a mass of information. The problem is to extract that part of it that is relevant to the questions to be answered by the project, in the simplest and most understandable way. This essentially involves checking for pertinent patterns and anomalies in the data. This is the basic role of statistical models: to simplify reality in a reasonable and useful way, a way that you can empirically check with the data. No model is ever "true," but some models are more useful than others for a given data and questions. Irrespective of whether the model constituted by a quantitative or qualitative data, the model can provide:

- a parsimonious description or summary of results, highlighting important features
- a basis for prediction of future observations
- biological or social insight into the processes under study
- a test of a prior theoretical relationship
- comparison of results from different studies
- measures of precision of quantities of interest.

Six Sigma emphasizes the model in terms of $Y = f(X)$. It mathematically summarizes the fact that the output from a business process is a function of the decisions made by the process owners. A best model is that which is free of all irregularities and inconsistencies in the data leaving very little chance for assignable causes (man, machine, method, materials, and processes) variation. A general model building involves the following steps:

1. Studying the important descriptive statistics, in order to become familiar with the data
2. Developing a reasonable model from the results of step 1 and from previous knowledge
3. Fitting the model to the data
4. Checking the goodness of fit of the model
5. Going back to step 2, if necessary
6. Using the model to draw appropriate conclusions.

The purpose of modeling is not to get the best fit to the data, but to construct a model that is not only supported by the data but also consistent with previous knowledge, including earlier empirical research, and that also has a good chance of describing future observations reasonably well. Generally, a probability model and a regression model are sought for describing these types of situations (see Lindsey [4] for details).

2.7.1 Probability Model-Based Project

It is expected that the main response variable (Y) under study should be specified in the protocol. In most cases, it is directly observable, but in some experimental trials, it may be constructed—for example, the difference between the responses at baseline, before the intervention began, and the final response after a certain length of treatment. In statistical models, we consider the response variable to arise at random in a certain sense: That is, we cannot predict in advance exactly what response each respondent will give so that random fluctuations are not reproducible. This variability arises primarily from differences among human beings, in contrast to studies in physics or chemistry where measurements error is predominant.

When the value of a variable is subject to random variation, or when it is the value of a randomly chosen member of a population, then it is called a random variable. A description of the possible values of a random variable and of their corresponding probabilities of occurrence is the probability distribution or the probability model. A probability distribution is a mathematical function that smoothes the histogram of observations in an informative way, while retaining and highlighting the basic shape. A probability distribution is defined for both qualitative and quantitative data. Binomial (for binary responses), Poisson (for counts), geometric, hyper-geometric, etc., are probability models for qualitative and discrete data, and normal, exponential, beta, gamma, uniform, Weibull, lognormal distributions, etc., are the probability models for continuous and quantitative data. Most probability distributions have one or more unknown and unobservable *parameters* (not explanatory variables). Most distributions that have a parameter that indicates the size of the responses, generally the mean and some that have a second parameter related to the shape of the distribution, may be called the variance. Also there are distributions having three parameters, namely location, scale, and shape parameters.

A detailed discussion on various probability models is taken up in different phases of Six Sigma in subsequent chapters. The relevance of such models with respect to the deliverables of DMAIC phases is presented in the respective chapters.

2.7.2 Regression Model-Based Project

The probability distribution describes the random variability in the response variable. However, in many studies these variables can come with *systematic* changes in the response under certain conditions. They are called the explanatory variables

(X's). This situation can also be translated into a statistical model by looking at how the probability distribution of the response, or more exactly the parameters in it, change under these conditions. This process may need some general assumptions to be made on the variables in the model. For simplicity, we assume that the mean of the distribution changes with the conditions of interest and variance to remain constant under all conditions. Further by assuming a linear relationship between the response and explanatory variables, the model building becomes an easy job. But in reality, this will not happen every time and hence the necessity of regression model.

The two conditions put together will give the standard (multiple) linear regression model, whereby some function of the mean changes with the conditions:

$$g(\mu_i) = \beta_0 + \beta_1 x_{i1} + \beta_2 x_{i2} + \cdots \tag{2.1}$$

where μ_i is the mean for the ith subject, x_{ij} is the observation of the jth explanatory variable for that subject, and β_j is the corresponding unknown parameter, the *regression coefficient*, to be estimated. This model that combines some probability distribution with a linear regression has come to be known as a *generalized linear model*. These and other regression models are discussed in the analyze phase of the DMAIC methodology later.

2.8 Quantitative Project Management

A quantitative project management involves:

- Establishing and maintaining the project's quality and process performance objectives
- Identifying suitable sub-processes that compose the project's defined process based on historical stability and capability data found in process performance baselines or models
- Selecting sub-processes within the project's defined process to be statistically managed
- Monitoring the project to determine whether the project's objectives for quality and process performance are being satisfied, and identifying appropriate corrective action
- Selecting measures and analytic techniques to be used in statistically managing selected sub-processes
- Establishing and maintaining an understanding of the variation of selected sub-processes using selected measures and analytic techniques
- Monitoring the performance of selected sub-processes to determine whether they are capable of satisfying their quality and process performance objectives, and identifying corrective action
- Recording statistical and quality management data in the organization's measurement repository.

Organizational Process Performance: Process performance is a measure of actual results achieved by following a process. Process performance is characterized by process measures (e.g., effort, cycle time, and defect removal effectiveness) and product measures (e.g., reliability, defect density, capacity, response time, and cost). The common measures for the organization consist of process and product measures that can be used to characterize the actual performance of processes in the organization's individual projects. By analyzing the resulting measurements, a distribution or range of results can be established that characterize the expected performance of the process when used on any individual project. Data mining is one such topic which can be incorporated at this stage to analyze the existing pool of data and segregate for arriving at a process capability trend of the past which would enable us gauge the changes in the future.

Creating Organizational Baselines: The expected process performance can be used in establishing the project's quality and process performance objectives and can be used as a baseline against which actual project performance can be compared. This information is used to quantitatively manage the project. Each quantitatively managed project, in turn, provides actual performance results that become a part of baseline data for organizational process assets.

2.9 Project Risk Assessment

In project work, any event that could prevent the project realizing the expectations of stakeholders is a *risk*. A risk that happens becomes an issue that must receive prompt attention to maintain the project schedule on time. There are risks to all projects, and *risk management* is a method of managing a project that focuses on identifying and controlling the areas or events that have the potential of creating and causing unwanted changes leading to unwanted results. Because of the complexity of risks, it is impossible to derive a universal process for managing all risks in a project. Three important risks associated with any project are as follows:

- *Business risks*—the viability and context of the project
- *Project risks*—associated with the technical aspects of the work to achieve the required outcomes
- *Process risks*—associated with the project process, procedures, tools and techniques employed to control the project.

All projects inherently contain risk by default. The success depends on how well you manage the risks throughout the project. An efficient project manager should have the potential to identify and evaluate potential risks and the capacity to resolve any issues arising from risks that can happen anytime during the project.

PROBABILITY OF OCCURRENCE	IMPACT ON PROJECT		
	LOW (0.1 – 0.29)	MEDIUM (0.3 – 0.64)	HIGH (0.65 – 1.0)
LOW (0.1 – 0.29)	Low	Medium	High
MEDIUM (0.3 – 0.64)	Medium	High	Unacceptable
HIGH (0.65 – 1.0)	Medium	High	Unacceptable

Fig. 2.1 Risk ranking matrix

2.9.1 Quantifying the Risk

Once a list of risk is derived, work with the team using their experience to decide for each risk: the probability of occurrence and the impact on the project if it does happen. The probability of occurrence can be assessed on a scale of 0–1, where close to zero is considered as low and most unlikely to happen and 1 being very high and essentially to happen. The impact can be calculated as 0.1–0.29 being low—some effect on schedule, little effect on costs; 0.3–0.64 being considered as medium effect—less serious effect on the schedule, some effect on costs and 0.65–1 being high impact— significant effect on the schedule and project costs. Once a set of risks has been assessed for impact and probability of occurrence, one can rank them using a matrix with the parameters of *probability* and *impact* on the project as shown in Fig. 2.1.

The course of action can be initiated as follows:

- *Low risk*—Not expected to have any serious impact on the project. Review regularly for ranking and monitor.
- *Medium risk*—Significant impact on the project with possible impact on other projects. Not expected to affect a project milestone. Review at each project meeting and assess ranking. Monitor regularly to ensure it does not turn into a HIGH risk.
- *High risk*—Major impact on the project schedule and costs. Serious consequent impact on other related projects. Likely to affect a project milestone. Must be monitored regularly and carefully. Review possible mitigation actions you can take to reduce the ranking or minimize the impact.
- *Unacceptable risk*—The project cannot proceed without some immediate actions to reduce this risk ranking to lower the probability of occurrence, either with alternative strategies or making significant decisions about cost, schedule, or scope.

Clearly, any project allowed to proceed with many unacceptable risks is likely to be speculative, with serious potential for failure. By identifying such risks in this process, one can alert the sponsor and management to what you consider may be a safer alternative strategy. Once risks to the project have been identified action plans can be derived. A close monitoring is a key activity toward achieving success. If risks happen, they become issues that have a time-related cost impact. Unresolved issues do not disappear; they just accumulate and threaten to drown the whole project.

2.10 Critical Evaluation of a Project

The critical evaluation of a project is achieved through the analysis of critical path method (CPM) and project (or program) evaluation and review technique (PERT). The fundamental purpose is to enable one to find the shortest possible time in which to complete the project. This is done by the inspection of the *network diagram* or *logic diagram*. Enter the duration on to your notelets (nodes) in the network diagram for each key stage. Begin at the START nodes and trace each possible route or path through the diagram to the FINISH nodes, adding the duration of all the key stages in the path. The path that has the longest duration is the "critical path" of the project and takes the least time to complete the project. All the key stages on the critical path must, by definition, finish on time or the project schedule will slip.

Example 2.1 Consider a sample project with 14 activities and duration is as follows (Table 2.1):

The PERT method of critical path planning and scheduling is the most commonly used technique for project management control. It is based on representing the activities in a project by boxes (or nodes) that contain essential information calculated about the project. The inter-dependencies between the activities are represented by arrows to show the flow of the project through its various paths in the logic diagram. The PERT diagram is identical to the logic diagram, where each

Table 2.1 A sample project

Activity	Immediate predecessors	Duration (months)
A	–	2
B	–	6
C	–	4
D	B	3
E	A	6
F	A	8
G	B	3
H	C, D	7
I	C, D	2
J	E	5
K	F, G, H	4
L	F, G, H	3
M	I	13
N	J, K	7

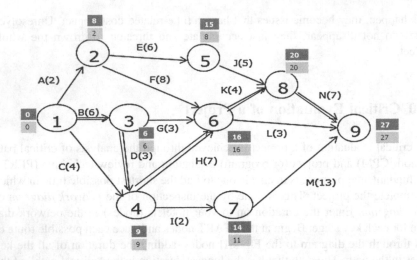

Fig. 2.2 A network diagram

notelet for a key stage representing a node. The analysis of the logic diagram or
network diagram is a simple logical process extending the initial calculation you
made earlier to locate the critical path (see Fig. 2.2 for the network diagram for the
above project). The two steps involved in the critical path calculations are as
follows:

1. Adding durations from start to finish—the *forward pass*
2. Subtracting the durations from finish to start—the *backward pass*

The critical path is that which corresponds to the events, where the time to start
and finish coincides. For the above project, the critical path is 1–3–4–6–8–9. Notice
that CPM is used for deterministic durations (that is, each activity is given a fixed
duration), whereas PERT is used for probabilistic durations (that is, each activity is
assumed to follow a logical probability distributions). PERT was developed primarily to simplify the planning and scheduling of large and complex projects. It
was developed for the US Navy Special Projects Office in 1957 to support the US
Navy's Polaris nuclear submarine project [5]. The advantages of PERT are as
follows:

- PERT chart explicitly defines and makes visible dependencies (precedence
  relationships) between the work breakdown structure (WBS) elements
- PERT facilitates identification of the critical path and makes this visible
- PERT facilitates identification of early start, late start, and slack for each activity
- PERT provides for potentially reduced project duration due to better understanding of dependencies leading to improved overlapping of activities and
  tasks where feasible
- The large amount of project data can be organized and presented in diagram for
  use in decision making.

See also Kerzner [2] and Klastorin [3] for details.

Since project schedules change on a regular basis, CPM allows continuous monitoring of the schedule, which allows the project manager to track the critical activities, and alerts the project manager to the possibility that non-critical activities may be delayed beyond their total float, thus creating a new critical path and delaying project completion.

In order to allow managers to prioritize activities for the effective management of project completion, and to shorten the planned critical path of a project by pruning critical path activities, one may use the techniques such as *fast tracking* (i.e., performing more activities in parallel), and/or by *crashing* the critical path (i.e., shortening the durations of critical path activities by adding resources). *Crash duration* is a term referring to the shortest possible time for which an activity can be scheduled [1]. It is achieved by shifting more resources toward the completion of that activity, resulting in decreased time spent and often a reduced quality of work, as the premium is set on speed. Crash duration is typically modeled as a linear relationship between cost and activity duration; however, in many cases a convex function or a step function is more applicable.

2.11 Role of Computing Technology in Project Management

Computing technology is a valuable aid in project management. Apart from providing computational support in a whole range of network scheduling calculations, it has made possible the generation and distribution of online reports for effective monitoring and control. Since a large number of activities and individuals are involved in a project, keeping everyone up to date and involved is itself a difficult task. This has been made easier through e-mail, the intranet, and the internet. This simplifies the coordination between the head office and multiple sites working on different environments. Moreover, it encourages the practice of green management of infrastructure and environment, which is the need of the hour. With these kinds of easier and cost-effective methods, one can enhance the efficiency of the project and save huge amount of manpower and finances. Plenty of computer softwares are available these days for enabling project management. Two such important softwares are the Microsoft Project and Primavera. The major advantages of such softwares are as follows:

- easy sorting and listing of activities
- easy updating and new listings of project progress over the life cycle
- advanced analysis and reporting can be done automatically
- decision making can be done effectively as per the resources available
- modification and alterations of constraints can be done in accordance with project priority.

2.12 Launch and Execution Process

The launch and execution process of a business project involves the following
activities:

- Derive the key stage work plans—use WBS, critical path, etc., to decide
  duration and float
- Establish the milestone schedule—use to resolve the risks and issues
- Create a communication plan to
  - understand current progress of the active tasks
  - identify the problems encountered
  - identify the technical difficulties being encountered
- Decide meetings schedule
- Derive change request process
- Hold launch meeting
- Initiate project execution.

2.13 Closure of the Project

Project completion is signified by:

- All finished tasks
- Agreed deliverables completed
- Testing completed
- Training materials prepared
- Equipment installed and operating
- Documentation manuals finished
- Process procedures finished and tested
- Staff training finished.

2.14 The Climate for Success

A successful Six Sigma project depends on its climate of success. Success is
defined as "attainment" of object, or of wealth, fame, or position. Synonyms of
success are victory, accomplishments, achievement, prosperity, attainment, fruition,
winning, etc. Success depends on who is measuring the project—sponsor, project
manager, project team, resource managers, customers, etc. How each one contribute
to success or failure is key to your management of the project.

The climate for success is generally influenced by organizational culture, organizational structure, and business strategy. However, the cultural influences of success depend on the following:

- morale
- mutual trust, support and respect for decisions, openness, and integrity
- risk taking and optimism—recognition of risks and sharing in success
- freedom of action—through accountability, pride, and participation in decision making
- commitment—a sense of belonging, avoiding confusion with clear responsibilities
- collaboration—shared benefits, teamwork and mutual assistance, and minimizing stress
- training—opportunities to learn both on and off the job.

Along with a favorable climate, the sponsor support is also crucial for success project management. They can be achieved through

- Responding rapidly to issues requiring senior management decisions
- Sustaining the agreed priority of the project in the organization
- Ensuring the project stays focused on the organizations strategic needs
- Building a working relationship with the customer
- Influencing the peer-group to provide cross-organization resources and services on time for the project
- Demonstrating concern for success by visible leadership
- Influencing other stakeholders in the approval and sign-off of the phases of the project.

The watch for potential failure is a continuous activity that must be the responsibility of everyone involved not just the project manager. Risk management processes are an essential and integral part of project management and will help reduce the probability of failure. Creating the platinum version of the product or services is ambitious and often more complex.

2.15 Relevance for Managers

Managing a project of various size and volume requires every stakeholder's attention and involvement. This commitment should flow throughout the project period starting from project conception to post-project evaluation. For a Six Sigma project to be relevant, it is essential to have a proper organizational culture, organizational structure, business strategy, and stakeholder confidence. A Six Sigma project is generally classified into two categories: project involving model uncertainties (probability models) and prediction models (regression-based models). These models are generally handy for managers, as they can facilitate both inferential and predictive results.

This chapter also discusses the issues related to risk assessment (project risk, process risk and, business risk), quantifying the risks and propose methods for minimizing the risks. The two important tools for evaluating the project, namely PERT and CPM are also included for a better understanding of the project. These tools are technically sound and support many management issues of project management. The role of computing technology for effective monitoring and control of the project is also included here. In fact, modern day businesses and Six Sigma projects demand the use of computing and information technology in all areas of its implementation.

Exercise

2.1. What is a project? Discuss various types of projects and their characteristics.
2.2. What is SWOT analysis?
2.3. Distinguish between a qualitative project and quantitative project.
2.4. Discuss various project phases associated with a project study.
2.5. What are the methods of managing stakeholders in a project? How do they impact the overall success of the project?
2.6. Discuss various features and characteristics of a Six Sigma project.
2.7. What are the benefits of model-based projects?
2.8. What are the characteristics of a quantitative project management?
2.9. What are the risks associated with a project? How do they quantify?
2.10. Discuss various methods of evaluating a project.
2.11. Distinguish between CPM and PERT.
2.12. Discuss the role of computing technology in project management.

References

1. Hendrickson, C., Tung, A.: Advanced Scheduling Techniques. Project Management for Construction. cmu.edu. Prentice Hall (2008)
2. Kerzner, H.: Project Management: A Systems Approach to Planning, Scheduling, and Controlling, 8th edn. Wiley, New York (2003)
3. Klastorin, T.: Project Management: Tools and Trade-offs, 3rd edn. Wiley, New York (2003)
4. Lindsey, J.K.: Revealing Statistical Principles. Arnold Publishers, New York (1999)
5. Malcolm, D.G., Roseboom, J.H., Clark, C.E., Fazar, W.: Application of a technique for research and development program evaluation. Oper. Res. 7(5), 646–669 (1959)
6. Montgomery, D.C., Woodall, W.H.: An overview of Six Sigma. Int. Stat. Rev. 76(3), 329–346 (2008)
7. Muir, A.: Lean Six Sigma Way. McGraw-Hill, New York (2006)

8. Snee, R.D., Rodebaugh Jr, W.F.: The project selection process. Qual. Prog. **35**(9), 78–80 (2002)
9. Thompson Jr, A.A., Strickland III, A.J.: Strategic Management, 10th edn. McGraw-Hill, New York (1998)
10. Young, T.: Successful Project Management. Kogan Page, London (2010)

Chapter 3
Six Sigma Process

3.1 Introduction

A process is a structured, measured set of activities also called inputs (X's) designed
to produce a specified output (Y's) for a particular customer or market. It implies a
strong emphasis on how work is done within an organization. A business process is
a set of logically related tasks performed to achieve a defined business outcome.

Processes are generally identified in terms of beginning and end points and
organization units involved, particularly the customer unit. Examples of processes
are as follows: developing a new product, ordering goods from a supplier, creating a
marketing plan, processing and paying an insurance claim, and so on.

Processes may be defined based on three dimensions:

- *Entities*: Processes take place between organizational entities: inter-organiza-
tions, inter-functional, or interpersonal
- *Objects*: Processes result in manipulation of physical or informational objects
- *Activities*: Processes could involve two types of activities: managerial (e.g.,
develop a budget) and operational (e.g., fill a customer order)

A work within an organization is completed by a single process or various
processes. Issuance of a purchase order is completed with a transaction process.
Admission of a patient in a hospital is completed with a registration process. Since
business processes contain several major steps, there are opportunities for breakdowns. That is, steps may not be completed the same way each time the process is
initiated. This inconsistency is termed *variation*.

One of the principle goals of Six Sigma or any other quality improvement
program is to reduce process variation, that is, to develop an approach to limit the
variation and more tightly focus on the process so as to produce the same results
over a long period of time. To reduce this process variation, you need a clear
understanding of what the process is and how it works. Mapping a process helps to
identify the flow of events in the process as well as the inputs and outputs in each

© Springer India 2015   39
K. Muralidharan, *Six Sigma for Organizational Excellence*,
DOI 10.1007/978-81-322-2325-2_3

step. The easy part is defining what goes into a process and the desired results. The hard part is trying to figure out the variables between input and output, which also called functions, say $f(x)$. A graphical representation of this mapping is shown in Fig. 3.1.

Thus, a Six Sigma process enables one to figure out which of the X's or variables in the business process and inputs have the biggest influence on the Y or the results and then use the changes in the overall performance of the process to adjust the business and keep it moving on a profitable path. In Six Sigma terminology, the X's and Y's have a variety of meanings. Some descriptions are as follows:

Y can mean

- a strategic goal
- any customer requirement
- profits
- customer satisfaction
- overall business efficiency

X can mean

- essential actions to achieve strategic goals
- quality of the work done by the business
- key influences on customer satisfaction
- process variables such as staffing, cycle time, amount of technology, etc., and
- quality of the inputs to the process from the customers or suppliers

Thus, the product results (Y) are function ($f(x)$) of many process variables (X's). The Six Sigma process identifies the process variables that cause variation in product results. Some of these process variables are critical to quality, are set at a certain value, and are maintained within a specified range (hence the name controllable variable). Other variables cannot be easily maintained around a certain value and are considered uncontrollable and are called *noise* variables (for more details, see [5]).

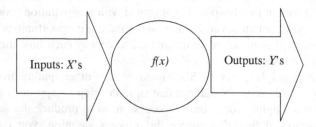

Fig. 3.1 A Six Sigma process

By using Six Sigma methods to understand the system and the variables, a company can learn to monitor and respond to the feedback from its processes, suppliers, employees, customers, and competitors to achieve higher levels of strength and performance.

Some examples of transaction processes are as follows: order generation, order fulfillment, open enrollment process, identifying customer requirements, engineering change process, collection of money, invoicing customers, scheduling logistics, equipment maintenance process, acquisition, admission process, inventory control, hospital management, and library management.

The quality of any process depends on the scope of the process and its deliverables. In the examples discussed above, there can be either a single or multiple deliverables and hence, dealing with a large process having multiple tasks will be monitored for any kind of uncertainties and failures. The selection of the processes is based on the critical success factors of the organization (see Fig. 3.2). Examples of critical success factors are reduction of product development time, increased perception of value, and higher yields. Candidate processes can then be ranked by assessing the importance of the process with regard to the critical factors and also the current process performance [2].

The relevance of each factor is entered in the body of the matrix (e.g., 5 is high relevance of the process to the critical success factor, 3 is moderate, and 1 is low relevance). The count is the total of these relevance ratings. The current performance is a rating of how the process is performing now (e.g., 1 is good performance, 3 is fair, and 5 is poor). The count or relevance is then multiplied by the current performance ratings to get the total score for the candidate process. Processes with the highest total would be likely selections for the formal process management (PM) approach.

Key Business processes	Critical Success Factors						Count	Current Performance	Total
	Product quality	Supplier quality	Empowered, skilled employees	Customer satisfaction	Lowest cost of poor quality	Lowest delivered cost			
	1	2	3	4	5	6			
A									
B									
C									
D									
E									
...									
...									

Fig. 3.2 Critical success factors (*Source* Gryna et al. [2], p. 198)

Documentation is the other area of concern where quality gets diluted for want of reliable data. Thus, a quality process is that which minimizes the non-value-added activities in the organization thereby increasing customer satisfaction. Two of the best tools for enabling this are the statistical process control (SPC) and statistical quality control (SQC) techniques. The "control" in both the above tools involves a universal sequence of steps: choosing the control subject, choosing a unit of measure, setting a goal, creating a sensor, measuring performance, interpreting the difference between actual performance and the goal, and taking action on the difference. A Six Sigma process does all these activities in a structured way.

SPC is the method of data gathering and analysis to monitor processes, identify performance issues, and determine variability existing in the process. It is also a strategy for reducing variation in products, deliveries, processes, materials, attitudes, and equipments. The SQC on the other hand is also an improvement technique carried out on analyzed data for understanding the performance of the process. The issues related to capability analysis, yield analysis, sigma-level calculations are performed as a part of SQC. Both these techniques help us to identify the controllable variables which affect the quality of the process.

A Six Sigma process is best described by the following components:

- Process characterization
- Process perception
- Process capability
- Process performance
- Process improvement
- Process control

Below we describe each component in detail with respect to a Six Sigma philosophy.

3.2 Process Characterization

A process involves the underlying characteristics of a good PM. PM is the totality of defined and documented processes, monitored on an ongoing basis, which ensure that measures are providing feedback on the flow and function of a process. The key measures include financial, process, people, innovation, etc. The objective of PM is thus to reduce the costs involved, defects, and cycle time; propose solutions; and suggest modifications and changes and the improvement necessary to the overall process. The improvement in a process can be suggested in the following way:

- Structured PM

    - Voice of customer—an organization's efforts to understand the customers' needs and expectations and to provide product and services that truly meet them
    - Voice of process—The 6σ spread between the upper and lower control limits as determined from a process

- Structured problem solving

  - Total quality management—overall quality improvement
  - Six Sigma—structured problem solving
  - Kaizen—continuous improvement
  - 5S—housekeeping and waste reduction.

- Structured documentation

  - Design of experiments
  - SQC techniques
  - SPC techniques.

- Structured content generation

  - Failure mode effect analysis
  - Root cause analysis
  - Quality function deployment
  - Value stream analysis.

Thus, PM is a system for identifying the critical business processes of an enterprise, focusing on "customer" requirements to reduce work in process and decisions in process. Understanding the inputs and outputs of cross-functional processes further improves the clarity of the project, and one can make assessment about the speed and flexibility of the ongoing project. It challenges us to:

- Do only those things that provide "value added": from the "customer's point of view"
- Move from managing functions to managing work processes
- Move from managing results (outcomes) to managing the processes that produces them
- Move from piece meal problem solving to system-wide improvement

Information about the process: One of the initial steps to understand or improve a process is to gather information about the important activities involved in a process. This can be done through a process map or flowcharts. Process mapping creates a picture of the activities that take place in a process. According to Oakland [4], it is important to consider the sources of information about the processes through some of the key issues as follows:

- Defining supplier–customer relationships
- Defining the process
- Standardizing procedures
- Designing a new process or modifying an existing one
- Identifying complexity or opportunities for improvement

The best way of understanding a process is to have a broad statement defining area of concern or opportunity, including its impacts and benefits to the organization. This statement should contain the details of those activities which are not going to be a part of the process and the potential improvements generated by

Fig. 3.3 Perceptions of a process

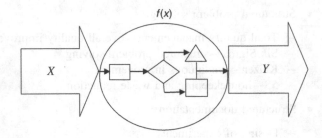

including a particular activity. It should also link to the business strategies and the tangible benefits in relation to the quality of the process.

3.3 Process Perception

Process perception is the method of deeper investigation into and understanding of how work is being done to identify inconsistencies, disconnects, or problem areas that might cause or contribute to the problem. To understand this, it is better to draw a process map and then identify the possible causes of problems and redundancies. In this journey, one can identify the value-added and non-value-added process activities and hence, an overall perception of the process can be brought forward.

A detailed process map is specific about what happens in a process. It often captures decision points, rework loops, complexity, etc. For instance, the four different perspectives on a process are as follows:

- *Perceived process*: What you think the process is
- *As is process*: What the process really is
- *Should be process*: What the process should be
- *Could the process*: What the process could be

Figure 3.3 describes a diagrammatic presentation of a perceived process. The other methods of describing process activities are flowcharts, value analysis mapping (VAM), and value stream mappings (VSM). These concepts are discussed in detail in various chapters below.

3.4 Process Capability

Process capability is the determination of whether a process, with normal variation, is capable of meeting customer requirements or measure of the degree a process is/ is not meeting customer requirements, compared to the distribution of the process. IF VOC and VOP are known, and then, the process capability can be expressed as the ratio of VOC to VOP. A process capability index [3, 4] is a measure relating the actual performance of a process to its specified performance, where processes are considered to be a combination of the equipments, materials, people, plant, methods, and the environment. The minimum requirement for a process to be capable is that three times process standard deviations each side of the process mean are contained within the specification limits. This means that about 99.7 % of the output will be within the tolerance. A process capability index may be calculated, only when the process is under statistical control. They are simply a means of indicating the variability of a process relative to the product specification tolerance.

The calculation of process capability depends on the type of data under investigation. For attribute data system, the capability is defined in terms of nondefective/defective, pass/fail, or good/bad as the case may be. They are calculated in terms of defects per million opportunities (DPMO), whereas for continuous data system, the capability is defined in terms of defects under the curve and outside of the specification limits. Some of the frequently used capability indices are process potential (capability) index and process performance index. Detailed discussions on these topics are presented in Chap. 8.

3.5 Process Performance

Process performance is a measure of actual results achieved by following a process. Process performance is characterized by process measures (e.g., effort, cycle time, and defect removal effectiveness) and product measures (e.g., reliability, defect density, capacity, response time, and cost). The common measures for the organization consist of process and product measures that can be used to characterize the actual performance of processes in the organization's individual projects [1]. By analyzing the resulting measurements, a distribution or range of results can be established that characterize the expected performance of the process when used on any individual project. Data mining or some exploratory data analysis (EDA) is one such topic which can be incorporated at this stage to analyze the existing pool of data to understand the pattern of variation in the raw data. This will help for arriving at a process capability trend of the past which would enable us gauge the changes in the future.

Recall that Six Sigma measurement focuses on tracking and reducing defects in a process. One can implement much simpler computational knowledge for understanding defect measures for variation. They can be applied to any process for which there is a performance standard or requirement, whether for continuous or discrete data. One can even track the performance and improvement across the projects and processes. A detailed discussion on these topics is also presented in Chap. 8.

3.6 Process Improvement

The term process improvement refers to a strategy of developing focused solutions so as to eliminate the root causes of business performance problems. Process improvement effort seeks to fix a problem while leaving the basic structure of the work process intact. In Six Sigma philosophy, the emphasis is on finding and targeting solutions to address the "vital few" factors (the X's) that cause the problem or effect (the Y's). Thus, the vast majority of Six Sigma projects are centered on process improvement efforts.

Process improvements are often achieved through specific opportunities, commonly called problems, being identified or recognized. A focus on improvement opportunities should lead to the creation of teams whose membership is determined by their work on and detailed knowledge of the process, and their ability to take improvement action. The teams must then be provided with good leadership and the right tools to tackle the job.

By using reliable methods, creating a favorable environment for team-based problem solving, this can be achieved. Some of the graphical tools such as histogram, Pareto analysis, scatter diagram, and box plots discussed in previous chapters are very useful in problem identification and solving. Even the use of SPC and control charts constitutes the ideal way of monitoring current process performance, predicting future performance and suggesting the need for corrective action. A detailed discussion on process improvement is presented in Chap. 10.

3.7 Process Control

A process control involves the inspection of the items for analyzing quality problems, and improving the performance of the production process, it is sometime termed as quality inspection. Inspection means checking of the material, components, or components itself at various stages in manufacturing. It often involves a decision on product acceptance, regulating manufacturing process, rating overall product quality, measuring inspection accuracy, and sorting out the faulty or defective items. The control and inspection act includes the interpretation of a specification, measurement of the product, and comparison of specification and measurement. The objectives of inspection are as follows:

- *Receiving inspection*: Inspection of incoming materials and purchased parts to ensure that they are according to the required specifications
- *In process inspection*: Inspection of raw materials as it undergoes processing from one operation to another
- *Finished good inspection*: To inspect the final finished product to detect the defects and its sources

- *Gauge maintenance*: Control and maintenance of measuring instruments and inspection gauges
- *Decision of salvage*: It is necessary to take decision on the defective parts. Some of these parts may be acceptable after minor repairs

Thus, a Six Sigma process summarizes the fact that the output from a business process is a function of the decisions made by the process owners. Hence, before embarking for a complete control of the process, we need to achieve a certain level of quality through inspection. At this stage, the people associated with the quality will be able to bring the process to operate within a predictable range of variation, that is, to bring a process to a stable and consistent level. This is ultimately achieved by the SQC and SPC techniques. Both these techniques involve the measurement and evaluation of variation in a process and the efforts made to limit or control such variation. In its most common applications, SPC helps an organization or process owner to identify possible problems or unusual incidents so that action can be taken promptly to resolve them. In Chap. 11, we will discuss various control methods of improving processes and quality in details after introducing the concepts of process quality in the next chapter and in some later sections.

3.8 Relevance or Managers

This chapter remains as one of the core subject of project management, where we show that every project is a sequence of processes and associated activities. A Six Sigma process is best described by components such as process characterization, process perception, process capability, process performance, process improvement, and process Control. A business organization, therefore, makes all effort to improve their processes to have a better return on investment. The technical language of a process in terms of $y = f(x)$, stressing the importance of various perceptions of y and x, is also presented in this chapter. This language of the process is the foundation to any quantitative model of a project. The two major techniques of assessing variation, namely SQC and SPC and its connection to process performance, are suitably done in a peripheral manner in this chapter, leaving a detailed study in the later chapters.

Exercises

3.1. What is a process? What are the dimensions of a process?
3.2. Distinguish between a dependant and independent variable. State some examples of each.
3.3. How does a management process differ from a Six Sigma process? Give examples for a Six Sigma process.

3.4. Discuss various components of a Six Sigma process stating the importance of each in successful project management.

3.5. Distinguish between VOC and VOP. How are they captured in a project study?

3.6. Distinguish between process capability and process performance. State various methods of evaluating the capability and performance of a process.

3.7. Distinguish between SPC and SQC. How are they used for identifying variations in a process?

References

1. Eileen, F.: CMMI for Services (CMMI-SVC): A Tutorial. Software Engineering Institute, Carnegie Mellon University, Pittsburgh (2010)
2. Gryna, F.M., Chua, R.C.H., Defeo, J.: Juran's Quality Planning and Analysis for Enterprise Quality. Tata McGraw-Hill, New Delhi (2007)
3. Montgomory, D.C.: Introduction to Statistical Quality Control. Wiley, India (2003)
4. Oakland, J.S.: Statistical Process Control. Elsevier, New Delhi (2005)
5. Sanders, D., Ross, B., Coleman, J.: The process map. Qual. Eng. 11(4), 555–561 (1999)

Chapter 4
Understanding Variation

Understanding variation is the best way to improve the quality of a process or product. We all know that variation is the cause of all defects and is inherent in all processes in varying amount. To make a product defect-free, it is essential to minimize the variation of that process. Hence, quantifying the amount of variation in a process is the first and the critical step toward improvement. This will further enhance the understanding of the types of causes and decide the course of action to reduce the variation. This action will have the lasting improvement on the quality of the product [4].

Looking at variation helps management to understand fully the real performance of its business and its processes. It has been common for organizations to measure and describe their efforts and results in terms of "averages": average cost, average cycle time, average inventory, etc. But averages actually can hide problems by covering up variation. For instance, a supplier has an actual average delivery time of 4 working days to deliver a product. Based on this, a customer is promised a delivery of his order within 6 working days from the date the order was placed. But it does not reveal the fact that due to wide variation in the processes involved; more than 15 % of orders are delivered after more than 6 days! With the same amount of variation, simple statistical calculations will show that it will be necessary to achieve an average of 2 days just to get all the orders delivered within 6 days. On the other hand, by significantly reducing the variation, with an average time of only 5 days, there would be no delivery beyond 6 days. In fact, it is far less expensive to achieve a five-day average delivery time, than to achieve a two-day average. With better control on variation, both supplier and the customers benefit by not having to provide for unpredictable wide variation.

According to Montgomery [3], quality is inversely proportional to variability. That is less the variability more the quality. Since variability can only be described by statistical terms, statistical methods play a central role in quality improvement efforts. The statistical tools and techniques offer a wide range of quality improvement tools for the detection of variations in the process. In the remaining part of this chapter, we will explore those important statistical techniques to study and detect variation.

© Springer India 2015
K. Muralidharan, *Six Sigma for Organizational Excellence*,
DOI 10.1007/978-81-322-2325-2_4

4.1 Types of Variation

As we know the amount of variation in a process tells us what that process is actually capable of achieving (tolerance), whereas specifications tell us what we want a process to be able to achieve. The first step in understanding variation should always be to plot a process data in time order. The time order can be hourly, daily, weekly, monthly, quarterly, yearly, or any suitable time period.

Generally, two types of variations are encountered in a process. They are special cause or assignable cause variations and the random or irregular type of variations. The goal of any process, then, will be to identify the causes and eliminate them to make an unstable process into a stable process. A process with only common cause variation is called *stable*.

4.1.1 Special Cause Variation

The variations which are relatively large in magnitude and viable to identify are the special causes of variation or assignable causes of variation. They may come and go sporadically. Quite often, any specific evidence of the lack of statistical control gives a signal that a special cause is likely to have occurred. Hence, such variations are local to the process and are unstable. The best way of dealing with this kind of variation is as follows:

- Get timely data so that special cause signal can identify easily
- Take immediate action to remedy any damage
- Search immediately for a cause; find out what was different on that occasion; and isolate the deepest cause that can affect the process
- Develop a long-term remedy that will prevent that special cause from recurring, or if results are good, retain that lesson
- Use early warning indicators throughout your operation. Take data at the early stage of the process so that changes can be identified as soon as possible.

4.1.2 Common Cause Variations

This type of variation is the sum of the multitude of effects of a complex interaction of random or common causes. It is common to all occasions and places and always present in some degree. Variation due to common causes will almost always give results that are in statistical control. These kinds of variations are normally stable and can be controlled with a proper treatment.

The ideal way of analyzing the special cause variation and common cause variation is the control chart techniques. A control chart is a device intended to be used at the point of operation, where the process is carried out and by the operators of that process. For every control chart, there are three zones, namely the stable zone (shows the presence of common causes of variation), warning zone, and action zone (both show the presence of special causes of variation). A statistically controlled chart will be called a normal process whose process means (a measure of accuracy) and standard deviation (a measure of precision) will work according to the tolerance level suggested by the customer.

A control chart for means (or $\overline{X}$-chart) is a graphical presentation of sample means against the sample numbers. There are three control lines (CLs) in the graph: visually a lower control limits (LCL), a CL corresponds to the overall mean, and the upper control limit (UCL). They are constructed under the assumption of normality. Any points outside the LCL or UCL are an indication of out of control process.

As an illustration, consider an example on measurements of inner radius of a pipeline cable, where 10 samples each of size 5 are drawn at regular intervals from a manufacturing process. The sample mean ($\overline{x}$) and their ranges (R) are given below:

Sample no.	1	2	3	4	5	6	7	8	9	10
Mean ($\overline{x}$)	49	45	48	53	39	47	46	39	51	45
Range (R)	7	5	7	9	5	8	8	6	7	6

It is proposed to have a quality check on a regular basis to ascertain the causes of variations. The control chart corresponds to the above data is presented in Fig. 4.1. Since none of the sample means falls outside the control limits, and the points are randomly distributed within the control limits, the process is said to be statistically in control. Whatever variations now remains in the process are purely due to random causes.

An $\overline{X}$-chart helps to detect the changes in the process average. If the concern is the variability in the process, then it can be first checked using the control chart for range (or R-chart). Thus, a control chart is a simple method of monitoring variations

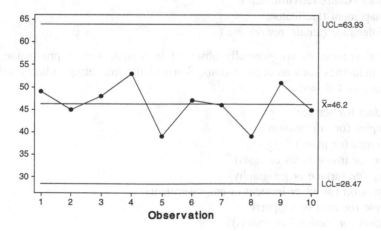

Fig. 4.1 Control chart for inner radius of cable

in an ongoing process for its consistency and quality. They generally do not involve rigorous technical computations and hence is easy to implement. Further discussion on charting techniques will be taken up again in Chap. 10 in detail.

4.2 Causes of Variations

There is a certain amount of variability in every product; consequently, no two things are exactly alike. For instance, a process is bound to vary from product to product, materials to materials, operation to operation, service to service, machine to machine, man to man, shift to shift, day to day, etc. However, if the variation is very large, the customer may perceive the unit to be undesirable and unacceptable. As mentioned earlier, quantifying the information is the first step toward the improvement of quality. The quantification is done on the basis of the characteristics under study. The characteristics can be either qualitative (or attributes) or quantitative (or variables) in nature. We will describe various tools and techniques for analyzing these characteristics in the chapters to follow.

Causes can be identified through brainstorming sessions (see Sect. 9.2 for details). These groups can then be labeled as categories of the fishbone diagram (discussed more in Sect. 9.9). They will typically be one of the traditional categories mentioned above but may be something unique to the application in a specific case. Causes can be traced back to root causes with the 5 Whys technique.

Typical categories of causes are as follows:

- Machine (technology)
- Method (process)
- Material (includes raw material, consumables, and information.)
- Man (physical work, brain work, etc.)
- Measurement (inspection)
- Mother Nature (environment)
- Management (facilitator)
- Maintenance (repair, rework etc.)

The above causes are generally observed in manufacturing, production, and service industries such as organizations. Some of the marketing industry specific causes are as follows:

- Product (or service)
- Samples (or information's)
- Schemes (or plans)
- Price (or investments or costs)
- Place (or surface or geography)
- Promotion (or advertisement or empowerment)
- People (or man or support)
- Process (or methods or means)
- Physical evidence (or visible lapses)

4.3 Need for Measuring Variation

With respect to a Six Sigma project, there are many significant notions of measuring variation. They include as follows:

- Measuring variability determines the reliability of an average by pointing out as to how far an average is representative of the entire data.
- Helps to determine the nature and cause of variation in order to control the variation itself.
- Enable comparisons of two or more distributions with regard to their variability.

Note that, a better improved quality product always pays less rework, fewer mistakes, fewer delays, and better use of time and materials. Most often, a major part of the organization's time will be utilized for finding and correcting mistakes and variations. The entire demand and supply of quality have undergone a sea change for the last two decades with the advent of scientific innovations putting pressure on both customers and producers. Although reasons can be sighted from many angles, here are some of the potential reasons:

- The growth of foreign markets and the shrinking nature of transportation.
- The discriminating attitude of consumers as they are offered more choices and preferences.
- Consumer's sophistication leads to the demand for new and better products and services.
- Increasing global competition and free market concepts due to the erosion of economic and political boundaries.
- Innovative concepts over the non-conventional aspects of quality and its practice.

The necessity of quality increases according to the growth of the organization, its output, and its customers requirements. The dimensionality of the quality characteristics therefore includes the cost, innovative products, time to market, lead time, employee morale, competitive environment, safety, engineering techniques including process, and product specifications. For a detailed discussion on these and other quality dimensions, see Garvin [1] and Montgomery [3]. Since customer being the sole assessor of the finished product, it is essential to have their voice heard in every aspects of the quality of an organization. Even the quality measurements and the corresponding metrics are decided based on customer's voice. A structured approach of addressing this aspect of organizational quality is the voice of customer (VOC) and voice of process (VOP). Both VOC and VOP strongly advocate the use of statistics and will be discussed in various chapters to follow.

For industrial and commercial organizations, which are relevant only if they provide satisfaction to the consumer, competitiveness in quality is not only central to profitability, but also crucial to business survival. The consumer should not be required to make a choice between price and quality, and for manufacturing or service organizations to continue to exist, they must learn how to manage quality.

In today's tough and challenging business environment, the development and implementation of a comprehensive quality policy is not merely desirable, and it is essential [2, 5].

In addition, in an organization, the quality of the product depends on the following approaches practiced. They are of passive or reactive type and proactive or preventive type. Setting acceptable quality levels for the product, inspecting to measure compliance in relation to the process, etc., are reactive type approaches. The proactive or preventive type include ensuring design quality in products and processes, identifying sources of variation in processes and materials, and finally monitoring the process performance.

These discussions emphasize the fact that the quality characteristics depend on the design and engineering of the product. Once such a designed product is ready, it is evaluated on the basis of specifications. For a manufactured product, the specifications are the desired measurements for the quality characteristics on the components and subassemblies that make up the product, as well as the desired values for the quality characteristics in the final product. A value of a measurement that corresponds to the desired value for those quality characteristics is called the target value for that characteristic. These target values are usually bounded by a range of values that most typically will be close to the nominal value suggested to and vary within that range. Thus, the smallest allowable value for quality characteristics is called the lower specification limit (LSL) and the largest allowable value for quality characteristics is called the upper specification limit (USL). Specification limits are usually the result of the engineering design process for the product.

In the next two subsections, we will discuss the importance of measurement variation and various types of measures of variations. The necessity of measurement system analysis (MSA) is also discussed for a better understanding of the variation.

4.4 Measurement Variations

For any decision making, one should have a good measurement and the associated knowledge of the variations to statistically remove them. The knowledge of change or fluctuation of a specific characteristic which determines how stable or predictable the process, may be affected by man/people, machine/equipment, methods/procedures, measurements, materials, and the mother earth (or environment) and many other variations. Knowledge of these is the best way to investigate the reasons of variations and eliminate them.

A measure also helps us to

- Understand a decision on

    - Meeting standards and specifications
    - Detection/reaction oriented activities
    - Short-term results

- Stimulate continuous improvement like
    - Where to improve?
    - How much to improve?
    - Is improvement cost effective?
    - Prevention oriented
    - Long-term strategy etc.

Recall the famous quote by Taguchi as "If you cannot measure, you cannot improve!" Thus, to identify and define the various components of measurement error and to list and discuss the sources of error, we should have a good MSA in place.

4.5 Measurement System Characteristics

A measurement system consists of measuring devices, procedures, definitions, and people. To improve a measurement system, we need to evaluate how well it works (by asking "how much of the variation we see in our data is due to the measurement system?") and evaluate the results and develop improvement strategies. The common problems with measurements are as follows:

- *Bias or inaccuracy*: The measurements have a different average value than a "standard" method
- *Imprecision*: Repeated readings on the same material vary too much in relation to current process variation
- *Not reproducible*: The measurement process is different for different operators, or measuring devices or labs. This may be either a difference in bias or precision
- *Unstable measurement system over time*: Either the bias or the precision changes over time
- *Lack of resolution*: The measurement process cannot measure to precise enough units to capture current product variation

The desired measurement characteristics for continuous variables are as follows:

- *Accuracy*: The measured value has little deviation from the actual value. Accuracy is usually tested by comparing an average of repeated measurements to a known standard value for that unit
- *Repeatability*: The same person taking a measurement on the same unit gets the same result
- *Reproducibility*: Other people (or other instruments or labs) get the same result as you get when measuring the same item or characteristic
- *Stability*: Measurements taken by a single person in the same way vary little over time
- *Resolution*: There is enough resolution in the measurement device so that the product can have many different values

- *Linearity*: Measurement is not "true" and/or consistent across the range of the "gage"

The actions associated with the above measurement problems are as follows:

Accuracy/Bias Actions

- Calibrate when needed/scheduled
- Use operations instructions
- Review specifications
- Review software logic
- Create Operational Definitions

Repeatability and Reproducibility actions

- Repair, replace, and adjust equipment
- Follow standard operating practices (SOP)
- Impart Training

Stability Actions

- Change/adjust components
- Establish "life" time frame
- Use control charts
- Use/update current SOP

Resolution Actions

- Measure to as many decimal places as possible
- Use a device that can measure smaller units
- Live with it, but document that the problem exists
- Larger sample size may overcome problem
- Priorities may need to involve for other considerations such as engineering tolerance and process capability
- Cost and difficulty in replacing device

Linearity Actions

- Use only in restricted range
- Rebuild
- Use with correction factor/table/curve
- Sophisticated study required and will not be discussed in this course

The Gage study is carried to identify the amount of variations present in continuous measurement variables, whereas, for discrete data, the variation is assessed through "bias" study, which is done using a kappa analysis. A detailed study on these aspects will be taken further when we discuss the measure phase of Six Sigma project (see Chap. 8, Sect. 8.7 for details).

4.6 Measures of Variations

The extent of degree to which a numerical data tend to spread about an average value is called the dispersion or variation of the data. Precision is the amount of accuracy you plan to do in a process. As seen in the previous section, the accuracy of a process relates to its ability to attain the target value (or the average value). The precision actually links to the type of scale or amount of detail of your operational definition, but it can have an impact on your sample size of the data set. That is why for more sample observations; the spread is exhibited in lower denominations. When the number of observations (sample size) is large, then the frequency polygon approximates to a symmetric curve (or bell shaped) called the normal curve, and hence, the distribution of the data will be a *normal distribution*. Figure 4.2 shows normal curves with same mean and different spreads.

A detailed discussion on normal distribution is taken up again in Chap. 8. There are number of measures available for measuring variation (dispersion) in a data. Range, quartile deviation (QD), mean deviation (MD), standard deviation (SD), coefficient of variation (CV), etc., are some of them and are discussed below.

Range: The Range of a set of values is the difference between the largest and smallest numbers in the set. If $x_{(1)}$ and $x_{(n)}$ are the smallest and largest observations of a given set of data, then

$$\text{Range} = x_{(n)} - x_{(1)} \tag{4.1}$$

Example 4.1 Find the Range of the data set 28, 54, 45, 30, 44, 20, 38, 37, 49, 50, and 27.

Solution Here, $x_{(1)} = 20$ and $x_{(n)} = 54$. Therefore,

$$\text{Range} = 54 - 20 = 34$$

The range offers a measure of scatter which can be used widely, owing to its simplicity. However, there are some problems associated with the use of this measure: (i) The value of the range depends on the number of observations in the

Fig. 4.2 Normal curves with different spreads

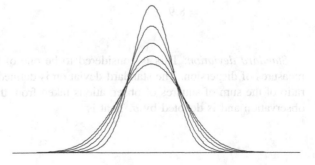

sample. The range will tend to increase as the sample size increases and (ii) calculation of the range uses only a portion of the data obtained. The value remains the same despite changes in the values lying between the lowest and the highest values.

Quartile deviation or semi-inter-quartile range: The QD is defined as half of the difference between the third quartile and first quartile. That is

$$QD = \frac{Q_3 - Q_1}{2} \tag{4.2}$$

where Q_1 and Q_3 are defined as the first and third quartiles, respectively, and they explain 25 and 75 % of the observations. The quartile that explains the 50 % observations is called the second quartile, Q_2, and is also called the median of the observations.

Mean deviation: The MD or average deviation of a set of numbers $x_1, x_2, \ldots, x_n$ is defined by the ratio of absolute difference of observations from mean to the number of observations. That is

$$MD = \begin{cases} \dfrac{\sum_{i=1}^{n} |x_i - \bar{x}|}{n}, & \text{for continuous data} \\ \dfrac{\sum_{i=1}^{n} f_i |x_i - \bar{x}|}{\sum_{i=1}^{n} f_i}, & \text{for grouped frequency distribution} \end{cases} \tag{4.3}$$

where $\bar{x}$ is the arithmetic mean of the given data. MD can also be defined in terms of median as well as mode in a similar way.

Example 4.2 Find MD about mean of the data: 32, 45, 43, 65, 25, 44, 55, 28, 38, and 40.

Solution Here

$$\bar{x} = \frac{\sum_{i=1}^{10} x_i}{10} = 41.5$$

Therefore,

$$MD_{\bar{x}} = \frac{|32 - 41.5| + |45 - 41.5| + \cdots + |40 - 41.5|}{10}$$

$$= 8.9$$

Standard deviation: This is considered to be one of the best and widely used measures of dispersion. The standard deviation is defined as the square root of the ratio of the sum of squares of observations taken from the mean to the number of observation and is denoted by σ. That is

$$\sigma = \begin{cases} \sqrt{\dfrac{\sum_{i=1}^{n}(x_i - \bar{x})^2}{n}}, & \text{for continuous data} \\[4mm] \sqrt{\dfrac{\sum_{i=1}^{n}f_i(x_i - \bar{x})^2}{\sum_{i=1}^{n}f_i}}, & \text{for grouped frequency distribution} \end{cases} \qquad (4.4)$$

Sometimes, the standard deviation of a sample's data is defined with $(n - 1)$ replacing n in the denominators of the expressions (4.4) as the resulting value represents a better estimate of the standard deviation of a population from which the sample is taken. For large values of n (say $n > 30$), the difference is negligible. In Chap. 5, we have given detailed estimation procedures of standard deviation under normal and non-normal cases.

Variance: The variance of a set of data is defined as the square of the standard deviation and is given by σ^2. The variance is never negative and can be zero only if all the data values are the same.

Example 4.3 Obtain the variance of 40, 33, 42, 36, 51, 31, 51, 45, 41, and 50.

Solution Here $n = 10$, and

$$\bar{x} = \frac{\sum_{i=1}^{10} x_i}{10} = 42$$

Therefore,

$$\sigma^2 = \frac{\sum_{i=1}^{n}(x_i - \bar{x})^2}{n}$$

$$= \frac{(40 - 42)^2 + (33 - 42)^2 + \cdots + (50 - 42)^2}{10}$$

$$= 53.11$$

Example 4.4 A particular intelligence test is given to the selected students and their performances, as determined by the test scores are noted as below:

30	33	37	35	39	43	32	24	28	38
32	47	41	51	41	54	32	31	46	15
38	26	50	40	38	44	21	45	31	37
40	33	42	36	51	31	51	45	41	50
46	50	26	15	23	41	38	40	37	40
68	41	30	52	52	61	53	48	21	28

Discuss what percentage of scores falls within 1-σ, 2-σ, and 3-σ limits?

Solution Here $n = 60$, and

$$\bar{x} = \frac{\sum_{i=1}^{60} x_i}{60} = 38.81667$$

and

$$\sigma = \sqrt{\frac{\sum_{i=1}^{n} (x_i - \bar{x})^2}{n}} = 10.68896$$

Therefore,

$$(\bar{x} - \sigma, \bar{x} + \sigma) = (28.1277, \ 49.5956): \text{ covers } 66.67\,\% \text{ observations}$$
$$(\bar{x} - 2\sigma, \bar{x} + 2\sigma) = (17.4387, \ 60.1946): \text{ covers } 93.33\,\% \text{ observations}$$
$$(\bar{x} - 3\sigma, \bar{x} + 3\sigma) = (6.7498, \ 70.8836): \text{covers } 100\,\% \text{ observations}$$

Thus, in general, for a moderate normal process, one can observe about 68 % sample values (observations) falls in 1-σ limit, 95 % sample values falls in 2-σ limits, and 99.73 % sample values in 3-σ limits. This is explained in Fig. 4.3. The figure shows the probability of sample values correspond to various standard deviations from mean of a standard normal random variate.

Relative dispersion: The measures discussed so far are absolute measures. However, a dispersion of 8 in. in measuring a distance of 100 ft is quite different in effect from the same variation of 8 in. in a distance of 30 ft. A measure of this effect is calculated on the basis of *relative dispersion*, which is defined as

$$\text{Relative dispersion} = \frac{\text{absolute dispersion}}{\text{average}}$$

Coefficient of variation: It is defined as the ratio of standard deviation to the mean. CV plays a very important role in deciding the consistency or stability in a data set when compared with other data sets. Since CV is usually expressed in percentages, it is easy to compare the CV of different sets to assess the uniformity of

Fig. 4.3 Proportion of samples covered by different standard deviations

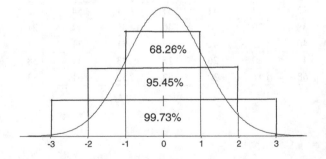

the data set. Less the CV, better the consistency. Symbolically, CV can be computed as

$$CV = \frac{\sigma}{\bar{x}} \cdot 100 \qquad (4.5)$$

Note that, the CV is independent of the units used, and hence, it can be used to compare different distributions.

Example 4.5 The outputs of measurements are taken from three processes to evaluate the consistency of performances. Ten measurements from each process are shown below. Decide which process is consistent (stable or uniform) in the production?

P-1	74	64	88	76	58	69	73	66	60	73
P-2	65	64	68	74	72	73	69	68	69	74
P-3	70	74	81	65	69	70	74	56	76	66

Solution To judge the consistency of the process, we compute the CV for each of the three processes. The mean and standard deviations of the three processes are as follows:

	P-1	P-2	P-3
Mean ($\bar{x}$)	70.1	69.6	70.1
Standard deviation (σ)	8.761659	3.565265	6.887186
Coefficient of variation ($\sigma/\bar{x}$)	0.12498	0.05122	0.098248

Thus, the CV for the three processes is 12.49, 5.12, and 9.82 %, respectively. Since the CV for the second process is low, the process P-2 is considered to be stable.

Covariance: If $x_1, x_2, \ldots, x_n$ and $y_1, y_2, \ldots, y_n$ are two pairs of observations, then the covariance between x and y is defined as

$$COV(x, y) = \frac{1}{n} \sum_{i=1}^{n} (x_i - \bar{x})(y_i - \bar{y}) \qquad (4.6)$$

Covariance helps us to find association between two variables. The extent of degree of association between two variables can be measured using the correlation coefficient and is given as

$$r = \frac{\text{COV}(x,y)}{\sigma_x \sigma_y}$$

$$= \frac{\sum_{i=1}^{n} (x_i - \bar{x})(y_i - \bar{y})}{\sqrt{\sum_{i=1}^{n} (x_i - \bar{x})^2} \sqrt{\sum_{i=1}^{n} (y_i - \bar{y})^2}} \tag{4.7}$$

The value of r ranges from -1 to 1. A value of $r = 0$ implies the variables are uncorrelated. We will study more on these concepts in Chap. 9.

Example 4.6 Consider the following data on measurements of number of bugs in particular software:

33.1639	31.9869	31.3154	34.1101	30.7042	30.8357
33.7121	32.2232	32.1460	32.7959	34.2559	34.3373
32.7708	32.1333	33.1144	31.7009	32.2436	30.8757
33.7269	32.1078	34.0607	32.2813	33.2847	30.5529
31.4589	32.9697	34.1473	32.4434	31.9802	31.4253

The histogram with normal fit of the data is shown in Fig. 4.4.
The descriptive statistics according to Excel software is:

Summary statistics of bug counts	
Mean	32.49546667
Standard error	0.205764748
Median	32.26245
Mode	32.1054
Standard deviation	1.127019939
Sample variance	1.270173943
Kurtosis	−0.953573096
Skewness	0.05827139
Range	3.7844
Minimum	30.5529
Maximum	34.3373
Sum	974.864
Count	30

4.7 Relevance for Managers

Variation (or amount of dispersion or spread) is inherent of any process. Understanding variation is therefore crucial for understanding the process stability and consistency, and hence the quality of the process. The dispersion, as a measure of

Fig. 4.4 Histogram with
normal fit

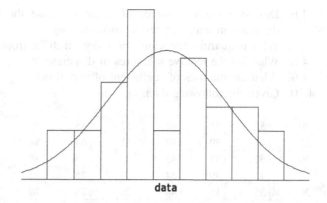

data

variation, is an important characteristic to understand the basics of any process, as it gives us additional information that enables us to judge the reliability of our measure of the central tendency (measure of location or target). Foregrounding this fact, we discussed in detail various types of variation, causes of variation, and methods of measuring variation in this chapter. Among various measures of dispersion, the standard deviation (σ) is considered to be a good measure of variation, as it is an absolute measure of dispersion that expresses variation in the same units as the original data. For different other estimators of σ, we recommend readers to proceed to the next chapter.

This chapter also connects the spread of a distribution to the normal curve, which is the basis of all sampling inferences of data. There are two basic reasons why normal distribution occupies such a prominent place in statistics. First, it has some properties that make it applicable to a great many situations in which it is necessary to make inferences by taking samples. Second, the normal distribution comes close to fitting the actual observed frequency distribution of many phenomena, including human characteristics (weights, heights, IQs), outputs from physical processes (dimensions, yields, etc.), and other measures of interest to managers in both public and private sectors. Failure to understand the significance of a normal curve ultimately leads to poor understanding of the state of the art. Hence, better understanding of normal distribution enhances knowledge of variation and vice versa.

Exercises

4.1. Elicit the statement: "Variation is the enemy of quality".
4.2. What are the types of variation encountered in a process? How are they caused?
4.3. Discuss the control chart method of detecting variation in a process.
4.4. What are the characteristics of a good measurement system?
4.5. Explain the practical significance of normal distribution in controlling the variation.

4.6. Discuss various measures of variation. State the importance of standard deviation in management decision making.

4.7. What is quartile deviation? How does it differ from mean deviation?

4.8. What are the relative measures of dispersion?

4.9. What are the uses of coefficient of variation?

4.10. Given the following data:

70	55	51	42	57	45	60	47	63	53
33	65	39	82	55	64	58	61	65	42
50	52	53	45	45	25	36	59	63	39
65	45	49	54	64	75	42	41	52	35
30	35	15	48	26	20	40	55	46	18

Compute

(i)   Mean and standard deviation

(ii) What percent of observations are covered under $(\bar{x} \pm \sigma), (\bar{x} \pm 2\sigma)$ and $(\bar{x} \pm 3\sigma)$

(iii) Range, quartile deviation, mean deviation about mean

(iv) Coefficient of variation

4.11. Following data show the demand corresponding to various dimensions of piston rings:

Dimension of piston rings (in mm)	Demand
10.50–19.50	5
20.50–29.50	45
30.50–39.50	42
40.50–49.50	38
50.50–59.50	32
60.50–69.50	29
70.50–79.50	26
80.50–89.50	20

(i)   Draw a histogram and assess the normality of the data

(ii) Compute mean and standard deviation

(iii) What percent of observations is covered under $(\bar{x} \pm \sigma), (\bar{x} \pm 2\sigma)$ and $(\bar{x} \pm 3\sigma)$

(iv) Compute quartile deviation and mean deviation about mean

(v)   Compute coefficient of variation

4.12. Two methods were used in a study of the latent heat of fusion of ice. Both A and B were conducted with the specimens coded to −0.72 °C. The following data represent the change in total heat from −0.72 °C to water 0 °C in calories per gram of mass:

Method A: 79.98, 80.02, 80.04, 80.04, 80.03, 80.04, 79.97, 80.05, 80.03, 80.02
Method B: 80.02, 79.94, 79.98, 79.97, 79.97, 80.03, 80.03, 79.95, 79.97, 80.00

Compute

(i) Coefficient of variation and decide which method is stable in maintaining the temperature.
(ii) Correlation coefficient between the two methods and decide what extent the methods are related.

4.13. A company is considering installing new machines to assemble its product. The company is considering two types of machines, but it will buy only one type. The company selected eight assembly workers and asked them to use these two types of machines to assemble products. The following table gives the time taken (in minutes) to assemble one unit of the product on each type of machine for each of these 8 workers.

Machine I: 23   26   19   24   27   22   20   18
Machine II: 21   24   23   25   24   28   24   23

Compute

(i) Coefficient of variation and decide which machine is consistent in assembling the unit.
(ii) Correlation coefficient between the two machines and decide whether there is any relationship between them.

4.14. What are the importance of normal distribution? State the managerial significance of normal distribution.

References

1. Garvin, D.A.: Competing in the Eight Dimensions of Quality. Harvard business Review, Sept–Oct 1987
2. Gryna, F.M., Chua, R.C.H., Defeo, J.A.: Juran's Quality Planning and Analysis for Enterprise Quality, 5th edn. Tata McGraw-Hill, New Delhi (2007)
3. Montgomory, D.C.: Introduction to Statistical Quality Control. Wiley, India (2003)
4. Muralidharan, K., Syamsundar, A.: Statistical Methods for Quality, Reliability and Maintainability. PHI Publications, New Delhi (2012)
5. Oakland, J.S.: Statistical Process Control. Elsevier, New Delhi (2005)

Chapter 5
Sigma Estimation

In the previous chapter, we have seen the importance of a process and variation associated with a process. The necessity of estimating standard deviation is always an important aspect in statistical study including variation estimation, sample size determination; control charts preparations, and sigma-level estimations. The importance of sigma estimation is the central part of Six Sigma study where process improvements are usually talked in terms of sigma level and its estimation. Various methods such as sample range and sample standard deviations have been used to estimate standard deviation. The obvious limitation of all these methods is their natural sensitivity regarding normality assumption of sample observations. In this chapter, we provide some robust estimates which are computationally simple and can be implemented as per the requirement in practical problems.

5.1 Introduction

When the process is operating under optimal conditions, the measurements taken on the process are typically modeled as independent draws from a normal distribution with mean μ and standard deviation σ^2. As mentioned in the previous chapter, the $\overline{X}$ and R charts are designed to detect mean shifts, changes in variability, and outliers. To set up $\overline{X}$ and R charts, one must either have a prior knowledge of μ and σ or an estimate of these parameters by observing the process over a period of time. Such estimates are usually made on the basis of a group of subsamples collected during an initialization period. The estimation of σ in the control chart context is obviously not a simple matter, given the possible occurrence of disturbances during the initialization period. The naïve approach is to ignore the possibility of such disturbances during the initialization period and to use the average subsample range as the basis for estimating σ.

There are several robust estimators that have been developed in recent years, including the M-estimator proposed by Huber [5], the A-estimators described by Hoaglin et al. [4], the S-estimator of Rousseau and Yohai [16], and the T-estimators of Yohai and Zamar [22]. Some distribution-based estimators are

© Springer India 2015
K. Muralidharan, *Six Sigma for Organizational Excellence*,
DOI 10.1007/978-81-322-2325-2_5

available in Simonoff [20] and Lax [7]. For other estimators and their comparisons, one may refer to Tatum [21], Schrader and McKean [17], Schrader and Hettmansperger [18], Rocke [13, 14], Muralidharan and Neha [9], and the references contained therein.

In the next section, we review some computationally simple estimators of standard deviation from the Six Sigma process improvement perspectives. A comparison of the estimators in terms of efficiency and mean square error is presented in the last section.

5.2 Some General Estimators of Standard Deviation

The sample standard deviation for the observations $X = (X_1, X_2, \ldots, X_n)$ is given by

$$s(X) = \sqrt{\frac{1}{n-1} \sum_{i=1}^{n} (X_i - \overline{X})^2} \tag{5.1}$$

where $\overline{X} = \frac{1}{n} \sum_{i=1}^{n} X_i$ is the sample mean. Equation (5.1) can also be written as follows:

$$s(X) = \sqrt{\frac{1}{n(n-1)} \sum_{i<j} (X_i - X_j)^2} \tag{5.2}$$

The sample variance $s^2(X)$ is the minimum variance unbiased estimator for the variance parameter σ^2 provided the data are normal. It has been found that this estimator is sensitive with respect to outliers and hence is not robust. The formula (5.2) shows that the sample variance is proportional to the average of the squared inter-point distances $X_i - X_j$.

Note that, the numerical difference between (5.1) and (5.2) is only material in the case of samples containing very few observations and that in such cases, no reliance can be placed on the precise value of any single estimate. It is then of more importance to have a means of determining upper and lower limits between which σ may lie within an expected confidence. Further, the choice between (5.1) and (5.2) may be assessed from the viewpoint of the practical worker, where the most convenient form of definition of higher order sample moment coefficients may be invoked. In the case where an estimate of σ is to be made from a number of small samples and therefore can be obtained with precision, it is essential to avoid the bias, which, in the case of a single small sample, is completely overlaid by the general uncertainty. A correction factor may be used to make the estimators unbiased in such cases (see Davies and Pearson [3]).

For any random sample drawn from a distribution with finite variance, the sample variance will be an unbiased estimator of σ^2. It follows that the sample standard deviation is a biased estimator of the underlying population standard deviation, but this bias has an analytic expression when the data are Gaussian. The expected value of the sample standard deviation in the case of Gaussian data is as follows:

$$E(s(X)) = \sigma \left[\left(\frac{2}{n-1} \right)^{1/2} \Gamma \left(\frac{n}{2} \right) \right] / \Gamma \left(\frac{n}{2} - \frac{1}{2} \right) \tag{5.3}$$

where $\Gamma(x)$ is the gamma function. If samples are normally distributed, then for $n = 20$, the sample standard deviation has expected value 0.9869σ.

Note that, $\Gamma(n)$ is defined as follows:

$$\Gamma(n) = \int_0^\infty x^{n-1} e^{-x} dx$$

and is computed through the recursive formula

$$\Gamma(n) = (n-1)\Gamma(n-1)$$
$$= (n-1)!$$

Also,

$$\Gamma \left(\frac{n-1}{2} \right) = \left(\frac{n-1}{2} \right) \left(\frac{n-3}{2} \right) \cdots \frac{3}{2} \frac{1}{2} \Gamma \left(\frac{1}{2} \right)$$

With

$$\Gamma \left(\frac{1}{2} \right) = \int_0^\infty x^{-1/2} e^{-x} dx$$

$$= \sqrt{2} \int_0^\infty e^{-y^2/2} dy, \text{ by taking } x = y^2/2$$

$$= 2\sqrt{\pi} \frac{1}{\sqrt{2\pi}} \int_0^\infty e^{-y^2/2} dy$$

$$= \sqrt{\pi} \frac{1}{\sqrt{2\pi}} \int_{-\infty}^\infty e^{-y^2/2} dy$$

$$= \sqrt{\pi}$$

A vague estimator of process standard deviation can be defined in terms of range as the difference between the highest and the lowest observation. Unfortunately, this estimator is highly influenced by the extreme values and hence is less useful. The most commonly calculated robust scale estimator is probably the inter-quartile range (IQR) which measures the difference between a distribution's upper and lower quartiles. For a continuous random variable X, with cumulative distribution function $F_X(x)$, the IQR is defined as follows:

$$IQR(X) = F_X^{-1}(0.75) - F_X^{-1}(0.25) \qquad (5.4)$$

where $F_X^{-1}(y) = x$ solves the equation $F_X(x) = y$. The sample equivalent of (5.4) is given as follows:

$$IQR(X) = Q_3(X) - Q_1(X) \qquad (5.5)$$

where $Q_1(X)$ and $Q_3(X)$ are the sample lower quartile and upper quartiles, respectively, given by

$$Q_1(X) = X_{(l)} - (l^* - l)\left[X_{(l)} - X_{(l+1)}\right]$$

where $X_{(i)}$ is the ith order statistic of the observations X and $l^* = 1 + \frac{1}{4}(n - 1)$ and $l = [l^*]$ and

$$Q_3(X) = X_{(u)} - (u^* - u)\left[X_{(u)} - X_{(u+1)}\right]$$

where $X_{(u)}$ is the ith order statistic of the observations X and $u^* = 1 + \frac{3}{4}(n - 1)$ and $u = [u^*]$. An estimate of standard deviation can be considered as the half of IQR also called the quartile deviation (QD). For moderately skewed distribution, this is approximately equals to 0.8σ.

Sometimes, the sample IQR is used as a simple technique of outlier detection, since it ignores the most extreme 25 % of each tail observations. This is the reason why box plots or whisker plots are generally used for detecting inconsistent observations. They are commonly drawn with observations 1.5 times the IQR above the upper quartile or below the lower quartile shown as points rather than included in the whiskers. The points that lie outside this range are considered to be potential inliers (lower side) and outliers (upper side), and for a random sample from the normal distribution with n large, we would expect only 0.7 % of the observations to be labeled in this way. Further discussion of this technique can be found in Hoaglin et al. [4].

An estimator similar to (5.2) is given by *Gini's mean difference* as given by

$$G(X) = \frac{2}{n(n - 1)} \sum_{i<j} |X_i - X_j| \qquad (5.6)$$

The difference here is that the squared inter-point differences are being replaced by absolute differences. For reasons outlined above for the sample standard deviation, this statistic is also not very robust. However, use of the absolute value rather than the square reduces the impact of large differences. This statistic forms the basis of robust estimation of risk in a strand of the financial literature for quantifying risk. However, Shalit and Yitzhaki [19] employ Gini's mean difference as an alternative measure of assessing risk. They are motivated mainly by the theoretical results it facilitates, rather than robustness.

A most common robust estimator of standard deviation in advanced use is given by the median absolute deviation (MAD) as follows:

$$MAD(X) = \text{median}_i |X_i - \text{median}_j(X_j)| \qquad (5.7)$$

If a large random sample is drawn from a normal distribution with variance σ^2, we expect $E[MAD(X)] = 0.6745\,\sigma$. The MAD is often used to give an auxiliary estimate of scale for other more complicated scale estimators, and for n large, it is commonly scaled so that it is asymptotically unbiased for the standard deviation σ for normal data. Rousseau and Croux [15] present two estimators as alternatives to the MAD, commonly referred to as S_n and Q_n, respectively, given as follows:

$$S_n = \text{median}_i \{\text{median}_j |X_i - X_j|\} \qquad (5.8)$$

and

$$Q_n = \{|X_i - X_j|; i < j\}_{(k)} \qquad (5.9)$$

which is the kth largest of the $|X_i - X_j|$ for $i < j$, where $k = \left(\begin{array}{c} [n/2] + 1 \\ 2 \end{array} \right)$. Thus, Q_n is the kth order statistic of the $\left(\begin{array}{c} n \\ 2 \end{array} \right)$ inter-point distances and hence is approximately the lower quartile of the inter-point distances. In the case where $n = 20$, $k = 55$ with $\left(\begin{array}{c} n \\ 2 \end{array} \right) = 190$ and hence k is the $100\,(55/190) = 28.9$th percentile of the ordered inter-point distances. The choice of k can be manipulated for optimizing the performance of Q_n. The unbiased estimators of S_n and Q_n are 1.1926σ and 2.2219σ, respectively (see Rousseau and Croux [15]).

We now define a score estimator for scale as follows:

$$S_c = \sqrt{\frac{1}{n} \sum_{i=1}^{n} \frac{1}{(1 + Z_i^2)} (X_i - M)^2} \qquad (5.10)$$

where $Z_i = \frac{X_i - M}{\sigma_0}$, M is a an auxiliary estimate of location, and σ_0 is an auxiliary estimate of scale. This estimator is a variant of A-estimator proposed by Lax [7].

The bi-weight A-estimator was well presented by Iglev [6] and was shown in a major study by Lax [7] to perform well compared to other robust univariate scale estimators. In the numerical calculations given below, the auxiliary estimate of M is assumed as the median of the sample and for σ_0, we assume the range. The efficiency can be improved, if the estimate for M is assumed as the mean.

In the next few examples, we compute the estimators discussed above and compare the performance of some of the estimators in terms of efficiency and mean square errors. Four different cases are discussed.

Example 5.1 The following data represent the life in years of 40 similar car batteries recorded to the nearest tenth of a year. The batteries were guaranteed to last 3 years. The observations are as follows:

2.2, 4.1, 3.5, 4.5, 3.2, 3.7, 2.6, 3.0, 3.4, 1.6, 3.1, 3.3, 3.8, 3.1, 3.7, 4.7, 2.5, 4.3, 3.4, 3.6, 2.9, 3.3, 3.1, 3.9, 3.3, 3.1, 3.7, 4.4, 3.2, 4.1, 3.4, 1.9, 4.7, 3.8, 3.2, 2.6, 3.9, 3.0, 3.5, and 4.2.

The estimates of scale parameter are presented in Table 5.1. The efficiency of the estimators is compared with the estimator given in (5.2).

Example 5.2 Here, we consider a simulated data set from $N(\mu, \sigma^2)$, where $\mu = 10$ and $\sigma = 3$. The observations are as follows:

12.2048, 11.9512, 8.5437, 6.3199, 7.0613, 8.8032, 9.7866, 6.6632, 8.7732, 10.4411, 13.0881, 11.5009, 15.3758, 11.0275, 12.7046, 13.0466, 9.8773, 13.8293, 12.7339, 7.7721, 14.6086, 2.9966, 9.1037, 6.6746, 6.1749, 8.2308, 8.6594, 9.2295, 10.4622, 6.2878, 5.0673, 9.7718, 6.5258, 9.1063, 6.8511, 9.6787, 10.1887, 14.6555, 7.2044, and 10.5923.

The estimates of scale parameter are presented in Table 5.2. The efficiency in Table 5.2 is compared with the population parameter σ.

Example 5.3 A Monte Carlo study (normal data)

Table 5.3 estimates are correspond to 10,000 simulations from $N(\mu, \sigma^2)$, where $\mu = 10$ and $\sigma = 3$. A sample size $N = 30$ is generated for each simulations. The mean square error (MSE) of the estimators is calculated using the formula

Table 5.1 Estimates for the battery life data

Formula	(5.1)	(5.2)	(5.6)	(5.7)	(5.8)	(5.9)	(5.10)
Estimates	0.70281	0.70281	0.8766	0.8923	1.0598	1.5730	0.6397
Efficiency (%)	100	100	80.17	78.76	66.31	44.68	109.18

Table 5.2 Estimates for the simulated data

Formula	(5.1)	(5.2)	(5.6)	(5.7)	(5.8)	(5.9)	(5.10)
Estimates	2.8614	2.8614	3.245	3.1349	3.5419	4.8381	2.6178
Efficiency (%)	100	100	88.17	91.28	80.78	59.14	109.31

Table 5.3 Estimates based on Monte Carlo study (normal data)

Formula	(5.1)	(5.2)	(5.6)	(5.7)	(5.8)	(5.9)	(5.10)
Estimates	3.0696	3.0696	3.1511	3.1203	3.4812	4.2234	3.0012
MSE	0.1731	0.2172	0.2327	0.2297	1.3568	1.8467	0.0913
Efficiency (%)	100	100	97.41	98.38	88.18	72.68	102.79

Table 5.4 Estimates based on Monte Carlo study (non-normal data)

Formula	(5.1)	(5.2)	(5.6)	(5.7)	(5.8)	(5.9)	(5.10)
Estimates	41.1606	41.1606	49.0867	46.7808	67.2345	72.8649	42.1606
MSE	2600.204	2612.356	3305.84	3027.77	4237.89	5671.21	2578.48
Efficiency (%)	100	100	83.85	87.98	61.22	56.49	97.62

$$\text{MSE} = \frac{1}{N} \sum_{i=1}^{N} (\hat{\sigma}_i - \sigma)^2,$$

where $\hat{\sigma}_i$ is the value of the estimator in the ith simulation and N is the simulation size (trials). The efficiency in Table 5.3 is compared with the population parameter σ.

Example 5.4 A Monte Carlo study (non-normal data)

Here, we run a simulations for 10,000 times, where 29 observations are drawn independently from $N(\mu, \sigma^2)$, where $\mu = 10$ and $\sigma = 3$, and a single observation is drawn from $N(\mu, \sigma^2)$, where $\mu = 10$ and $\sigma = 300$ (this is called one-wild sample, see Randal [12] and Morgenthaler and Tukey [8] for details). The above type of sampling is done rarely and in robustness studies. The estimates and MSE are presented in Table 5.4. The efficiency in Table 5.4 is compared with the population parameter $\sigma = 3$.

It is observed from tables that the estimators (5.1) and (5.2) are almost equal in all the cases. Therefore, the use of these estimators (5.1) and (5.2) is only a choice of conformability. One advantage of the pair wise difference measures is that, as they do not involve measures of location, they allow a measure of spread to be defined before a measure of location. This seems the natural order because in statistics, variability or spread is part of the problem, and summary measures of location are part of the solution; it seems natural to define the problem precisely before defining the solution precisely.

The other estimator that remains robust is the estimator defined in (5.10), which is found to be more efficient than any other estimator. The probable reason may be due to its accountability on measure of location and variability simultaneously. This is included in the score function. Even though the range is used as the auxiliary estimator for σ, the overall variability is well controlled by the score function and the squared deviation in the numerator. However, the choice is again left to the investigator for the auxiliary estimators for M and σ_0. Among the other estimators, the estimators given in (5.7) and (5.6) also perform equally efficient in most of the cases. The estimators (5.8) and (5.9) remained poor irrespective of the sampling scheme. One can also see that the MSE of the robust estimator is always less than the other estimators.

5.3 Estimation of Standard Deviation Through Control Charts

The determination of whether a process is in or out of control is greatly facilitated by the use of control charts, which are determined by two numbers—the upper and lower control limits. To employ such a chart, the data generated by the manufacturing process are divided into subgroups and subgroup statistics—such as the subgroup mean and subgroup standard deviation are computed. When the subgroup statistics does not fall within the upper and lower control limits, we conclude that the process is out of control. A detailed discussion of the control charts for variables and attributes will be taken up in later chapters. We now use this technique to compute the estimate of standard deviation for further use.

5.3.1 Default Method Based on Individual Measurements

When each subgroup sample contains a single observation (i.e., $n_i = 1$), the process standard deviation σ is estimated as $\hat{\sigma} = \frac{\bar{R}}{d_2(2)}$, where $\bar{R}$ is the average of the moving ranges of consecutive measurements taken in pairs. This is the method used to estimate σ for individual measurements and moving range charts. If the measurements are not in pairs, then a default estimate for σ is as follows:

$$\hat{\sigma} = \frac{\bar{R}}{d_2(n)} \qquad (5.11)$$

where $\bar{R}$ is the average of the moving ranges, n is the number of consecutive individual measurements used to compute each moving range, and the unbiasing factor $d_2(n)$ is defined so that if the observations are normally distributed, the expected value of R_i is $E(R_i) = d_2(n_i)\sigma$.

A relatively simple estimators are obtained through sample median. They are the MAD estimator and the median moving range (MMR) estimator (see Boyles 1997 for details). The MAD estimator for σ is computed as follows:

$$\hat{\sigma} = \frac{1}{0.6745} \, \text{median}\{|x_i - \tilde{x}|, \ 1 \le i \le N\}, \qquad (5.12)$$

where $\tilde{x}$ is the sample median. Similarly, the MMR estimate for σ is computed as $\hat{\sigma} = \frac{\tilde{R}}{0.954}$, where $\tilde{R}$ is the median of the non-missing moving ranges. We now discuss the sigma estimation for subgroups of observations.

5.3.2 Sigma Estimation for Subgroups

When control limits are computed from the input data, then σ is estimated using subgroup standard deviations or through subgroup ranges. If standards are unknown, then an estimate for σ is as follows:

$$\hat{\sigma} = \frac{1}{N} \sum_{i=1}^{N} \frac{R_i}{d_2(n_i)} \tag{5.13}$$

where N is the number of subgroups for which $n_i \geq 2$ and $R_i = \max_{1 < j < n_i} (x_{ij}) - \min_{1 < j < n_i} (x_{ij})$ is the sample range of the observations $x_{i1}, x_{i2}, \ldots, x_{in_j}$ in the ith subgroup. The unbiasing factor $d_2(n_i)$ is defined so that, if the observations are normally distributed, the expected value of R_i is $d_2(n_i)\sigma$. Thus, $\hat{\sigma}$ is the unweighted average of N unbiased estimates of σ.

If standards are known, then a default estimate for σ is as follows:

$$\hat{\sigma} = \frac{1}{N} \sum_{i=1}^{N} \frac{s_i}{c_4(n_i)} \tag{5.14}$$

where s_i, the sample standard deviation of the ith subgroup, is obtained as follows:

$$s_i = \sqrt{\frac{1}{n-1} \sum_{j=1}^{n_i} (x_{ij} - \bar{x}_i)^2}$$

and

$$c_4(n_i) = \frac{\Gamma(n_i/2)\sqrt{2/(n_i - 1)}}{\Gamma((n_i - 1)/2)}$$

here $\Gamma(.)$ denotes the gamma function discussed earlier, and $\bar{x}_i$ denotes the ith subgroup mean. If the observations are normally distributed, the expected value of s_i is $c_4(n_i)\sigma$. Thus, $\hat{\sigma}$ is the un weighted average of N unbiased estimates of σ. Table 5.5 presents the values of $c_4(n_i)$ for $n = 2, 3, \ldots, 10$.

Table 5.5 The value of $c_4(n_i)$

n_i	$c_4(n_i)$
2	0.7978849
3	0.8862266
4	0.9213181
5	0.9399851
6	0.9515332
7	0.9593684
8	0.9650309
9	0.9693103
10	0.9726596

We now discuss some other estimators of σ satisfying good statistical properties. One such estimator is the minimum variance linear unbiased estimate (MVLUE). Below, we describe the MVLUE of σ when subgroup ranges and standard deviations are given (Burr [1, 2]; Nelson [10, 11]).

5.3.3 MVLUE Method Based on Subgroup Ranges

The MVLUE is a weighted average of N unbiased estimates of σ of the form $R_i/d_2(n_i)$, and it is computed as follows:

$$\hat{\sigma} = \frac{1}{\sum_{i=1}^{N} f_i} \sum_{i=1}^{N} \frac{f_i R_i}{d_2(n_i)} \tag{5.15}$$

where $f_i = \left[\frac{d_2(n_i)}{d_3(n_i)}\right]^2$. The unbiasing factor $d_3(n_i)$ is defined so that, if the observations are normally distributed, the expected value of σ_{R_i} is $d_3(n_i)\sigma$. The MVLUE assigns greater weight to estimates of σ from subgroups with larger sample sizes, and it is intended for situations where the subgroup sample sizes vary. If the subgroup sample sizes are constant, the MVLUE reduces to the default estimate.

5.3.4 MVLUE Method Based on Subgroup Standard Deviations

This estimate is a weighted average of N unbiased estimates of σ of the form $s_i/c_4(n_i)$, and it is computed as follows:

$$\hat{\sigma} = \frac{1}{\sum_{i=1}^{N} h_i} \sum_{i=1}^{N} \frac{h_i s_i}{c_4(n_i)} \tag{5.16}$$

where $h_i = \frac{[c_4(n_i)]^2}{1-[c_4(n_i)]^2}$. The MVLUE assigns greater weight to estimates of σ from subgroups with larger sample sizes, and it is intended for situations where the subgroup sample sizes vary. If the subgroup sample sizes are constant, the MVLUE reduces to the default estimate. One can also use the $s_i's$ and their corresponding degrees of freedom to obtain the following estimate for σ as follows:

$$\hat{\sigma} = \frac{1}{c_4(n)} \frac{\sqrt{\sum_{i=1}^{N}(n_i-1)s_i^2}}{\sqrt{\sum_{i=1}^{N}n_i - N}} \tag{5.17}$$

The weights are the degrees of freedom $n_i - 1$. If the unknown standard deviation σ is constant across subgroups, the root-mean-square estimate is more efficient than the MVLUE. However, in process control applications, it is generally not assumed that σ is constant, and if σ varies across subgroups, the root-mean-square estimate tends to be more inflated than the MVLUE.

5.4 Relevance for Managers

As discussed in the previous chapter, for better understanding of process variation, the knowledge of sigma (σ) (or population standard deviation) is essential. Sigma estimation is crucial for any sample-based research study. Further, for the computation of sample size, tests of significance, confidence interval estimation, sampling study, control charts, and design of experiments, the knowledge of sigma is vital. In order to make use of the knowledge of sigma, generally the sample standard deviation is used as the simplest estimate of the standard deviation. Through an extensive literature survey, we brought out the importance of various estimators of σ and their practical relevance for decision makers. The challenge for other efficient estimators is also felt in many research studies and in decision making. Hence, this chapter offers various computationally simple and efficient estimators for σ in a ready to use form. The comparative study of the estimators for normal and non-normal processes is given on the basis of simulation study.

Since Six Sigma is a set of tools to drive process improvement through defect reduction and the minimization of variation, it is apparent that a proper measurement for variation should be in place. This is best suggested through the evaluation of sigma. Also, this chapter has immediate use in the next chapter on sample size determination. In subsequent chapters, we explore the managerial significance of sigma in various statistical computations discussed in practical situations.

Exercises

5.1 What are the managerial significance of sigma (σ) value. How is the value of σ influence the decision making?

5.2 The number of ATM transactions of a particular bank per day was recorded at 15 locations in a large city. The data are given as follows:

135, 149, 129, 150, 130, 112, 126, 156, 198, 167, 98, 170, 183, 195, and 138.

Compute

(i)   The average and median transactions
(ii)   s(X), IQR(X), G(X), and MAD(X)
(iii) S_n and Q_n and compare the values
(iv)   A score estimator with auxiliary estimate of location as mean

5.3 The following table shows the diameters in centimeters of a sample of 50 ball bearings manufactured by a company:

1.738	1.729	1.743	1.74	1.736	1.741	1.735	1.731	1.726	1.373
1.728	1.737	1.736	1.735	1.724	1.733	1.742	1.736	1.739	1.735
1.745	1.736	1.742	1.740	1.728	1.738	1.725	1.733	1.734	1.732
1.733	1.730	1.732	1.730	1.739	1.738	1.739	1.727	1.735	1.744
1.32	1.737	1.731	1.746	1.735	1.735	1.729	1.734	1.730	1.740

(i)   Check whether the data follow a normal process
(ii)   Compute various estimators of sample standard deviations
(iii) Compute S_n and Q_n and compare the values
(iv)   Compute Gini's coefficient
(v)   Suggest a suitable MVLU estimator for standard deviation

5.4 From a manufacturing process, a sample of 120 piston rings are taken and their diameters are observed as given in the data below

Ring diameter (in mm)	No. of piston rings
23.980–23.985	6
23.985–23.990	10
23.990–23.995	19
23.995–24.000	23
24.000–24.005	22
24.005–24.010	22
24.010–24.015	13
24.015–24.020	5

(i) Compute various estimators of sample standard deviations
(ii) Compute S_n and Q_n and compare the values
(iii) Compute Gini's mean difference and median absolute deviation about mean and compare the values.

5.5 Discuss a method of estimating sample standard deviation using control chart
5.6 Show that standard deviation is independent of change of origin and dependant of change of scale.

References

1. Burr, I.W.: Control charts for measurements with varying sample sizes. J. Qual. Technol. **1**, 163–167 (1969)
2. Burr, I.W.: Statistical quality control methods. Marcel Dekker, New York (1976)
3. Davies, O.L., Pearson, E.S.: Methods of estimating from samples the population standard deviation. J. Roy. Stat. Soc. **1**(1), 76–93 (1934)
4. Hoaglin, D., Mosteller, F., Tukey, J. (eds.): Understanding Robust and Exploratory Data Analysis. Wiley, New York (2000)
5. Huber, P.J.: Robust estimation of a location parameter. Ann. Math. Stat. **35**, 73–101 (1964)
6. Iglev, B.: Robust scale estimators and confidence intervals for location. In: Hoaglin, D.C., Mosteller, F., Tukey, J.W. (eds.) Understanding Robust and Exploratory Analysis, pp. 404–431. Wiley, New York (1983)
7. Lax, D.A.: Robust estimators of scale: finite-sample performance in long-tailed symmetric distributions. J. Am. Stat. Assoc. **80**, 736–741 (1985)
8. Morgenthaler, S., Tukey, J.W. (eds.): Configural Polysampling: A Route to Practical Robustness. Wiley, New York (1991)
9. Muralidharan, K., Neha, R.: Estimation of process standard deviation. Int. J. Comput. Math. Numer. Simul. **5**(1), 179–186 (2012)
10. Nelson, L.S.: Standardization of shewhart control charts. J. Qual. Technol. **21**, 287–289 (1989)
11. Nelson, L.S.: Shewhart control charts with unequal subgroup sizes. J. Qual. Technol. **26**, 64–67 (1994)
12. Randal, J.: A reinvestigation of robust scale estimation in finite samples. Ph.D. dissertation (2003)
13. Rocke, D.M.: Robust statistical analysis of inter laboratory studies. Biometrika **70**, 421–431 (1983)
14. Rocke, D.M.: Robustness and balance in the mixed model. Biometrics **47**, 303–309 (1991)
15. Rousseau, P.J., Croux, C.: Alternatives to the median absolute deviation. J. Am. Stat. Assoc. **88**, 1273–1283 (1993)
16. Rousseau, P., Yohai, V.: Robust regression by means of s-estimators. In: Robust and Nonlinear Time Series Analysis. Lecture Notes in Statistics, vol. 26. Springer, Berlin (1984)
17. Schrader, R.A., McKean, J.W.: Robust analysis of variance. Commun. Stat. A: Theory Methods **6**, 879–894 (1977)
18. Schrader, R.A., Hettmansperger, T.P.: Robust analysis of variance based upon a likelihood ratio criterion. Biometrika **67**, 93–101 (1980)
19. Shalit, H., Yitzhaki, S.: Mean-Gini portfolio theory and the pricing of risky assets. J. Finance **39**, 1449–1468 (1984)
20. Simonoff, J.S.: Outlier detection and robust estimation of scale. J. Stat. Comput. Simul. **27**, 79–92 (1987)

21. Tatum, L.G.: Robust estimation of the process standard deviation for control charts. Technometrics **39**(2), 127–141 (1997)
22. Yohai, V.J., Zamar, R.: High breakdown-point estimates of regression by means of the minimization of an efficient scale. J. Am. Stat. Assoc. **83**, 406–413 (1988)

Chapter 6
Sample Size Determination

A *Sample* size refers to the number of observational units. A sample size can only be calculated for some specific aspect (value of interest) of the *population* (the entire set of units) to be estimated, usually related to some important response variable. The value of interest may be mean, proportion, variance, correlation, etc. The procedure of drawing a sample from a population is called *random sampling* or simply *sampling*. Since sample forms the basics of all statistical decision making, it is essential to have a suitable sample size for any investigation. Such a sample size will depend on many factors such as population size, type of data (continuous or discrete), *precision* required, and confidence level. Figure 6.1 shows the sequence of activities taking place from the data collection to decision making [4].

6.1 Accuracy and Precision

The size of sample that will require attaining a given precision for such an estimate depends on the variability of the population and on the extent to which it is possible to reduce the different components of this variability in the random sampling error. The measured value that has little deviation from the actual value is called *accuracy*. Accuracy is usually tested by comparing an average of repeated measurements to a known standard value for that unit. Mean, mode, median, etc., are all accuracy measures. The *standard error* (standard deviation of the sample observations), a crude measure of the precision of an estimate obtained from a sample, is accurate enough to make the sample size calculation as it is a function of sample size.

Let us represent the population value of interest as μ, and let its sample estimate be $\bar{x}$. If σ is the standard deviation of the population, then the standard error is $\frac{\sigma}{\sqrt{n}}$, where n is the sample size. Note that, approximately one-third of the observable random variability in an estimate will be greater than the standard error and one-twentieth greater than twice the standard error [2]. An estimate will be biased with respect to the population of interest if the study fails to include certain units because of coverage or no-response errors. In the same way, the standard error can only

© Springer India 2015
K. Muralidharan, *Six Sigma for Organizational Excellence*,
DOI 10.1007/978-81-322-2325-2_6

81

Fig. 6.1 Data collection to
decision making

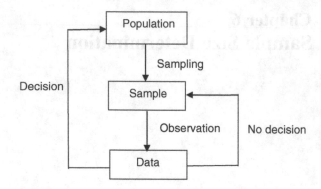

measure variability among samples due to not including all of the population in the observed sample; it does not take into account non-coverage and non-response. Lack of accuracy immediately places in question the value of any results and conclusions, whereas lack of precision generally only increases the uncertainty surrounding the exact values calculated.

Precision is how narrow you want the range to be for an estimate of a characteristic. For example, we quite often make the following statements about a process: The estimate of cycle time must be within two days, the estimate of the percent defective is within 3 %, the estimate of the proportion should be within 5 % of the total, the estimate of the standard deviation can vary between ±2 of the process mean, etc. So, what we mean by these statements is the amount of precision that can be allowed in the process specified according to the value of interest. Now, consider the following statements: The 95 % confidence interval (CI) of the cycle time is 40 ± 2 days [or the CI is (38, 42)], the 99 % CI of the sample proportion is 0.6 ± 0.15 [or the CI is (0.45, 7.5)], the 98 % approximate CI of the population variance is 12 ± 2.5 h [or the CI is (9.5, 14.5)], etc. That is, we are specifying the allowable precision (or margin of error) admissible in each situation along with the characteristics of the value of interest.

Note that, a CI in general can be written a

$$\text{Confidence interval} = \text{Accuracy measure} \pm \text{Precision}$$
$$= \text{Point estimate} \pm \text{critical value } X \text{ standard error} \qquad (6.1)$$

and therefore, precision is equal to half the width of a CI. For example, a 95 % confidence interval = (38, 42) for cycle time (in days) means we are 95 % confident that the interval from 38 to 42 days contains the average cycle time. Therefore, the width of the CI = 4 days, and hence, the precision is 2 days (i.e., the estimate is within ±2 days). Thus

$$\text{Precision} = \frac{\text{UCL} - \text{LCL}}{2}, \qquad (6.2)$$

where UCL and LCL are the upper confidence limit and lower confidence limit, respectively.

Suppose the characteristics of interest is the process average, say μ, and if the process follow a normal distribution with known standard deviation, say σ, then the 95 % CI for μ is $\bar{x} \pm 1.96 \frac{\sigma}{\sqrt{n}}$, which is equivalent to $\bar{x} \pm k \frac{1}{\sqrt{n}}$, where $k = 1.96\sigma$.

Similarly, the 95 % CI for population proportion p is $\hat{p} \pm 1.96\sqrt{\frac{\hat{p}(1-\hat{p})}{n}}$, which is equivalent to $\hat{p} \pm k_1 \frac{1}{\sqrt{n}}$, where $k_1 = 1.96\sqrt{\hat{p}(1-\hat{p})}$. Thus, precision is a function of the sample size and standard error. Moreover, precision (d) is inversely proportional to the square root of sample size n. That is

$$d \propto \frac{1}{\sqrt{n}} \qquad (6.3)$$

Hence, the knowledge of precision can ease the calculation of sample size. Note that, lesser the value of d, more precise the estimator will be. So for good precision, we need to have a large sample size. As such, there is no clear-cut answer about how much precision you need; the answer depends on the business impact of using the estimate. Each situation is unique and should not be influenced by someone else's decision. But it is always possible to make a wild guess about precision. Also note that, to improve precision, you need to increase sample size (which incurs more cost). The converse may not be true.

The next important aspect of sample size calculation is to get some knowledge about the standard deviation. You need to have some idea of the amount of variation in the data because as the variability increases, the necessary sample size increases. There are many options for this:

- Find an existing data and calculate $\hat{\sigma} = s$ or
- Use a control chart (for individuals) from a similar process and get $\hat{\sigma} = s$ or
- Collect a small sample and calculate $\hat{\sigma} = s$ or
- Take an educated guess about s based on your process knowledge and memory of similar data.

See also Chap. 5 for various methods of sigma estimation. Now, let us discuss the sample size computation in more detail and answer its statistical importance in estimation and inferential studies.

6.2 Sample Size When Characteristic of Interest Is Mean

As seen above, if the characteristics of interest is the process mean, say μ, then the $(1 - \alpha)100 \%$ CI for μ is $\bar{x} \pm z_{\alpha/2} \frac{\sigma}{\sqrt{n}}$. Here, α is called the significance level and $z_{\alpha/2}$ is the cut-off (or percentile) point of the area corresponds to $(\alpha/2)$ of the standard normal curve. Specifically, if the confidence level is sought for 95 %, then $\alpha = 0.05$ and $z_{\alpha/2} = 1.96$. Therefore,

$$d = 1.96 \frac{\sigma}{\sqrt{n}}$$

$$\Rightarrow n = \left(\frac{1.96\hat{\sigma}}{d}\right)^2 = \left(\frac{1.96s}{d}\right)^2 \qquad (6.4)$$

The practitioners working on industrial applications generally consider the approximate value of the sample size as $n \cong \left(\frac{2\hat{\sigma}}{d}\right)^2 \cong \left(\frac{2s}{d}\right)^2$. Even in most of the Six Sigma applications, the black belts consider this simplified formula. This adjustment will only increase a few samples extra and will not exaggerate the results in any way.

For many biological, psychological, and social science research, the researchers may look for more precision in their estimators as getting sample size may not be a difficult problem for them. In that case, the precision may be improved upon by considering higher levels of standard errors. That is, if the formula derived in (6.4) is corresponding to a precision of one standard error of the mean, then the formula corresponding to a precision of two standard errors (margin of error reduced by half) and three standard errors (margin of error reduced by one-third) are, respectively, given by

$$\frac{d}{2} = 1.96 \frac{\sigma}{\sqrt{n}}$$

$$\Rightarrow n = \left(\frac{2 \times 1.96\hat{\sigma}}{d}\right)^{2^2}$$

$$= \left(\frac{2 \times 1.96s}{d}\right) \qquad (6.5)$$

and

$$\frac{d}{3} = 1.96 \frac{\sigma}{\sqrt{n}}$$

$$\Rightarrow n = \left(\frac{3 \times 1.96\hat{\sigma}}{d}\right)^{2^2}$$

$$= \left(\frac{3 \times 1.96s}{d}\right) \qquad (6.6)$$

Note that, for more precise estimates, more sample size is required and vice versa. The sample size formula for different confidence level may be obtained similarly. The other value of α generally used in statistical research is $\alpha = 0.01$ (with $z_{\alpha/2} = 2.58$) and $\alpha = 0.10$ (with $z_{\alpha/2} = 1.645$). These values may be inserted

Table 6.1 Sample size for mean correspond to one precision

Level of confidence (%)	d = 1		d = 2		d = 4		d = 6		d = 10	
	s = 5	s = 10	s = 5	s = 10	s = 5	s = 10	s = 5	s = 10	s = 5	s = 10
90	67	272	17	68	4	17	2	8	1	3
95	96	384	24	96	6	24	3	11	1	4
98	136	543	34	136	8	34	4	15	1	5
99	166	666	42	166	10	42	5	18	2	7

Table 6.2 Sample size for mean correspond to two precision

Level of confidence (%)	d = 1		d = 2		d = 4		d = 6		d = 10	
	s = 5	s = 10	s = 5	s = 10	s = 5	s = 10	s = 5	s = 10	s = 5	s = 10
90	269	1089	68	272	17	68	8	30	3	11
95	384	1537	96	384	24	96	11	43	4	15
98	543	2172	136	543	34	136	15	60	5	22
99	666	2663	166	666	42	166	18	74	7	27

Table 6.3 Sample size for mean correspond to three precision

Level of confidence (%)	d = 1		d = 2		d = 4		d = 6		d = 10	
	s = 5	s = 10	s = 5	s = 10	s = 5	s = 10	s = 5	s = 10	s = 5	s = 10
90	605	2450	153	613	38	153	17	68	6	25
95	864	3457	216	864	54	216	24	96	9	35
98	1222	4886	305	1222	76	305	34	136	12	49
99	1498	5991	374	1498	94	374	42	166	15	60

suitably in the formula discussed above as per the requirement. Tables 6.1, 6.2, and 6.3 present the value of sample size according to various levels of confidence, the estimate of sigma and precision levels (all entries are corresponds to the actual $z_{\alpha/2}$ values).

Example 6.1 Extensive monitoring of a computer time-sharing system has suggested that response time to a particular editing command is normally distributed with standard deviation 25 ms. A new operating system has been installed, and we wish to estimate the true average response time μ for the new environment. Assuming that response times are still normally distributed with $s = 25$, what sample size is necessary to ensure that the resulting 95 % CI has a width of (at most) 10?

Solution Given

$$s = 25 \text{ and } d = \frac{\text{UCL} - \text{LCL}}{2} \frac{10}{2} = 5$$

Therefore,

$$n \cong \left(\frac{1.96 \times 25}{5}\right)^2 = 96, \text{ Correspond to one standard error precision}$$

$$n \cong \left(\frac{3.92 \times 25}{5}\right)^2 = 384, \text{ Correspond to two standard error precision}$$

$$n \cong \left(\frac{5.88 \times 25}{5}\right)^2 = 864, \text{ Correspond to three standard error precision}$$

Assuming that the width of the CI is still at most 10, then the sample size corresponding to 99 % CI is obtained as follows:

$$n \cong \left(\frac{2.58 \times 25}{5}\right)^2 \cong 166, \text{ Correspond to one standard error precision}$$

$$n \cong \left(\frac{5.16 \times 25}{5}\right)^2 \cong 666, \text{ Correspond to two standard error precision}$$

$$n \cong \left(\frac{7.74 \times 25}{5}\right)^2 \cong 1498, \text{ Correspond to three standard error precision}$$

6.3 Sample Size When Characteristic of Interest Is Proportion

As seen above, if the characteristics of interest is the population proportion p, then the 95 % CI for p is $\hat{p} \pm 1.96\sqrt{\frac{\hat{p}(1-\hat{p})}{n}}$. Therefore,

$$d = 1.96\sqrt{\frac{\hat{p}(1-\hat{p})}{n}}$$

$$\Rightarrow n = \frac{(1.96)^2\hat{p}(1-\hat{p})}{d^2} \tag{6.7}$$

Similarly, the formula correspond to a precision of two (margin of error reduced by half) and three standard errors (margin of error reduced by one-third) are, respectively, obtained as follows:

$$n = \frac{(2 \times 1.96)^2\hat{p}(1-\hat{p})}{d^2} \tag{6.8}$$

and

$$n = \frac{(3 \times 1.96)^2 \hat{p}(1 - \hat{p})}{d^2} \tag{6.9}$$

The sample size formula for proportion for different confidence level may be obtained similarly. Tables 6.4, 6.5, and 6.6 present the value of sample size for estimating proportion according to various level of confidence, the estimate of proportion and precision levels (all entries correspond to the actual $z_{\alpha/2}$ values).

Example 6.2 Suppose the proportion is about 80 % and the CI is suggested for a precision of 4 %, determine the necessary sample size.

Solution Given

$$\hat{p} = 0.8 \quad \text{and} \quad d = 0.04$$

Then

$$n \cong \frac{4(0.8)(0.2)}{(0.04)^2} = 400, \text{Correspond to one standard error precision}$$

$$n \cong \frac{16(0.8)(0.2)}{(0.04)^2} = 1600, \text{Correspond to two standard error precision}$$

$$n \cong \frac{36(0.8)(0.2)}{(0.04)^2} = 3600, \text{Correspond to three standard error precision}$$

Example 6.3 For a binary response study variable, the value of p will be the expected average probability under two treatments. Suppose the investigator wants to detect a difference in probability of response of 10 %, between 85 % under control and 95 % under active treatment. Determine the sample size corresponds to two standard error precision.

Solution Given

$$\hat{p} = 0.9 \quad \text{and} \quad d = 0.1$$

Then, the sample size corresponds to two standard error precision is obtained as follows:

$$n \cong \frac{16(0.9)(0.1)}{(0.1)^2} = 144$$

Table 6.4 Sample size for proportion correspond to one precision

Level of confidence (%)	d = 5 %			d = 10 %			d = 20 %			d = 30 %		
	p = 0.2	p = 0.5	p = 0.7	p = 0.2	p = 0.5	p = 0.7	p = 0.2	p = 0.5	p = 0.7	p = 0.2	p = 0.5	p = 0.7
90	172	269	226	43	67	56	11	17	14	5	7	6
95	246	384	323	61	96	81	15	24	20	7	11	9
98	347	543	456	87	136	114	22	34	29	10	15	13
99	426	666	559	107	166	140	27	42	35	12	18	16

Table 6.5 Sample size for proportion correspond to two precision

Level of confidence (%)	$d = 5\%$			$d = 10\%$			$d = 20\%$			$d = 30\%$		
	$p = 0.2$	$p = 0.5$	$p = 0.7$	$p = 0.2$	$p = 0.5$	$p = 0.7$	$p = 0.2$	$p = 0.5$	$p = 0.7$	$p = 0.2$	$p = 0.5$	$p = 0.7$
90	689	1076	904	172	269	226	43	67	56	19	30	25
95	983	1537	1291	246	384	323	61	96	81	27	43	36
98	1390	2172	1824	347	543	456	85	136	114	39	60	51
99	1704	2663	2237	426	666	559	107	166	140	47	74	62

Table 6.6 Sample size for proportion correspond to three precision

Level of confidence (%)	d = 5 %			d = 10 %			d = 20 %			d = 30 %		
	p = 0.2	p = 0.5	p = 0.7	p = 0.2	p = 0.5	p = 0.7	p = 0.2	p = 0.5	p = 0.7	p = 0.2	p = 0.5	p = 0.7
90	1549	2421	2033	387	605	508	97	151	127	43	67	56
95	2213	3457	2904	553	864	726	138	216	182	61	96	81
98	3127	4886	4104	782	1222	1026	195	305	257	87	136	114
99	3834	5991	5032	959	1498	1258	240	374	315	107	166	140

That is, 72 in each treatment group. For a better estimate, one may use 64 instead of 16 in the formula, if covariates are not considered in the study. This will improve the reliability and consistency of the measurements of response variables.

6.4 Sample Size When Characteristic of Interest Is Counts

If the characteristics of interest is some average counts, say μ, then $\hat{\sigma}$ may be estimated as $\sqrt{\hat{\mu}}$, then as developed in Sect. 6.2, we obtain the sample size as follows:

$$n = \left(\frac{1.96}{d}\right)^2 \hat{\mu}, \text{ Correspond to one standard error precision} \tag{6.10}$$

$$n = \left(\frac{2 \times 1.96}{d}\right)^2 \hat{\mu}, \text{ Correspond to two standard error precision} \tag{6.11}$$

and

$$n = \left(\frac{3 \times 1.96}{d}\right)^2 \hat{\mu}, \text{ Correspond to three standard error precision} \tag{6.12}$$

Tables 6.7, 6.8, and 6.9 present the value of sample size for estimating counts according to various level of confidence, the estimate of count and precision levels (all entries are correspond to the actual $z_{\alpha/2}$ values).

Example 6.4 If suppose the average number of literate people in a family is 3 and the interval suggested is within a precision of 0.4. Obtain the sample size to decide the population characteristic of literate people approximately.

Solution Given

$$\hat{\mu} = 3 \quad \text{and} \quad d = 0.4$$

Then

$$n = \left(\frac{1.96}{0.4}\right)^2 3 = 72, \text{ Correspond to one standard error precision}$$

$$n = \left(\frac{2 \times 1.96}{0.4}\right)^2 3 = 288, \text{ Correspond to two standard error precision}$$

$$n = \left(\frac{3 \times 1.96}{0.4}\right)^2 3 = 648, \text{ Correspond to three standard error precision}$$

Table 6.7 Sample size for counts correspond to one precision

Level of confidence (%)	d = 5 %			d = 10 %			d = 20 %		
	$\mu = 2$	$\mu = 4$	$\mu = 6$	$\mu = 2$	$\mu = 4$	$\mu = 6$	$\mu = 2$	$\mu = 4$	$\mu = 6$
90	2152	4303	6455	538	1076	1614	134	269	403
95	3073	6147	9220	768	1537	2305	192	384	576
98	4343	8686	13,029	1086	2172	3257	271	543	814
99	5325	10,650	15,975	1331	2663	3994	333	666	998

Table 6.8 Sample size for counts correspond to two precision

Level of confidence (%)	d = 5 %			d = 10 %			d = 20 %		
	$\mu = 2$	$\mu = 4$	$\mu = 6$	$\mu = 2$	$\mu = 4$	$\mu = 6$	$\mu = 2$	$\mu = 4$	$\mu = 6$
90	8607	17,213	25,820	2152	4303	6455	538	1076	1614
95	12,293	24,586	36,879	3073	6147	9220	768	1537	2305
98	17,372	34,745	52,117	4343	8686	13,029	1086	2172	3257
99	21,300	42,601	63,901	5325	10,650	15,975	1331	2663	3994

Table 6.9 Sample size for counts correspond to three precision

Level of confidence (%)	d = 5 %			d = 10 %			d = 20 %		
	$\mu = 2$	$\mu = 4$	$\mu = 6$	$\mu = 2$	$\mu = 4$	$\mu = 6$	$\mu = 2$	$\mu = 4$	$\mu = 6$
90	19,365	38,730	58,095	4841	9683	14,524	1210	2421	3631
95	27,660	55,319	82,979	6915	13,830	20,745	1729	3457	5186
98	39,088	78,176	117,264	9772	19,544	29,316	2443	4886	7329
99	47,926	95,852	143,778	11,982	23,963	35,945	2995	5991	8986

6.5 Sample Size When Characteristic of Interest Is Difference of Means

As we know, if the characteristics of interest is the difference of population means $\mu_1 - \mu_2$, then for equal sample sizes, the 95 % CI for $\mu_1 - \mu_2$ is $(\bar{x}_1 - \bar{x}_2) \pm 1.96 \sqrt{\frac{\sigma_1^2}{n} + \frac{\sigma_2^2}{n}}$. Therefore,

$$d = 1.96\sqrt{\frac{\sigma_1^2}{n} + \frac{\sigma_2^2}{n}}$$

$$\Rightarrow n = \frac{(1.96)^2(\sigma_1^2 + \sigma_2^2)}{d^2}, \tag{6.13}$$

where σ_1^2 and σ_2^2 are the variances of populations under consideration. Similarly, the formula corresponds to a precision of two (margin of error reduced by half) and three standard errors (margin of error reduced by one-third) are, respectively, obtained as follows:

$$n = \frac{(2 \times 1.96)^2(\sigma_1^2 + \sigma_2^2)}{d^2} \tag{6.14}$$

and

$$n = \frac{(3 \times 1.96)^2(\sigma_1^2 + \sigma_2^2)}{d^2} \tag{6.15}$$

The sample size formula for difference of means for different confidence level may be obtained similarly.

Example 6.5 The following information are given for estimating the difference of means: $\sigma_1 = 2.0, \sigma_2 = 2.5, \alpha = 0.01$ and $d = 1.2$. Decide the sample size.

Solution The sample size is obtained as follows:

$$n = \frac{(2.58)^2(\sigma_1^2 + \sigma_2^2)}{d^2}$$

$$= \frac{(2.58)^2(2^2 + 2.5^2)}{1.2^2}$$

$$= 47.38 \cong 47$$

6.6 Sample Size When Characteristic of Interest Is Difference of Proportions

If characteristics of interest is difference of population proportions $P_1 - P_2$, then for equal sample sizes, the 95 % CI for $P_1 - P_2$ is $(p_1 - p_2) \pm 1.96\sqrt{\frac{p_1(1-p_1)}{n} + \frac{p_2(1-p_2)}{n}}$. Therefore,

$$d = 1.96\sqrt{\frac{p_1(1 - p_1)}{n} + \frac{p_2(1 - p_2)}{n}}$$

$$\Rightarrow n = \frac{(1.96)^2[p_1(1 - p_1) + p_2(1 - p_2)]}{d^2} \tag{6.16}$$

where p_1 and p_2 are the sample proportions of the study populations. Similarly, the formula corresponds to a precision of two (margin of error reduced by half) and three standard errors (margin of error reduced by one-third) are, respectively, obtained as follows:

$$n = \frac{(2 \times 1.96)^2 [p_1(1 - p_1) + p_2(1 - p_2)]}{d^2} \qquad (6.17)$$

and

$$n = \frac{(3 \times 1.96)^2 [p_1(1 - p_1) + p_2(1 - p_2)]}{d^2} \qquad (6.18)$$

The sample size formula for difference of proportion for different confidence level may be obtained similarly.

Example 6.6 A marketing specialist of a car manufacturing company wants to estimate the difference between the proportion of those customers who prefer a domestic car and those who prefer an imported car. A historical study estimates 60 % of the customers preferred domestic car and 40 % preferred imported car. How large a sample should be considered to see the difference in proportion of customer's preference will have a margin of error of 2.5 % with a confidence coefficient of 99 %?

Solution Given $p_1 = 0.60, p_2 = 0.40, \alpha = 0.01, z_{\alpha/2} = 2.58$ and $d = 2.5\%$. Therefore, the sample size is obtained as

$$\begin{aligned}
n &= \frac{(2.58)^2 [p_1(1 - p_1) + p_2(1 - p_2)]}{d^2} \\
&= \frac{(2.58)^2 [0.6 \times 0.4 + 0.4 \times 0.6]}{(0.025)^2} \\
&= 5112.11 \cong 5112
\end{aligned}$$

When designing experiments intended for testing statistical hypothesis, one should consider both the desired level of significance (α) and the desired power of the test ($1 - \beta$) for the sample size calculations. The value α is the probability of wrongly rejecting a true hypothesis [also called P(Type-I error)], whereas β is the probability of accepting a wrong hypothesis [also called Type-II error)]. Generally, one can control both of these quantities by selection of the number n of replicates, the power with fixed α increasing as n increases. One may refer to Odeh and Fox [6] and Muralidharan [3] for a detailed study on sample size choice for experiments dealing with linear models and tests of hypothesis.

Bratcher et al. [1] provides tables of sample sizes in the analysis of variance models. The other useful references can be found in Stein [7] and Neyman and Pearson [5] and references contained therein. See Chap. 9 for a detailed study on tests of hypothesis and other decision-making tools.

6.7 Relevance for Managers

If all statistical analysis (quantitative research) address the issue of sampling and sample-based study, then the knowledge of sample size is crucial for any investigation. Studying samples is easier than studying the whole population. This chapter is therefore unique in addressing this concept. This study takes care of sample size computation for various characteristics such as mean, proportion, counts, difference of means, and difference of proportions. In many chapters to follow, the knowledge of sample size is an integral part in facilitating the inference thereafter. For example, in designing a control chart, we must specify both the sample size to use and the frequency of sampling. Another example is the study of hypothesis testing, where the sample size is crucial for increasing the power of the test.

For any meaningful inference, a project study is expected to have a large sample size. A large sample size generally improves the accuracy and precision of the estimate. Getting more precise estimators needs proper sampling frame and sampling units. So, the frequently asked questions such as "what is the ideal sample size" and "how large a sample size should be?" can be answered in the context in which the study is carried out. To some extent, we have simplified this issue in this chapter. Without delving too much on the technical aspects such as decision errors, significance level, and confidence coefficient, we have studied the relevance of sample size through an applied statistician perspective.

Exercises

6.1. Extensive monitoring of a computer time-sharing system has suggested that response time to a particular editing command is normally distributed with standard deviation 25 ms. A new operating system has been installed and one wishes to estimate the true average response time μ for the new environment. Assuming that response times are still normally distributed with $\sigma = 25$, what sample size is necessary to ensure that the resulting 95 % C.I has a width of (at most) 10?

6.2. A manufacturer of college textbooks is interested in estimating the strength of the bindings produced by a particular binding machine. Strength can be measured by recording the force required to pull the pages from the binding. If the force is measured in pounds how many books should be tested to estimate the average force required to break the binding within 0.11b with 95 % confidence. Assume that σ is known to be 0.8. What will be the size correspond to two precision?

6.3. Eleven rabbits were examined for the total fatty acid content of their plasma, with the following results (in mg/100 ml):

169, 168, 154, 156, 172, 163, 169, 175, 150, 167 and 166.

Find a 90 % confidence interval for the mean of the population from which this sample has been taken, giving the conditions and assumptions necessary for this calculation, to be valid. An experimenter wishes to estimate the true fatty acid content of the plasma to within ±1 % of the true value, with 90 % confidence. How large a sample would you recommend?

6.4. A researcher wants to estimate the mean of a population by taking a random sample and computing the sample mean. He wants to be 95 % confident that margin of error of the estimate is 2. If the population standard deviation is 20, how many observations need to be included in the sample? Estimate the same for two and three precisions, respectively?

6.5. A newspaper reporter is preparing a story on automobile inspection stations and the reporter wants to estimate the average cost of a state inspection (excluding the cost of repairs). The reporter believes that costs will vary from $10 to $25. How many inspection stations should be surveyed if the reporter wants the margin of error of the estimate to be at most $1 with 95 % confidence?

6.6. In a manufacturing process, an estimate is sought for the difference between two population means. If suppose the population standard deviations are known to be 3 and 3.5, respectively. How large a sample should be taken so that with probability 99 % the estimate is within 1.5 units of the true value of $\mu_1 - \mu_2$.

6.7. The means of two large samples are 67.5 and 68, respectively. If the two samples have same standard deviation as 2.5, then decide the sample sizes, such that the difference in population mean will have a probability 95 % with a margin of error within 2 units. Estimate the same for two and three precisions, respectively?

6.8. A vaccination drive is proposed to carry out to assess the support of people who favor the new medicine for a flue. How large a sample should be taken in order to estimate the proportion of people with a margin of error of 3 % with 95 % probability? Estimate the same for two and three precisions, respectively?

6.9. A marketing specialist of a tea manufacturing company wants to estimate the difference between the proportions of those customers who prefer two particular brands of tea. A historical study estimate 45 % of the customers preferred the first brand and 50 % preferred the second brand. With a confidence coefficient of 99 %, how large a sample should be considered to see the difference in customer's preference is within a margin of error of 2.5 %?

6.10. Discuss the relevance of standard deviation in the determination of sample size.

References

1. Bratcher, T.L., Moran, M.A., Zimmer, W.I.: Tables of sample sizes in the analysis of variance. J. Qual. Technol. **2**, 156–164 (1970)
2. Lindsey, J.: Revealing Statistical Principles. Arnold Publishers, London (1999)
3. Muralidharan, K.: On sample size determination. In: International Journal of Computational Mathematics and Numerical Simulations, (to appear) (2014)
4. Muralidharan, K., Syamsundar, A.: Statistical Methods for Quality, Reliability and Maintainability. PHI Publications, New Delhi (2012)
5. Neyman, J., Pearson, E.S.: The use and interpretation of certain test criteria for purpose of statistical inference. Part-I, Biometrika **20a**, 175–240 (1928)
6. Odeh, R.E., Fox, M.: Sample Size Choice: Charts with Experiments with Linear Models. Marcel Dekker, INC, New York (1991)
7. Stein, C.: A two-sample test for a linear hypothesis whose power is independent of the variance. Ann. Math. Statist. **16**, 243–258 (1945)

Chapter 7
Define Phase

7.1 Project Charter

A project charter is a document stating the purposes of the project. A project charter at the define phase sets the stage for a successful Six Sigma project by helping you to answer four critical questions:

1. What is the *problem* or opportunity on which we will focus?
2. What *results* do you want to accomplish and by when (usually the goals)?
3. Who is the *customer* served and/or impacted by this process and problem?
4. What is the *process* we are investigating?

In documenting project goals and parameters, improvement teams can help ensure that their work meets with the expectations of their organization leaders and project sponsor. Thus, a project charter serves as an informal contract that helps the team stay on track with the goals of the enterprise.

7.1.1 The Problem Statement

This is a concise and focused description of "what's wrong"—either the pain arising from the problem or the opportunity that needs to be addressed. In some cases, the problem statement can be a distilled version of the project rationale, but usually, a team will need to define their issue much more specifically, since even the best project rationale statements will be pretty broad.

A problem statement generally serves the purposes such as:

- Validating the project rationale
- Consensus formation of project teams
- Assessing the clarity of the supporting data
- Establish a baseline measure against which progress and results can be tracked.

© Springer India 2015   99
K. Muralidharan, *Six Sigma for Organizational Excellence*,
DOI 10.1007/978-81-322-2325-2_7

A problem statement never implies a cause or a solution, but identifies a visible deficiency in a planned outcome. A mission statement is based on the problem statement but provides direction to the project team. A goal or other measure of project completion and a target date should be defined as a mission for a better result. Like problem statement, mission statement should not imply cause or solution.

7.1.2 The Goal (or Result) Statement

A *goal* is a desired result to be achieved in a specified time. A goal statement defines "relief" in terms of concrete results. Goal statement structure can be standardized into three elements:

1. A description of what is to be accomplished.
2. A measurable target for desired results. The target should quantify the desired cost saving, defect elimination, or time reduction, etc., in percentages or actual numbers.
3. A project deadline and/or time frame for results. The data set in the early part of the project may need to be revised later, but establishing a deadline helps to rally resources and commitment and shorten project cycle time.

A goal statement should be SMART:

- *Specific*: A goal statement should not be very general and philosophical, narrow it to the goal
- *Measurable*: Unless the team has measurable goals, it will not know whether it is making progress or it has succeeded
- *Achievable*: The target achievable through planning and execution can also be aggressive in nature
- *Relevant*: The goal must be specifically linked to the strategic goals of the enterprise
- *Timely*: The goal must make sense in the time frame in which the team must work

Defeo and Barnard [1] identify seven areas in which goals are minimally required:

- Product performance
- Competitive performance
- Quality improvement
- Cost of poor quality
- Performances of business processes
- Customer satisfaction
- Customer loyalty and retention

According to Gryna et al. [2], the quality goals are identified through a "company-wide assessment of quality." These studies identify the strengths, weaknesses, opportunities, and threats. The other inputs to help formulate goals include the following:

- Pareto analysis of repetitive external alarm signals (failures, complaints, returns, etc.)
- Pareto analysis of repetitive internal alarm signals (scrap, rework, sorting, testing, etc.)
- Proposals from key insiders—managers, supervisors, professionals, etc.
- Proposals from suggestion schemes
- Field study of users' needs and costs
- Data on performance of products versus competitors
- Comments of key people outside the company (customers, vendors, critics, etc.)
- Findings and comments of government regulators, reformers, etc.

Table 7.1 depicts a project charter for optimization in electrical power consumption in a batch plant. The business problem here is to identify the areas where power consumption can be decreased or optimized and also to identify the areas of improvement. The goal statement clearly shows the company's current requirement to reduce the electrical power consumption by 1000 kWh per day from 3100 kWh. This reduction will lead to savings of Rs. 80 lakhs per annum for the year 2014–2015.

7.1.3 Customer Identification

The person or organization that receives an output from a process is called customer. They are the people who are affected by the product or process. Hence, the person or organization that is the end user of a product or service is also a consumer. The three broad categories of customers are as follows:

- External customers—A person or organization that receives a product, a service, or information, but is not part of the organization supplying it
- Internal customers—A person or process within the organization that receives output from another person or process within the same organization. They can be both management and stakeholders
- Suppliers as customers—A person or process outside the organization that provides the input to an organization

Among the category of each, customers are further classified according to their defined needs and requirements. Hence, how you define your strategy and differentiate your customers will have a big impact on the accuracy of your data and the resources needed to establish the voice of customer system. Broadly, they can be classified as follows:

Table 7.1 A project charter for optimization in electrical power consumption

Project charter			
General project informations			
Project name	Optimization in electrical power consumption in batch plant		
Project manager	Dr. XYZ	Start date: 10.01.2013	End date: 10.07.2013
Organizational unit	Power alignment plant		
Project essentials			
Business case	The electrical power consumption due to batch plant operation is Rs. 53 lakhs for the year 2013. Thereby, it requires focus for identification of areas where power consumption can be decreased/optimized and also identify areas of improvement		
Problem statement	The electrical average power consumption of batch plant is 3100 kWh per day for year 2013 which is in terms of Rs. 110 lakhs		
Goal statement	To reduce the electrical power consumption by 1000 kWh per day from 3100 kWh (base level that is considered the average per day of the year 2013), which will lead to a savings of Rs 80 lakhs per annum for the year 2014–2015		
Project KPOVs	1. Monthly power consumption at batch plant		
	2. Pull/no. of batches		
Assumptions	Size of the batch should be uniform throughout the year		
Constraints	Nil		
Project scope (in scope/out scope)	In scope: number of batch, pull		
	Out scope: water heater usage, batch recipe		
	Key milestone/ timelines	Start date	Completion date
Project schedule	Review and scope	10.01.2013	16.01.2013
	Plan and kickoff	17.01.2013	20.01.2013
	Define phase	21.01.2013	28.01.2013
	Measure phase	29.01.2013	15.02.2013
	Analyze phase	16.02.2013	19.03.2013
	Improve phase	21.03.2013	30.04.2013
	Control phase	01.05.2013	10.06.2013
	Summary and closure	11.06.2013	10.07.2013
Project team	Mr. L. Patel (Champion), Dr. R. Ravi (MBB), Mr. Rahul (BB), Mr. Vijay Kumar, Mr. M. Pawan, Mr. Peter Marcos, Mrs. S. Maiti, Mr. K. Malhotra		

- Current, happy customers—those customers who are always loyal to the company and their products
- Current, unhappy customers (that includes both those who complain and those who do not)—those customers who are indifferent to product and services
- Lost customers—those customers who are one time visitors

- Competitors customers—those customers who enjoy services as per convenience and shift loyalty
- Prospective customers—i.e., those who have not purchased from you or your competitors, but are potential buyers of your products/services

According to Harrington and Anderson [3], the characteristics of satisfied customers are as follows:

- The customer understands through the producer's actions that he or she is valued as an individual or organization
- The customer understands through actions or proactive plans that the producer will be effective in solving any reliability problem if and when it occurs
- The customer is comfortable with the producer's activities that monitor and communicate product performance
- The customer feels convertible in establishing some degree of partnership and sharing responsibility for business investments and benefits
- The customer understands through experience that the producer always strives to exceed the specified reliability

Why is the customer important? In light of today's speed of change, it is most likely that the organizations that do not give importance to their customers are going to suffer a lot in terms of productivity. Over and above, an organization should keep a tag on creating and maintaining a comprehensive system to gather and use customer input and market trends in tandem with time. Further,

- Customer identity brings a real awakening to a business and its leaders as well
- Making improvement should be treated as a continuous effort
- It imposes strategic impact on the accuracy of data
- It helps to identify the trends happening in the market
- It helps to address your weaknesses and motivate you for new efforts

Failing to disseminate customer-focused knowledge throughout the organization can also be a serious weakness for some and handy for some. Still the efforts to gather various inputs and understand customer requirements are always a smart approach.

How to define customer needs and requirements? Customers define their needs as per their output requirements and service requirements. The output requirements are the features and/or characteristics of the final product or service that are delivered to the customer at the end of the process, whereas service requirements are guidelines for how the customer should be treated or served during the execution of the process itself. Specifically, a customer's requirements depend on

- Identifying the output or service situations
- Identifying the customer or customer segment
- Reviewing available data on customer needs, expectations, comments, complaints, etc.
- Drafting a requirement statement
- Validating the requirement
- Refining and finalizing the requirement statement

See Pande et al. [4], for more details.

While identifying core process and key customer, it is always advised to focus on activities that directly add value to customers and involve a mix of people for opinion. See that you are not overloading the process with lots of inputs and outputs without making drastic changes in your core processes.

7.1.4 Process Models

This is the most important part of the activity involved at the define stage. A process is a set of linked or related activities that take inputs and transform them into outputs with the purpose of producing a product, information, or service for external or internal customers. Therefore, it is essential to establish a communication link between the customer requirements and the process parameters. This is where the need for process orientation arises. This can be done through graphically outlining the sequence of a process, showing how steps in a process relate to each other, identifying bottlenecks, pinpointing redundancies, and locating waste in the process, etc. In a way, these activities will help us to achieve two things: (i) document of how the process works and (ii) to identify the value-added (VA) and non-valued activities. One of the best methods of describing the process orientation and mapping is the suppliers, input, process, outputs and customer (SIPOC) model described below.

The supplier–customer relationship can best be described using a SIPOC diagram which is one of the most useful and often used techniques of process management and improvement. It is used to present an "at-a-glance" view of work flows. SIPOC is an acronym for suppliers, input, process, outputs and customer.

- *Supplier*—the person or group providing key information, materials, or other resource to the process
- *Input*—the "thing" provided
- *Process*—the set of steps that transforms and ideally adds value to the input
- *Output*—the final product of the process
- *Customer*—the person, group, or process that receives the output

A SIPOC diagram (sometimes called high-level process map) is an extended version of the input-process-output (IPO) diagram where supplier–customer interface is brought into entire improvement activities of the organization. Such an involvement often results in joint economic planning, joint technological planning, and cooperation during contract execution. It also generates customer confidence and healthy working atmosphere in the organization. A SIPOC diagram is given in Fig. 7.1.

Depending upon the organization and the project, the process model may vary. For example, instead of identifying suppliers first, for a customer-driven organization, the critical process area may start from the customer level and proceed to supplier level. In those cases, the model may be reverted to customer, output,

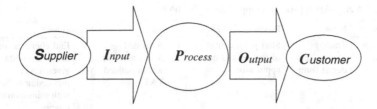

Fig. 7.1 A SIPOC diagram

process, input, and supplier (COPIS) to show the process mapping. In both the cases, it is the process of thinking that produces excellence in the output unit. Some of the process-thinking intangibles are as follows:

- Customer thinking (customer specifications)
- Process thinking (leading indicators)
- Statistical thinking (analysis of variation)
- Casual thinking [usually of the form: $y = f(x_1, x_2, \ldots, x_n)$]
- Experimental thinking (data-driven hypothesis testing)
- Control thinking (process quality control)
- Common language/way of thinking about problems
- Accountability thinking

A typical SIPOC model for an assembly and packaging unit is presented in Table 7.2.

In Table 7.3, we present a high-level process map, established through a SIPOC diagram. This kind of diagram exposes missing links, bottlenecks, unnecessary steps, and redundancies in the process. A SIPOC diagram is an essential tool for initiating the measure phase of DMAIC. This also links the customer requirements with the project charter for creating an efficient process map. In Sect. 7.5 below, we embark on these issues.

Table 7.2 A SIPOC model

Suppliers	Inputs	Process	Outputs	Customers
1. Valve manufacturers (external and internal) 2. Body manufacturers (external and internal) 3. Planning shop 4. Packaging materials suppliers 5. Turning shop 6. Production planning	1. Valves 2. Bodies 3. Packaging materials 4. Miscellaneous consumables 5. Production schedules	1. Assembly of materials 2. Departmental orientation 3. Layout activities in sequence 4. Material preparation 5. Coding and segregating the materials 6. Preparing the material for quality inspection	1. Assembled and packaged valves 2. Production summaries 3. Materials' usage reports 4. Leak test report	1. Warehouse (internal) 2. End users (external) 3. Production scheduling 4. Purchasing 5. Quality assurance

Source Muralidharan and Syamsundar [12]

Table 7.3 A high-level process mapped through SIPOC

Suppliers		Input	Process (high level)	Output		Customers	
1	Partners clients	Trading account and demat form	Start point collection of trading and demat application	1	Activated demat account within defined TAT	1	End client: account activation within given TAT, processing of form with minimum queries
		Supporting documents	Operational activities			2	Partner: account activation within TAT, processing of form with minimum queries, update of every status
			1	Application and document verification			
			2	Inward entry			
			3	In person verification			
			4	Receipt at HO			
			5	Second-level verification			
			6	Activation of demat account			
			7	Activation of trading account			
			8	Activation/ rejection of application			

According to Pande et al. [4], the benefits of SIPOC are the following:

- It displays a cross-functional set of activities in a single and simple diagram
- It uses a framework applicable to processes of all sizes
- It helps to maintain a "big picture perspective" to which additional detail can be added.

7.2 Defining Team Roles

The two general types of teams are *informal* and *formal*. *Formal teams* have a specific goal or goals linked to the organization's plans. The goal is referred to as the mission or statement of purposes of the team. The formal team may also have a charter that includes its mission statement, a listing of team members, and a statement of support from management. *Informal teams* usually have none of these documents and may have a more fluid membership depending on their needs. The other teams are as follows:

- *Virtual teams*: made up of people in different locations who may never meet in person. Instead, the team may meet using conferencing facilities, or they may conduct all communication through written words via e-mail. These teams are used when expertise is separated geographically or temporally
- *Process improvement teams*: formed to find and propose changes in specific processes. These teams are usually cross-functional, involving representation from various groups likely to be impacted by the changes
- *Self-directed teams* and *work group teams*: usually have a broader and more ongoing mission involving day-to-day operations. They typically are in a position to make decisions about safety, quality, maintenance, scheduling, personal, and so forth

Teams seem to work best when various members have assigned jobs and roles and others understand the team rules and dynamics. The roles associated with various personnel with a Six Sigma projects can be summarized as follows: (Pande et al. [4], Kubiak and Benbow [5]).

The leadership group or council. They are the top management team who understand the relevance of Six Sigma project, who can delegate and initiate projects. Their job profile includes the establishment of roles and responsibilities of other team members, selecting projects and allocating resources, reviewing the progress of ongoing projects, assessing progress, identifying strengths and weaknesses in the effort, sharing best practices throughout the organization, and applying the lessons learned to their own individual management styles.

The Sponsor or Champion or the Process Owner. They are the authorizing entity of the project. They prepare objectives, goals, and scope necessary for a project. They find resources for the projects; help to smooth out issues and overlaps that arise between teams, or with people outside the team; work with process owners to ensure a smooth handoff at the conclusion of an improvement projects; and apply their gained knowledge of process improvement to their own management tasks. They are the people who take on a new, cross-functional responsibility to manage an "end-to-end" set of steps that provide value to an internal or external customer. He or she receives the "handoff" from improvement teams or becomes the owner of new and newly designed processes.

The Facilitator. They are the people who chair the meeting and sessions. A facilitator helps the team leader keep the team on track and ensures that all members have an equal opportunity to speak, contribute, and discuss. He summarizes progress using visual aids as appropriate, provides methods for reaching decisions, and helps resolve conflicts.

The Six Sigma Coach. The coach provides expert advice and assistance to process owners and Six Sigma improvement teams, in areas that may range from statistics to change management to process design strategies. The coach is the technical expert, though the level of expertise will vary from business to business based on how the roles are structured and the level of complexity of the problem. Their responsibility also includes resolving team member disagreements and conflicts, gathering and analyzing data about team activities, and helping teams promote and celebrate their successes.

The Team Leader or Project Leader. The team leader is the individual who takes primary responsibility for the work and the results of a Six Sigma project. Most team leaders focus on process improvement or design/redesign, but they also can take on efforts tied to voice of customer systems, measurements, or process management. Some of his or her specific responsibilities are as follows:

- Reviewing/clarifying the project rationale with the sponsor
- Developing and updating the project charter and implementation plan
- Helping or selecting to select the project team members
- Identifying and seeking resources and information
- Maintaining the project schedule and keeping progress moving toward final solutions and results
- Supporting the transfer of new solutions or processes to ongoing operations, while working with functional managers and/or the process owner
- Documenting final results and creating a executive summary of the project

The Team members. They are the people who drive the project to its developmental end. They may also be termed as the brains and muscle behind the measurements, analysis, and improvement of a process. They communicate ideas, and expertise listens to all ideas and completes action assignments as scheduled.

Note that, these "generic" roles are not all mandatory, as there can be some overlap among these responsibilities as it is. Since Six Sigma philosophy advocates the use of "belt" system (or titles) for the personal involved in a project, the roles and responsibilities also will be assigned as per the qualification of "belts" gained by the individual. In Table 7.4, we provide the variations in generic roles and "belts."

Green Belt (GB). Green belts are close to process operations and work directly with shop floor operators and service delivery personnel. They are responsible for data collection, make initial interpretations, and begin to formulate recommendations that are fed to black belts.

Black Belts (BB). A team leader trained in the DMAIC process and facilitation skills, responsible for guiding an improvement project to completion. They are responsible to train and mentor GB as well as lead improvement projects using specified methodologies. BBs generally perform rigorous statistical analysis with

Table 7.4 Role versus titles

Generic role	"Belt" or other title
Leadership council	Quality council, Six sigma steering committee
Sponsor	Champion, process owner
Implementation leader	Six Sigma director, quality leader, master black belt
Coach	Master black belt or black belt
Team leader	Black belt or green belt
Team member	Team member or green belt
Process owner	Sponsor or Champion

Source Pande et al. [4]

additional data and input from other sources and make recommendations to MBBs and project Champions.

Master Black Belt (MBB). They are responsible to train full time and to mentor BB as well as lead the strategy personals involved with a Six Sigma project. It is their responsibility to ensure that the improvement activities are in place and in right perspectives. An MBB can be a trained BB or a consultant outside the organization. They should have a thorough expertise in decision-making subjects and engineering concepts. The responsibilities also cover the implementation of DMAIC or DMADV and Design for Six Sigma (DFSS) concepts.

The appointments of officials and representatives are altogether the organizations own matter of policy. In some situations, team composition will change over the life cycle of the project. A team must be provided with the resources needed to complete team activities and assignments. This includes ensuring that team member participation is viewed as part of a team member's job and that adequate time is allotted to complete the assignments. Team members and team dynamics bring resources to bear on a project that people working individually would not be able to produce. The best teams with the best intentions will perform sub-optimally, however, without careful preparation. There seems to be no standard list of preparation steps, but at a minimum, the following items should be performed [5]:

- Set clear purposes and goals that are directly related to the project charter. Teams should not be asked to define their own purpose, although intermediate goals and timetables can be generated
- Team members should be given some basic team-building training. Without an understanding of how a team works and the individual behavior that advances team progress, the team will often get caught in personality concerns and turf wars
- A schedule of team meetings should be published early, and team members should be asked to commit to attend all meetings
- Teams can succeed only if management wants them to. Team members must know that they have the authority to gather data, ask difficult questions, and think outside the box
- Team meetings should end with a review of activities individual team members will complete before the next meeting. Minutes highlighting these activities should follow each meeting
- All teams should have a sponsor who has a vested interest in the project. The sponsor reviews the team's progress, provides resources, and removes organizational roadblocks.

7.3 Managing the Project Team

As discussed earlier, a project team can vary at any point of time with minor changes in the responsibilities and roles. A strong leadership is necessary to head such a team. An important part of team leadership is generating and maintaining a high level of motivation toward project completion. The best way of showing recognition is through letters of appreciation sent to individuals and public expression of appreciation via meetings and newsletters. During team meetings, facilitators and other team members can enhance motivation by setting an example of civility and informality, relaxing outside roles, and exhibiting a willingness to learn from one another. Teamwork may be enhanced by celebrating the achievement of a project milestone. According to Kubiac and Benbow [5], a project team is said to go through stages such as:

- *Forming*: Members struggle to understand the goal and its meaning for them individually
- *Storming*: Members express their own opinions and ideas, often disagreeing with others
- *Norming*: Members begin to understand the need to operate as a team rather than as a group of individuals
- *Performing*: Team members work together to reach their common goal
- *Adjourning*: A final team meeting is held, during which management decisions regarding the project are discussed and other loose ends are tied up
- *Recognition*: The team's contribution is acknowledged

The other important aspects of team management are as follows:

Communication. Lack of communication is one of the most frequently noted causes of team's failure. Serious effort toward improving communication should be made at each stage of team's development.

Initiating the team. The team's initiation should be in the form of a written document from an individual at the executive level of the organization. A separate document from each member's immediate supervisor is also helpful in recognizing that the team member's participation is important for the success of the team. Team norms are often established at the first meeting. They provide clear guidelines regarding what the team will and will not tolerate, and often define the consequences of violating the norms. Announcement of team meetings including time, location, agenda, and such must be made as early as possible. Minutes or summary of the meeting should be provided immediately after the session. Following the project completion, final reports and follow-up memos regarding disposition of team proposals should be provided to each team member.

Team dynamics. Once teams have been formed, they must be built, because a true team is more than a collection of individuals. A team begins to take on a life of its own that is greater than the sum of its parts. The building of team begins with well-trained members who understand roles and responsibilities of team members and how they may differ from roles outside the team. The best model for team leadership is the coach who strives to motivate all members to contribute their best. Productive teams occur when team coaches facilitate progress while recognizing and dealing with obstacles. The coaching function may be performed by team leaders or team facilitators or both.

Time management. Time is the most critical resource for any team. Although meetings and presentations are considered to be non-value-added (NVA) activities, time management is an essential part of any project management. It is up to the team leader and the facilitator to make every minute count, although every team member has this responsibility as well. These things can be planned through Gantt chart, affinity diagrams, and time line preparations.

According to Muir [6], the GRPI (described below) tool helps manage a successful team over the length of the project. The roles and responsibilities can continue as input in the GRPI model. The GRPI model may also be useful as a diagnostic tool when the team is not working well and you are not sure what is wrong.

The GRPI model periodically assesses the status of the project with respect to:

- *Goals*—The goal should be SMART as described above and included in the project charter. Each member of the team should understand and be able to explain the project's goal in their own terms
- *Roles and responsibilities*—Team members should know what parts they are associated with and for which others are responsible for
- *Processes and procedures*—This is the area where a strong communication is required
- *Interpersonal relationships*—It may be necessary to itemize the rules of conduct for the team. Like communication, it is very essential for establishing transparency in the entire duration of the project

Initially, the goals, roles, and processes are the most important to define, but as the project progresses, the interpersonal state of the team can quickly become the greatest concern for continued success.

7.4 Planning Tools

Planning a project facilitates establishment of flow in the project. The planning process facilitates the prerequisites for achieving and sustaining improvements in business performance. According to DeFeo and Barnard [1], the six types of breakthroughs as prerequisites for achieving business performances are as follows: (1) leadership, (2) organization, (3) current performance, (4) management, (5) adaptability, and (6) culture. The interaction, dynamics, and leadership are all important for increasing the performance of a business activity. The idea for a project is often the result of a good leadership and experience of the organization, customer base, or marketplace. This technical knowledge is often considered as the perfect qualification for leading a project team to a successful outcome.

Planning may be divided into three categories:

1. *Strategic activities*: A strategic planning process helps the management to study the current state of the process in which the organization operates, envisions the ideal future state, and plans the path for realizing the project. Strategic planning must be a priority of top management. For strategic planning to be successful, senior management must be actively engaged in the process. It is not something

that can be delegated. Further, successful strategic planning must create a line of sight from strategic to tactical to operational activities. This does not mean that individuals executing activities at the lowest level of the organization must understand the details of the strategic plan.

2. *Tactical activities*: Tactical plans are designed to support strategic plans. TRIZ (theory of inventive problem solving), PUGH analysis (a decision matrix, when a single option must be selected from several and multiple decisions), etc., are some of the tactical planning of the processes.

3. *Operational activities*: The planning process is really operational planning directed at product and process planning. This activity is in contrast to strategic planning, which establishes long-term goals and the approach to meet those goals.

In the context of the Juran Trilogy, DFSS and other design methodologies are for planning, whereas Six Sigma, Lean, and such methodologies are for improvement. For a detailed discussion on these methodologies from planning perspectives, one may see Gryna et al. [2]. Two of the visual methods of planning processes are described below.

7.4.1 Gantt Chart

Gantt chart or time line are a type of bar chart used in process/project planning and control to display planned work and finished work in relation to time. The information from project network diagrams such as Project Evaluation and Review Techniques (PERT) and critical path method (CPM) helps to prepare Gantt charts. A sample of Gantt chart is shown in Fig. 7.2.

Gantt charts were considered extremely revolutionary when they were first introduced by James [7] in recognition of Henry Gantt's contributions. The Henry Laurence Gantt Medal is awarded for distinguished achievement in management and in community service. This chart is also used in information technology to represent data that has been collected. Modern Gantt charts also show the dependency (i.e., precedence network) relationships between activities and can be used to

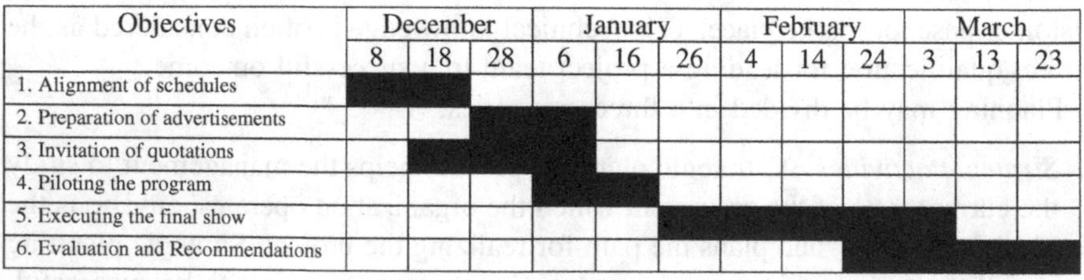

Objectives	December			January			February			March		
	8	18	28	6	16	26	4	14	24	3	13	23
1. Alignment of schedules	■											
2. Preparation of advertisements		■										
3. Invitation of quotations			■									
4. Piloting the program				■								
5. Executing the final show						■	■	■				
6. Evaluation and Recommendations										■	■	■

Fig. 7.2 Example of a Gantt chart

S. No	Activity	Hours	11 March, 2012 to 22 April 2012	25 April, 2012 to 12 May 2012	15 May, 2012 to 31 May 2012	1 June, 2012 to 20 June 2012
1	Identification of failure		⟹			
2	Dismantling			⟹		
3	Repair / replacement				⟹	
4	Reassembly					⟹
5	Alignment / Testing					⟹

Fig. 7.3 A typical corrective maintenance work schedule for repair on failure

show current schedule status. A maintenance schedule of an industrial process can also be depicted in the form of Gantt charts as shown in Fig. 7.3.

Thus, Gantt chart and activity diagrams are easy to understand and can track the project activities at any phases. They are sometimes used to convey project information involving changes that are easier to draw or photograph than to explain in words. A relatively new diagram most often used in workplace management (or *Gemba* organizations) is the affinity diagram, which is described below.

7.4.2 Affinity Diagram

Affinity diagrams are the tools used to organize information and help achieve order out of the chaos that can develop in a brainstorming session. Large amount of data, concepts, and ideas are grouped on the basis of their natural relationship to one another. It is more a creative process than a logical process. It helps to produce numerous possible answers to an open problem initiated through a brainstorming session. This is considered to be one of the seven Japanese management and planning tools. The other related tools are as follows:

- *Tree diagram*—Also called systematic diagram and helps dissect a problem into sub-problems and causes
- *Process decision program chart*—Assess alternative processes to help select the best process
- *Matrix diagram*—Shows the presence or absence of relationships among collected pairs of elements in a problem situation
- *Interrelationship diagram*—Helps identify the key factors for inclusion in a tree diagram
- *Prioritization matrix*—Evaluates options through a systematic approach of identifying weighing and applying criteria to the options

- *Arrow diagram*—Also called activities network diagram. It helps to analyze the sequence of tasks necessary to complete a project and determines the critical tasks to monitor and to execute the project efficiently

For recent literatures on these creative methods of quality management, one may refer to Gryna et al. [2] and Wardsworth et al. [8] and references contained therein.

7.5 Process Map and Flowchart

Process maps and flowcharts are among the two most essential tools of Six Sigma, in which improving, designing, measuring, and managing processes are the primary focus. A process map/flowchart involves a series of tasks (rectangles) and decisions/reviews (diamonds), connected by arrows to show the flow of work. As you build process maps for your Six Sigma projects, you are likely to find that some of the most enlightening information comes right in the actual "map creation" session, as people start to hear about how work is done and processes managed in other parts of the business. When a process is documented and validated, you can analyze it for some of the following specific problem areas:

- *Disconnects*: Points where handoffs from one group to another are poorly handled or where a supplier and customer have not communicated clearly on one another's requirements.
- *Bottlenecks*: Points in the process where volume overwhelms capacity, slowing the entire flow of work. Bottlenecks are the "weak link" in getting products and services to customers on time and in adequate quantities.
- *Redundancies*: Activities that are repeated at two points in the process also can be paralleled to activities that duplicate the same result.
- *Rework loops*: Places where a high volume of work is passed "back" up the process to be fixed, corrected, or repaired.
- *Decisions/Inspections*: Points in the process where choices, evaluation, checks, or appraisal intervene-creating potential delays. These activities tend to multiply over the life of a business and/or process.

The purpose of the process mapping is to document the process steps necessary for taking the transactions from the customer into the business processes and then bringing it back to the customer. A detailed process map also helps to identify the VA and NVA activities in a process. At this stage, our focus should be on understanding the nonces of VA/NVA activities, not the classification of such activities. In the measure phase, we discuss this aspect in detail and accordingly prepare the VA/NVA process map.

A flowchart depicts in a pictorial form, the sequence in which conditions are to be tested and process activities carry out a particular task. Figure 7.4 shows the flow

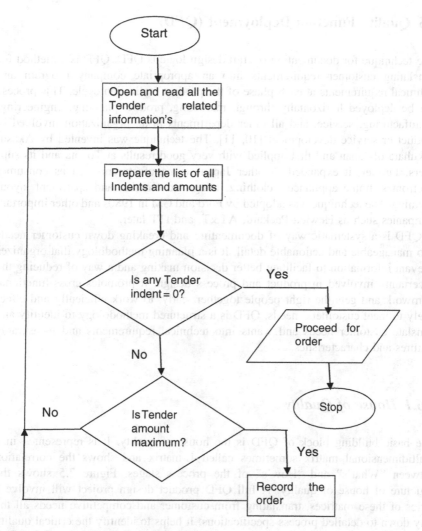

Fig. 7.4 A flowchart for tender identification

diagram for picking the largest tender in an order process. The diagram enables one to sequentially carryout the highest tender amount along with the tender identification from a large number of database of tenders. Flow diagrams use four symbols for events: a diamond for a decision-making event, a small square for an activity, a group of consecutive squares for a rework loop, and a paper symbol for a document or database (see also Salvendy [9]).

7.6 Quality Function Deployment (QFD)

One technique for documenting overall design logic is QFD. QFD is a method for translating customer requirements into an appropriate company program and technical requirements at each phase of the product realization cycle. This process can be deployed horizontally through marketing, product planning, engineering, manufacturing, service, and all other departments in an organization involved in product or service development [10, 11]. The technique was invented by Akashi Fukuhara of Japan and first applied with very good results at Toyota and its suppliers. Further, it expanded to other Japanese manufacturers such as consumer electronics, home appliances, clothing, integrated circuits, and apartment layout planning. The technique was adopted by Ford and GM in 1980s and other important companies such as Hewlett-Packard, AT&T, and ITT later.

QFD is a systematic way of documenting and breaking down customer needs into manageable and actionable detail. It is a planning methodology that organizes relevant information to facilitate better decision making and a way of reducing the uncertainty involved in product and process design. It promotes cross-functional teamwork and gets the right people together, early, to work efficiently and effectively to meet customers' needs. QFD is a structured methodology to identify and translate customer needs and wants into technical requirements and measurable features and characteristics.

7.6.1 House of Quality

The basic building block of QFD is the house of quality. It is represented in a multidimensional matrix sometimes called L-matrix and shows the correlation between "What's" and "How's" of the process stages. Figure 7.5 shows the structure of house of quality. A full QFD product design project will involve a series of these matrices, translating from customer and competitive needs all the way down to detailed process specifications. It helps to identify the critical quality characteristics. The characteristics critical to quality (CTQ) are nothing but a product feature or process step that must be controlled to guarantee that you deliver what the customer wants. For a detailed description of QFD, one may refer to Sanchez et al. [10], ReVelle et al. [11], Pande et al. [4], Muir [6], Gryna et al. [2], and Kubiak and Benbow [5].

As seen in the Fig. 7.5, QFD matrix consists of several parts. The matrix is formed by first filling in the customer requirements (area 1), which are developed from analysis of VOC. The technical requirements are established in response to the customer requirements and place in area 2. The relationship area 3 displays the connection between the technical requirements and the customer requirements. The comparison between the competitors for the customer requirement is shown in area 4 (optional), and the area 5 provides an index to documentation concerning

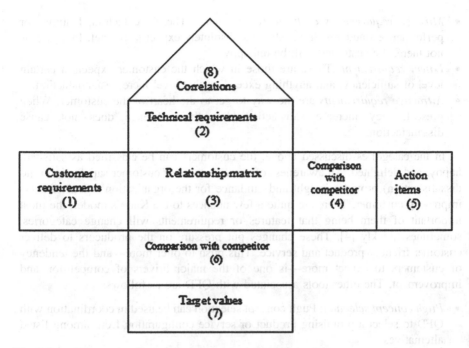

Fig. 7.5 A QFD matrix. *Source* Muralidharan and Syamsundar [12]

improvement activities. Area 6, like area 4, is again optional and it plots the comparison with the competition for the technical requirements. Area 7 lists the target values for the technical requirements. Area 8 shows the corelationships between the technical requirements. A positive correlation indicates that both technical requirements can be improved at the same time. A negative correlation indicates that improving one of the requirements will worsen the other.

The completed matrix can provide a database for product development, serve as a basis for planning product or process improvements, and suggest opportunities for new or revised product or process introductions. If a matrix has more than 25 customer voice lines, it tends to become unmanageable. In such situation, a convergent tool such as the *affinity diagram* (see Sect. 7.4.2 above) and *Kano's model* may be used to condense the list.

7.6.2 Kano's Model

Kano's model of customer satisfaction identifies several types of requirements that impact customer satisfaction. It separates customer requirements into several categories such as:

- *Must be requirements or basic requirements*: These are factors, features, or performance standards that customers absolutely expect to be met. In case it is not there, the customers will be unhappy.
- *Latent requirement*: These are those in which the customer expects a certain level of sufficiency, and anything exceeding that level increases satisfaction.
- *Attractive requirements* are the way to get to the heart of the customer. When present, they increase satisfaction, but their absence does not cause dissatisfaction.

In the categories discussed above, the customers can be classified as satisfied, happy, and delighted. An awareness of these features of customer satisfaction (and dissatisfaction) provides insight and guidance for the organization's goal of ever-improving customer. There are quite a few nuances to the Kano's model, the most important of them being that features or requirements will change categories, sometimes quickly [4]. These changes put pressure on the producers to deliver customer friendly product and service. This push to offer more—and the tendency of customers to expect more—is one of the major drivers of competition and improvement. The other tools associated with QFD are as follows:

- *Pugh concept selection*: Pugh concept selection can be used in coordination with QFD to select a promising product or service configuration from among listed alternatives.
- *Modular Function Deployment*: Modular Function Deployment uses QFD to establish customer requirements and to identify important design requirements with a special emphasis on modularity.
- *Hoshin Kanri process*: The QFD-associated Hoshin Kanri process somewhat resembles management by objectives (MBO), but adds a significant element in the goal setting process, called "catchball." Use of these Hoshin techniques by US companies such as Hewlett-Packard has been successful in focusing and aligning company resources to follow stated strategic goals throughout an organizational hierarchy. Since the early introduction of QFD, the technique has been developed to shorten the time span and reduce the required group efforts.

There are few minor differences between the applications of QFD in Modular Function Deployment as compared to house of quality, for example, the term "customer attribute" is replaced by "customer value," and the term "engineering characteristics" is replaced by "product properties." But the terms have similar meanings in the two applications.

7.7 Understanding Defects, DPU, and DPMO

A Six Sigma project is assessed through its level of sigma. Generally, at define stage, one will not be interested in sigma-level calculations, but for baseline performance check, it may be necessary to know the importance of defects and its connections with sigma-level calculations. This will be warranted during the measure phase of the Six Sigma project, as the purpose of executing project will be decided on the basis of initial sigma level. The sigma level can be computed through capability analysis, control chart techniques, defect per million opportunities (DPMO) calculations, Z-score analysis, etc.

Calculation of sigma level through DPMO involves concept of unit, opportunity, and defects. *Unit* is any product, part, assembly, process, or service for which quality characteristics are desired. *Opportunity* is a VA feature of a unit that should meet specifications proposed by the customer. *Defect* is the characteristic of the product that fails to meet customer requirements, and *defectives* are the total number of units containing different types of defects. Thus, total opportunity (TOP), defect per unit (DPU), and defect per million opportunities (DPMO) are defined as follows:

$$TOP = \text{Number of units checked} * \text{Number of opportunities of failure} \quad (7.1)$$

$$DPU = \frac{\text{No. of Defects}}{\text{No. of units checked}} \quad (7.2)$$

and

$$DPMO = \frac{DPU}{\text{No. of opportunity per unit}} * 10^6 \quad (7.3)$$

Then,

$$\sigma_{LT} = \Phi^{-1}\left(1 - \frac{DPMO}{10^6}\right) \quad (7.4)$$

and

$$\sigma_{ST} = \sigma_{LT} + 1.5\sigma \quad (7.5)$$

where σ_{LT} and σ_{ST} are the long-term and short-term sigma levels, respectively, and $\Phi^{-1}(.)$ is the inverse standard normal distribution.

As an illustration, consider 250 units are selected from a production process and each one is tested for 20 possible opportunities. Then, the TOP $= 250 * 20 = 5000$. Suppose 75 defects are observed in the process, then the DPU is obtained as DPU $= \frac{75}{250} = 0.3$. Hence, the defects per million opportunities are calculated as DPMO $= \frac{0.3}{20} \times 10^6 = 15000$. That is the process fails to meet specification for 15,000 opportunities out of 1,000,000 (1 million) opportunities. This DPMO is corresponding to a short-term sigma level of 3.67σ as per Appendix Table A.12.

One may refer to Chap. 12 of this book, for a detailed discussion on defects, and the sigma-level estimation of processes under normal and non-normal distributions.

7.8 Incorporating Suggestions, Improvements, and Complaints

While defining a Six Sigma project, it is necessary to consider all kinds of suggestions, improvement, and complaints from all the stakeholders (customers, suppliers, employees, investors, and communities). This will decide the process to follow, the kind of training to impart, the kind of people to include, and the model to pursue. Complaints should not be treated as a negative aspect of the project; instead, it should be treated as one of the essential requirements for improving the project. A Six Sigma project may impact stakeholders in the following ways:

- Process inputs may be altered, which changes the requirements to suppliers
- Process procedures may be changed that impact operators and managers
- Process outputs may be altered, which impact customers
- Changes to tooling, preventive maintenance schedules, and so forth, impact the suppliers or those materials and services

7.9 Readying for the Next Phase

The success of any project depends on the stakeholders inputs which come in the form of information or data. A strong VOC system enables one to achieve this. To better understand customer needs, expectations, and requirements, an appropriate data should be in place. Surveys, interviews and focus group, observations, etc., are the methods of observing data. Statistically speaking, the most valid procedure for collecting customer data is to randomly select a reasonably large representative group of customers and obtain complete and accurate data on each one. Written surveys can be conducted on a randomly selected group of customers or potential customers, asking for their responses. They are called focus groups and have special rights and privileges in the organization.

Focus groups are an attempt to improve the depth and accuracy of the responses. They generally provide more accurate informations and have the ability to probe deeper into issues or concerns on a real-time basis. In this case, the sample may not be random and may be too small. Personal interviews permit a higher response rate than written surveys; however, the respondents tend to be self-focused, and therefore, sample may not be random and sample size may be too small. For a valid analysis and conclusion, the data collected should be reliable and objective and designed to shed light on customer requirements and market demands. This is to be ensured while defining the measures and metrics. This is discussed in the next chapter.

7.10 Define Check Sheets

The define check sheet includes the following parameters:

- Ensure that the project is worthwhile and supported by the business leaders
- Prepare a clearly defined project rationale explaining the potential impact of the project on customers, profits, and its relationship on the company's business strategies
- Prepare a clearly defined problem statement and possible symptoms identified
- Prepare a goal statement defining the results we are seeking from the project with a measurable target
- Prepare other key elements of a DMAIC team charter, including a list of constraints and assumptions, and a review of roles and responsibilities
- Review the project charter with the sponsor and ensure their support
- Identify the primary customer and key requirements of the process being improved and created by a SIPOC diagram of the areas of concern
- Prepare a detailed process map of areas of the process for measurement and analysis

7.11 Relevance for Managers

The relevance of this chapter is to propose the body language of a Six Sigma project. Any improvement project generally includes a project charter, a problem statement, and a goal statement preferably in the SMART format. Other relatively important aspects are the scoping of the project and the schedule of the project. This information will then pave the way for the development of the process models and micro- and macro-level process maps. The development of SIPOC diagram is crucial to these steps. The team roles and responsibilities for managing the project are then finalized for a smooth planning and execution of the project.

Since Six Sigma project is more structured than any other management project, a time line is drawn for its completion. Therefore, all planned and finished works should be in relation to feasibility of time and resources available for a project. Adherence to time is most crucial for executives and managers who are associated with a project. Although most of the planning tools are familiar to them, one has to have complete involvement in the knowledge discovery of the project. Another important aspect of Six Sigma project is the documentation and accountability of the project. This is achieved through the QFD technique. Another objective of Six Sigma project is to reduce defects in a project, for which the knowledge of defects per unit, defects per million opportunities, and yield is necessary. This is further used in the computation of sigma-level estimation of the process. This chapter, thus, introduces the first phase of the DMAIC philosophy.

Exercises

7.1. What are the objectives of a problem statement? Describe various characteristics of a problem statement.

7.2. Discuss how a project charter is different from a problem statement.

7.3. Describe various types of customers and their priority demands.

7.4. Define a process and state various components of a process.

7.5. What is SIPOC? State all the benefits of a SIPOC model.

7.6. Discuss various roles and responsibilities associated with a Six Sigma project.

7.7. What is a team? How a team is prepared for facilitating a project management?

7.8. Discuss three categories of planning stages.

7.9. Distinguish between a Gantt chart and an affinity diagram. How are they used for planning a project?

7.10. Distinguish between a process map and a flowchart?

7.11. What is QFD? Discuss various components of QFD.

7.12. How a Kano's model is effectively used in quality improvement?

7.13. What is DPU, TOP, and DPMO? How are they used in finding the sigma level of a process?

7.14. A process produces 10,000 pencils. Three types of defects can occur. The number of occurrences of each type of defects is blurred printing—36, wrong dimensions—118, and rolled ends—11. Calculate DPU and DPMO. Also obtain the sigma level.

References

1. Defeo, J.A., Barnard, W.: Juran's Six Sigma Breakthrough and Beyond. McGraw-Hill, New York (2003)
2. Gryna, F.M., Chua, R.C.H., Defeo, J.: Juran's Quality Planning and Analysis for Enterprise Quality. Tata McGraw-Hill, New Delhi (2007)
3. Harrington, H.J., Anderson, L.C.: Reliability Simplified. McGraw Hill, New York (1999)
4. Pande, P.S., Newuman, R.P., Cavanagh, R.R.: The Six Sigma way. Tata McGraw Hill, New Delhi (2003)
5. Kubiak, T.M., Benbow, D.W.: The Certified Six Sigma Black Belt Handbook, 2nd edn. Dorling Kindersley Pvt Ltd., India (2010)
6. Muir, A.: Lean Six Sigma way. McGraw Hill, New York (2006)
7. James, J.M.: Gantt charts: a centenary appreciation. Eur. J. Oper. Res. **149**(2), 430–437 (2003)
8. Wardsworth, H.M., Stephens, K.S., Godfrey, A.B.: Modern Methods for Quality Control and Improvement. Wiley, New York (2001)
9. Salvendy, G.: Handbook of Industrial Engineering, 2nd edn. Wiley, New York (1992)
10. Sanchez, S.M., Ramberg, J.S., Fiero, J., Pignatiello Jr, J.J.: Quality by design. In: Kusaiak, A. (ed.) Concurrent Engineering: Automation, Tools, and Techniques (Chapter 10). Wiley, New York (1993)
11. ReVelle, J.B., Moran, W., Cox, C.A.: The QFD Handbook. Wiley, New York (1998)
12. Muralidharan, K., Syamsundar, A.: Statistical Methods of Quality, Reliability and Maintainability. PHI, New Delhi (2012)

Chapter 8
Measure Phase

8.1 Initiating Measure Phase

This is one of the most difficult and critical phases of a Six Sigma project management. This is where we connect customer requirements to the process voice. This phase entails selecting product characteristics, i.e., dependent variables, mapping the respective processes, making the necessary measurement, recording the results, and estimating the short- and long-term process and capabilities. One of the main objectives of this phase is to establish standards for performance that are based on actual customer input, so that process effectiveness and capability can be accurately measured. While compiling the project metrics, attention will be given on critical quality variables like the following [8]:

- *Critical to quality* (CTQ): They describe the requirements of quality in general but lack the specificity to be measurable. CTQs can be physical measurements like cycle time, height, weight, pressure, intensity, and radius.
- *Critical to cost* (CTC): They are similar to CTQs but deal exclusively with the impact of cost on the customer. However, CTQs and CTCs may be similar, yet stated by the customer for different reasons
- *Critical to process* (CTP): CTPs are typically the key process input variables (or independent variables)
- *Critical to safety* (CTS): CTSs are stated customer needs regarding the safety of the product or process. Though identical to the CTQ and CTC, it is identified by the customer preference to quality
- *Critical to Delivery* (CTD): CTDs represent those customers with stated needs regarding delivery. On-time delivery is always appreciated

Projects aligned with CTxs, and subsequently critical customer requirements, have the biggest impact on the customer and the business directly. Hence, measure phase is important in understanding the long-term variabilities in each of the quality characteristics as discussed above. Process mapping, VA/NVA assessment, data

© Springer India 2015   123
K. Muralidharan, *Six Sigma for Organizational Excellence*,
DOI 10.1007/978-81-322-2325-2_8

collection plan, gage R&R, and check sheets are some of the techniques through which a good measurement system can be achieved. Even quality function deployment (QFD) plays a major role in selecting critical product characteristics.

8.2 Process Mapping

In Chap. 7, we have described various process models and the SIPOC diagram, where we have found that the supplier–customer relationship is well established through a SIPOC diagram. This diagram helps the process manager to identify the core requirements needed for processes like the following:

- the key materials, information, or products provided for the process
- the absolutely essential components required for the work to be performed
- whether the components are consumed or used during the process, or passed through to the customer as an output
- the provider of those outputs, etc.

During the measure stage, this information helps to identify the work in progress (WIP), work in queue (WIQ), cycle time, and takt time. One can also use these tools to identify the yield, throughput, value-added/non-value-added time, and setup time required to process a particular activity in the process. Thus, the customer requirements can very well be explained through the input–output processes of SIPOC.

A key step in understanding processes is the development of process maps and flowcharts. The format of these documents will vary with the situation, but in general, flowcharts show each step in a process, decision points, inputs, and outputs. Process map usually contains additional information about the steps, including inputs and outputs, costs, setup time, cycle time, inventory, types of defects that can occur, probability of defects, and other appropriate information. Process maps and flowcharts enable a broader perspective of potential problems and opportunities for process improvement.

A SIPOC diagram displays a cross-functional set of activities in a single and simple diagram and helps maintain a "big-picture" perspective. A process mapping at this stage is very important for the identification of value-added/non-value-added activities (VA/NVA). This is the stage in which we will begin to identify the customer requirements for the process in terms of its interdependencies coming through supplier requirements. Figure 8.1 is such a process, where the supplier and customer are linked through a process.

As described in the figure, voice of customer (VOC) and voice of process (VOP) are the two important interfaces which link the inputs and outputs through the process. Since quality is defined by the customer, changes to a process are usually made to increase satisfaction of internal and external customers. The knowledge of VOC and VOP also helps to measure the capability as

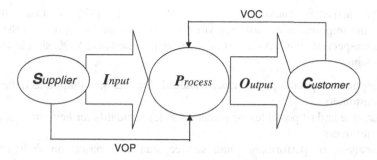

Fig. 8.1 Mapping supplier and customer through a process

$$\text{Process capability} = \frac{\text{VOC}}{\text{VOP}} \tag{8.1}$$

Almost every process can be improved by reducing variation. To do so, it is necessary to always do things the same way. The best way to ensure that things are done the same way is to have written procedures and work instructions. A written procedure is like standard operating procedures (SOP) and can be used anytime during the execution of a project. Work instructions are generally product specific and often include graphics showing discrete steps including defects to avoid, safety hazards, tooling required, and any information that ensures the job will be performed in a standard way. When generating work instructions and procedures, the people closest to the process must be involved. If more than one shift or workstation is affected, the appropriate people should participate. The most difficult part of documenting instructions and procedures is maintaining up-to-date versions. As workers find improvements or engineers change designs, the documentation must be updated to avoid problems, issues, defects, or obsolescence. Another useful tool for analyzing processes is known as the value stream map (VSM), which is discussed below in Sect. 8.4.

8.2.1 Voice of Customer

The voice of customer or voice of client (VOC) is a continuous process of collecting customer views on quality and can include customer needs, expectations, satisfaction, and perception. The emphasis is on in-depth observing, listening, and learning. A customer is the person who is instrumental in shaping the organization's future and producing good quality product. For any process management and improvement, it is essential to know what do customers really want, and how do their needs, requirements, and attitudes change over a period of time. This systematic approach of understanding customers and their needs is nothing but VOC. It is all about the strategy and system for continually tracking and upgrading customer requirements, competitor activities, market changes, etc. Pande et al. [17] and Oakland [16] describe the importance of customer voice in terms of gathering data as the most important aspect of VOC. As a part of improvement activity, VOC should include the following:

- description of specific performance standards for each key output, as defined by the customers
- observable and (if possible) measurable service standards for key interfaces with the customers
- an analysis of performance and service standards based on their relative importance to customers and customer segments and their impact on business strategy
- analyzing and prioritizing requirements as per the business strategy.

The characteristics of a VOC Metrics are as follows:

- *Credibility*:
  - How widely accepted is the measure?
  - Does it have a good track record of results?
  - Is it based on a scientifically and academically rigorous methodology?
  - Will management trust it?
  - Is there proof that it is tied to financial results?

- *Reliability*:
  - Is it a consistent standard that can be applied across the customer life cycle and multiple channels?

- *Precision*:
  - Is it specific enough to provide insight?
  - Does it use multiple related questions to deliver greater accuracy and insight?

- *Accuracy*:
  - Is the measurement right?
  - Is it representative of the entire customer base, or just an outspoken minority?
  - Do the questions capture self-reported importance or can they derive importance based on what customer says?
  - Does it have an acceptable margin of error and realistic sample sizes?

- *Actionability*:
  - Does it provide any insight into what can be done to encourage customers to be loyal and to purchase?
  - Does it prioritize improvements according to biggest impacts?

- *Predictability*: Can it project the future behaviors of the customer based on their satisfaction?

One of the core requirements of a VOC system will be the ability to identify customer requirements while catching trends, thus helping to keep ahead of changes in market preferences, to be aware of new challenges, and so on. In this process, if one does not have the access to correct information, one will miss new opportunities and become vulnerable to new competitors. Although luring the customers is not ideal as people do these days, it is sometimes necessary to retain one's customer base by offering sales offers, loyalties, and discounts. This will sometimes help them to share their confidential views and sentiments with one, by which one can understand their requirements.

A model being used by a growing number of companies to analyze requirements is based on the work of Noriaki Kano, a Japanese engineer and consultant. In the most common application of *Kano analysis* (discussed in detail in Sect. 7.6.2), the customer's requirements are grouped into basic, variable, and latent requirements. The objective of these requirements is to make the customers happy and to understand the pulse of the market.

In some segment of the market like shopping malls and mega-stores, delighting a customer have become a route to success. They load the customers with lots of offers and discounts and thus increase the purchasing power of the customers. This push to offer more, and the tendency of customers to expect more, is one of the major drivers of competition and improvement. As your business develops a more objective and complete picture of customer requirements, you can also apply a concept like Kano analysis to get a better idea of what the various features and capabilities mean in terms of your customers satisfaction and your competitive edge.

8.2.2 *Voice of Process*

At many stages in a process, it is necessary for customers to determine their needs or give their reaction to proposed changes in the process. For this, it is often useful to describe the edges or boundaries of the process. This is what VOP does. A clearly stated VOP enables one to carry out the documentation of the inputs and outputs of the process. This is done through two important process analysis strategies as follows:

1. *Process analysis*: Deeper investigation into and understanding of how work is being done to identify inconsistencies, disconnects, or problem areas that might cause or contribute to the problem.
2. *Data analysis*: Use of measures and data to discern patterns, tendencies, or other factors about the problem that either suggest or prove/disprove possible causes.

As seen earlier, a process is a set of linked or related activities that take inputs and transform them into outputs with the purpose of producing a product, information, or service for external or internal customers [14]. Therefore, it is essential to establish a communication link between the customer requirements and the process parameters. This is where the need for *process orientation* arises. This can be done through graphically outlining the sequence of a process, showing how steps in a process relate to each other, identifying bottlenecks, pinpointing redundancies, and locating waste in the process. In a way, these activities will help us to achieve two things: (i) document how the process works and (ii) to identify the value-added and non-valued activities. One of the best methods of describing the process orientation and mapping is the SIPOC model already described in the define phase (Chap. 7).

8.3 Adding Value Through Customer Service

There are many ways to add value to a product/process or service. They generally depend on customer needs and requirements. It is also not necessary that every activity will add value to the service. There are the activities that add value to the customers and that do not. The activities that add value to the customer are called value-added activities (VA). For this, customers are willing to pay for it. It physically changes the product or service. Hence, just moving things around is usually not value adding. So it is better to do things correctly for the first time. Some examples of value-added activities are entering order, ordering materials, preparing drawing, assembling, legally mandated testing, packaging, shipping to customer, etc.

The activities that generally do not contribute any value to the process activities and output are called non-value-added activities (NVA). Also they are not essential to produce output. For example, preparation/setup, control/inspection, defects, errors, omissions, over-production, processing, inventory, transporting, motion, waiting, delays, sorting, counting, recording, obtaining approvals, testing, reviewing, copying, filing, revising/reworking, tracking, etc., are non-value-added activities.

According to Bovee and Thill [2], some of the possible ways to add value to a product are the following:

- Be flexible
- Tolerate customer errors
- Give personal attention
- Provide helpful information
- Increase convenience.

Similarly, for bank account-related services, the value can be added through the following ways:

- Let customer design their own checks
- Let there be no fees for check leaves and overdrafts
- Help customers with individual tax questions
- Publish a brochure on financial planning
- Install ATMs as per convenience of the people.

The best way of documenting the VA/NVA activities is by using process mapping. It is as good as documenting the process steps necessary for taking the transaction from the customer into the business processes and then bringing it back to the customer. During the analysis process, each process step will be classified as to whether the process step adds value to the service or product. The VA/NVA process map can then be used to classify and document whether a process is well designed to deliver service to the customer in an efficient manner. The final result of a VA/NVA process mapping analysis is a list of non-value-added process steps that can be eliminated or automated to decrease the total cycle time. See Muir [12] for more details.

The methods for collecting information on customer needs are many and varied. Some of the common methods include focus groups, observations at customer sites, executive interactions with customers, special customer surveys, analysis of complaints, participation at trade shows, and comparing products with those of competitors. Even monitoring Internet messages is also a good method to find out what customers are saying about a product [4].

8.4 Value Stream Mapping

The *value stream* consists of all activities required to bring a product from conception to commercialization. It includes detailed design, order taking, scheduling, production, and delivery. Understanding the value stream allows one to see value-added steps, non-value-added but needed steps, and non-value-added steps. A value stream mapping (VSM) is a visual representation of every process in the material and information flow. It helps to see and understand the flow of material and information as a product makes its way through the value stream.

A VSM normally puts the activities in their hierarchical form. For instance, a customer-focused production process will have customer information, process information, raw and finished materials, information flow, and timeline to finish the product as the hierarchy. Typical value stream mapping tips are as follows:

- Collect information while walking the actual paths of material and information yourself
- Walk the path once to get a feel of the flow and sequence of processes; then, go back and start collecting information
- Begin at the end. Start closest to the customer, and you will focus on things linked most closely to the customer
- Bring a stopwatch, see it, and time it for yourself, and do not trust information from the "system".

The value stream mapping summarizes the delay and execution times for an individually identifiable entity as it moves through the process and reflects the process at a particular time. All normal backlogs are reflected in the delay times. Inventory levels at each step can also be recorded and noted on the value stream map between process steps.

8.5 Data Collection Plan

Data are a numerical expression of an activity and are essential for all kinds of decision making [14]. Data collection is an integral part of any Six Sigma project. It involves the basic definitions for the concepts to be investigated, inquiries to communicate, delineation of the environment in which the data will be collected, designing of the process involved, etc. During the data collection, we must consider the diverse tasks of assignment and recruitment of staff, ways of increasing response rates, costs and bias, sources under alternative collection approaches, and proper training of the personnel. The effect of each of these on accuracy, monetary costs, and time constraints must be evaluated. Finally, the collection phase must be supervised as well as planned. A properly collected data invariably help us to know and quantify the status of the study, helps to monitor the ongoing process, helps to decide acceptance or rejection of the outputs produced, and help us to decide the course of action.

Data can be classified based on how they are collected: They are primary data and secondary data. The *primary data* are those which are collected afresh and for the first time and thus happen to be original in character. Such data are published by authorities who themselves are responsible for their collection. The collection of primary data involves a great deal of deliberation and expertise. Depending on the nature of information, one can adopt any of the following methods of primary data collection: observation method, questionnaire method, mailed questionnaire method, telephone interview, door-to-door survey, etc. Some of these methods need lots of skill and knowledge and are very expensive and time-consuming.

On the other hand, the *secondary data* are those which have already been collected and processed by some other agency. They are information which has previously been collected by some organization to satisfy its own need, but it is being used by the department under reference for an entirely different reason. For example, the information collected by National Informatics Center (NIC) throughout India is disseminated for researchers, demographers, and other social organizations for planning and research. Thus, the information collected by NIC is primary for them and the same data for the researchers, demographers, and other organizations will be a secondary data. Even when an organization interested to collect primary data, it is necessary to take help of various types of secondary data to design and analysis. Some of the sources of secondary data are the publications of central, state, and local governments; international bodies; journals of trade, economics, commerce, engineering; books; magazines; and newspapers.

As for any project investigation, it is imperative to have a defined problem (goal) and the data set thereof. So a complete understanding of all elements associated with a research is essential for making the right decision. Two such elements are unit of analysis and characteristics of interest.

8.5.1 Unit of Analysis

The individuals or objects whose characteristics are to be measured are called the unit of analysis [14]. They may be persons, groups of persons, business establishments, transactions, monetary units, inanimate objects, cycle time or just about objects, or activity a person can name. Basically, it talks about "What objects am I interested in?"

As an example, consider a manufacturer of small electrical motors who wishes to ascertain the extent to which its potential customers know the company exists. The potential customers are basically business entities. But the units of population (universe) could also be defined as purchasing departments, production departments, engineering departments, or particular individuals within one or more departments. Again we come to all the pervasive question of what alternative actions are being considered by the manufacturer. In terms of these actions, who should be aware of the manufacturer's existence? Is the company considering specific acts that might increase awareness levels for certain groups? These are the sorts of questions that should be considered in specifying the appropriate units of analysis.

8.5.2 Characteristics of Interest

The characteristics of interest identify the whereabouts of the units that are of concern to the decision maker. These characteristics fall into two categories: the independent variables (usually denoted as the variable X) and the dependant variables (usually denoted as the variable Y). The dependant variables are those of interest for their own sake. For example, in marketing, often we refer to behavior or attitude toward a firm's offering. Examples are purchases, awareness, opinions, or profits associated with consumer behavior attitudes. The independent variables included in the problem definition are those characteristics thought to be related to the dependant variables. These variables may either be within the control of the firm (*endogenous*)—such as advertising, pricing, and grading—or beyond the control of the firm (*exogenous*). Exogenous variables (X's) of potential interest cover a multitude of possibilities, varying from competitor and government actions to economic conditions to individual consumer characteristics.

The characteristics of interest are crucial to the management. Its value will serve as the basis for choosing the correct variables among alternative actions. The unit of analysis establishes the source for the information. In many cases, the unit employed is dictated by convenience rather than the proper problem definition. Therefore, it is better to have an exhaustive understanding about the difference between the two for a better investigation. For example, a manufacturer of drugs is interested in rupee value sales of a particular generic drug. The manufacturer wants to know rupee sales for a group of five states during the period of July 2000–June 2001. In this example, "rupee value sales" is the characteristics of interest, because this is the concern to the drug company. The unit of analysis identifies "on what" or "on whom" the characteristic of interest is measured. Hence, the unit of analysis for the drug company may be the "individual drug store." Hence, the research (investigation) is to collect rupee sales from the individual drug stores, and then determine the total sales in the five states by adding or collating together the information of all stores.

The identification of unit of analysis and characteristics of interest are crucial for any Six Sigma project. The personals associated with such projects should be clear with the distinction as well as the importance of using those in the data collection plan. A good measurement in such situations defines the necessary tools for good analysis and interpretations.

8.5.3 Data Types

Data types are generally classified into two main categories. They are *discrete* (or attribute or categorical) type and *continuous* type. Number of telephone calls, number of defectives, number of calls on hold past 30 min, number of under-specified chips, number of delayed boarding incidents, number of empty (or full) tank, out-of-specifications chips, out-of-specifications screws, number of units ordered, number of non-delivered items, out-of-specifications wire, etc., are examples for discrete data. Some of the examples for *continuous* data are cycle time, delay time, pressure, width of a chip, radius of a screw, cost per unit, thickness of the wire, holding time/incoming call, temperature, intensity, minutes to board plane, quantity of gas in a tank, etc.

Attribute data are further classified as nominal and ordinal, whereas continuous type is classified into interval and ratio type (see Fig. 8.2 for details). The attribute data are usually measured in discrete forms. For example, unit's delivered/day, rating of services, number of typographical errors, number of prescriptions, number of claims in a dispute, fill rate, number of accidents, etc., are all discrete measurements.

Nominal scale: It is a system of assigning number symbols to events in order to label them. It provides convenient ways of keeping track of objects, people, and events, for example, customer types, document types, and defect types. Nominal data classify occurrences into unordered categories; for example, color of product is acceptable or not, a customer survey response on a satisfaction questionnaire is either "yes" or "no" type, and a customer contact is considered acceptable or not.

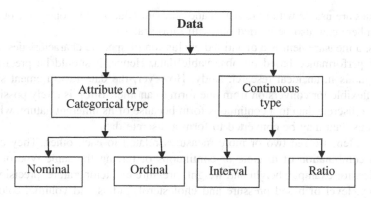

Fig. 8.2 Data types

The counting of members in each group is the only arithmetic operation possible in these cases. Hence, *mode* is the best measure of central tendency for nominal data sets, and *chi-square test* is used to test the association for significance.

Ordinal scale: Ordinal data classify occurrences into ordered categories; for example, a customer response survey on a questionnaire is either "excellent, very good, good, average, poor" or "strongly disagree, moderately disagree, neutral, moderately agree, strongly agree," etc.; a sales lead is considered "unlikely," "likely," or "very promising"; an IT failure is classified as being of "low," "medium," or "high" severity; and a package to be delivered is categorized as acceptable, downgraded, or reject.

The ordinal scale places events in order and ranks. So quite often, we encounter uses of ordinal data sets in market research, for example, severity ratings, priority ratings, and customer satisfaction ratings. In fact, a questionnaire-based sample survey usually involves these kinds of measures, and hence, *median* is used as the suitable measure of central tendency, and quartiles or percentiles are used for measuring dispersion. Sometimes, when it can be justified, the scale can be used as if it were continuous. The tests of hypothesis are restricted to *nonparametric tests* only.

Interval scale: This is the most powerful measurement than ordinal scale. Interval scales have meaningful differences but no absolute zero, so ratios are not useful. For example, temperature, pressure, speed, cycle time, rate of flow, productivity, defect density, and test time are all continuous variables measured in interval scale. *Mean* is the most appropriate measure of central tendency for this kind of measure, while standard deviation is the most widely used measure of dispersion. A lot of statistical tools are available for studying these kinds of data sets.

Ratio scale: Ratio scale represents the actual amounts of variables. They have meaningful differences, and an absolute zero exists. For example, length measured in inches is a ratio scale, as zero length is defined as having no length and 30 in. length is twice as long as 15 in. Measurement on any continuous scale such as weight, height, distance, etc., is an example of ratio scale. Generally, all statistical techniques are usable with ratio scales and all manipulations that one carry out with real numbers can also be carried out with ratio scales.

Thus, a measurement is a quantified evaluation of specific characteristics and/or level of performance based on observable data. Hence, it should be precise and unambiguous in an ideal research study. However, the data measurement should also be flexible for conversion from one form to another. But it is rarely possible to convert a discrete data to a continuous form because of its inherent nature, whereas continuous data may be converted to form a discrete data.

Very often, we see two or more measures related to each other. They can be either a combination of discrete or continuous or having the same type of characteristics, for example, height and weight, pressure and temperature, precision and efficiency, level of blood pressure and cholesterol, radius and volume, color and odor, sales and expenditure, thickness and pressure, length and width, and performance and IQ. Establishing correlations and associations will be the prime concern in these types of variables. Similarly, there can be variables depending on many other associated measures, and hence, the study becomes multidimensional. Many important techniques like multiple correlation, multiple regression, principle component analysis, factor analysis, and classification techniques are the useful tools for studying the relationship between two or more variables. Some of these techniques will be discussed in detail in Chap. 9 of analyze phase.

8.6 Cycle Time, Takt Time, Execution Time, and Delay Time

The time required to complete one cycle of an operation in a process is called the *cycle time*, which includes the actual work time and wait time. Cycle time is also described in terms of takt time sometimes. A *takt time* is the available production time divided by the rate of customer demand. Operating under takt time sets the production pace to customer demand. If cycle time for every operation in a complete process can be reduced to equal takt time, products can be made in single-piece flow. The time actually spent doing something to the product or service as it flows on its way to the customer is called *execution time*, whereas the time the product or service spends waiting for something to be done is called the *delay time*. The *setup time* or *changeover time* is the time required to convert from producing one product to producing a different one on, say, a particular machine, process step, and so forth. Another time associated with speed delivery is the *just in time*.

During the measure phase, it is important to know the distinction between the times associated with the process. In most cases, the ultimate goal of studying process flow is to implement a one-piece flow instead of a batch-and-queue operation. This is necessary, because the measurement and modeling depends entirely on this. This also helps in the elimination of duplicate or unnecessary tasks and eliminates rework or duplicate works during the entire duration of the process. Therefore, time analysis is very important during a process to identify the *idle time*, during a process. Since cycle time improvement has been a priority, time analysis has been a tremendous benefit to cutting process turnaround to minutes instead of hours, days instead of months. In fact, many successful Six Sigma projects are focused on cycle time reduction, because of these reasons.

8.7 Measurement System Analysis

Since the data collection plan is in place and the measurements are available, the next step is to check the validity and reliability of the measurements. The real purpose of measurement system analysis (MSA) is to force the project team to consider the data as reflecting not one, but two processes. The largest sources of variation in the measurement come from poorly defined and misunderstood operational definitions of the components of transactional processes and assumptions made about data derived from historical systems that were never designed for the purposes of data collection. The project focus must always remain on reducing defects as defined by the customer, but be careful not to lose sight of the measurement process for the sake of the business process.

During the measure phase of the project, one must resist the temptation to start analyzing the data and concentrate on understanding it. As seen earlier, the measurements are generally classified into two types, namely continuous and discrete types. Therefore, the assessment also should be done suitably to achieve the stated purposes. In drawing conclusions about measurement error, it is worthwhile to study the causes of variation in observed values. If observed variation in the process is due to many causes, and if the causes act independently, then the observed variation can be expressed as

$$\sigma_{observed} = \sqrt{\sigma^2_{cause\ A} + \sigma^2_{casue\ B} + \cdots + \sigma^2_{cause\ N}} \qquad (8.2)$$

It is valuable to find the numerical values of the components of observed variation because the knowledge may suggest where effort should be addressed to reduce variation in the product. A separation of the observed variation into product variation plus other causes of variation may indicate important factors other than the manufacturing process. Thus, it is found that the *measurement error* is a large percentage of the total variation; this finding must be analyzed before proceeding with a quality improvement program. Finding the components (e.g., instrument, operator, methods) of this error may help to reduce the measurement error, which in turn may completely eliminate a problem (see [5] for more details).

Note that every measurement system consists of the following key components:

- Measurement instrument
- Appraiser(s) (also known as operators)
- Methods or procedures for conducting the measurement
- Environment.

If observations from an instrument are used to measure a series of different units of product, then the total variation can be attributed to (i) the variation due to the measuring method and (ii) the variation in the product itself. This is expressed as

$$\sigma_{total} = \sqrt{\sigma_{product}^2 + \sigma_{measurement}^2} \tag{8.3}$$

The components of measurement error are further broken down into repeatability and reproducibility (R&R). *Repeatability* concerns variation due to measurement gauges and equipment, whereas *reproducibility* concerns variation due to human (or appraisers) who use the gauges and equipment. Studies to estimate these components are often called "gage R&R" studies.

Below, we discuss the gage R&R study for assessing bias in continuous and discrete (or attribute) measurements separately. The measurement system analysis for continuous variables uses the following methods:

- AIAG method: It is an organization called Automotive Industry Action Group (AIAG) and provides the tabular procedure with specific guidelines for identifying variations associated with various components.
- ANOVA method: It is called analysis of variance (ANOVA) and uses rigorous statistical method of identifying variations associated with various components.
- Control chart method: It is a graphical procedure of identifying variations associated with various components.

8.7.1 Assessing Bias in Continuous Measurements

An important method of testing the effectiveness of a continuous measurement is known as *gage R&R*. It is carried out to assess the accuracy, repeatability, and reproducibility of a continuous measurement system. It is typically used in manufacturing or other applications where "gages" or devices are used to measure important physical characteristics that are continuous, for example, thickness, viscosity, strength, and pressure.

In a *gage R&R* study of measurement system, multiple operators measure multiple units a multiple number of times. For instance, 3 operators each measure 7 units twice in a particular time period. A blindness approach is extremely desirable, so that the operator do not know that the part being measured is part of a special test. At a minimum, they should not know which of the test parts they are currently measuring. Then, analyze the variation in the study results to determine how much of it comes from differences in the operators, techniques, or the units themselves.

8.7.1.1 Gage R&R Study: Computational Aspects

- Select units or items for measuring that represent the full range of variation typically seen in the process. Measurement systems are often more accurate in some parts of the range than in others, so you need to test them over the full range.
- Have each operator measure those items repeatedly. In order to use MINITAB to analyze the results, each operator must measure each unit the same number of times. It is extremely desirable to randomize the order of the units and not let the operator know which unit is being measured.
- Look at the total variation in the items or units measured.
- Then, estimate the proportion of the total variation that is due to

    1. *Part-to-part variation*: physical or actual differences in the units being measured.
    2. *Repeatability*: inconsistency in how a given person takes the measurement (lots of inconsistency = high variation = low repeatability).
    3. *Reproducibility*: Inconsistency in how different people take the measurement (lots of inconsistency = high variation = low reproducibility).
    4. *Operator–part interaction*: An interaction that causes people to measure different items in different ways (e.g., people of a particular height may have trouble measuring certain parts because of lighting, perspective, etc.).

If there is excessive variation in repeatability or reproducibility (relative to part-to-part variation), you must take action to fix or improve the measurement process. The goal is to develop a measurement system that is adequate for your needs.

Acceptance criteria for MSA

1. Gage R&R <10 %—The measurement system is excellent
2. Gage R&R 10–30 %—The measurement system is acceptable
3. Gage R&R >30 %—The measurement system is not acceptable

Interpretation of MSA Results: It is expected that repeatability and reproducibility error should be low. If suppose the repeatability is high, then it is concluded that (i) instrument and method of measurement are improper. An overall system improvement is required, if R&R error is low. Similarly, if reproducibility error is high, then it is concluded that a strong training and instruction is necessary to train the operator and the inspector.

Example 8.1 Consider an experiment in which three operators were asked to make two measurements on each of 10 different parts. The results are shown below:

Part	Operator 1				Operator 2				Operator 3			
	Measurements		Mean	Range	Measurements		Mean	Range	Measurements		Mean	Range
	A-1	A-2	$\bar{x}_A$	R_A	B-1	B-2	$\bar{x}_B$	R_B	C-1	C-2	$\bar{x}_C$	R_C
1	31.83	31.67	31.75	0.16	32.55	32.07	32.31	0.48	31.99	32.23	32.11	0.24
2	30.31	30.47	30.39	0.16	30.63	30.39	30.51	0.24	30.15	30.15	30.15	0.0

(continued)

Part	Operator 1				Operator 2				Operator 3			
	Measurements		Mean	Range	Measurements		Mean	Range	Measurements		Mean	Range
	A-1	A-2	$\bar{x}_A$	R_A	B-1	B-2	$\bar{x}_B$	R_B	C-1	C-2	$\bar{x}_C$	R_C
3	29.51	29.83	29.67	0.32	29.83	29.67	29.75	0.16	29.51	29.83	29.67	0.32
4	29.43	29.27	29.35	0.16	29.83	29.35	29.59	0.48	29.27	29.51	29.39	0.24
5	29.91	30.23	30.07	0.32	30.07	29.67	29.87	0.4	29.83	29.27	29.55	0.56
6	30.47	30.31	30.39	0.16	30.87	30.87	30.87	0.0	30.31	30.31	30.31	0.0
7	28.79	28.79	28.79	0.0	29.03	28.71	28.87	0.32	28.23	28.47	28.35	0.24
8	29.99	29.67	29.83	0.32	30.07	29.91	29.99	0.16	29.67	29.83	29.75	0.16
9	28.55	28.63	28.59	0.08	28.95	28.79	28.87	0.16	28.55	28.23	28.39	0.32
10	29.35	29.11	29.23	0.24	29.83	29.59	29.71	0.24	29.59	29.51	29.55	0.08
	$\bar{\bar{x}}_A = 29.806, \bar{R}_A = 0.192$				$\bar{\bar{x}}_B = 30.034, \bar{R}_B = 0.264$				$\bar{\bar{x}}_C = 29.722, \bar{R}_C = 0.216$			

Solution We have

$$\bar{\bar{R}} = \frac{1}{3}\left(\bar{R}_A + \bar{R}_B + \bar{R}_C\right)$$

$$= \frac{1}{3}(0.192 + 0.264 + 0.216) = 0.22$$

Hence the estimate of gage repeatability is

$$\hat{\sigma}_{\text{repeatability}} = \frac{0.22}{1.128} = 0.2$$

where the value of d_2 is obtained from Appendix Table A.11 for subgroup size 2. The gage reproducibility is the variability that arises because of difference among the three operators. To estimate the gage reproducibility, let

$$\bar{\bar{x}}_{\text{max}} = \max(\bar{\bar{x}}_A, \bar{\bar{x}}_B, \bar{\bar{x}}_C) = 30.034$$
$$\bar{\bar{x}}_{\text{min}} = \min(\bar{\bar{x}}_A, \bar{\bar{x}}_B, \bar{\bar{x}}_C) = 29.722$$
$$R_{\bar{\bar{x}}} = \bar{\bar{x}}_{\text{max}} - \bar{\bar{x}}_{\text{min}} = 0.312$$

Hence, the estimate of gage reproducibility is

$$\sigma_{\text{reproducibility}} = \frac{R_{\bar{\bar{x}}}}{d_2}$$

$$= \frac{0.312}{1.693} = 0.184$$

Therefore, the total gage variation is

$$\hat{\sigma}^2_{\text{gage}} = \hat{\sigma}^2_{\text{repeatability}} + \hat{\sigma}^2_{\text{reproducibility}}$$
$$= (0.2)^2 + (0.184)^2 = 0.073$$

The ANOVA method of gage R&R study carried out using the MINTAB software is shown below. Note that the estimate of gage repeatability is 0.18672 and gage reproducibility is 0.1825, which are comparable with the output obtained through the manual calculations. Also the estimate of total gage variation is 0.06805 which is roughly equal to 0.073 as obtained above.

The above estimates are useful for finding the process capability of the process. According to Montgomery [9], the ratio $6\sigma_{gage}$ to the total tolerance band is often called the *precision-to-tolerance ratio* (or P/T ratio), a measure of process capability. From a process model point of view, it is simply the ratio of VOC to VOP. Hence, if the process specification for the above example is LSL = 20 and USL = 40, then

$$\text{Process capability} = \frac{\text{VOC}}{\text{VOP}} = \frac{6\hat{\sigma}_{gage}}{\text{USL} - \text{LSL}}$$

$$= \frac{6(0.2709)}{40 - 20} = 0.0812$$

Since the P/T ratio is less than 10 %, it is concluded that the process has adequate capability. In the absence of specification limits, one can calculate the process capability as the percentage of the product characteristic variability as

$$\text{Process capability} = \frac{\sigma_{gage}}{\sigma_{product}} \times 100,$$

Since

$$\sigma_{total}^2 = \sigma_{product}^2 + \sigma_{gage}^2$$

we estimate the value of σ_{total}^2 and σ_{gage}^2 from the whole sample of 60 observations ignoring the operator differences. These values are $\sigma_{total}^2 = 0.947$ and $\sigma_{gage}^2 = 0.076$, respectively. Hence,

$$\sigma_{product}^2 = \sigma_{total}^2 - \sigma_{gage}^2 = 0.8713$$

Therefore, the process capability is

$$\text{Process capability} = \frac{\sqrt{0.076}}{\sqrt{0.8713}} \times 100 = 29.5341$$

This capability ratio is more meaningful than the P/T ratio discussed above, as it does not depend on the specification limits (Fig. 8.3).

MINITAB Gage R&R Study - ANOVA Method

Two-Way ANOVA Table With Interaction

Source	DF	SS	MS	F	P
Parts	9	52.8017	5.86686	108.248	0.000
Operator	2	1.0426	0.52128	9.618	0.001
Parts * Operator	18	0.9756	0.05420	1.540	0.144
Repeatability	30	1.0560	0.03520		
Total	59	55.8758			

Gage R&R

Source	VarComp	%Contribution (of VarComp)
Total Gage R&R	0.06805	6.56
Repeatability	0.03520	3.39
Reproducibility	0.03285	3.17
Operator	0.02335	2.25
Operator*Parts	0.00950	0.92
Part-To-Part	0.96878	93.44
Total Variation	1.03683	100.00

Source	StdDev (SD)	Study Var (6 * SD)	%Study Var (%SV)
Total Gage R&R	0.26087	1.56522	25.62
Repeatability	0.18762	1.12570	18.43
Reproducibility	0.18125	1.08753	17.80
Operator	0.15282	0.91692	15.01
Operator*Parts	0.09746	0.58478	9.57
Part-To-Part	0.98426	5.90559	96.66
Total Variation	1.01825	6.10949	100.00

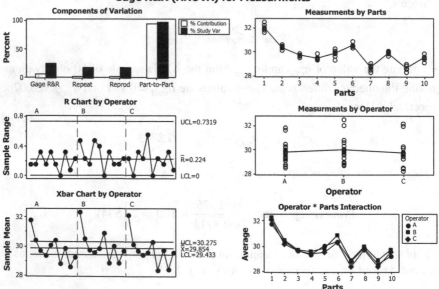

Gage R&R (ANOVA) for Measurments

Fig. 8.3 Gage R&R study

Analysis interpretation: From the gage R&R (ANOVA), it is inferred that the parts and operators are highly significant, whereas the interaction is not significant. From the gage R&R output, we can see that the total variation explained is 6.56 % which is much less than the part-to-part variation which is 93.44 %. Hence, it is concluded that much of the variation is due to parts. Since the total variation is less than 10 %, the measurement system is acceptable.

Note that the study variance as per numerical calculation is $6\hat{\sigma}_{gage} = 6(0.2709) = 1.6255$, which according to the MINITAB output is 1.56522. Similarly, the percentage contribution according to numerical calculation is roughly 7 % $\left(\hat{\sigma}^2_{gage} = 0.073\right)$ as compared to 6.56 % in the output table. These differences are negligibly small and will not affect the overall conclusion about the process.

Again, from the graph panel, it is observed that the percentage variation due to part to part is very high as compared to other variations. A similar conclusion can be drawn from the measurements by parts graph as the parts are not aligned in a straight line. The *R*-chart and the graph of measurement by operator show that the Operator B measured the observations slightly above the average. In the $\overline{X}$ chart by operator, most of the points in the $\overline{X}$ chart are outside the control limits, indicating variation is mainly due to differences between parts. The Operator* Part Interaction graph is a visualization of the *p* value (0.144 in this case)—indicating no significant interaction between each part and operator.

8.7.2 Assessing Bias of Attribute Data

Note that the attribute data (nominal or ordinal data) often result from subjective decisions by observers or raters. This subjectivity frequently results in bias, repeatability, and reproducibility problems. The more one can remove subjectivity from the rating process, the better. Studies of bias require knowledge of the true measure of an item or occurrence. As with continuous data, nominal and ordinal measurement systems are evaluated in terms of stability, bias, repeatability, and reproducibility. One such measure for assessing the above problem for nominal data is *kappa*.

Nominal measurement studies, which address repeatability and reproducibility, are often called *concordance studies*. In such studies, intra-rater agreement measures repeatability (within rater), and inter-rater agreement measures the combination of repeatability and reproducibility (between rater). And the probability of

Table 8.1 Kappa agreement table

Kappa	Strength of agreement
<0.00	Poor or none
0.00–0.20	Slight
0.21–0.40	Fair
0.41–0.60	Moderate
0.61–0.80	Substantial
0.81–1.00	Almost perfect

overall agreement among raters, or within a rater, can be estimated using a confidence interval. The measure kappa (denoted by κ) is useful for measuring both intra-rater and inter-rater rater agreement for nominal data and is defined as

$$\kappa = \frac{\text{no. of observed agreement} - \text{no. of expected agreement}}{\text{total observations} - \text{no. of expected agreement}} \tag{8.4}$$

The strength of agreement is evaluated based on the value of κ and is presented in the Table 8.1.

Kappa can be generalized to situations where there are more than two ratings by one rater, or where there are several raters. Kappa is also easily generalized to a situation where the response has more than two categories.

Example 8.2 The data in Table 8.2 show the results of three appraisers responded twenty items subjected to sampling inspection. The quality of the item is classified as "good" or "bad." Study the agreement between the appraisers.

Solution The MINITAB analysis of attribute agreement analysis output is shown below. The percent agreement table shows that Appraiser B, C, and A have matched for 55, 50, and 40 %, respectively (see also Fig. 8.4). The Fleiss' kappa statistic assumes the level of disagreement among responses is the same. Since the kappa values are negative in each case, it indicates that the measurement system is unacceptable.

Table 8.2 Attribute agreement analysis data

Sample	Appraiser	Response	Sample	Appraiser	Response	Sample	Appraiser	Response	Sample	Appraiser	Response
1	A	Good	6	A	Good	11	A	Good	16	A	Bad
1	A	Good	6	A	Good	11	A	Good	16	A	Good
1	B	Bad	6	B	Bad	11	B	Good	16	B	Good
1	B	Good	6	B	Good	11	B	Bad	16	B	Good
1	C	Bad	6	C	Bad	11	C	Bad	16	C	Good
1	C	Good	6	C	Good	11	C	Bad	16	C	Good
2	A	Bad	7	A	Good	12	A	Bad	17	A	Bad
2	A	Good	7	A	Good	12	A	Good	17	A	Good
2	B	Good	7	B	Good	12	B	Good	17	B	Good
2	B	Good	7	B	Good	12	B	Bad	17	B	Good
2	C	Good	7	C	Good	12	C	Good	17	C	Good
2	C	Good	7	C	Good	12	C	Good	17	C	Bad
3	A	Bad	8	A	Bad	13	A	Bad	18	A	Bad
3	A	Good	8	A	Good	13	A	Good	18	A	Good
3	B	Bad	8	B	Bad	13	B	Bad	18	B	Good
3	B	Bad	8	B	Bad	13	B	Bad	18	B	Bad
3	C	Good	8	C	Good	13	C	Good	18	C	Good
3	C	Bad	8	C	Bad	13	C	Bad	18	C	Good
4	A	Good	9	A	Good	14	A	Good	19	A	Good
4	A	Good	9	A	Good	14	A	Good	19	A	Good
4	B	Good	9	B	Good	14	B	Good	19	B	Good
4	B	Good	9	B	Good	14	B	Bad	19	B	Good
4	C	Good	9	C	Good	14	C	Good	19	C	Bad
4	C	Bad	9	C	Bad	14	C	Bad	19	C	Bad

(continued)

Table 8.2 (continued)

Sample	Appraiser	Response	Sample	Appraiser	Response	Sample	Appraiser	Response	Sample	Appraiser	Response
5	A	Bad	10	A	Bad	15	A	Bad	20	A	Bad
5	A	Bad	10	A	Bad	15	A	Good	20	A	Bad
5	B	Good	10	B	Good	15	B	Good	20	B	Good
5	B	Bad	10	B	Bad	15	B	Good	20	B	Bad
5	C	Good	10	C	Good	15	C	Good	20	C	Good
5	C	Good	10	C	Good	15	C	Good	20	C	Good

Attribute Agreement Analysis for Response

Within Appraisers

```
Assessment Agreement

Appraiser  # Inspected  # Matched  Percent      95 % CI
A                   20          8    40.00  (19.12, 63.95)
B                   20         11    55.00  (31.53, 76.94)
C                   20         10    50.00  (27.20, 72.80)

# Matched: Appraiser agrees with him/herself across trials.

Fleiss' Kappa Statistics

Appraiser  Response     Kappa  SE Kappa         Z  P(vs > 0)
A          Bad      -0.250000  0.223607  -1.11803     0.8682
           Good     -0.250000  0.223607  -1.11803     0.8682
B          Bad       0.040000  0.223607   0.17889     0.4290
           Good      0.040000  0.223607   0.17889     0.4290
C          Bad      -0.098901  0.223607  -0.44230     0.6709
           Good     -0.098901  0.223607  -0.44230     0.6709
```

Between Appraisers

```
Assessment Agreement

# Inspected  # Matched  Percent      95 % CI
         20          0     0.00  (0.00, 13.91)

# Matched: All appraisers' assessments agree with each other.

Fleiss' Kappa Statistics

Response      Kappa  SE Kappa         Z  P(vs > 0)
Bad      -0.0453333  0.0577350  -0.785196    0.7838
Good     -0.0453333  0.0577350  -0.785196    0.7838
```

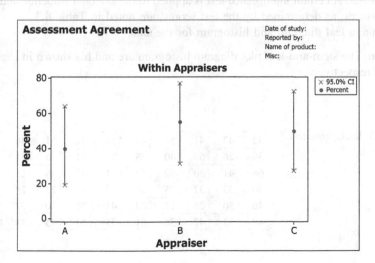

Fig. 8.4 Graphical results of attribute agreement analysis

8.8 Descriptive Statistics

The descriptive statistics play a very important role in studying the basic characteristics of a process data. They enable one to understand the pattern of variations in terms of accuracy measures or location measures like mean, median, and mode (also called measures of central tendency) and precision measures like range, quartile deviation, standard deviation, and variance (also called measures of dispersion). The objective of compressing the information into a descriptive statistics is to find a representative value which can be used to locate and summarize the entire set of varying values. This one value can be used to make many decisions concerning the entire set. The measures of central tendency also enable us to compare two or more sets of data to facilitate comparison. Which measure is good and applicable depends on the purpose and objective of the study. But it is expected that the measure should be

- Easy to understand
- Simple to compute
- Based on all observations
- Uniquely defined
- Capable of further algebraic treatment
- Simple to apply

Two of the best ways to describe a given data set is the histograms and stem-and-leaf plot methods. A *histogram* is a visual display of the frequency distribution. They are rectangles in varying sizes drawn adjacent to each other according to the frequency of each observation. A *stem-and-leaf diagram* is another way to obtain an informative visual display of a data set, where each number consists of at least two digits. As an illustration, consider the following example:

Example 8.3 A certain intelligence test is applied to the selected students, and their performances, as determined by the test scores, are noted in Table 8.3.
     Prepare a leaf diagram and histogram for the data.

Solution The stem-and-leaf plot diagram histogram are and his shown in Figs. 8.5 and 8.6 respectively.

Table 8.3 Intelligence test scores

32	47	41	51	41	54	32	31	46	15
38	26	50	40	38	44	21	45	31	37
68	41	30	52	52	61	53	48	21	28
30	33	37	35	39	43	32	24	28	38
46	50	26	15	23	41	38	40	37	40
40	33	42	36	51	31	51	45	41	50

Fig. 8.5 Stem-and-leaf plot of test scores

Stem	leaf	Count
1	55	2
2	1134	4
2	6688	4
3	0011122233	10
3	5677788889	10
4	000011111234	12
4	556678	6
5	0001112234	10
6	18	2

Fig. 8.6 Histogram of intelligence test scores

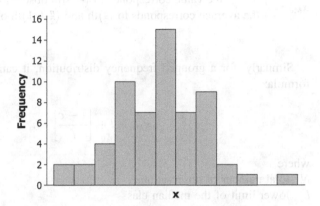

8.8.1 Measures of Accuracy

Arithmetic Mean: The most commonly used accuracy measure in Six Sigma project is the arithmetic mean (AM) or simply mean or process average. It is defined as the ration of sum of all observations to the total number of observations. If $x_1, x_2, \ldots, x_n$ are the set of n observations, then symbolically the mean is defined as

$$\bar{x} = \frac{x_1 + x_2 + \cdots + x_n}{n}$$

$$= \frac{1}{n} \sum_{i=1}^{n} x_i \tag{8.5}$$

If suppose the sample observations $x_1, x_2, \ldots, x_n$ occur with various frequencies, say $f_1, f_2, \ldots, f_n$, then the corresponding mean is obtained as

$$\bar{x} = \frac{x_1 f_1 + x_2 f_2 + \cdots + x_n f_n}{\sum_{i=1}^{n} f_i},$$

$$= \frac{1}{N} \sum_{i=1}^{n} x_i f_i \tag{8.6}$$

where $\sum_{i=1}^{n} f_i = N$.

Median: Median is that value which divides the distribution into two equal parts. Mostly, this measure is computed for characteristics which are measured in counts and numbers. Median is that value of the middle observation when the series is arranged in the order of the size of magnitude. Thus, 50 % of the observations in the distributions are above the value of median and 50 % above the median value. If the number of observations is odd, then the median is equal to exact middle observation. If the number of observations is even, then the median will be the average of the two middle observations. Symbolically, it can be calculated using the following formula:

$$x_{Me} = \begin{cases} \text{the value corresponds to } \left(\frac{n+1}{2}\right) \text{th observation,} & \text{if } n \text{ is odd} \\ \text{the average corresponds to } \left(\frac{n}{2}\right) \text{th and } \left(\frac{n}{2}+1\right) \text{th observation} & \text{if } n \text{ is even} \end{cases}$$

(8.7)

Similarly, for a grouped frequency distribution, it can be calculated using the formula:

$$x_{Me} = l + \left[\frac{\frac{N}{2} - c}{f} \right] * h,$$

(8.8)

where
N total number of observations
l lower limit of the median class
c cumulative frequency preceding to the median class
f frequency of the median class
h class interval of the median class.

The quantity cumulative frequency for a sample of numerical data corresponding to a number x is the total number of observations that are less than or equal to x. A diagram representing grouped numeric data in which cumulative frequency is plotted against the upper class boundary and the resulting points are joined by straight line segments is called a *frequency polygon*.

Quartiles, Deciles, and Percentiles: Like median, quartiles, deciles, and percentiles are the related positional measures of central tendency. *Quartiles* are those values which divide the total data into four equal parts. The first quartile, Q_1, is the value such that 25 % of the observations are smaller and 75 % of the observations are larger. The second quartile Q_2 is the median, and the third quartile Q_3 is the value such that 25 % values are larger and 75 % observations are smaller.

Mode: The mode is defined as the most frequently occurring value in a data set. It is that value which occurs most often with the highest frequency. For the raw data 23, 34, 26, 28, 30, 30, 27, 29, and 30, the mode is 30 as it occurs frequently. For a frequency distribution or grouped data, the model class is the class with the maximum frequency. It is calculated using the formula:

$$x_{Mo} = l + \left[\frac{f_1 - f_0}{2f_1 - f_0 - f_2}\right] * h, \tag{8.9}$$

where
l   lower limit of the modal class
f_1 frequency of the modal class
f_0 frequency preceding to the modal class
f_2 frequency succeeding to the modal class
h   class interval of the median class

8.8.2 Measures of Symmetry and Shape

The measure of central tendency and variation do not reveal all the characteristics of a given set of data. Two distributions may have the same mean and standard deviation (see Chaps. 4 and 5 for a detailed discussion on measures of variation) but may differ widely in the shape of their distribution. That is, the data are either symmetrical or not symmetrical. If the distribution is not symmetrical, it is called asymmetrical or skewed. Thus, skewness refers to the lack of symmetry in a distribution. A symmetric distribution has skewness zero, and hence, it is called a bell-shaped distribution (or normal distribution). In this case, mean, median, and mode all will coincide at one point (see Fig. 8.7). For a statistically controlled process, these accuracy measures must be same. Any departure from this will indicate the presence of variation and will be the primary concern for improvement.

If the distribution is not symmetric, then the other two possibilities are either a long tail on the left side or on the right side. If the observations are tailed toward the lower side, the skewness is negative. It happens when the mean is decreased by some extremely low values, thus making mean < median < mode (see Fig. 8.8). Similarly, if the values are tailed toward the right side, the skewness is positive. It occurs when the mean is increased by some unusually high values, thereby making, mean > median > mode (see Fig. 8.9).

Fig. 8.7 A symmetric distribution

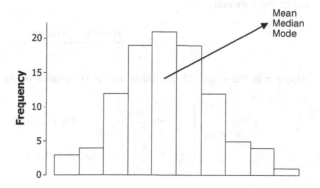

Fig. 8.8 A negatively skewed distribution

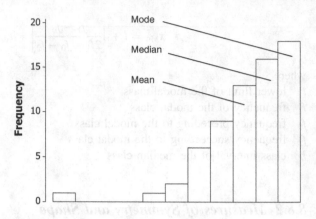

Fig. 8.9 A positively skewed distribution

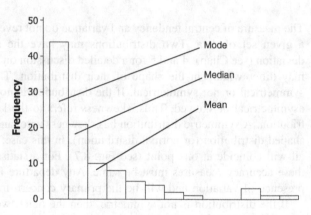

A simple measure of skewness is computed using the formula

$$S_k = \frac{\text{mean} - \text{mode}}{\sigma} \qquad (8.10)$$

In case mode is not well defined for a data set, the skewness may be computed using the formula:

$$S_k = \frac{3(\text{mean} - \text{median})}{\sigma} \qquad (8.11)$$

where σ is the standard deviation and is computed using the formula

$$\sigma = \begin{cases} \sqrt{\dfrac{\sum_{i=1}^{n}(x_i - \bar{x})^2}{n}}, & \text{for continuous data} \\[3mm] \sqrt{\dfrac{\sum_{i=1}^{n} f_i(x_i - \bar{x})^2}{N}}, & \text{for grouped frequency data} \end{cases} \qquad (8.12)$$

The standard deviation is a measure of spread or scatter in the population expressed in the original unit. For various estimators of sigma, one may refer to Chap. 5 for details. If $S_k = 0$, then the distribution is normal, and if $S_k > 0$, the distribution is positively skewed; otherwise, it is negatively skewed. When the distributions are given with open-ended classes having extreme values, then the skewness can be computed using the positional measures based on quartiles as

$$S_k = \frac{Q_1 - 2Q_2 + Q_3}{Q_3 - Q_1} \tag{8.13}$$

Skewness can also be computed from the given observations as follows:

$$S_k = \begin{cases} \dfrac{\left[\sum_{i=1}^{n}(x_i-\bar{x})^3\right]^2}{\left[\sum_{i=1}^{n}(x_i-\bar{x})^2\right]^3}, & \text{for continuous data} \\[4mm] \dfrac{\left[\sum_{i=1}^{n}f_i(x_i-\bar{x})^3\right]^2}{\left[\sum_{i=1}^{n}f_i(x_i-\bar{x})^2\right]^3}, & \text{for grouped frequency data} \end{cases} \tag{8.14}$$

The shape of the distribution can be further investigated using the height (or peakedness) of the frequency curve of the distribution. A measure of peakedness is called kurtosis. A normal distribution has skewness 0 and kurtosis equal to 3, and hence, it is called mesokurtic curve. A normal distribution, therefore, plays an important role in identifying the amount of variation present in an ongoing measurement system. Even a normal probability plot is used as a quality tool for describing the variation present in the process.

The kurtosis can be calculated using the following formula:

$$K_u = \begin{cases} \dfrac{\sum_{i=1}^{n}(x_i-\bar{x})^4}{\left[\sum_{i=1}^{n}(x_i-\bar{x})^2\right]^2}, & \text{for continuous data} \\[4mm] \dfrac{\sum_{i=1}^{n}f_i(x_i-\bar{x})^4}{\left[\sum_{i=1}^{n}f_i(x_i-\bar{x})^2\right]^2}, & \text{for grouped frequency data} \end{cases} \tag{8.15}$$

We now illustrate the numerical computation of the characteristics given in various equations above for the example on intelligent scores discussed in Table 8.3.

Example 8.4 The descriptive statistics corresponding to the data on intelligent scores (Table 8.2) is computed as follows:

Class	f	x	xf	$(x_i - \bar{x})$	$f(x_i - \bar{x})^2$	$f(x_i - \bar{x})^3$	$f(x_i - \bar{x})^3$
10–20	2	15	30	−25.33	1283.56	−32,516.74	823,757.43
20–30	8	25	200	−15.33	1880.89	−28,840.30	442,217.88
30–40	20	35	700	−5.33	568.89	3034.07	16,181.73
40–50	18	45	810	4.67	392.00	1829.33	8536.89
50–60	10	55	550	14.67	2151.11	31,549.63	462,727.90
60–70	2	65	130	24.67	1216.89	30,016.59	740,409.28
Total	60		2420		7493.33	−995.56	2,493,831.11

Then

$$\text{Mean} = \bar{x} = \frac{1}{N} \sum_{i=1}^{n} f_i x_i$$

$$= \frac{2420}{60} = 40.33$$

$$\text{Variance} = \frac{1}{N} \sum_{i=1}^{n} f_i (x_i - \bar{x})^2$$

$$= \frac{7493.33}{60} = 124.89$$

$$\text{Skewness} = \frac{\left[\sum_{i=1}^{n} f_i (x_i - \bar{x})^3 \right]^2}{\left[\sum_{i=1}^{n} f_i (x_i - \bar{x})^2 \right]^3}$$

$$= \frac{(-995.56)^2}{(7493.33)^3} = 0.00$$

$$\text{Kurtosis} = \frac{\sum_{i=1}^{n} f_i (x_i - \bar{x})^4}{\left[\sum_{i=1}^{n} f_i (x_i - \bar{x})^2 \right]^2}$$

$$= \frac{2493831.11}{(7493.33)^2} = 0.04$$

The histogram with normal curve is shown in Fig. 8.10.

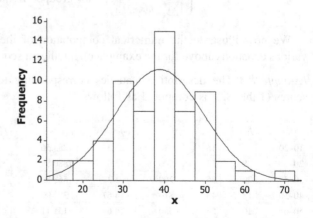

Fig. 8.10 Histogram with normal curve for intelligent scores

8.9 Describing Sources of Variation

The change or fluctuation of a specific characteristic from the normal condition is called variation. Variation affects every process, which in turn affects the quality of the product directly or indirectly. Only a quantitative measurement like accuracy measures, measures of dispersion, and measures of skewness and kurtosis is not enough to explain the amount of variation, as the factors affecting variation may not be exclusively shown in a descriptive measure every time. Therefore, we use various charts and graphical plots to achieve this. In this section, we discuss some of the quality tools used for identifying variations in the data (see also Chap. 4, for a detailed discussion on the sources and types of variations). In the management decision-making process, they are usually called 7-QC tools, as they are very easy to study and interpret the variations. Some of the most frequently used tools in Six Sigma project is discussed below.

8.9.1 Pareto Chart

Pareto chart is a quality tool used for understanding the pattern of variations in attribute or categorical type data sets. It is based on Pareto principle introduced by Vilfredo Pareto's research stating that the vital few (20 %) of the causes have a greater impact that the trivial many (80 %) causes with a lesser impact. It uses data with columns arranged in descending order, with highest occurrences (highest bar) shown first, and uses a cumulative line to track percentages of each category/bar, which distinguishes the 20 % of items causing 80 % of the problems.

Thus, dividing data into categories, Pareto chart judges the relative impact of various parts of a problem (quantifying the problem) and tracks down the biggest contributor(s) to a problem. A Pareto chart helps to decide where improvement efforts will have the biggest payoff. A Pareto chart can be used only when the problem under study can be broken down into categories and the number of occurrences can be counted for each category. The Pareto chart is illustrated through the following example.

Example 8.5 The data below show the downtime associated with a particular manufacturing process of a type company. There are about nine patterns of failures (D1–D9) observed for a period of one year. The number of times each failure occurred is summarized below.

Failure pattern	D1	D2	D3	D4	D5	D6	D7	D8	D9
Failure counts	274	59	19	43	3	4	6	10	5

The Pareto chart associated with the above data is given in Fig. 8.11.

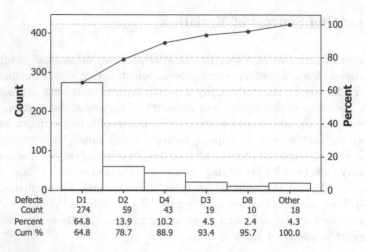

Defects	D1	D2	D4	D3	D8	Other
Count	274	59	43	19	10	18
Percent	64.8	13.9	10.2	4.5	2.4	4.3
Cum %	64.8	78.7	88.9	93.4	95.7	100.0

Fig. 8.11 A Pareto chart

How to read a Pareto chart:

- Type of data used to create the chart—is the chart based on valid data?
- Relative heights of the bars (including height of the Y-axis)—making sure the Pareto Principle applies.
- Size of the "other" category—making sure that another category cannot be formed from some of the "other" data.

Some interpretations of the Pareto chart are the following:

1. If the Pareto principle holds and a few categories are responsible for most of the problems, then begin work on the largest bar(s) and start analyzing for causes.
2. When all the bars are roughly the same height and/or many categories are needed to account for most of the problems, the data need to be interpreted in another way.

Thus, using Pareto chart, the areas of concern can be identified and efforts can be made to correct those defects that account for the largest percentage of defects. In Fig. 8.10, the bar height reflects the frequency or impact of the causes. The graph is distinguished by the inclusion of a cumulative percentage line that identifies the vital few opportunities for improvement.

8.9.2 Control Charts

In Chap. 4, we had discussed the sources of variation and the types of variations associated with a measurement system in detail. As said, the best way of analyzing the special cause variation and common cause variation is through the control chart techniques. Control chart at this phase helps project people to identify the assignable causes (man, machine, materials, methods, and management) and chance causes of variation (environment, technology, etc.) present in the process.

A statistically controlled chart will be called a normal process whose process means (a measure of accuracy) and standard deviation (a measure of precision) will work according to the tolerance level suggested by the customer. This is called statistical process control (SPC). SPC involves the measurement and evaluation of variation in a process, and the efforts made to limit or control such variation. Thus, SPC helps an organization or process owner to identify possible problems or unusual incidents so that action can be taken promptly to resolve (or to control) them. Figure 8.12 is the control chart which corresponds to the measurements discussed in Example 8.1. A detailed discussion on charting techniques, the steps involved in making a process in control, and other issues concerned with out-of-control processes, etc., will be taken up later in the control phase in Chap. 11.

8.9.3 Cause and Effect Diagram

A cause and effect (C&E) diagram also called *Ishikava diagram* or *Fishbone diagram* introduced by Kaoru Ishikawa in 1976 [7] is to generate in a structured manner maximum number of ideas regarding possible causes for a problem by using brainstorming technique. It is used when there are a large number of potential

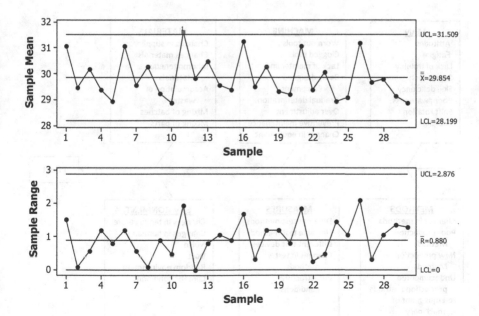

Fig. 8.12 Control chart of part measurements

causes that it is difficult to focus the analysis and when there is a lack of clarity about the relationship between different potential causes. It also helps

- To stimulate thinking during a brainstorm of potential causes
- To understand relationships between potential causes
- To track which potential causes have been investigated, and which proved to contribute significantly to the problem.

How to prepare a C&E diagram?

- Clarify the problem
- Gather members for discussion
- Conduct brainstorming session
- Group the causes (man, material, machine, method, environment, etc.)
- Draw the cause and effect chart
- Check for missing information
- Determine importance of significance of causes.

A general C&E diagram depicting the major causes and its effect is shown in Fig. 8.13.

How to identify Potential Causes?

- Define the "effects" for a cause and effect diagram (CED) clearly
- Prepare a 5 M, process, or stratification cause and effect diagram (SCED) for each effect. One may choose to use a combination of these

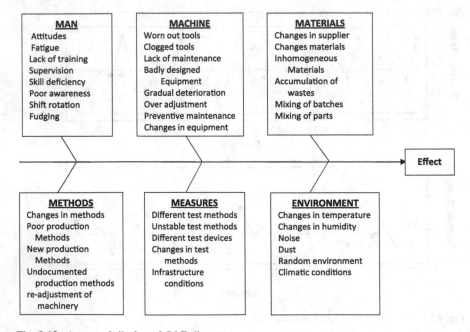

Fig. 8.13 A general display of C&E diagram

- Prepare a time line analysis (TLA) if the problem was not always present. Identify "what changed, when"?
- Perform a comparative analysis (CA) to determine whether the same or a similar problem existed in related products or processes
- Identify several potential causes. Develop a plan for investigating each cause and update the action plan (AP)
- Evaluate a potential cause against the problem description.

Some frequently occurring causes are material failure, assembly error, incorrect speeds/feeds, incorrect tooling, inadequate venting, misalignment, improper torque, inadequate gauging, overload capacity, out of tolerance, tool damage/burrs, clamping, packaging damage, handling damage, improper tool setup, improper surface preparation, inadequate control system, damaged part, inadequate gating, and inadequate holding etc.

A typical C&E diagram for the effect of "high petrol consumption" is given in Fig. 8.14.

How to Analyze Potential Causes?

- Use an alternative process to analyze each potential cause

  - *Hypothesis generation*: How does the potential cause result in the problem?
  - *Design*: What type of data can most easily prove or disapprove the hypothesis?
  - *Preparation*: Obtain materials and prepare a check sheet

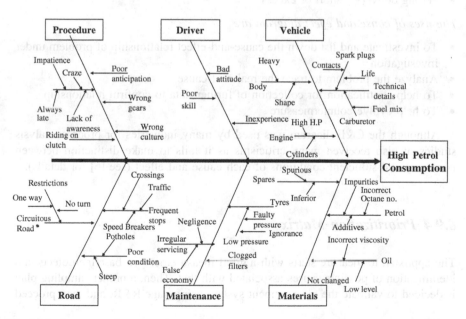

Fig. 8.14 Cause and effect diagram for high petrol consumption (*Source* Six Sigma MBB lecture notes, ISI Bangalore)

- *Data collection*: Collect the data
- *Analysis*: Use simple, graphical methods to display data

- Investigate several potential causes independently.
- Use an action plan to manage the analysis process for each potential cause being studied.
- Validate root causes
- Clearly state root cause(s) and identify data which suggest a conclusion

The potential effect(s) of failure are

- Poor coordination
- Abrupt use of materials
- Air/water leakage
- Brake failure
- Contamination
- Customer dissatisfaction
- Delay in delivery/payment
- Excess interest
- Poor brake efficiency
- Poor valve performance
- Oil carryover
- Inadequate service life
- Unable to assemble
- Wrong material
- Wrong delivery—short or excess

The uses of cause and effect diagram are

- To investigate and list down the cause-and-effect relationship of problem under investigation
- Analyze the problem to trace the real root cause
- To help stratification for collection of further data to confirm relationship
- To help evolve countermeasure

Although the C&E diagram was used by many industries for process analysis studies, it also received many criticisms as it fails to make distinction between necessary and sufficient conditions of each cause and effect (see [6] for details).

8.9.4 Prioritization Matrix

The approach to measure starts with a good measurement on baseline defects and identification of possible causes associated with that. Then, a proper sampling plan is decided to validate the measurement system using gage R&R, and then proceed with the analysis of the data. In this process, selecting the "right" measures is key to gain a good understanding of the problem. This is achieved through two techniques: *prioritization matrix* and failure mode effect analysis (FMEA). Here, we discuss prioritization matrix first and take up FMEA in the next phase of Six Sigma.

Prioritization matrix is used when

- There are too many variables that might have an impact on the output of the process
- Collecting data about all possible variables would cost too much time and money
- Team members have different theories about what happens in the process

Therefore, we develop a prioritization matrix to

- Focus the data collection effort
- Formulate theories about causes and effects and then
- Identify the critical few variables that need to be measured and analyzed

A prioritization matrix is constructed in the following way:

1. List out all the variables
2. Rank order and weight the output variables
3. List all input and process variables
4. Evaluate the strength of the relationship between output and input/process variables (correlation factor)
5. Cross-multiply weight and correlation factor
6. Highlight the critical few variables

A typical prioritization matrix of a manufacturing process is presented in Table 8.4. As per the table, the critical process variable is temperature and the critical input variables are die and lubricant, as their total score is high.

Table 8.4 A prioritization matrix

	Output variables	Strength	Surface quality	Diameter	Cast/lay	Total
	Weight	8	8	7	5	
Process variables	Speed	2	4	5	2	93
	Temperature	9	8	2	5	175
	Pressure	5	5	2	1	99
	...	...	...	...	...	...
Input variables	Die	5	8	8	2	170
	Lubricant	8	7	7	2	179
	Payoff	5	5	5	3	130
	Wire	4	4	9	3	142
	...	...	...	...	...	...

8.10 Dealing with Uncertainty: Probability Concepts

Uncertainty is a part and parcel of the human life. Decision making in such cases is facilitated through formal and precise expressions for the uncertainties involved. Probability theory provides us with the ways and means to attain such a task for uncertainties involved in different situations [3, 11, 19]. Probability, in common parlance, connotes the chance of occurrence of an event or happening. These occurrences are either the result of an experiment or an investigation.

The term *experiment* is used in probability theory in a broader sense than in other branches of sciences. Any action, whether drawing an item form a consignment, or measurement of a product's dimension to ascertain quality, or the launching of a new product in the market, deciding a product good or bad, rating a service, tossing of a coin constitute an experiment in the probability theory.

Any experiments in probability theory have three things in common:

1. There is always uncertainty about the outcomes
2. It is possible to specify the outcomes in advance
3. There are two or more outcomes of each experiment

The set of all possible outcomes of an experiment is defined as the *sample space* denoted as S. Each outcome is thus visualized as a sample point in the sample space. For example, the set {success, failure} defines the sample space of a launching experiment. Similarly, the set {head, tail} defines the sample space of a coin tossing experiment. An *event* in probability theory constitutes one or more possible outcomes of an experiment. Thus, an event is a subset of the sample space. There are many ways to define probability of an event. A useful and practical definition of probability is defined through axioms. These axioms are fundamental to probability theory and hence are important for any decision making under uncertainty. The *axioms* are as follows:

(i)   If A is an event, then $0 \leq P(A) \leq 1$
(ii) $P(S) = 1$
(iii) If A and B are two *mutually exclusive events*, then,

$$P(A \text{ or } B) = P(A) + P(B)$$

However, if the events are not mutually exclusive, then the probability of occurrence of either A or B is given by

$$P(A \text{ or } B) = P(A) + P(B) - P(A \text{ and } B)$$

That is, notationally, it is written as

$$P(A \cup B) = P(A) + P(B) - P(A \cap B), \qquad (8.16)$$

which is called the addition theorem of probability. The quantity $P(A \cap B)$ is called the joint occurrence of the events A and B. The result can be extended for any number of events.

If two or more events together define the total sample space, the events are said to be collectively *exhaustive events*. Further, if we are interested in calculating probabilities when some partial information concerning the result of the experiment is available, or in calculating them in light of additional information, then we use *conditional probability*.

For example, consider the experiment of throwing a dice twice. If we denote (i, j) as the outcome corresponding to first and second dice, respectively, then

$$S = \{(i,j), i = 1,2,3,4,5,6 \text{ and } j = 1,2,3,4,5,6\}$$

Suppose that the first dice shows 4. Then, given this information, what is the probability that the sum of the two dice equals 8? To calculate this probability, we define the event A and B and the associated outcomes as follows:

A: The first dice shows 4, then $A = \{(4,1), (4,2), (4,3), (4,4), (4,5), (4,6)\}$

and

B: The sum of the dice is 8, then $B = \{(2,6), (3,5), (4,4), (5,3), (6,2)\}$

The probability of A and B are 6/36 and 5/36, respectively. Note that one outcome is common to both A and B, which is the set $\{(4, 4)\}$. Therefore, $P(A \cap B) = 1/36$. The required probability is

$$P(\text{sum of the face is } 8 | \text{first face shows } 4) = P(B|A)$$
$$= P(A \cap B)/P(A)$$
$$= 1/6.$$

Definition 8.1 If A and B are two events, then the conditional probability of A given B is defined as

$$P(A|B) = \frac{P(A \cap B)}{P(B)}, \quad P(B) \neq 0 \tag{8.17}$$

or

$$P(A \cap B) = P(A|B)P(B)$$
$$= P(B|A)P(A)$$

If events A and B are *independent*, then $P(A|B) = P(A)$ and $P(B|A) = P(B)$. Hence,

$$P(A \cap B) = P(A)P(B), \qquad (8.18)$$

which is called the *multiplication theorem* of probability. For any experiment, the multiplication probability can depend on the sampling procedures used in the experiment. In *sampling with replacement*, the object that was drawn at random is placed back to the given set and the set is mixed thoroughly; then, the next item is drawn randomly. In *sampling without replacement,* the object that was drawn is put aside, and then, the next sample is drawn at random.

Example 8.6 Disks of polycarbon plastic from a supplier are analyzed for scratch and shock resistance. The results from 100 disks are summarized as follows:

Shock resistance	Scratch resistance	
	High	Low
High	70	10
Low	16	4

A disk high is selected at random. What is the probability that

(i)   the disk has high shock resistance
(ii) the disk has high shock resistance given that it has high scratch resistance
(iii) the disk has high shock resistance and also has low scratch resistance
(iv) Are the events' high shock resistance and high scratch resistance independent?

Solution Let us denote

A: the event that disk has high shock resistance

B: the event that a disk has high scratch resistance

Then, the required probabilities are

(i) $P(A) = \frac{80}{100} = 0.80$

(ii) $P(A|B) = \dfrac{P(A \cap B)}{P(B)}$: According to conditional probability definition

$$= \frac{70}{100} \Big/ \frac{86}{100} = \frac{70}{86} = 0.814$$

(iii) $P(A \cap B') = \frac{10}{100} = 0.10$ (directly observed from the table)

Using the definition of conditional probability, this can be obtained as

$$P(A \cap B') = P(A)P(B'|A)$$

$$= P(B' \cap A) = \frac{10}{100} = 0.10$$

(iv) We have $P(A \cap B) = \frac{70}{100} = 0.70$, $P(A) = \frac{80}{100} = 0.80$ and $P(B) = \frac{86}{100} = 0.86$. Therefore,

$$P(A) \times P(B) = 0.80 \times 0.86 = 0.688$$

Since $P(A \cap B) \neq P(A) \times P(B)$, the events A and B are not independent.

8.10.1 Principles of Counting

The situation in the above problem can be simplified by using the *principle of counting*. The two general principles of counting are permutations and combinations. A *permutation* of given things (elements or objects) is an arrangement of these things in a row in some order. For example, for three letters p, q, and r, there are $3! = 1 \times 2 \times 3 = 6$ permutations: *pqr, prq, qpr, qrp, rpq, rqp*. Further,

(a) The number of permutations of n different things taken all at a time is

$$n! = 1 \times 2 \times 3 \ldots n \tag{8.19}$$

(b) If n given things can be divided into c classes of alike things differing class to class, then the number of permutations of n different things taken all at a time is

$$\frac{n!}{n_1! n_2! \ldots n_c!}, n_1 + n_2 + \cdots + n_c = n, \tag{8.20}$$

and n_i is the number of things in the ith class.

(c) The number of permutations of n different things taken r at a time without repetition is

$$n(n-1)(n-2)\ldots(n-r+1) = \frac{n!}{(n-r)!} = {}^nP_r \tag{8.21}$$

In a permutation, the order of the selected things is essential. In contrast, a *combination* of given things means any selection of one or more things without regard to the order. There are two kinds of combinations, as follows:

(a) The number of different combinations of n different things, r at a time, without repetitions is

$$\frac{n(n-1)(n-2)\ldots(n-r+1)}{1\times 2\ldots r}=\frac{n!}{r!(n-r)!}={}^nC_r=\binom{n}{r} \tag{8.22}$$

(b) The number of different combinations of n different things, r at a time, with repetitions is

$$\binom{n-r+1}{r} \tag{8.23}$$

Example 8.7 In a lot of 8 items, 2 are defective. Find the number of different samples of 3, where:

(i)   the number of samples containing no defective
(ii) the number of samples containing 1 defective
(iii) the number of samples containing 2 defectives.

Solution Since the lot contains 8 items (2 defective and 6 non-defective), different samples of 3 can be obtained as a combination of selecting 3 items out of 8 items. Hence

$$\text{The number of different samples of } 3 = \binom{8}{3}$$
$$= \frac{8\times 7\times 6}{1\times 2\times 3}=56$$

Further

(i) the number of samples containing no defective $=\binom{6}{3}$
$$=\frac{6\times 5\times 4}{1\times 2\times 3}=20$$

(ii) the number of samples containing 1 defective $=\binom{2}{1}\binom{6}{2}$
$$=\frac{2\times 6\times 5}{1\times 2}=30$$

(iii) the number of samples containing 2 defectives $=\binom{2}{2}\binom{6}{1}=6$

Some other frequently used combinatorial identities in probability theory are as follows:

(i) $\binom{n}{k} + \binom{n}{k+1} = \binom{n+1}{k+1}, k \geq 0$ is an integer

(ii) $\sum_{s=0}^{n-1} \binom{k+s}{k} = \binom{n+k}{k+1}$

(iii) $\sum_{k=0}^{r} \binom{p}{k} + \binom{q}{r-k} = \binom{p+q}{r}$

(iv) $(a+b)^n = \sum_{k=0}^{n} \binom{n}{k} a^k b^{n-k}$

(v) $n! \cong \sqrt{2\pi} n^{n+1/2} e^{-n}$

Here, the identity given in (iv) is the binomial expansion and (v) is called the *Stirling's approximation* of n. There are number of uses of the identities in probability theory and modeling. In some of the sections in various chapters, we use these quantities as a part of probability models and analysis purposes without any detailed proof.

An important theorem, which makes use of conditional probability, is the *Bayes' theorem*. The Bayes' theorem offers the revision of probability estimates with added information in random experiments.

Definition 8.2 *Bayes's theorem*: If suppose $A_1, A_2, \ldots, A_k$ are k mutually exclusive and collectively exhaustive events and B is another event defined in the context of the same experiment associated with the occurrence of $A_1, A_2, \ldots, A_k$, then the conditional probability of A_i given B is given by

$$P(A_i|B) = \frac{P(A_i)P(B|A_i)}{\sum_{j=1}^{k} P(A_j)P(B|A_j)} \tag{8.24}$$

Bayes' theorem is useful in computing probabilities of various hypothesis $A_1, A_2, \ldots, A_k$ that have resulted in the event B. Below, we give an example to illustrate the Bayes' theorem.

Example 8.8 A manufacturing company uses three machines to produce a particular item. It is found that 60 % of the items are produced through machine 1, 20 % through machine 2, and remaining through machine 3. It is found that 5, 3, and 2 % of the produced items are defectives reported through machines 1, 2, and 3, respectively. During a quality check, one of the items is selected from a machine and found to be defective. What is the probability that the defective item was drawn from

(a) Machine 1?
(b) Machine 1 or Machine 2?

Solution As per the Bayes' theorem, we first define the event and corresponding probabilities as follows: Let

$$A_1: \text{drawing Machine-1}$$
$$A_2: \text{drawing Machine-2}$$
$$A_3: \text{drawing Machine-3}$$

Such that

$$P(A_1) = 0.60, \quad P(A_2) = 0.20, \quad P(A_3) = 0.20$$

Note that the events A_1, A_2, and A_3 are mutually exclusive and collectively exhaustive, as $\sum_{i=1}^{3} P(A_i) = 1$. Now define

$$B: \text{The item drawn is defective}$$

Since event B is associated with the events A_1, A_2, and A_3, we have

$$P(B|A_1) = 0.05, \quad P(B|A_2) = 0.03, \quad P(B|A_3) = 0.02$$

Then

(a) To find the probability that defective item was drawn from machine 1, we need to find $P(A_1|B)$. That is

$$P(A_1|B) = \frac{P(A_1)P(B|A_1)}{\sum_{j=1}^{3} P(A_j)P(B|A_j)}$$

$$= \frac{0.60 \times 0.05}{0.60 \times 0.05 + 0.20 \times 0.03 + 0.20 \times 0.02}$$

$$= 0.75$$

(b) To find the probability that defective item was drawn from machine 1 or machine 2, we need to find the probability $P(A_1|B) + P(A_2|B)$, where

$$P(A_2|B) = \frac{P(A_2)P(B|A_2)}{\sum_{j=1}^{3} P(A_j)P(B|A_j)}$$

$$= \frac{0.20 \times 0.03}{0.60 \times 0.05 + 0.20 \times 0.03 + 0.20 \times 0.02}$$

$$= 0.15$$

Therefore,

$$P(A_1|B) + P(A_2|B) = 0.75 + 0.15 = 0.90$$

In the discussions made so far, we assumed the probability as a measurable quantity of an uncertain event. The axiomatic approach provides a quantitative measure for probability when events are mutually exclusive. In the discussion that follows, we will now treat probability as a function which assigns probability value P to each sample point of an experiment abiding by the above axioms. As the events are interest to us, the computation of probability through any of the definition above turns out to be tedious in many cases. Hence, there arises the need for probability models, generally called *probability distributions* to study the characteristics of the random phenomena associated with a random experiments. In various sections below, we will study the importance of probability models and show how uncertainty of experimental events is accounted for in measurement systems.

8.11 Random Variables and Expectation

When the value of a variable is subject to random variation, or when it is the value of a randomly chosen member of a population, it is called a *random variable*. A random variable is a function that assigns a real number to each outcome in the sample space of a random experiment. A random variable is denoted by an uppercase letter such as X, and the value of it is denoted by the lowercase letter such as x. A random variable can be discrete or continuous.

8.11.1 Discrete Random Variables

A *discrete random variable* is a random variable with a finite (or countably infinite) range. For example, number of defective pieces in a lot, number of scratches on a surface, number of telephone calls recorded during a day, number of industrial accidents, and number of typographical errors in a page. Thus, discrete random variables are generally recorded in terms of counts and the measurements are limited to integers. A description of the possible values of a random variable and of their corresponding probabilities of occurrence is the *probability distribution*.

Definition 8.3 Let X be a discrete random variable with possible values $x_1, x_2, \ldots$. Then, the probability mass function (pmf) is a function such that

(a) $f(x_j) \geq 0$

(b) $f(x_j) = P(X = x_j), \quad j = 1, 2, \ldots,$

(c) $\sum_{j=1}^{n} f(x_j) = 1$

$$(8.25)$$

Thus, for a discrete random variable, the distribution is often specified by just a list of the possible values along with the probability of each. As an illustration, consider the following coin tossing experiments:

Suppose a coin is tossed three times. The distribution of the *"number of tails"* can be obtained as follows:

The possible outcomes of the experiment here is

$$S = \{HHH, HHT, HTH, HTT, THH, THT, TTH, TTT\}$$

Let the random variable X be "number of tails." Then, X can take values 0, 1, 2, and 3 with corresponding probability of occurrences as

$$P(X = 0) = 1/8, P(X = 1) = 3/8, P(X = 2) = 3/8 \text{ and } P(X = 3) = 1/8$$

Thus, the probability distribution of "the number of tails" is in Table 8.4.

The graph of *pmf* of "number of tails" is shown in Fig. 8.15. Note that the pmf is a line graph, where each line represents a probability, corresponding to the outcomes.

Sometimes, we must be interested in finding the probabilities fewer or greater than a given value. This is realized through the cumulative distribution function of the random variable. A distribution function is defined as follows:

Definition 8.4 The cumulative distribution function or simply the *distribution function $F(x)$* of a discrete random variable is defined as

$$F(x) = P(X \le x) = \sum_{x_i \le x} f(x_i) \tag{8.25}$$

For a discrete random variable, $0 \le F(x) \le 1$, and $F(x) \le F(y)$, if $x \le y$.

Fig. 8.15 *pmf* of "number of tails"

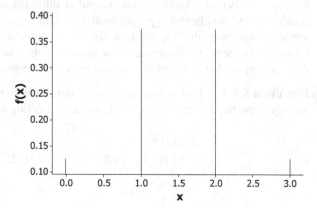

Example 8.9 Show that the following function is a pmf

$$f(x) = \frac{1 + 2x}{25}, \quad x = 0, 1, 2, 3, 4$$

Also find $P(X \le 2), P(X \ge 1), P(1 \le X < 4)$.

Solution The distribution table can be obtained as

X	0	1	2	3	4	Total
$P(X = x)$	$\frac{1}{25}$	$\frac{3}{25}$	$\frac{5}{25}$	$\frac{7}{25}$	$\frac{9}{25}$	1
$P(X \le x)$	$\frac{1}{25}$	$\frac{4}{25}$	$\frac{9}{25}$	$\frac{16}{25}$	$\frac{25}{25}$	

Since $\sum_{x=0}^{4} P(X = x) = 1$, the given function is a *pmf*. A plot of this *pmf* is shown in Fig. 8.16.

Also

$$P(X \le 2) = P(X = 0) + P(X = 1) + P(X = 2)$$
$$= \frac{9}{25}$$
$$P(X \ge 1) = P(X = 1) + P(X = 2) + P(X = 3) + P(X = 4)$$
$$= \frac{24}{25}$$
$$P(1 \le X < 4) = P(X = 1) + P(X = 2) + P(X = 3)$$
$$= \frac{15}{25}$$

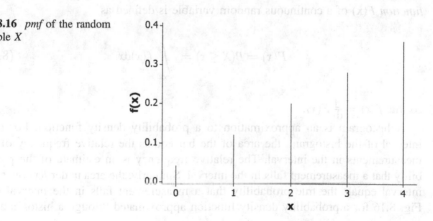

Fig. 8.16 *pmf* of the random variable X

8.11.2 *Continuous Random Variables*

A *continuous random variable* is a random variable with an interval (either finite or infinite) of real numbers for its range, for example, temperature, pressure, radius, voltage, speed, weight, and height. The range of a continuous random variable includes all values in an interval of real numbers.

Definition 8.5 Let X be a continuous random variable; then, the *probability density function (pdf)* is a function such that

$$(a)\ f(x) \geq 0$$

$$(b)\ \int_{-\infty}^{\infty} f(x)\mathrm{d}x = 1 \tag{8.27}$$

$$(c)\ P(x_1 \leq X \leq x_2) = \int_{x_1}^{x_2} f(x)\mathrm{d}x, \quad \text{for all } x_1 < x_2$$

A probability density function provides a simple description of the probabilities associated with a random variable. As long as $f(x)$ is nonnegative and $\int_{-\infty}^{\infty} f(x)\mathrm{d}x = 1$, $0 \leq P(x_1 < X < x_2) \leq 1$, so that the probabilities are properly restricted. Since, $P(X = x) = 0$ for a continuous random variable, and for any x_1 and x_2,

$$P(x_1 \leq X \leq x_2) = P(x_1 < X \leq x_2) = P(x_1 \leq X < x_2) = P(x_1 < X < x_2), \tag{8.28}$$

holds.

Definition 8.6 The cumulative distribution function or simply the *distribution function $F(x)$* of a continuous random variable is defined as

$$F(x) = P(X \leq x) = \int_{-\infty}^{x} f(x)\mathrm{d}x, \tag{8.29}$$

so that $f(x) = \frac{\mathrm{d}}{\mathrm{d}x}F(x)$.

A histogram is an approximation to a probability density function. For each interval of the histogram, the area of the bar equals the relative frequency of the measurements in the interval. The relative frequency is an estimate of the probability that a measurement falls in the interval. Similarly, the area under $f(x)$ over any interval equals the true probability that a measurement falls in the interval (see Fig. 8.16 for a probability density function approximated through a histogram).

The two important quantities that characterize any probability distribution is the expectation and variance of the random variable. The *expected value* of a random variable X, which is denoted by $E(X)$, may be interpreted as the long-term average value of X. The expectation can be defined as

$$E(X) = \begin{cases} \sum_j x_j P(X = x_j), & \text{if } X \text{ is discrete} \\ \int_{-\infty}^{\infty} x f(x) dx, & \text{if } X \text{ is continuous} \end{cases} \tag{8.30}$$

If X and Y are two random variables, then the following expectation rules hold:

(i) $E(X + Y) = E(X) + E(Y)$
(ii) $E(X - Y) = E(X) - E(Y)$
(iii) $E(aX + b) = aE(X) + b$
(iv) $E(aX + bY) = aE(X) + bE(Y)$
(v) $E\left(\sum_{j=1}^{n} a_j X_j\right) = \sum_{j=1}^{n} a_j E(X_j)$

where a, b, and a_j are all constants. Further, if X and Y are independent, then

$$E(XY) = E(X) * E(Y).$$

Similarly, the *variance* of a random variable is defined as the expectation of the squared difference between the random variable and its expectation. That is

$$Var(X) = E\left[\{X - E(X)\}^2\right] = E(X^2) - \{E(X)\}^2 \tag{8.31}$$

According to the variable X is discrete or continuous, the variance can be defined as

$$Var(X) = \begin{cases} \sum_j x_j^2 P(X = x_j) - \{E(X)\}^2, & \text{if } X \text{ is discrete} \\ \int_{-\infty}^{\infty} x^2 f(x) dx - \{E(X)\}^2, & \text{if } X \text{ is continuous} \end{cases} \tag{8.32}$$

The variance is often denoted by σ^2. For the experiment discussed for the number of tails in three tosses of a coin (see Table 8.5), the mean and variance of is obtained as

$$E(X) = \sum_j x_j P(X = x_j) = \frac{12}{8} = 1.5$$

Table 8.5 Distribution of "number of tails"

X	0	1	2	3	Total
$P(X = x)$	$\frac{1}{8}$	$\frac{3}{8}$	$\frac{3}{8}$	$\frac{1}{8}$	1

and

$$\text{Var}(X) = \sum_j x_j^2 P(X = x_j) - \{E(X)\}^2$$

$$= \frac{24}{8} - \left(\frac{12}{8}\right)^2 = 0.75$$

respectively.

Example 8.10 Suppose that X is a continuous random variable whose probability density function is given by

$$f(x) = \begin{cases} C(4x - 2x^2), & 0 < x < 2 \\ 0, & \text{Otherwise} \end{cases}$$

Find

(i)   the value of C
(ii) $P(X > 1)$
(iii) $E(X)$ and $Var(X)$

Solution

(i)   Since $f(x)$ is a density function, we must have $\int_{-\infty}^{\infty} f(x)dx = 1$. Therefore,

$$C \int_0^2 (4x - 2x^2)dx = 1$$

$$\Rightarrow C \left[2x^2 - \frac{2}{3}x^3\right]_{x=0}^{x=2} = 1$$

$$\Rightarrow C = 3/8$$

The graph of $f(x)$ is shown in Fig. 8.17.

Fig. 8.17 Histogram approximated to $f(x)$

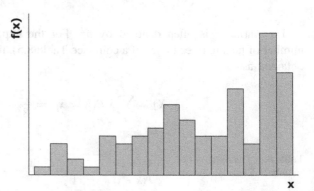

(ii)   Now

$$f(x) = \begin{cases} \frac{3}{8}(4x - 2x^2), & 0 < x < 2 \\ 0, & \text{Otherwise} \end{cases}$$

Hence

$$P(X > 1) = \int_1^\infty f(x)dx = \frac{3}{8} \int_1^\infty (4x - 2x^2)dx = \frac{1}{2}$$

(iii) $E(X) = \int_0^2 xf(x)dx = \frac{3}{8}\int_0^2 x(4x - 2x^2)dx = 1$

and

$$E(X^2) = \int_0^2 x^2 f(x)dx = \frac{3}{8}\int_0^2 x^2(4x - 2x^2)dx = \frac{6}{5}$$

Therefore,

$$\text{Var}(X) = E(X^2) - \{E(X)\}^2 = \frac{1}{2}$$

Example 8.11 If $f(x) = kx^2, 0 \le x \le 1$ and 0 otherwise. Find k. Find c_1 and c_2 such that $P(x \le c_1) = 0.1$ and $P(x \le c_2) = 0.9$.

Solution Since $\int_{-\infty}^\infty f(x)dx = 1$, the value of k is such that

$$\int_0^1 kx^2 dx = 1 \Rightarrow k = 3/8.$$

Now, the value of c_1 and c_2 is such that

$$P(x \le c_1) = 0.1 \Rightarrow \int_0^{c_1} \frac{3}{8}x^2 dx = 0.1$$

$$\Rightarrow c_1 = 0.9283$$

and

$$P(x \leq c_2) = 0.9 \Rightarrow \int\limits_{0}^{c_2} \frac{3}{8} x^2 \mathrm{d}x = 0.9.$$

$$\Rightarrow c_2 = 1.9310$$

8.11.3 Jointly Distributed Random Variables

For a given experiment, we are often interested not only in the probability distribution function of individual random variables but also in the relationship between two or more random variables. For instance, an engineer may be interested in the relationship between the shear strength and the diameter of a spot weld in a fabricated sheet steel specimen. To specify such relationship between two random variables, we define the joint cumulative probability distribution function of X and Y by

$$F_{XY}(x, y) = P(X \leq x, Y \leq y) \tag{8.33}$$

A knowledge of the joint distribution function enables one to compute the probability of any statement concerning the values of X and Y. For instance, the marginal (or cumulative) distribution of X and Y is given by

$$F_X(x) = P(X \leq x)$$
$$= P(X \leq x, Y \leq \infty)$$
$$= F(x, \infty)$$

Similarly,

$$F_Y(y) = F(\infty, y)$$

If X and Y are jointly continuous, then the joint probability function of X and Y can be obtained as

$$f_{XY}(x, y) = \frac{\mathrm{d}^2}{\mathrm{d}x\mathrm{d}y} F(x, y) \tag{8.34}$$

Thus

Definition 8.7 A *joint probability density* for the continuous random variables X and Y, denoted as $f_{XY}(x, y)$, satisfies the following properties:

(a) $f_{XY}(x, y) \geq 0$ for all x, y

(b) $\displaystyle \int_{-\infty}^{\infty} \int_{-\infty}^{\infty} f_{XY}(x, y) dx dy = 1$ (8.35)

(c) $P[(X, y) \in C] = \displaystyle \iint_{(x,y) \in C} f_{XY}(x, y) dx dy$

In the definition, C is a region of two-dimensional space. Thus, $f_{XY}(x, y)$ is defined over all of two-dimensional space by assuming that $f_{XY}(x, y) = 0$ for all points for which $f_{XY}(x, y)$ is not specified. The double integral of $f_{XY}(x, y)$ over a region C provides the probability that (X, Y) assumes a value in C. This integral can be interpreted as the volume under the surface $f_{XY}(x, y)$ over the region C.

The above definition of joint probability density function helps us to define the *marginal probabilities* of X and Y as

$$f_X(x) = \int_{-\infty}^{\infty} f_{XY}(x, y) dy \tag{8.36}$$

and

$$f_Y(y) = \int_{-\infty}^{\infty} f_{XY}(x, y) dx \tag{8.37}$$

respectively. The conditional probabilities of the random variables are defined as follows:

Definition 8.8 The conditional probability X given Y is defined as

$$f_{X|Y}(x|y) = \frac{f_{XY}(x, y)}{f_Y(y)}, \quad f_Y(y) \geq 0 \tag{8.38}$$

and the conditional probability Y given X is defined as

$$f_{Y|X}(y|x) = \frac{f_{XY}(x, y)}{f_X(x)}, \quad f_X(x) \geq 0 \tag{8.39}$$

In case X and Y are both discrete random variables whose possible values are $x_1, x_2, \ldots,$ and $y_1, y_2, \ldots,$ respectively, we define the *joint probability mass function* of X and Y as

$$f_{XY}(x, y) = P\{X = x, Y = y\}, \tag{8.40}$$

Such that $\sum_x \sum_y f_{XY}(x,y) = 1$. The corresponding marginal pmf of X and Y is obtained as

$$f_X(x) = \sum_y f_{XY}(x,y) \tag{8.41}$$

and

$$f_Y(y) = \sum_x f_{XY}(x,y) \tag{8.42}$$

Some examples concerning joint probability distributions are discussed below:

Example 8.12 The joint density function of X and Y is given by

$$f(x,y) = \begin{cases} 2e^{-(x+2y)}, & 0<x<\infty, 0<y<\infty \\ 0, & \text{Otherwise} \end{cases}$$

Compute

(i) $P(X > 1, Y < 1)$
(ii) $P(X < Y)$
(iii) $P(X < k)$
(iv) Mean and variance of Y
(v)   Are X and Y independent?

Fig. 8.18 Joint density plot of (X, Y)

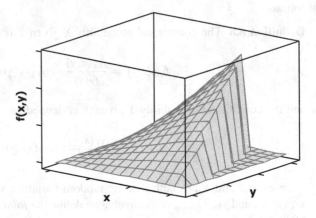

Solution The plot of the joint density is shown in Fig. 8.18. Further

(i) $P(X > 1, Y < 1) = \int\limits_{0}^{1} \int\limits_{1}^{\infty} f(x,y)dxdy$

$$= \int\limits_{0}^{1} \int\limits_{1}^{\infty} 2e^{-(x+2y)}dxdy$$

$$= e^{-1}(1 - e^{-2})$$

(ii) $P(X < Y) = \iint\limits_{(x,y):x<y} f(x,y)dxdy$

$$= \int\limits_{0}^{\infty} \int\limits_{0}^{y} 2e^{-(x+2y)}dxdy$$

$$= \frac{1}{3}$$

(iii) $P(X < k) = \int\limits_{0}^{k} \int\limits_{0}^{\infty} f(x,y)dxdy$

$$= \int\limits_{0}^{k} \int\limits_{0}^{\infty} 2e^{-(x+2y)}dxdy$$

$$= 1 - e^{-k}$$

(iv) Mean of $Y = E(Y) = \int\limits_{0}^{\infty} yf_Y(y)dy,$

where

$$f_Y(y) = \int\limits_{0}^{\infty} f(x,y)dx$$

$$= \int\limits_{0}^{\infty} 2e^{-(x+2y)}dx$$

$$= 2e^{-2y}, \quad 0 < y < \infty$$

Therefore,

$$E(Y) = \int\limits_{0}^{\infty} y f_Y(y) \mathrm{d}y$$

$$= \int\limits_{0}^{\infty} 2y e^{-2y} \mathrm{d}y$$

$$= \frac{1}{2}$$

and

$$E(Y^2) = \int\limits_{0}^{\infty} y^2 f_Y(y) \mathrm{d}y$$

$$= \int\limits_{0}^{\infty} 2y^2 e^{-2y} \mathrm{d}y$$

$$= \frac{1}{2}$$

Hence

$$\text{Variance of } Y = \text{Var}(Y)$$
$$= E(Y^2) - [E(Y)]^2$$
$$= \frac{1}{4}$$

(v)   Now

$$f_X(x) = \int\limits_{0}^{\infty} f(x, y) \mathrm{d}y$$

$$= \int\limits_{0}^{\infty} 2e^{-(x+2y)} \mathrm{d}y$$

$$= e^{-x}, \quad 0 < x < \infty$$

Note that

$$f_x(x)f_y(y) = 2e^{-(x+2y)}$$
$$= f_{XY}(xy)$$

Hence, the variables X and Y are independent.

Example 8.13 A product is classified according to the number of defects it contains and the factory that produces it. Let X_1 and X_2 be the random variables that represent the number of defects per unit and the factory number, respectively. The entries in Table 8.6 represent the joint probability mass function of a randomly chosen product:

(a) Find the marginal probability distribution of X_1 and X_2.
(b) Find $E(X_1), E(X_2), \text{Var}(X_1), \text{Var}(X_2), \text{Cov}(X_1, X_2)$ and $r(X_1, X_2)$.
(c) Are the variables independent?

Solution Let us denote

X_1: the random variables that represent the number of defects per unit

X_2: the random variables that represent the factory number.

Then

(a) For marginal distribution of X_1, we find probabilities for each outcome of X_1 for varying values of X_2. Thus

$$\text{For } X_1 = 0: f_{x_1}(0) = P(X_1 = 0, X_2 = 1) + P(X_1 = 0, X_2 = 2)$$
$$= \frac{1}{8} + \frac{1}{16} = \frac{3}{16}$$

$$\text{For } X_1 = 1: f_{x_1}(1) = P(X_1 = 1, X_2 = 1) + P(X_1 = 1, X_2 = 2)$$
$$= \frac{1}{16} + \frac{1}{16} = \frac{2}{16}$$

Table 8.6 Joint probability mass function

X_1	X_2	
	1	2
0	1/8	1/16
1	1/16	1/16
2	3/16	1/8
3	1/8	1/4

For $X_1 = 2$: $f_{X_1}(2) = P(X_1 = 2, X_2 = 1) + P(X_1 = 2, X_2 = 2)$

$$= \frac{3}{16} + \frac{1}{8} = \frac{5}{16}$$

For $X_1 = 3$: $f_{X_1}(3) = P(X_1 = 3, X_2 = 1) + P(X_1 = 3, X_2 = 2)$

$$= \frac{1}{8} + \frac{1}{4} = \frac{6}{16}$$

Similarly, for marginal distribution of X_2, we find probabilities for each outcome of X_2 for varying values of X_1. Thus

For $X_2 = 1$: $f_{X_2}(1) = P(X_1 = 0, X_2 = 1)$
$$+ P(X_1 = 1, X_2 = 1) + P(X_1 = 2, X_2 = 1) + P(X_1 = 3, X_2 = 1)$$

$$= \frac{1}{8} + \frac{1}{16} + \frac{3}{16} + \frac{1}{8} = \frac{8}{16}$$

For $X_2 = 2$: $f_{X_2}(2) = P(X_1 = 0, X_2 = 2) + P(X_1 = 1, X_2 = 2)$
$$+ P(X_1 = 2, X_2 = 2) + P(X_1 = 3, X_2 = 2)$$

$$= \frac{1}{8} + \frac{1}{16} + \frac{3}{16} + \frac{1}{8} = \frac{8}{16}$$

Table 8.7 shows the joint and marginal distributions of (X_1, X_2).

(b) $E(X_1) = \sum_{x_0=0}^{3} x_1 f_{X_1}(x_1)$

$$= 0 \times \frac{3}{16} + 1 \times \frac{2}{16} + 2 \times \frac{5}{16} + 3 \times \frac{6}{16}$$

$$= \frac{30}{16}$$

$E(X_1^2) = \sum_{x_1=0}^{3} x_1^2 f_{X_1}(x_1)$

$$= 0 \times \frac{3}{16} + 1^2 \times \frac{2}{16} + 2^2 \times \frac{5}{16} + 3^2 \times \frac{6}{16}$$

$$= \frac{76}{16}$$

Table 8.7 Joint and marginal distributions

X_1	X_2		
	1	2	$f_{X_1}(x_1)$
0	1/8	1/16	3/16
1	1/16	1/16	2/16
2	3/16	1/8	5/16
3	1/8	1/4	6/16
$f_{X_2}(x_2)$	8/16	8/16	1

Therefore,

$$\text{Var}(X_1) = E(X_1^2) - [E(X_1)]^2 = 1.2343$$

Now

$$E(X_2) = \sum_{x_2=1}^{2} x_2 f_{X_2}(x_2)$$

$$= 1 \times \frac{8}{16} + 2 \times \frac{8}{16}$$

$$= \frac{3}{2}$$

$$E(X_2^2) = \sum_{x_2=1}^{2} x_2^2 f_{X_2}(x_2)$$

$$= 1^2 \times \frac{8}{16} + 2^2 \times \frac{8}{16}$$

$$= \frac{5}{2}$$

Therefore,

$$\text{Var}(X_2) = E(X_2^2) - [E(X_2)]^2 = 0.25$$

Also

$$E(X_1 X_2) = \sum_{x_1=0}^{3} \sum_{x_2=1}^{2} x_1 x_2 f_{X_1 X_2}(x_1 x_2)$$

$$= 0 \times 1 \times (1/8) + 0 \times 2 \times (1/16) + \cdots + 3 \times 2 \times (1/4)$$

$$= \frac{47}{16} = 2.93$$

Therefore,

$$\text{Cov}(X_1, X_2) = E(X_1 X_2) - E(X_1)E(X_2)$$
$$= 2.93 - 1.875 \times 1.5 = 0.125$$

Hence

$$r(X_1, X_2) = \frac{\text{Cov}(X_1, X_2)}{\sqrt{\text{Var}(X_1)\text{Var}(X_2)}}$$

$$= \frac{0.125}{\sqrt{1.2343 \times 0.25}}$$

$$= 0.225$$

(c) Since $\text{Cov}(X_1, X_2) \neq 0$, the variables are not independent.

Note Independence of two random variables can also be proved, if any one of the following relationship holds:

(i) $E(X_1 X_2) = E(X_1)E(X_2)$
(ii) $f_{XY}(xy) = f_X(x)f_y(y)$
(iii) $\text{Cov}(X_1, X_2) = 0$

Also note that $\text{Cov}(X_1, X_2) = 0$ may not imply the independence of the variables.

8.12 Probability Models

In the discussions carried out so far, it is apparent that a random phenomenon can be very well explained through few characteristics of a probability distribution. A probability distribution may be characterized by either a discrete or a continuous random variable. The distribution of a discrete variable is called a discrete probability distribution, and the distribution that corresponds to a continuous variable is continuous probability distribution. Some of the most frequently used discrete probability distributions are Bernoulli distribution, Poisson distribution, geometric distribution, hypergeometric distribution, rectangular distribution, etc. And some of the continuous uncertainty models are exponential distribution, normal distribution, gamma distribution, Weibull distribution, triangular distribution, uniform distribution, etc. We study some of those important distributions frequently used in Six Sigma projects.

8.12.1 Binomial Distribution

A binomial distribution is defined on the basis of a Bernoulli *process*. Any uncertain situation or experiment that is marked by the following three properties is known as a Bernoulli process.

1. There are only two mutually exclusive and collectively exhaustive outcomes in the experiment.

2. In repeated observations of the experiment, the probabilities of occurrence of these events remain constant.
3. The observations are independent of one another.

The random variables that may be of interest in these situations are the following: the number of success or failure in a specified number of trials, the number of good or bad lots in a consignment, the number of units accepted or rejected, and the number of people who agree or disagree with quality of the product. Given the knowledge on the probability of a success (say p) in a trial and if the experiment is observed n times, then the random phenomena of the distribution of number of successes, say r, follow a binomial distribution, denoted as $BD(n, p)$, having the density function:

$$P(X = x) = \binom{n}{x} p^x q^{n-x}, \quad x = 0, 1, 2, \ldots, n; \ 0 \le p \le 1, p + q = 1 \qquad (8.43)$$

where $\binom{n}{x} = \frac{n!}{x!(n-x)!}$.

The binomial distribution has mean np and variance npq. Figure 8.19 shows binomial distributions for various values of n and p. The binomial distribution function can be computed using the following relationships

$$P(X \le i) = \sum_{x=0}^{i} \binom{n}{x} p^x q^{n-x}, \quad i = 0, 1, 2, \ldots, n \qquad (8.44)$$

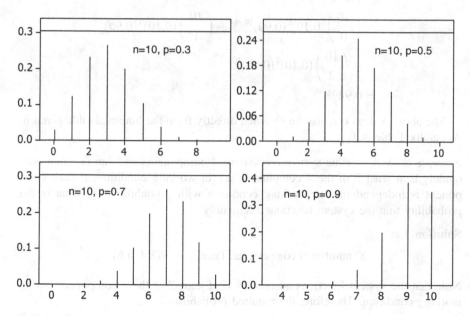

Fig. 8.19 Binomial distributions

and the binomial probabilities can be easily computed using the recurrence relationship

$$P(X = x+1) = \frac{p(n-x)}{(1-p)(x+1)} P(X = x), \tag{8.45}$$

For $x = 0, 1, \ldots, n$, one can obtain the successive probabilities. The relationship shown in (8.45) is also called the recurrence relationship between probabilities. The binomial probability sums shown in (8.44) can also be found from the Appendix Table A.1 for various combinations of n and p.

Example 8.14 A machine produces 10 % defective items. Ten items are selected at random. Find the probability that

(i) 5 are defective and
(ii) Not more than 2 items are defective.

Solution Let

$$X\text{: the number of defective. Then, } X \sim BD(10, 0.10)$$

(i) $p(X = 5) = \binom{10}{5}(0.10)^5(0.9)^5$

$$= 0.0015$$

(ii) $P(X \le 2) = \sum_{r=0}^{2} \binom{10}{r}(0.10)^r(0.90)^{10-r}$

$$= \binom{10}{0}(0.10)^0(0.90)^{10-0} + \binom{10}{1}(0.10)^1(0.90)^{10-1}$$

$$+ \binom{10}{2}(0.10)^2(0.90)^{10-2}$$

$$= 0.9298$$

The above values can also be obtained directly from the binomial table given in Appendix Table A.1.

Example 8.15 A satellite system consists of 4 components and can function adequately if at least 2 of the 4 components are in working condition. If each component is independently in working condition with probability 0.6, what is the probability that the system functions adequately?

Solution Let

$$X\text{: number of components. Then, } X \sim BD(4, 0.6)$$

Note that the system functions adequately if at least 2 of the 4 components are in working condition. Therefore, the required probability is

$$P(X \geq 2) = \sum_{x=2}^{4} \binom{4}{x} (0.6)^x (0.4)^{4-x}$$

$$= \binom{4}{2}(0.6)^2(0.4)^2 + \binom{4}{3}(0.6)^3(0.4)^1 + \binom{4}{4}(0.6)^4(0.4)^0$$

$$= 0.8208$$

Alternatively,

$$P(X \geq 2) = 1 - P(X \leq 1)$$

$$= 1 - \sum_{x=0}^{1} \binom{4}{x}(0.6)^x(0.4)^{4-x}$$

$$= 1 - 0.1792 = 0.8208 \text{ (From Appendix Table A.1)}$$

8.12.2 Poisson Distribution

The Poisson distribution is used to describe a number of processes, including the distribution of telephone calls going through a switchboard, the demand of patients for service at a health institution, the number of industrial accidents, and the arrival of trucks and cars at a tollbooth and the number of accidents at an intersection. In all the examples, the outcome is described by a discrete random variable that takes on integer values 0, 1, 2, ... and so on. The conditions leading to a Poisson probability distribution are as follows:

- The experiment consists of counting the number of times a particular event occurs during a given unit of time or in a given area or volume or weight or distance.
- The probability that an event occurs in a given unit of time is same for all the units.
- The number of events that occur in one unit of time is independent of the number that occurs in other units.
- The mean number of events in each unit will be denoted by λ, say

If X is the random variable that follows the Poisson law, then the probability of exactly x occurrences in a Poisson distribution is calculated with the mass function

$$P(X = x) = \frac{e^{-\lambda}\lambda^x}{x!}, \quad x = 0, 1, 2 \ldots \tag{8.46}$$

where λ is a positive constant. The Poisson distribution has mean λ and variance λ. If we note that $P(X = 0) = e^{-\lambda}$, then successive probabilities can be calculated by using the recurrence relation

$$P(X = r) = \frac{\lambda}{r} f(x = r - 1) \tag{8.47}$$

For $r = 1, 2, \ldots$, one can obtain the successive probabilities of Poisson distribution. If λ is not an integer, the mode is the value of the integer r for which $r - 1 < \lambda < r$. If λ is an integer, then $P(X = \lambda - 1) = P(X = \lambda)$ and both λ and $\lambda - 1$ are modes. See Appendix Table A.2 for cumulative Poisson probabilities for various values of λ. Figure 8.20 shows Poisson distributions for various mean values.

If $\lambda < 1$, the graph of the probability function decreases steadily, whereas if $\lambda > 1$, the graph increases steadily to the value at the mode, then decreases steadily tending to 0 as $r \to \infty$. Since binomial distribution and Poisson distribution share a lot of similarities, it is possible to establish a relationship between the two. When n is very large and p is very small, binomial distribution tends to Poisson distribution. This is seen as follows:

Suppose that X is a binomial random variable with parameters (n, p) and let $\lambda = np$. Then

$$P(X = r) = \frac{n!}{r!(n-r)!} p^r (1-p)^{n-r}$$

$$= \frac{n!}{r!(n-r)!} \left(\frac{\lambda}{n}\right)^r \left(1 - \frac{\lambda}{n}\right)^{n-r}$$

$$= \frac{n(n-1)\ldots(n-r+1)}{n^r} \frac{\lambda^r}{r!} \left(1 - \frac{\lambda}{n}\right)^{n-r}$$

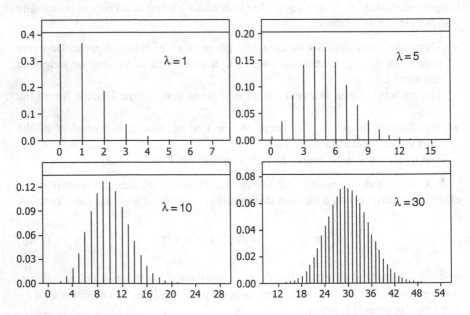

Fig. 8.20 Poisson distributions for various means

For large n and small p,

$$\left(1 - \frac{\lambda}{n}\right)^n \approx e^{-\lambda}, \frac{n(n-1)\ldots(n-r+1)}{n^r} \approx 1 \text{ and } \left(1 - \frac{\lambda}{n}\right)^r \approx 1$$

Hence, for large n and small p,

$$P(X = r) \approx e^{-\lambda}\frac{\lambda^r}{r!},$$

which is Poisson with parameter λ.

Some examples of random variables that usually obey to a good approximation with Poisson law are as follows:

- Number of errors per 100 invoices
- Number of surface defects in a casting
- Number of faults of insulation in a specified length of cable
- Number of visual defects in a bolt of cloth
- Number of spare parts required over a specified period of time
- The number of absenteeism in a specified period of time
- The number of death claims in a hospital per day
- The number of breakdowns of a computer per month
- The PPM of toxicant found in water or air emission from a manufacturing plant.

Further, for large n, a Poisson distribution can be approximated to a normal distribution. For instance, for large n, if X is according to Poisson random variable and $\lambda \to \infty$, then $\frac{X - \lambda}{\sqrt{\lambda}}$ follows a standard normal distribution.

Example 8.16 If the prices of new cars increase on an average of four times every 3 years, then in a randomly selected period of 3 years, find the probability of

(a) No price hikes
(b) Two price hikes
(c) Five or more hikes

Solution Here, $\lambda = 4$

(a) P(no price hikes) $= P(X = 0)$
$$= e^{-\lambda} = 0.018$$

(b) P(Two price hikes) $= P(X = 2)$
$$= \frac{e^{-\lambda}\lambda^2}{2!} = 0.1465$$

(c) P(Five or more hikes) $= P(X \geq 5) = 1 - P(X \leq 4)$

$$= 1 - \sum_{r=0}^{4} \frac{e^{-\lambda}\lambda^r}{r!}$$

$$= 1 - e^{-4}\left(1 + 4 + 8 + \frac{32}{3} + \frac{32}{3}\right)$$

$$= 0.3712$$

Alternatively,

$$P(X \geq 5) = 1 - P(X \leq 4)$$
$$= 1 - 0.629 = 0.371 \text{ (From Appendix Table A.2)}$$

Example 8.17 Suppose that the probability of an item produced by a certain machine will be defective is 0.05. If the produced items are sent to the market in packets of 20, find the number of packets containing at least, exactly, and at the most 2 defective items in a consignment of 1000 packets using

(i) Binomial distribution and
(ii) Poisson approximation to binomial distribution

Solution Let us define the random variable as

X: the item is defective. Then given $p = 0.05$ and $n = 20$.

Now, we find the probabilities associated with each outcome assuming binomial and Poisson distribution:

(i) according to *binomial distribution*,

$$P(\text{at least 2 defective items}) = P(X \geq 2)$$

$$= \sum_{x=2}^{20} \binom{20}{x}(0.05)^x(0.95)^{20-x}$$

$$= 1 - [P(X = 0) + P(X = 1)]$$

$$= 0.2642$$

$$P(\text{exactly 2 defective items}) = P(X = 2)$$

$$= \binom{20}{2}(0.05)^2(0.95)^{20-2}$$

$$= 0.1887$$

$$P(\text{at most 2 defective items}) = P(X \leq 2)$$

$$= \sum_{x=0}^{2} \binom{20}{x}(0.05)^x(0.95)^{20-x}$$

$$= 0.9245$$

Therefore, in a consignment of 1000 packets

the packets having at least 2 defective items $= 1000 \times P(X = 2) = 264$

the packets having exactly 2 defective items $= 1000 \times P(X = 2) = 189$

and

the packets having at the most 2 defective items $= 1000 \times P(X \geq 2) = 925$

Now

(ii) according to *Poisson distribution*, $\lambda = np = 1$,

$$P(\text{at least 2 defective items}) = P(X \geq 2)$$

$$= \sum_{x=2}^{\infty} \frac{e^{-1} 1^x}{x!}$$

$$= 1 - [P(X = 0) + P(X = 1)]$$

$$= 0.2642$$

$$P(\text{exactly 2 defective items}) = P(X = 2)$$

$$= \frac{e^{-1}}{2!}$$

$$= 0.1839$$

$$P(\text{at the most 2 defective items}) = P(x \leq 2)$$

$$= \sum_{x=0}^{2} \frac{e^{-1} 1^x}{x!}$$

$$= 0.9197$$

Therefore, in a consignment of 1000 packets

the packets having at least 2 defective items $= 1000 \times P(X \geq 2) = 264$

the packets having exactly 2 defective items $= 1000 \times P(X = 2) = 184$

and

the packets having at the most 2 defective items $= 1000 \times P(X \leq 2) = 919$

Note that both the probability models gave identical results.

Example 8.18 If X and Y are independent Poisson variates such that $P(X = 1) = P(X = 2)$ and $P(Y = 2) = P(Y = 3)$. Find

(i)   the variance of $X - 2Y$
(ii) $P(X = 0)$
(iii) $P(0 < Y < 5)$
(iv) $P(0 < X + Y < 5)$

Solution Suppose X and Y are independent Poisson variates with mean λ and μ, respectively. Now

$$P(X = 1) = P(X = 2) \Rightarrow \frac{e^{-\lambda}\lambda}{1!} = \frac{e^{-\lambda}\lambda^2}{2!}$$

$$\Rightarrow \lambda = 2$$

and

$$P(Y = 2) = P(Y = 3) \Rightarrow \frac{e^{-\lambda}\mu^2}{2!} = \frac{e^{-\lambda}\mu^3}{3!}$$

$$\Rightarrow \mu = 3$$

Further

(i) $\text{Var}(X - 2Y) = \text{Var}(X) + 4\text{Var}(Y) = 14$
(ii) $P(X = 0) = \frac{e^{-2}}{0!} = 0.1353$
(iii) $P(0 < Y < 5) = \sum_{y=1}^{4} \frac{e^{-3}3^y}{y!} = 0.7655$

Now, X and Y are independent and $X \sim PD(2)$ and $Y \sim PD(3)$. Hence,

$$Z = X + Y \sim PD(5)$$

(iv) $P(0 < X + Y < 5) = P(0 < Z < 5)$

$$= \sum_{z=1}^{4} \frac{e^{-5}5^z}{z!} = 0.4338$$

8.12.3 Hypergeometric Distribution

Consider the following situations: If a box contains N screws, and M of them are defective, the probability of drawing a defective screw in a trial is $p = \frac{M}{N}$, and the probability of drawing a non-defective screw is $q = \frac{N-M}{N} = 1 - p$. Then, the probability of drawing x defectives in n trials is given by

$$P(X = x) = \binom{n}{x}\left(\frac{M}{N}\right)^x\left(1 - \frac{M}{N}\right)^{n-x}, \quad x = 0, 1, 2, \ldots, n \qquad (8.48)$$

which is nothing but the binomial probability law, as the sampling is done with replacement and the trials are independent.

Suppose the trials are not independent; that is, after each draw, the screws are not replaced in the mix; then, the probability law has the mass function:

$$P(X = x) = \frac{\binom{M}{x}\binom{N-M}{n-x}}{\binom{N}{n}}, \quad \max(0, n - N + M) \leq x \leq \min(n, M) \quad (8.49)$$

The probability law given in (8.49) is called the *hypergeometric distribution* and is denoted by $HGD(N, M, n)$. The graph of hypergeometric distribution for various values of (N, M, n) is given in Fig. 8.21.

The distribution has mean

$$\mu = n\frac{M}{N} \qquad (8.50)$$

and the variance

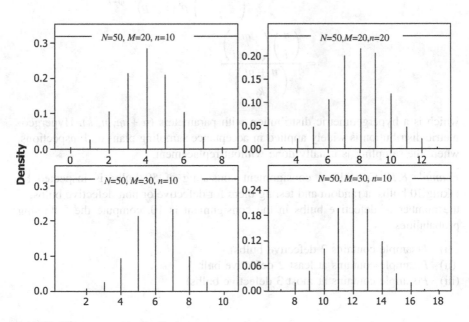

Fig. 8.21 Hypergeometric distribution

$$\sigma^2 = \frac{nM(N-M)(N-n)}{N^2(N-1)} \tag{8.51}$$

Note that mean μ of hypergeometric distribution is similar to the mean of a binomial random variable. Also, σ^2 differs from the result for a binomial variable only by the term $\frac{N-n}{N-1}$, which is often called the *finite population correction factor*. If n is small relative to N, the correction is small and the hypergeometric distribution is similar to the binomial distribution.

Result 8.1 If X and Y are independent binomial random variables having respective parameters (n, p) and (m, p), then the conditional pmf of X given that $X + Y = k$ is hypergeometric distribution.

The result can be proved as follows:

$$P(X = i | x + Y = k) = \frac{P(X = i, X + Y = k)}{P(X + Y = k)}$$

$$= \frac{P(X = i, Y = k - i)}{P(X + Y = k)}$$

$$= \frac{P(X = i)P(Y = k - i)}{P(X + Y = k)}$$

$$= \frac{\binom{n}{i} p^i (1-p)^{n-i} \binom{m}{k-i} p^{k-i}(1-p)^{m-(k-i)}}{\binom{n+m}{k} p^k (1-p)^{n+m-k}}$$

$$= \frac{\binom{n}{i}\binom{m}{k-i}}{\binom{n+m}{k}},$$

which is a hypergeometric distribution with parameters $(n + m, n, k)$. Hypergeometric distribution is widely applied in acceptance sampling plans and inspections, where the sampling is usually done without replacement.

Example 8.19 Consider a consignment consisting of 50 bulbs is inspected by taking 10 bulbs at random and testing them for defective or non-defective bulbs. If the number of defective bulbs in the consignment is 10, compute the following probabilities

  (i) P(sample contains 2 defective bulbs)
 (ii) P(sample contains at least 2 defective bulbs)
(iii) P(sample contains at most 3 defective bulbs)

Further, if the sample of 10 bulbs contains at the most 1 defective bulb, the consignment containing 50 bulbs is accepted; otherwise, it is rejected. Compute the probability for the consignment containing 50 bulbs to be accepted.

Solution As per the notation, we have $N = 50$, $M = 10$, $n = 10$. Then

(i) $P(X = 2) = \binom{10}{2}\binom{40}{8} / \binom{50}{10}$

$= 0.3368$

(ii) $\Pr\{X \geq 2\} = \sum_{x=2}^{10} \binom{10}{x}\binom{40}{n-x} / \binom{50}{10}$

$= 1 - \sum_{x=0}^{1} \binom{10}{x}\binom{40}{n-x} / \binom{50}{10}$

$= 1 - \left[\binom{10}{0}\binom{40}{10-0} / \binom{50}{10} + \binom{10}{1}\binom{40}{10-1} / \binom{50}{10} \right]$

$= 0.6513$

(iii) $\Pr\{X \leq 3\} = \sum_{x=0}^{3} \binom{10}{x}\binom{40}{n-x} / \binom{50}{10}$

$= \binom{10}{0}\binom{40}{10} / \binom{50}{10} + \binom{10}{1}\binom{40}{9} / \binom{50}{10}$

$+ \binom{10}{2}\binom{40}{8} / \binom{50}{10} + \binom{10}{3}\binom{40}{7} / \binom{50}{10}$

$= 0.9034$

Again, the lot containing 50 bulbs is accepted, if $P(X \leq 1)$. Therefore,

$$P(X \leq 1) = \sum_{x=0}^{1} \binom{10}{x}\binom{40}{n-x} / \binom{50}{10}$$

$$= \binom{10}{0}\binom{40}{10} / \binom{50}{10} + \binom{10}{1}\binom{40}{9} / \binom{50}{10} = 0.3487$$

Hence, the lot will be accepted if $P(X \leq 1) = 0.3487$.

Example 8.20 The environment protection agency has purchased 40 precision instruments to be used to measure the air pollution at various locations. Eight of these are randomly selected and tested for defects. If 4 of the 40 instruments are defective, what is the probability that the sample will contain no more than one defective instrument? Use binomial distribution approximation to find the same probability of the same and compare. Also use the Poisson distribution approximation to binomial distribution and compare the results.

Solution Given $N = 40$, $M = 4$, $n = 8$. To find

$$P(X \leq 1) = \sum_{x=0}^{1} \binom{4}{x}\binom{36}{8-x} \bigg/ \binom{50}{8}$$
$$= 0.828$$

For Binomial distribution, we assume, $p = \frac{M}{N} = 0.1$. Then, the required probability is

$$P(X \leq 1) = \sum_{x=0}^{1} \binom{8}{x}(0.1)^x(0.9)^{8-x}$$
$$= 0.813$$

Again for Poisson distribution, we assume $\lambda = np = 0.8$. Then, the required probability is

$$P(X \leq 1) = \sum_{x=0}^{1} \frac{e^{-0.8}(0.8)^x}{x!}$$
$$= 0.809$$

Note If N, M, and $N - M$ are large compared with n, then it does not matter too much whether we sample with or without replacement, and in this case, the hypergeometric distribution may be approximated by the binomial distribution. Hence, sampling with replacement is equivalent to sampling from an infinite set because the proportion of success remains constant for every trial in the experiment.

8.12.4 Normal Distribution

The distribution central to all Six Sigma processes and projects is the normal distribution. It is the most versatile of all the continuous probability distributions and is synonymous with quality control distribution. It is found to be useful in statistical inferences, in characterizing uncertainties in many real-life processes and in approximating other probability distributions. If X is a continuous random variable, then the probability density function of normal distribution is given by

$$f(x) = \frac{1}{\sqrt{2\pi}\sigma} \exp\left[-\frac{1}{2}\left(\frac{x-\mu}{\sigma}\right)^2\right], \quad -\infty \leq x \leq \infty \tag{8.52}$$

where μ = mean, σ = standard deviation, and π = 2.14159. A normal distribution with a specified mean and variance is denoted as $N(\mu, \sigma^2)$. Figure 8.22 shows

Fig. 8.22 Normal
distributions

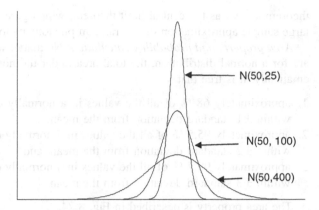

different normal curves for the same mean and different variances. Since all the
three central moments mean, median, and mode coincides, the distribution is
symmetric and is called bell shaped.

Standard normal distribution: If the variable X is expressed in terms of standard
units, say $Z = \frac{X-\mu}{\sigma}$, then the distribution of Z is a standard normal, having the form

$$f(z) = \frac{1}{\sqrt{2\pi}}\exp\left[-\frac{z^2}{2}\right] \tag{8.53}$$

A standard normal distribution has mean zero and variance one; hence, it is
usually denoted by $N(0, 1)$. The graph of a $N(0, 1)$ is presented in Fig. 8.23. The
importance of the standard normal curve lies in the fact that one does not have to
tabulate probability (areas) under the curve between two lines $x = a$ and $x = b$ for
different value of μ and σ. Only one table of area is sufficient for the standard
normal curve, as for this curve mean is zero and variance is one.

In Appendix Table A.3, the standard normal area for various values of z is provided. The normal distribution was introduced by the French mathematician Abra-
ham de Moivre in 1733 and was used by him to approximate probabilities associated
with binomial random variables when the binomial parameter n is large. This result
was later extended by Laplace and others and is now encompassed in a probability

Fig. 8.23 Standard normal
curve

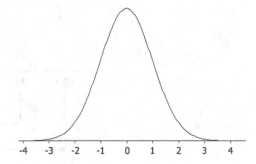

theorem known as the central limit theorem, which gives a theoretical base to the large sample approximation of any random phenomena to a normal distribution.

Area property and probability calculation: No matter what the values of μ and σ are for a normal distribution, the total area under the normal curve is one. Mathematically, it is true that

1. approximately 68 % of all the values in a normally distributed population lie within ± 1 standard deviation from the mean,
2. approximately 95.5 % of all the values in a normally distributed population lie within ± 2 standard deviation from the mean, and
3. approximately 99.7 % of all the values in a normally distributed population lie within ± 3 standard deviation from the mean.

The area property is described in Fig. 8.24.

Since the two tails of the normal probability distribution extend indefinitely and never touch the horizontal axis, it is not possible to have different tables for every possible normal curve. The standard normal distribution helps us to find area under any normal curve. The table given in Appendix A.3 helps us to determine the area and the ordinate of a normally distributed random variable. This is illustrated through some examples below:

Example 8.21 If X is a normal random variable with mean 3 and variance 16, find

(a) $P(X < 11)$
(b) $P(X > -1)$
(c) $P(2 < X < 7)$

Solution Here, $X \sim N(3, 16)$. Then

(a) $P(X < 11) = P\left(\dfrac{X - 3}{4} < \dfrac{11 - 3}{4}\right)$

$\qquad\qquad\quad = P(Z < 2\} = 0.9772 \qquad$ (see Fig. 8.23a)

Fig. 8.24 Area under a normal curve

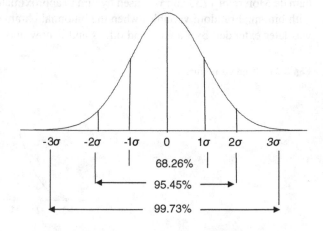

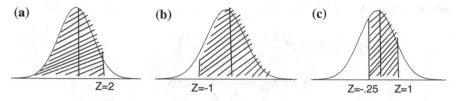

(a) Z=2   **(b)** Z=-1   **(c)** Z=-.25   Z=1

Fig. 8.25 a $P(Z<2\}$, b $P(Z>1\}$, c $P(-0.25<Z<1)$

(b) $P(X>-1) = P\left(\dfrac{X-3}{4} > \dfrac{-1-3}{4}\right)$

$\qquad\qquad = P(Z>-1\} = P(Z<1\} = 0.8413$ (see Fig. 8.23b)

(c) $P(2<X<7) = P\left(\dfrac{2-3}{4} < \dfrac{X-3}{4} < \dfrac{7-3}{4}\right)$

$\qquad\qquad = P(-0.25<Z<1)$

$\qquad\qquad = 0.4400$ (see Fig. 8.23c)

The above three probabilities are shown diagrammatically in Fig. 8.25. The shaded portions are the required probabilities.

Example 8.22 A production engineer finds that on an average a mechanic working in a machine shop completes a certain task in 15 min. The time required to complete the task is approximately normally distributed with a standard deviation of 3 min. Find the probabilities that the task is completed

(a) in less than 8 min
(b) in more than 9 min
(c) between 10 and 12 min.

Solution Let X be the time taken to complete the task. Then, $X \sim N(15, 9)$ and

(a) $P(X<8) = P\left(\dfrac{X-15}{3} < \dfrac{8-15}{3}\right)$

$\qquad\qquad = P(Z<-2.33) = 0.0099$ (from Appendix Table A.3)

(b) $P(X>9) = P\left(\dfrac{X-15}{3} > \dfrac{9-15}{3}\right)$

$\qquad\qquad = P(Z>-2) = 0.9772$ (from Appendix Table A.3)

(c) $P(10<X<12) = P\left(\dfrac{10-15}{3} < \dfrac{X-15}{3} < \dfrac{12-15}{3}\right)$

$\qquad\qquad = P(-1.67<Z<-1)$ (from Appendix Table A.3)

Example 8.23 Let X be normal with mean 14 and variance 16. Determine c such that $P(X \le c) = 95\,\%$, $P(X \le c) = 5\,\%$, and $P(X \le c) = 99.5\,\%$.

Solution Here, $X \sim N(14, 16)$. Then

$$P(X \leq c) = 0.95 \Rightarrow P\left(\frac{X - 14}{4} \leq \frac{c - 14}{4}\right)$$

$$\Rightarrow P\left(Z \leq \frac{c - 14}{4}\right) = 0.95$$

$$\Rightarrow \frac{c - 14}{4} = 1.645 \text{ (from Appendix Table A.3)}$$

$$\Rightarrow c = 20.579$$

$$P(X \leq c) = 0.05 \Rightarrow P\left(\frac{X - 14}{4} \leq \frac{c - 14}{4}\right)$$

$$\Rightarrow P\left(Z \leq \frac{c - 14}{4}\right) = 0.05$$

$$\Rightarrow \frac{c - 14}{4} = -1.645 \text{ (from Appendix Table A.3)}$$

$$\Rightarrow c = 7.42$$

$$P(X \leq c) = 0.995 \Rightarrow P\left(\frac{X - 14}{4} \leq \frac{c - 14}{4}\right)$$

$$\Rightarrow P\left(Z \leq \frac{c - 14}{4}\right) = 0.995$$

$$\Rightarrow \frac{c - 14}{4} = 2.576 \text{ (from Appendix Table A.3)}$$

$$\Rightarrow c = 24.30$$

8.12.5 Distributions Arising from the Normal

A normal probability model is vital for any decision making, and through standard normal distribution tables, one can estimate any probabilities associated with a practical situation. Interestingly, one can also standardize any distribution with its mean and standard deviation (basis of central limit theorem) and shown to follow normal distribution when the sample size is large. Some of the frequently used distributions derived through normal distribution are the chi-square, t-, and F-distributions. These distributions are called sampling distributions.

8.12.5.1 Chi-Square Distribution

If $Z_1, Z_2, \ldots, Z_n$ are independent standard normal random variables, then X defined by

$$X = Z_1^2 + Z_2^2 + \cdots + Z_n^2$$

is said to have a *chi-square distribution* with *n degrees of freedom*. In notation, we write it as $X \sim \chi^2_{(n)}$. The probability density function of chi-square distribution is given by

$$f(x) = \frac{1}{2^{n/2}\Gamma(n/2)} x^{n/2-1} e^{-x/2}, \quad x > 0 \tag{8.54}$$

The graph of the density function of $\chi^2_{(n)}$ for various values of n is given in Fig. 8.26. The chi-square distribution has mean n and variance $2n$. For $n \le 2$, the mode is at zero; otherwise, it is $n-2$. A chi-square distribution is a special case of a gamma distribution. The case $n = 2$ corresponds to the exponential distribution.

Chi-square distribution is one of the most widely used probability distributions in inferential statistics. This distribution is used both in hypothesis testing and in construction of confidence intervals. Specifically, the chi-square distribution is used as the test for goodness of fit of an observed distribution to a theoretical one, the independence of two criteria of classification of qualitative data, and in confidence interval estimation for a population standard deviation of a normal population. Many nonparametric statistical tests also use this distribution, such as Kruskal–Wallis test and Friedman's analysis of variance by ranks.

The percentage points of chi-square distribution (see Appendix Table A.4) are used as critical values in carrying out a χ^2 test based on the values of $\chi^2_{\alpha,n}$ for a specified value of significance value α. Thus, for any $\alpha \in (0, 1)$, the quantity $\chi^2_{\alpha,n}$ is defined to be such that

Fig. 8.26 Chi-square density with n degrees of freedom

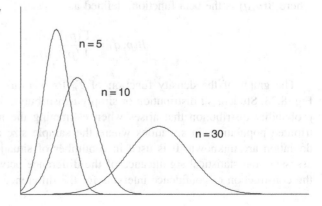

n = 5

n = 10

n = 30

Fig. 8.27 Percentage points
of chi-square distribution

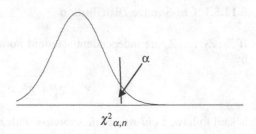

$$P(X \geq \chi^2_{\alpha,n}) = \alpha \tag{8.55}$$

This probability is shown in Fig. 8.27. In Appendix A.4, we provide $\chi^2_{\alpha,n}$ for a variety of values of n and α.

8.12.5.2 The *t*-Distribution

If Z and $\chi^2_{(n)}$ are independent random variables, with Z having standard normal distribution and $\chi^2_{(n)}$ having a chi-square distribution with n degrees of freedom, then the random variable T_n defined by

$$T_n = \frac{Z}{\sqrt{\chi^2_{(n)}/n}} \tag{8.56}$$

is said to have a *t*-distribution with n degrees of freedom. A *t*-distribution has the probability density function

$$f(t) = \frac{1}{\sqrt{n}B\left(\frac{1}{2},\frac{n}{2}\right)}\left(1 + \frac{t^2}{n}\right)^{-(n+1)/2}, \quad -\infty < t < \infty \tag{8.57}$$

where $B(p,q)$ is the beta function, defined as

$$B(p,q) = \frac{\Gamma(p)\Gamma(q)}{\Gamma(p+q)} \tag{8.58}$$

The graph of the density function of T_n for various values of n is given in Fig. 8.28. Student's *t*-distribution or simply *t*-distribution is a family of continuous probability distribution that arises when estimating the mean of a normally distributed population, in situations where the sample size and population standard deviation are unknown. It is used in a number of situations, including *t* test for assessing the statistical significance of the difference between two sample means, the construction of confidence intervals for the difference between two population

Fig. 8.28 Density functions of T_n

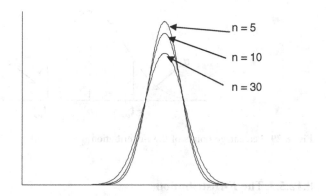

means, and in linear regression analysis. The t-distribution also arises in the Bayesian analysis of data from a normal family.

Note that $E(T_n) = 0$ and $\text{Var}(T_n) = \frac{n}{n-2}, n > 2$. Thus, the variance of T_n decreases to 1 and converges to standard normal random variable as $n \to \infty$. Hence, t-distribution is also symmetric like normal distribution.

The percentage points of t-distribution (see Appendix Table A.5) are used as critical values in carrying out a t test based on the values of t_n for a specified value of significance value α. For $\alpha \in (0, 1)$, the quantity $t_{\alpha,n}$ be such that

$$P(T_n \geq t_{\alpha,n}) = \alpha \tag{8.59}$$

Since the distribution is symmetric, $-T_n$ has the same distribution as T_n and so

$$\alpha = P(-T_n \geq t_{\alpha,n})$$
$$= P(T_n \leq -t_{\alpha,n})$$
$$= 1 - P(T_n > -t_{\alpha,n})$$

Therefore,

$$P(T_n \geq -t_{\alpha,n}) = 1 - \alpha$$

which implies

$$-t_{\alpha,n} = t_{1-\alpha,n} \tag{8.60}$$

Figure 8.29 shows the percentage points of t-distribution. In Appendix A.5, we provide the right tail area of $T_{\alpha,n}$ for a variety of values of n and α.

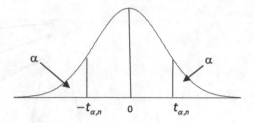

Fig. 8.29 Percentage points of the t-distribution

8.12.5.3 The F-Distribution

If $\chi^2_{(n)}$ and $\chi^2_{(m)}$ are independent chi-square random variables with n and m degrees of freedom, respectively, then the random variable $F_{n,m}$ defined by

$$F_{n,m} = \frac{\chi^2_{(n)}/n}{\chi^2_{(m)}/m} \tag{8.61}$$

is said to have F-distribution with n and m degrees of freedom and is denoted as $F_{n,m}$-distribution. The probability density function of $F_{n,m}$-distribution is given by

$$f(x) = \frac{B\left(\frac{n}{2}, \frac{m}{2}\right) n^{n/2} m^{m/2} x^{n/2-1}}{(m + nx)^{(n+m)/2}}, \quad x > 0, \tag{8.62}$$

where $x = F_{n,m}$. The distribution has mean and variance as

$$E(x) = \frac{m}{m-2}, \quad \text{provided that } m > 2 \tag{8.63}$$

and

$$\text{Var}(X) = \frac{2m^2(n+m-2)}{n(m-2)^2(m-4)}, \quad \text{provided that } m > 4 \tag{8.64}$$

If X has an $F_{1,m}$-distribution, then $\sqrt{X}$ has a t-distribution with m degrees of freedom. The graph of the density function of $F_{n,m}$ for various values of n and m is given in Fig. 8.30. The F-distribution arises frequently as the null distribution of a test statistic, most notably in the analysis of variance (ANOVA).

The percentage points of $F_{n,m}$-distribution (see Fig. 8.31) are used as critical values in carrying out the F-test based on the values of $F_{n,m}$ for a specified value of significance value α. Also for $\alpha \in (0, 1)$, the quantity $F_{\alpha,n,m}$ be such that

Fig. 8.30 $F_{n,m}$-distribution

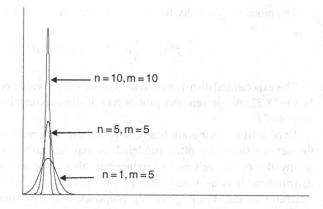

Fig. 8.31 Percentage points of F-distribution

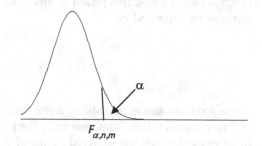

$$P(F_{n,m} > F_{\alpha,n,m}) = \alpha \tag{8.65}$$

In Appendix Tables A.6–A.10, we provide $F_{\alpha,n,m}$ for a variety of values of n, m, and α.

The χ^2-distribution, t-distribution, and F-distribution are all connected to each other by their theoretical developments as well. For instance, in chi-squared tests and in estimating variances, the χ^2-distribution enters the problem of estimating the mean of a normally distributed population and the problem of estimating the slope of a regression line via its role in t-distribution. The t-distribution enters all analyses of variance problems via its role in the F-distribution, which is the distribution of the ratio of two independent chi-squared random variables, each divided by their respective degrees of freedom.

8.12.6 Exponential Distribution

The exponential distribution is the most common product reliability model in use. It is found to be useful for characterizing uncertainty in a machine's life, breakdown of machines, life of an electronic unit, length of telephone calls, etc.

The probability density function of an exponential random variable X is stated as

$$f(x) = \frac{1}{\theta} e^{-x/\theta}, \quad x \geq 0, \quad \theta > 0 \tag{8.66}$$

The exponential distribution has mean θ, and the variance of the distribution is θ^2. In Fig. 8.32, we present the plot of $f(x)$ of the exponential distribution for various values of θ.

Exponential distribution has been found useful in many areas: In *queuing theory*, the service times are often modeled as exponentially distributed variables. Reliability theory and reliability engineering also make extensive use of exponential distribution. It is also very convenient because it helps to add failure rates in a reliability model. For engineering purposes, reliability function (or survival function) is defined as follows: the probability that a device will perform its intended function during a specified period of time under stated conditions. Mathematically, this may be expressed as

$$R(t) = \Pr(T > t) = \int\limits_{t}^{\infty} f(x) \mathrm{d}x \tag{8.67}$$

where $f(x)$ is the failure probability density function and t is the length of the period of time assumed to start from time zero. The hazard rate or instantaneous failure rate of an element which has survived t units of time is expressed as

$$h(t) = \lim_{\Delta \to 0} \frac{F(t + \Delta t) - F(t)}{\Delta P(T > t)}$$

$$= \frac{f(t)}{R(t)} \tag{8.68}$$

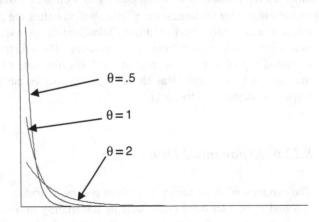

Fig. 8.32 Exponential distribution

The function $\Lambda(t) = \int_0^t h(x)\mathrm{d}x$ is called the cumulative hazard function. In repairable system models or maintenance system models, the function $h(t)$ and $\Lambda(t)$, respectively, are called the intensity function and cumulative intensity function or mean value function. For any lifetime random variable, the reliability function and hazard function are related through the relation

$$R(t) = \exp\left\{ - \int\limits_0^t h(x)\mathrm{d}x \right\} \tag{8.69}$$

The exponential distribution has survival function

$$R(t) = \int\limits_t^\infty \frac{1}{\theta} e^{-x/\theta}\mathrm{d}x$$

$$= e^{-t/\theta} \tag{8.70}$$

and hazard function as

$$h(t) = \frac{f(t)}{R(t)}$$

$$= \frac{1}{\theta} \tag{8.71}$$

Thus, for exponential distribution, the hazard function is constant.

The exponential distribution is, however, not appropriate for modeling the overall lifetime of organisms or technical devices, because the "failure rates" here are not constant; more failures occur for very young and for very old systems. In physics, if you observe a gas at a fixed temperature and pressure in a uniform gravitational field, the heights of the various molecules also follow an approximate exponential distribution.

Meeting reliability requirements has become a necessity for a good business to retain its customers. Sufficiently reliable data are required for this. The reasons for collecting reliability data include the following [9]:

- Assessing the characteristics of materials over a warranty period or during the product's design life
- Predicting product reliability
- Predicting product warranty costs
- Assessing the effect of a proposed design change
- Providing inputs needed for system failure risk assessment
- Assessing whether customer requirements and government regulations have been met

- Supporting programs to improve reliability through the use of laboratory experiments, including accelerated life tests
- Comparing components from two or more different manufacturers, materials, production periods, operating environments, and so on
- Checking the veracity of an advertising claim.

Example 8.24 The mileage (in thousands of miles) which car owners get with a certain kind of radial tyre is a random variable having an exponential distribution with mean 1/40 miles. Find the probabilities that one of these tyres will last

(a) at least 20,000 miles
(b) at most 30,000 miles

Solution Let X denote the mileage. Then, $X \sim \text{EXP}(40)$.
  Therefore,

(a) $P(X \geq 20) = e^{-20/40}$

    $= 0.6065$

(b) $P(X \leq 30) = 1 - e^{-30/40}$

    $= 0.5276$

Example 8.25 A certain kind of appliance requires repairs once every two years on an average. Assuming that the time between repairs is exponentially distributed, what is the probability that such an appliance will work at least three years without requiring repairs?

Solution Let X denote the repair time. Then, $X \sim \text{EXP}(2)$.
  Then, the required probability is

$$P(X \geq 3) = e^{-3/2}$$

$$= 0.2231$$

Properties of exponential distribution

1. One of the key properties of an exponential distribution is that it is memoryless, where we say that a nonnegative random variable X is *memoryless* if

$$P(X \geq t + s | X > s) = P(X > t), \quad \text{for all } t \geq 0 \qquad (8.72)$$

2. If $X_1, X_2, \ldots, X_n$ are independent exponential random variables having respective parameters $\theta_1, \theta_2, \ldots, \theta_n$, then $\min(X_1, X_2, \ldots, X_n) = X_{(1)}$ is exponential with parameter $\sum_{i=1}^{n} \theta_i$.

An exponential distribution can also be characterized through a Poisson process as follows:

Poisson Process: Suppose that the events are occurring at random time points, and let $N(t)$ denote the number of events that occur in the time interval $[0, t]$, and if the average occurrence of the event of interest is θ per unit time, then the number of occurrences in a given length of time x has a Poisson distribution with the pmf

$$P[N(t) = k] = \frac{e^{-\theta t}(\theta t)^k}{k!}, \quad k = 0, 1, \dots \tag{8.73}$$

Then, the time between any two consecutive occurrences will be exponential with parameter θ.

8.12.7 Gamma Distribution

A two-parameter gamma distribution is defined as

$$f(x; \theta, \beta) = \frac{1}{\theta^\beta \Gamma(\beta)} x^{\beta-1} e^{-x/\theta}, \quad x > 0 \tag{8.74}$$

The parameter θ is called the scale parameter, and β is the shape parameter. If β is an integer, then the distribution represents the sum of β independent exponentially distributed random variables, each of which has a mean of θ. The gamma distribution has mean $\theta\beta$ and variance $\theta\beta^2$. For $\beta = 1$, the distribution reduces to an exponential with mean θ. The graphical plot of the density functions for various values of β is given in Fig. 8.33.

Fig. 8.33 Gamma distributions for $\theta = 1$

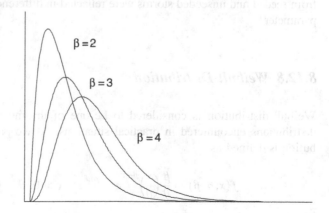

The quantity $\Gamma\beta$ in (8.71) is called the gamma function and is defined as

$$\Gamma(\beta) = \int_0^\infty \frac{1}{\theta^\beta} x^{\beta-1} e^{-x/\theta} dx$$

$$= \int_0^\infty y^{\beta-1} e^{-y} dy \qquad \text{(by letting } y = x/\theta)$$

$$= -e^{-y} y^{\beta-1} \Big|_{y=0}^{y=\infty} + (\beta-1) \int_0^\infty y^{\beta-2} e^{-y} dy \text{ (by integration by parts)}$$

$$= (\beta-1) \int_0^\infty y^{\beta-2} e^{-y} dy$$

$$= (\beta-1)\Gamma(\beta-1)$$
$$= (\beta-1)(\beta-2)\Gamma(\beta-2)$$
$$= (\beta-1)(\beta-2)(\beta-3)\ldots\Gamma(1)$$

$$= (\beta-1)! \qquad \text{Since } \Gamma(1) = \int_0^\infty e^{-y} dy = 1$$

The gamma function arises in various contexts of mathematical and statistical applications [1]. Unfortunately, the survival function and hazard function for gamma distribution do not have any closed form, because of the presence of incomplete gamma functions.

The gamma distribution is frequently used as a probability model for waiting periods; for instance, in life testing, the waiting period until death is a random variable that is frequently modeled with a gamma distribution. Gamma distributions were fitted to rainfall amounts from different storms, and differences in amounts from seeded and unseeded storms were reflected in differences in estimated β and θ parameter.

8.12.8 Weibull Distribution

Weibull distribution is considered to be one of the most important engineering distributions encountered in practical situations. A two-parameter Weibull distribution is defined as

$$f(x; \theta, \beta) = \frac{\beta}{\theta} \left(\frac{x}{\theta}\right)^{\beta-1} e^{-(x/\theta)^\beta}, \quad x > 0, \ \theta > 0, \ \beta > 0 \qquad (8.75)$$

Fig. 8.34 Weibull distribution for $\theta = 1$

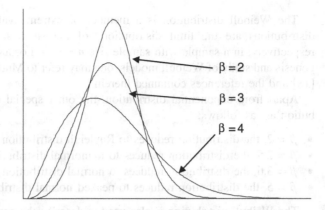

The parameter θ is called the scale parameter, and β is the shape parameter. For $\beta = 1$, the distribution reduces to exponential distribution. For $\beta < 1$, the failure rate decreases, and for $\beta > 1$, the failure rate increases. A graph of density function of Weibull distribution is given in Fig. 8.34.

The Weibull distribution has mean and variance as

$$\mu = \theta \, \Gamma\left(\frac{1}{\beta} + 1\right) \tag{8.76}$$

and

$$\sigma^2 = \theta^2 \left[\Gamma\left(\frac{2}{\beta} + 1\right) - \left\{ \Gamma\left(\frac{1}{\beta} + 1\right) \right\}^2 \right] \tag{8.77}$$

respectively. The hazard function and reliability function are respectively obtained as

$$h(t) = \frac{f(t)}{R(t)}$$

$$= \frac{\beta}{\theta} \left(\frac{t}{\theta}\right)^{\beta - 1} \tag{8.78}$$

and

$$R(t) = \int_t^\infty \frac{\beta}{\theta} \left(\frac{x}{\theta}\right)^{\beta - 1} e^{-(x/\theta)^\beta} \, \mathrm{d}t$$

$$= e^{-(t/\theta)^\beta} \tag{8.79}$$

The Weibull distribution is a member of extreme value distributions. These distributions are the limit distributions of the smallest or the greatest value, respectively, in a sample with sample size $n \to \infty$. For a complete account of the genesis and study of Weibull models, one may refer to Murthy et al. [15] and Rinne [18] and the references contained therein.

Apart from exponential distribution, the other special cases of Weibull distribution are as follows:

- $\beta = 2$, the distribution reduces to Rayleigh distribution
- $\beta = 2.5$, the distribution reduces to lognormal distribution
- $\beta = 3.6$, the distribution reduces to normal distribution
- $\beta = 5$, the distribution reduces to peaked normal distribution.

The Weibull distribution is also used to describe the particle size distribution of particles generated by grinding, milling, and crushing operations. Because of its frequent use in reliability, learning curves, error rates, etc., it is considered to be the most popular, modern distribution to be used in modeling performance data.

The *Rayleigh distribution* is often used to model the arrival of defects across a life cycle. By fitting a Rayleigh curve to historical data on defect arrivals, one may predict future defect arrivals with prediction intervals. The Rayleigh distribution has the density function

$$f(x; \theta) = \frac{2}{\theta} \left(\frac{x}{\theta}\right) e^{-(x/\theta)^2}, \quad x > 0, \theta > 0 \tag{8.80}$$

Some numerical characteristics of Weibull and Rayleigh distributions for $\theta = 10$ are shown in Table 8.8.

Example 8.26 During a model fitting of a lifetime data measured in hours, it is found that the data follow a Weibull distribution with parameters $\theta = 4000$ and $\beta = 0.5$. Then, find

(a) The mean time before the system breaks down
(b) $P(T \geq 5000)$
(c) $P(T \geq 10,000)$
(d) $P(T < 10,000)$

Table 8.8 Sample characteristics of Weibull distribution for $\theta = 10$

Characteristics	$\beta = 0.5$	$\beta = 0.8$	$\beta = 1$	$\beta = 2$	$\beta = 3$
Mean	19.9	11.332	10.058	8.854	8.914
Q_1	0.808	2.104	2.877	5.352	6.592
Q_2	4.76	6.293	7.003	8.331	8.854
Q_3	19.20	15.061	13.951	11.760	11.134
Standard deviation	46.00	14.335	10.050	4.627	3.239

Solution

(a) The mean time $= \mu = \theta\Gamma\left(\frac{1}{\beta}+1\right)$

$$= 40{,}000\Gamma\left(\frac{1}{0.5}+1\right)$$

$$= 4000\Gamma(3) = 8000$$

(b) $P(T \geq 5000) = e^{-(5000/4000)^{0.5}}$

$$= e^{-1.118}$$

$$= 0.3269$$

(c) $P(T \geq 10{,}000) = e^{-(10{,}000/4000)^{0.5}}$

$$= e^{-1.5811}$$

$$= 0.2057$$

(d) $P(T < 10{,}000) = 1 - e^{-(10{,}000/4000)^{0.5}}$

$$= 0.7943$$

8.12.9 Sampling Distributions

Many procedures in statistical inference are based on the use of the normal probability distribution. This is because of the *central limit theorem*. This theorem states that when a random sample of n observations is selected from a population (any population) with a mean of μ and a standard deviation of σ, for large n, the sampling distribution of the mean is approximately a normal distribution with a mean of μ and a standard deviation of $\sigma/\sqrt{n}$ (standard error of the mean). Since many of the estimators that are used to make inferences about the characteristics of a population are sums or means of sample measurements, we can expect the estimator to be approximately normally distributed in repeated sampling, when n is sufficiently large. And, of course, there is always the question, "What do you mean by n being sufficiently large?"

For practical considerations, people use the large sample and small sample concept (based on sample size value) to substantiate their findings. Therefore, any sample which is more than 30 is considered to be large; otherwise, the study is considered to be a small sample one. This assumption is reasonably good for all parametric-based inferences. As discussed in Chap. 6, the sample size determination is a general issue in statistical research. For an ideal sample size calculation for a particular characteristic, at least one must have the knowledge of the following aspects (see also [13]):

- a desired level of confidence $100(1 - \alpha)$ %, which determines the critical value, $Z_{\alpha/2}$
- an acceptable sampling error or precision or margin of error, say d, and
- the standard deviation, σ.

The value of a sample size plays a very important role in establishing the large sample approximation of a distribution. The central limit theorem can be formally stated as follows:

Central limit theorem: If $x_1, x_2, \ldots, x_n$ are n random variables which are independent and having the same distribution with mean μ and standard deviation σ, then as $n \to \infty$, the limiting distribution of the standardized mean

$$\frac{\bar{x} - \mu}{\sigma/\sqrt{n}} \to N(0, 1) \qquad (8.81)$$

In practice, if the sample size is sufficiently large, we need to know the population distribution because the central limit theorem (CLT) assures us that the distribution of $\bar{x}$ can be approximated by a normal distribution. A sample size larger than 30 is generally considered to be large enough for this purpose. Generally, in most of the practical cases, the sample sizes are higher than 30; hence, the sampling distribution of the mean can be approximated by a normal distribution. It is better to use the CLT when the population distribution is either unknown or known to be non-normal. If the population distribution is known to be normal, then $\bar{x}$ will also be normally distributed, irrespective of the sample size.

Below, we give some illustrations to explain the central limit theorem.

Example 8.27 If X is binomial with parameters $n = 150$ and $p = 0.6$, compute the value of $P(X \leq 80)$.

Solution Since n is very large, we use large sample approximation to compute the probability. We have $n = 150$, $p = 0.6$. Therefore, $\mu = np = 90$ and $\sigma^2 = 36$. Hence,

$$P(X \leq 80) = P\left(\frac{X - \mu}{\sigma} \leq \frac{80 - 90}{6}\right)$$
$$= P(Z \leq -1.67)$$
$$= 0.0475$$

Example 8.28 The average weight of a population of workers is 70 kg and the standard deviation 20 kg. Then

(a) If a sample of 36 workers is chosen, approximate the probability that their weight lies between 63 and 73 and
(b) Find the probability in part (a), when the sample is of size 64 and 81.

Solution Since we do not have any idea about the population from which the samples are taken, we use CLT to approximate the probability.

(a) Let X be the variable weight of the worker and $\overline{X}$ be its mean. Given $n = 36$ and $\sigma = 20$, then, according to CLT,

$$\overline{X} \sim N\left(\mu, \frac{\sigma^2}{n}\right)$$

i.e., $\overline{X} \sim N\left(70, \frac{400}{36}\right) = N(70, 11.11)$

Hence, the required probability is

$$P(63 < \overline{X} < 73) = P\left(\frac{63 - 70}{3.3} < \frac{\overline{X} - \mu}{\sigma/\sqrt{n}} \leq \frac{73 - 70}{3.3}\right)$$
$$= P(-2.12 < Z < 0.909)$$
$$= 0.7989$$

(b) If $n = 64$ then $\overline{X} \sim N(70, 6.25)$. Then, the required probability is

$$P(63 < \overline{X} < 73) = P\left(\frac{63 - 70}{2.5} < \frac{\overline{X} - \mu}{\sigma/\sqrt{n}} \leq \frac{73 - 70}{2.5}\right)$$
$$= P(-2.8 < Z < 1.2)$$
$$= 0.1125$$

Similarly, when $n = 100$, $\overline{X} \sim N(70, 2)$ and the required probability is

$$P(63 < \overline{X} < 73) = P\left(\frac{63 - 70}{2} < \frac{\overline{X} - \mu}{\sigma/\sqrt{n}} \leq \frac{73 - 70}{2}\right)$$
$$= P(-3.5 < Z < 1.5)$$
$$= 0.9329$$

Example 8.29 The lifetime of a certain electrical part is a random variable with mean 100 h and standard deviation 20 h. If 16 such parts are tested, find the probability that the sample mean is

(a) less than 104
(b) between 98 and 104 h.

Solution Given $n = 16, \mu = 100, \sigma = 20$. Then

(a) $P(\overline{X} < 104) = P\left(\dfrac{\overline{X} - 100}{20/4} < \dfrac{104 - 100}{20/4}\right)$

$\qquad\qquad\quad = P(Z < 4/5)$

$\qquad\qquad\quad = 0.7881$

(b) $P(98 < \overline{X} < 104) = P\left(\dfrac{98 - 100}{20/4} < \dfrac{\overline{X} - 100}{20/4} < \dfrac{104 - 100}{20/4}\right)$

$\qquad\qquad\qquad\qquad = P(-0.4 < Z < 0.8)$

$\qquad\qquad\qquad\qquad = 0.4435$

8.13 Capability Analysis

The method of determining the extent to which the long-term performance of an industrial process complies with engineering requirements or managerial goal is called capability analysis [10]. Often, it is required that the output should lie between upper and lower specification limits. It is important to recognize that the information about capability and the requirements come from different sources and are totally independent. The specification does not determine the capability of the process, and the process capability does not determine the requirements, but they do need to be known. Figure 8.35 shows the connection between the specification limits and tolerance of a given process.

A process capability index [16] is a measure relating the actual performance of a process to its specified performance, where processes are considered to be a combination of the equipment, materials, people, plant, methods, and the environment. The absolute minimum requirement is that three process standard deviations on each side of the process mean are contained within the specification limits. This means that about 99.7 % of the output will be within the tolerance. A process capability index may be calculated, only when the process is under statistical control.

Fig. 8.35 Specification limits and tolerance

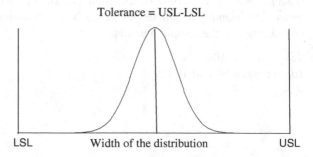

Tolerance = USL-LSL

LSL   Width of the distribution   USL

They are simply a means of indicating the variability of a process relative to the product specification tolerance.

The calculation of process capability depends on the type of data under investigation. For attribute data system, the capability is defined in terms of pass/fail or good/bad as the case may be. They are calculated in terms of defects per million opportunities (DPMO). On the other hand, for continuous data system, the capability is defined in terms of defects under the curve and outside the specification limits. Some frequently used capability indices and their potential uses are described below.

8.13.1 Process Potential Index (C_p Index)

It is defined as the ratio of maximum allowable characteristics ($USL - LSL$) to the normal variation of the process (6σ) and is given by

$$C_p = \frac{USL - LSL}{6\sigma} \tag{8.82}$$

If $USL - LSL = 2T$, then C_p index is just the ratio of $2T$ to 6σ. Here, the numerator is controlled by design engineering and the denominator is controlled by the process engineering. For large values of C_p, the process becomes capable, and for any value of C_p less than one, the process variation is greater than the specified tolerance. So, the process is incapable. Thus, C_p offers a simple comparison of total variation with tolerances.

8.13.2 Process Performance Index (C_pk Index)

One of the main drawbacks of the C_p index is that it does not take into account the centering of the process. C_{pk} accounts for both the process variation and the centering, which is widely used as a means of communicating the process capability. If we define

$$C_{pl} = \frac{\overline{X} - LSL}{3\sigma} \quad \text{and} \quad C_{pu} = \frac{USL - \overline{X}}{3\sigma}$$

then

$$C_{pk} = \text{Minimum of } \left(C_{pl}, C_{pu} \right) \tag{8.83}$$

A C_{pk} value of less than or equal to one means that the process variation and its centering are such that at least one of the tolerance limits will be exceeded and the

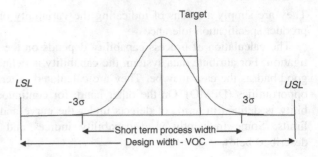

Fig. 8.36 Short-term capability process (*Source* Muralidharan and Syamsundar [14])

process is incapable. Both C_p and C_{pk} evaluate the short-term capability (see Fig. 8.36) of the process and they help us to

- predict whether rejections will take place on the higher side or on the lower side,
- take centering decisions,
- decide whether to consider broadening of tolerances,
- take decisions on whether to go in for new m/cs, and
- decide on the level of inspection required.

While interpreting the capability indices, one has to have a thorough knowledge of the source of data and the possible causes of variation. Some of the interpretations of C_{pk} are given below:

C_{pk} < 1: A situation in which the producer is not capable and there will inevitably be nonconforming output from the process

C_{pk} = 1: A situation in which the producer is not really capable (correspond to 3σ process)

C_{pk} = 1.5: Fair, but still nonconforming output will occur and the chances of detecting it are still not good enough.

C_{pk} = 1.67: Promising, nonconforming output will occur but there is a very good chance that it will be detected.

C_{pk} = 2: High level of confidence in the producer, provided that control charts are in regular use (correspond to 3σ process).

We shall now discuss some capability indices related to long-term variability. Long-term variability (see Fig. 8.37) is made up of short-term variability and

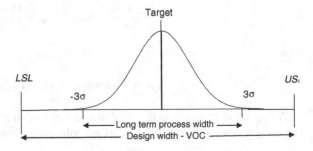

Fig. 8.37 Long-term capability process (*Source* Muralidharan and Syamsundar [14])

process drift. The long-term drift is determined by assuming 1.5σ shift relative to target. A rational subgroup concept is usually put in place to assess the long-term variability where we assume that the variability within the group is smaller than the variability from group to group ($\sigma_{total}^2 = \sigma_{within}^2 + \sigma_{between}^2$). This allows estimation of the pure short-term variability and the long-term drift of the process. Two such indices are P_p and P_{pk}.

P_p is based on the same equation as C_p with one exception; P_p employs the long-term standard deviation, whereas C_p employs the short-term standard deviation. It is defined as

$$P_p = \frac{USL - LSL}{6\sigma_{LT}} \qquad (8.84)$$

And P_{pk} is based on the same equation as C_{pk} with one exception; P_{pk} employs the long-term standard deviation, whereas C_{pk} employs the short-term standard deviation. It is given by

$$P_{pk} = \text{Minimum of } \left(P_{pl}\ P_{pu}\right), \qquad (8.85)$$

where

$$P_{pl} = \frac{\overline{X} - LSL}{3\sigma_{LT}} \text{ and } P_{pu} = \frac{USL - \overline{X}}{3\sigma_{LT}}$$

In both the above formulae, σ_{LT} is the long-term sigma value of the process discussed earlier.

Example 8.30 The thickness of a printed circuit board is an important quality parameter. Data on board thickness (in inches) are given in Table 8.9, for 25 samples of three boards each.

Table 8.9 Thickness measurements

Sample number	X_1	X_2	X_3
1	0.0629	0.0636	0.0640
2	0.0630	0.0631	0.0622
3	0.0628	0.0631	0.0633
4	0.0634	0.0630	0.0631
5	0.0619	0.0628	0.0630
6	0.0613	0.0629	0.0634
7	0.0630	0.0639	0.0625
8	0.0628	0.0627	0.0622
9	0.0623	0.0626	0.0633
10	0.0631	0.0631	0.0633

(continued)

Table 8.9 (continued)

Sample number	X_1	X_2	X_3
11	0.0635	0.0630	0.0638
12	0.0623	0.063	0.063
13	0.0635	0.0631	0.063
14	0.0645	0.064	0.0631
15	0.0619	0.0644	0.0632
16	0.0631	0.0627	0.063
17	0.0616	0.0623	0.0631
18	0.0630	0.0630	0.0626
19	0.0636	0.0631	0.0629
20	0.064	0.0635	0.0629
21	0.0628	0.0625	0.0616
22	0.0615	0.0625	0.0619
23	0.0630	0.0632	0.0630
24	0.0635	0.0629	0.0635
25	0.0623	0.0629	0.0630

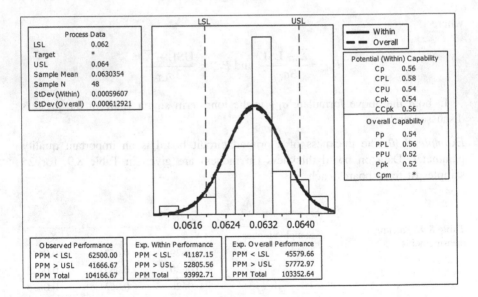

Fig. 8.38 Process capability analysis of thickness

A capability analysis using MINITAB is carried out, and the output is shown in Fig. 8.38. The output gives nonconformance rates, capability indices, and many other measures.

The analysis shows capability less than 1 (that is $C_{pk} < 1$), and the PPM total is also very high. Hence, the process is not capable. The sigma level corresponding to this PPM is roughly 2.5.

8.14 Baseline Performance Evaluation

The tools and methods of data collection are important in any type of business process measurement. Before getting into the root causes and analysis of the data, it is important to establish performance "baselines"—to determine how well processes are working currently—so that we can focus on and measure improvement. At the measure stage, the output measures will be looked at first, followed by the internal measurements that will affect the performance outputs.

At this stage, we explore the descriptive statistics of the data along with the process performance measures like Yield, DPMO, sigma, and the cost of poor quality (CoPQ) to identify the gray areas of concern. This will set the stage for preparing the *goal statement*, and then, one can look for improvement.

As an illustration, we can recall Example 8.25 on the thickness of a printed circuit board discussed above. The descriptive statistics are shown in Fig. 8.39. For a baseline performance evaluation, we use these statistics.

Note that the average thickness is average thickness is 0.062952 and the minimum thickness is 0.0613. This will help us to set a (typical) goal as

$$Goal = Baseline - 0.7(baseline - entitlement)$$
$$= 0.062952 - 0.7(0.062952 - 0.0613)$$
$$= 0.0618$$

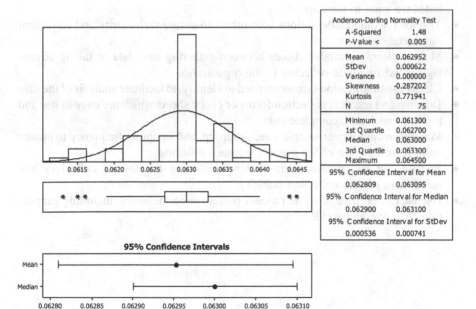

Fig. 8.39 Descriptive statistics of thickness

Thus, a goal statement offers a relief in terms of concrete results. The multiplying factor 0.7 is the leverage admissible. This value can change as per the process requirement. A typical goal statement has the following components:

- A *description of what is accomplished*. For instance, "reduce cycle time," "increase the availability," eliminate defect by, etc.
- A *measurable target for desired results*. The target should quantify the desired cost saving, defect elimination, or time reduction in percentages or actual numbers
- A *project deadline and/or timeframe for results*. The time frame decided in early part of the project may be revised at this time based on the goal statement. The completion time of the project should be clearly stated.

8.15 Measure Checklists

Before moving to the next phase of DMAIC, ensure the following measure checklist:

- Determine what we want to learn about our problem and process and where in the process can we get the answer
- Identify the types of measures we want to collect and have a balance between effectiveness/efficiency and input/process/output
- Develop clear and unambiguous operational definitions of the things or attributes we want to measure
- Test the operational definitions with others to ensure their clarity and consistent interpretation
- Make a clear, reasonable choice between gathering new data or taking advantage of existing data collected in the organization
- Clarify the stratification factors we need to identify, to facilitate analysis of the data
- Develop and test data collection forms or check sheets which are easy to use and provide consistent, complete data
- Identify an appropriate sample size, subgroup, and sampling frequency to ensure valid representation of the process we are measuring
- Prepare and test measurement system, including training of collectors and assessment of data collection stability
- Use data to prepare baseline process performance measures, including proportion defective and yield.

8.16 Relevance for Managers

Managing a Six Sigma project is as good as managing man, machines, materials, methods, environment, etc., as they are the sources of variations and problems in a process. A great deal of statistical thinking and temperament are vital for this. We cannot efficiently manage anything unless we have a mechanism to measure the performances and variations. This chapter helps to identify the critical-to-quality (CTQ) parameters of a process through two data capturing methods, viz. VOC and VOP. The mechanism also helps to differentiate the value-added and non-value-added activities in a process, established through a value-stream mapping and process mapping. This will enable necessity of data collection methods for further treatment and facilitation. Managers need skills to collect, analyze, and interpret statistical data relating to production, marketing, finance and human resources and R&D functions besides customer and public opinions on a continuing basis which would lead to operating and strategic decisions.

Statistical methods would be of immense value in exploring the environment in which they have been operating. The methods also facilitate better planning, organizing, evaluating, and controlling the business processes both qualitatively and quantitatively. The knowledge of various data types, namely nominal, ordinal, ratio, and scale; various time measurements such as cycle time, take time, execution time, and delay time helps the Six Sigma project personnel to identify proper statistical methods of applications. The necessity of carrying out measurement system analysis for variables and attributes, conducting descriptive statistics, etc., enhances better understanding of the variability of measurements. The most frequently used Pareto chart for prioritizing the sources of variation, control chart for analyzing the special cause variation and common cause variation, cause and effect diagram for identifying the causes for an effect, etc., are some of the basic tools used for increasing the visibility of process variations.

This chapter also discusses the issues related to uncertainty and the probability modeling of random phenomena, which often poses lots of challenges to Six Sigma project management. The practical significance of every probability models (discrete or continuous) is described in detail in reference to a Six Sigma improvement project. For instance, the use of central limit theorem (CLT) through which the sampling distributions of normal distributions (t, F, and χ^2) are derived is some of the core results of probability theory and has immense relevance in decision making under uncertainty.

As per engineering requirements, every process needs to be assessed in terms of its capability and performance before and after improvement. That will judge the worthiness of a Six Sigma project. Even viability of a project can be ascertained through this simple exercise. This chapter therefore includes the discussion of process capability and potential estimation and shows the method of setting initial goal for a project. For Six Sigma professionals, this baseline performance will be the stepping stone for further improvement and progress.

Exercises

8.1. What is process mapping? Describe the methods of identifying VA/NVA in a process.

8.2. Distinguish between VOC and VOP. How are they captured in a process?

8.3. Discuss various characteristics associated with a VOC and VOP.

8.4. What is value-stream mapping? Describe how a process flow is established through a VSM?

8.5. Distinguish between a primary data and secondary data. Give examples for each.

8.6. How a unit of analysis is different from characteristics of interest?

8.7. Describe various data types and their importance in data analysis.

8.8. Write short notes on cycle time, takt time, execution time, and delay time.

8.9. What is measurement system analysis? State various components of a good MSA.

8.10. What is gage R&R? Describe a method of assessing bias in a continuous measurement of a process.

8.11. An article in quality engineering 1997 presents data for a gage capability studies based on a measuring device known as an optical comparator; two operators took two measurements (in mm) of a particular dimension on 10 gear parts. The results are shown below:

Part	Operator 1		Operator 2	
	Test 1	Test 2	Test 1	Test 2
1	3.045	3.041	3.048	3.046
2	3.037	3.038	3.031	3.038
3	3.032	3.024	3.023	3.022
4	3.017	3.012	3.011	3.016
5	3.048	3.049	3.042	3.048
6	3.046	3.045	3.042	3.043
7	3.039	3.034	3.033	3.035
8	3.033	3.035	3.038	3.037
9	3.048	3.041	3.034	3.032
10	3.041	3.049	3.046	3.045

(i) Construct a repeatability range chart for the data. Are there any out-of-control points indicating operator difficulty?

(ii) Estimate gage repeatability and reproducibility.

(iii) Obtain an estimate for gage percent repeatability and reproducibility. What is assessment of the measurement system?

(iv) If the specifications for the measured dimension are 3.035 ± 0.05, what can you say about gage capability in terms of a P/T ratio?

8.12. Discuss a method of assessing bias in an attribute data.

8.13. The following table shows the results of three appraisers responded twenty items subjected to a lot-by-lot sampling inspection. The quality of the item is classified as "Accept" or "Reject." Carry out kappa analysis to assess the bias between the appraisers.

Sample	Appraiser	Response	Sample	Appraiser	Response	Sample	Appraiser	Response	Sample	Appraiser	Response
1	A	Reject	6	A	Reject	11	A	Reject	16	A	Reject
1	A	Reject	6	A	Accept	11	A	Accept	16	A	Accept
1	B	Accept	6	B	Accept	11	B	Accept	16	B	Accept
1	B	Accept	6	B	Accept	11	B	Reject	16	B	Reject
1	C	Reject	6	C	Accept	11	C	Accept	16	C	Accept
1	C	Accept	6	C	Accept	11	C	Accept	16	C	Accept
2	A	Accept	7	A	Accept	12	A	Reject	17	A	Reject
2	A	Reject	7	A	Reject	12	A	Accept	17	A	Accept
2	B	Accept	7	B	Reject	12	B	Reject	17	B	Accept
2	B	Accept	7	B	Reject	12	B	Reject	17	B	Reject
2	C	Accept	7	C	Accept	12	C	Accept	17	C	Accept
2	C	Accept	7	C	Accept	12	C	Accept	17	C	Accept
3	A	Accept	8	A	Accept	13	A	Reject	18	A	Reject
3	A	Accept	8	A	Accept	13	A	Accept	18	A	Accept
3	B	Reject	8	B	Reject	13	B	Accept	18	B	Reject
3	B	Accept	8	B	Accept	13	B	Reject	18	B	Accept
3	C	Reject	8	C	Reject	13	C	Accept	18	C	Accept
3	C	Reject	8	C	Reject	13	C	Reject	18	C	Reject
4	A	Reject	9	A	Reject	14	A	Accept	19	A	Accept
4	A	Accept	9	A	Accept	14	A	Accept	19	A	Accept
4	B	Accept	9	B	Accept	14	B	Accept	19	B	Reject
4	B	Accept	9	B	Accept	14	B	Reject	19	B	Accept
4	C	Reject	9	C	Reject	14	C	Accept	19	C	Reject
4	C	Reject	9	C	Reject	14	C	Reject	19	C	Reject

(continued)

Sample	Appraiser	Response	Sample	Appraiser	Response	Sample	Appraiser	Response	Sample	Appraiser	Response
5	A	Accept	10	A	Reject	15	A	Accept	20	A	Reject
5	A	Accept	10	A	Accept	15	A	Accept	20	A	Accept
5	B	Reject	10	B	Accept	15	B	Accept	20	B	Reject
5	B	Accept	10	B	Reject	15	B	Reject	20	B	Accept
5	C	Reject	10	C	Accept	15	C	Accept	20	C	Accept
5	C	Accept	10	C	Accept	15	C	Accept	20	C	Reject

8.14. What is the importance of descriptive statistics in managerial decision making?

8.15. Piston rings used in automobiles are coated with hard chrome plating. For a particular stage of production, the requirement of plating thickness is 195 ± 20 micrometers (μm). A random sample of 50 piston rings was obtained having the measurements:

18 9	181	183	191	180	182	187	188	189	189
177	191	192	180	184	190	179	198	197	179
171	188	165	191	176	172	199	185	183	183
187	189	186	191	185	172	182	184	185	201
189	187	186	183	178	173	172	193	184	183

(i) Prepare a leaf diagram and a histogram,
(ii) Obtain all the summary statistics (descriptive statistics) of the measurement, and
(iii) Compute skewness and kurtosis and decide the shape of the measurement.

8.16. The life in hours of 60 electric bulbs is shown below:

511	911	1177	1016	600	777	895	1199	806	950
749	1067	980	923	1314	1108	1137	1262	1027	1081
906	1230	1090	1242	803	1131	918	907	1061	1198
1240	1057	980	997	763	759	1394	991	1155	750
1111	1117	1143	808	948	857	962	813	1139	1127
922	817	1057	665	1171	936	1068	1163	934	515

(i) Prepare a frequency distribution,
(ii) Draw a histogram,
(iii) Compute all summary statistics, and
(iv) Check the normality of the data

8.17. What are the types of variation encountered in a manufacturing process? How are they controlled?

8.18. What is the importance of Pareto diagram? How is Pareto chart implemented?

8.19. The following defects are noted for respirator masks inspected during a given time period:

Dent	Pinhole	Strap	Dent	Pinhole
Pinhole	Discoloration	Dent	Discoloration	Strap
Strap	Dent	Strap	Strap	Pinhole
Strap	Pinhole	Discoloration	Dent	Discoloration
Discoloration	Strap	Discoloration	Dent	Pinhole

(continued)

Dent	Pinhole	Discoloration	Pinhole	Discoloration
Dent	Discoloration	Strap	Pinhole	Dent
Discoloration	Strap	Dent	Discoloration	Strap

Prepare a Pareto chart and interpret.

8.20. Discuss the importance of control charts.

8.21. Describe the importance of cause and effect diagram. How are they used in identifying potential causes of variations?

8.22. What are the effects of potential failures? How are they prioritized?

8.23. Discuss the importance of prioritization matrix. State the ways of constructing a prioritization matrix.

8.24. Define probability of an event. Discuss how it is used as a measure of uncertainty?

8.25. A batch of 140 semiconductor chips is inspected by choosing a sample of five chips. Assume 10 of the chips do not conform to customer requirements. (i) Find the number of different samples of 5. Find the number of samples containing, (ii) no defective, (iii) 1 defective, and (iv) 2 defectives.

8.26. A batch of 100 iron rods consists of 15 oversized, 25 undersized, and remaining rods of desired length. If two rods are drawn at random without replacement, what is the probability of obtaining (i) 2 rods of desired length, (ii) exactly 2 rods of desired length, (iii) 2 rods of undersized, and (iv) none of oversized.

8.27. Four cards are drawn at random from a pack of 52 cards. Find the probability that (i) two are kings and two are queens, (ii) they are a king a queen, a jack and an ace, (iii) two are black and two are red, and (iv) there are two cards of hearts and two cards of diamonds.

8.28. An urn contains 6 white, 4 red, and 9 black balls. If 3 balls are drawn at random, find the probability that (i) two of the balls drawn are white, (ii) one of each color, (iii) none is red, and (iv) at least one is white.

8.29. From 25 tickets, marked with first 25 numbers, one is drawn at random find the chance that (i) it is multiple of 5 or 7 and (ii) it is a multiple of 3 or 7.

8.30. If A and B are events such that $(P \cup B) = 3/4$, $P(A \cap B) = 1/4$ and $P(A') = 2/3$ Find (i) $P(A)$, (ii) $P(B)$, and (iii) $P(A \cap B')$.

8.31. If $P(A|B) = 0.4$, $P(B) = 0.8$, and $P(A) = 0.5$. Are the events A and B independent? Also determine $P(B|A)$.

8.32. A person applies for a job in two firms X and Y. The probability of his being selected in firm X is 0.7 and being rejected at Y is 0.5. The probability of at least one of his applications being rejected is 0.6. What is probability that he will be selected in one of the firms?

8.33. The probability that a randomly selected student from a college is a senior is 0.20 and the joint probability that the student is a computer science major and a senior is 0.03. Find the conditional probability that a student selected at random is a computer science major given that he/she is a senior.

8.34. An urn contains 3 red balls, 4 white balls, and 5 green balls. Four balls are selected at random from this urn.

   (i) Find the probabilities of the following events.

       A: All balls are of same color,
       B: All balls are of green color.

   (ii) Are A and B are independent? Justify your answer.

8.35. Customers are used to evaluate preliminary product designs. In the past 95 % of highly successful products received good reviews, 60 % of moderately successful products received good reviews, and 10 % of poor products received good reviews. In addition, 40 % of products have been highly successful, 35 % have been moderately successful, and 25 % have been poor products.

   (i) What is the probability that a product attains good review?
   (ii) If a new design attains a good review, what is the probability that it will be a highly successful product?

8.36. In a city school in Ahmedabad, there are three candidates for the position of principal. Mr. Patel, Mr. Shah, and Mr. Singh whose chances of getting the appointment are in the proportion 4:2:3, respectively. The probability that Mr. Patel if selected would introduce coeducation in the college is 0.3. The probabilities of Mr. Shah and Mr. Singh doing the same are, respectively, 0.5 and 0.8.

   (i) What is probability that there will be coeducation in college?
   (ii) If there is coeducation in the college, what is the probability that Mr. Shah is the principal.

8.37. The contents of three boxes are as follows:

   Box-1: 4 red, 5 white, and 2 yellow balls
   Box-2: 5 each of red, white, and yellow balls
   Box-3: 3 red, 6 white, and 5 yellow balls.
   A box is selected at random and then a ball is selected. It is found to be white. What is the probability that the ball selected is coming from

   (i) first box,
   (ii) first or second box, and
   (iii) not from the second.

8.38. The supervisor of customer relations for KLM airlines is studying the company's overbooking problem. He is concentrating on three late-night flights out of Mumbai airport. In the last year, 7, 8, and 5 % of the passengers on the Heathrow, Hong Kong, and Detroit flights, respectively, have been bumped. Further, 55, 20, and 25 % of the late-night KLM airline passengers at Mumbai airport take the Heathrow, Hong Kong, and Detroit flights, respectively. What is the probability that a bumped passenger was scheduled to be on the (a) Heathrow flight, (b) Hong Kong flight, and (c) Hong Kong or Detroit flight?

8.39. A machine tool is idle 15 % of the time. You request immediate use of the tool on five different occasions during the year. Assume that your request represents independent events. What is the probability that

- (i)   the tool is idle at the time of all of your requests?
- (ii) the tool is idle at the time of exactly four of your requests? and
- (iii) the tool is idle at the time of at least three of your requests?

8.40. Define mean and variance of a random variable. Obtain the mean and variance of number of heads in four tosses of an unbiased coin.

8.41. For the given random experiment, decide the variable is discrete or continuous:

- (i)   Number of molecules in a sample of gas,
- (ii)   Current in an electrical circuit,
- (iii) Outside diameter of a machined shaft,
- (iv)   Time until the projectile returns to earth,
- (v)   Number of times a transistor in a computer memory changes states,
- (vi)   Number of cracks exceeding 1.2 cm in 15 km of an interstate highway,
- (vii) Concentration of output from a reactor,
- (viii) Weight of an injection-molded plastic part,
- (ix)   Volume of gasoline that is lost to evaporation during the filling of tank, and
- (x)   Average age of a university student.

8.42. Suppose that a day's production of 850 manufactured parts contain 50 parts that do not conform to customer requirements. Two parts are selected at random, without replacement. Let the random variable X be the number of non-conforming parts in the sample. Obtain the distribution of X and also find the cumulative distribution, mean, and variance of the random variable X.

8.43. A continuous random variable X having values between 0 and 4 has a density function given by $f(x) = \frac{1}{2} - ax$ where a is a constant. Calculate (i) the value of a, (ii) the value of $P(1 < X < 2)$ and $P(X > 2)$, and (iii) the mean and variance of this distribution.

8.44. Suppose $f(x) = \frac{3}{4}(1 - x^2)$, $-1 \leq x \leq 1$. Find a such that $P(X \leq a) = 0.95$.

8.45. A small filling station is supplied with gasoline every Saturday afternoon. Assume that its volume X of sales in ten thousands of gallons has the probability density $f(x) = 6x(1 - x)$, $0 \leq x \leq 1$ and 0 otherwise. Determine the mean and variance.

8.46. A product is classified according to the number of defects it contains and the machine that produces it. Let X_1 and X_2 be the random variables that represent the number of defects per unit and the machine number, respectively. The entries in the table represent the joint possibility mass function of a randomly chosen product:

X_1	X_2	
	1	2
0	0.10	0.05
1	0.15	0.08
2	0.20	0.12
3	0.05	0.25

    (i) Find the marginal probability distribution of X_1 and X_2;

   (ii) Find $E(X_1)$, $E(X_2)$, $Var(X_1)$, $Var(X_2)$, $Cov(X_1, X_2)$, and $r(X_1, X_2)$;

  (iii) Are the variables independent ?; and

   (iv) Compute $E(X_1|X_2)$ and $Var(X_1|X_2)$.

8.47. Let $f(x,y) = k$, $4 \leq x \leq 10$, $0 \leq y \leq 5$ and zero elsewhere. Find k. Also find $P(X \leq 8, 3 \leq Y \leq 4)$ and $P(9 \leq X \leq 13, Y \leq 1)$.

8.48. The joint density function of X and Y is given by

$$f(x,y) = \frac{6}{7}(x^2 + xy/2), \quad 0<x<1, \ 0<y<2$$

    (i) Verify that this is indeed a joint density function,

   (ii) Compute the density function of X,

  (iii) Find $P(X > Y)$,

   (iv) Find mean and variance of Y,

    (v) Find Covariance between X and Y, and

   (vi) Find Correlation between X and Y.

8.49. What are the mean thickness and the standard deviation of transformer cores each consisting of 50 layers of sheet metal and 49 insulating paper layers if the metal sheets have mean thickness 0.5 mm each with a standard deviation of 0.05 mm and the paper layers have mean 0.05 mm each with a standard deviation of 0.02 mm, if covariance between the metal and paper is −0.001 mm?

8.50. Let X and Y be the diameters of a pin and a hole, respectively. Suppose (X, Y) has joint density: $f(x,y) = 625$, $0.98<x<1.02$, $1.00<y<1.04$ and 0 otherwise. Find

    (i) The marginal distributions,

   (ii) The probability that a pin chosen at random will fit a hole whose diameter is 1.00, and

  (iii) Are the variables independent?

8.51. A university found that 20 % of its students drop out without completing the introductory statistics course. Assume that 20 students have registered for the course this quarter.

   (i)   What is the probability that two or fewer will dropout?
   (ii) What is the probability that exactly four will dropout?
   (iii) What is the probability that more than three will dropout?
   (iv) What is the expected number of withdrawals?

8.52. The phone lines to an airline reservation system are occupied 50 % of the time. Assume that the events that the lines are occupied on successive calls are independent. If 10 calls are placed to the airline, what is the probability that

   (i)   For exactly three calls the lines are occupied?
   (ii) For at least one call the lines are not occupied?
   (iii) What is the expected number of calls in which the lines are all occupied?

8.53. With the usual notations, find p for a binomial random variable x, if $n = 6$ and if $9 \, P(X = 1) = P(X = 2)$.

8.54. The mean and variance of a binomial distribution are 3 and 2, respectively. Find the probability that the variate takes values.

   (i)   Less than or equal to 2 and
   (ii) Greater than or equal to 7.

8.55. The number of monthly breakdowns of computer is a random variable having Poisson distribution with mean equal to 1.8. Find the probability that this computer will function for a month

   (i)   Without a breakdown,
   (ii) With only one breakdown, and
   (iii) With at least one breakdown.

8.56. If the numbers of telephone calls coming into a telephone exchange between 9.00 a.m. and 10.00 a.m. and between 10.00 a.m. and 11.00 a.m. are independent and follow Poisson distributions with parameter 2 and 6, respectively, what is the probability that more than 5 calls come between 9 a.m. and 11 a. m.?

8.57. A contractor purchases a shipment of 100 transistors. It is his policy to test 10 of these transistors and to keep the shipment only if at least 9 of the 10 are in working condition. If the shipment contains 20 defective transistors, what is the probability that it will be kept?

8.58. Consider a lot consisting of 1000 bulbs is inspected by taking at random 10 bulbs and testing them for defective or non-defective. If the proportion of defective bulbs in a lot containing 1000 bulbs is 0.06, compute the probabilities stated in Example 8.2. State the approximation used in computing these probabilities.

8.59. If an auditor selects 5 returns from among 15 returns of which 9 contain illegitimate deductions, what is the probability that a majority of the selected returns contains illegitimate deductions?

8.60. $X \sim N (50, 10)$, determine the following probabilities

(i) $P(X<40)$ (ii) $P(40<X<65)$ (iii) $P(X > 55)$ (iv) $P(38 \leq X<62)$

8.61. Suppose $X \sim N (0, 1)$, find

(i) $P(X > 0.14)$ (ii) $P(X<2.12)$ (iii) $P(0.9<X<2)$ (iv) $P(-1.43<X<1.43)$

8.62. Assume the mean height of the student as 35.88 inches with variance 5.2 sq. inches. How many students in a class of 100 would you expect above 35 inches and between 30 to 37 inches tall?

8.63. Let $X \sim N (25, 10)$, find the values X corresponding to the following probabilities:

(i) $P(X > x) = 0.1251$ (ii) $P(X > x) = 0.9382$ (iii) $P(X<x) = 0.3859$
(iv) $P(X<x) = 0.8340$

8.64. In a normal distribution, 55 % of the observation are under 55 and 25 % are above 65. Find the mean and standard deviation.

8.65. The diameter of the dot produced by a printer is normally distributed with a mean diameter of 0.005 cm and a standard deviation of 0.001 cm.

(i)   What is the probability that the diameter of a dot exceeds 0.0065 cm?
(ii) What is the probability that a diameter is between 0.0035 and 0.0065 cm?
(iii) What standard deviation of diameters is needed so that the probability in part (ii) is 0.995?

8.66. In an entrance test for admission to MBA program, the number of candidates passing the test is 46 % and that of those getting distinction is 9 %. Estimate the average marks obtained by the candidate if the minimum marks for passing and obtaining distinction being 40 and 75, respectively.

8.67. The width of a slot on forging is normally distributed with mean 0.9 inch and standard deviation 0.004 inch. The specifications are 0.900 ± 0.005 inches. What percentage of forgings will be defective?

8.68. The mean of a normal distribution is 50 and 5 % of the values are greater than 60. Find the standard deviation?

8.69. In a normal population with mean 15.00 and S.D. 3.5, it is known that 647 observations exceed 16.25. What is the total number of observations in the population?

8.70. The time until recharge for a battery in a laptop computer under common conditions is normally distributed with a mean of 260 min and a standard deviation of 50 min.

   (i)   What is the probability that a battery lasts more than 4 h?
  (ii)   What are the quantiles (the 25 and 75 % values) of battery life? and
 (iii)   What value of life in minutes is exceeds with % probability?

8.71. The length of time (in minutes) that a certain lady speaks on the telephone is a random variable specified by pdf as

$$f(x) = \begin{cases} k\,e^{-0.2x}, & x \geq 0 \\ 0, & \text{Otherwise} \end{cases}$$

   (i) Evaluate A.
  (ii) What is the probability that the number of minutes she takes on phone is (a) more than 10, (b) less than 5, and (c) between 5 and 10?

8.72. A certain kind of appliance requires repairs on the average once every two years. Assuming that the times between repairs are exponentially distribution, what is the probability that such an appliance will work at least three years without requiring repairs?

8.73. If X ~ EXP(5), obtain (a) reliability functions at 2, 5, and 10 and (b) hazard functions at 2, 5, and 10. Also plot the reliability function and hazard function.

8.74. The life time of a mechanical assembly in a vibration test is exponentially distributed with mean of 400 h.

   (i)   What is the probability that an assembly on test fails in less than 100 h?
  (ii)   What is the probability that an assembly operates for more than 500 h before failure?
 (iii)   If an assembly has been on test for 400 h without a failure, what is the probability of a failure in the next 100 h?

8.75. The life of automobile voltage regulators has an exponential distribution with a mean life of 6 years. You purchase an automobile that is six years old, with a working voltage regulator, and plan to own it for six years. What is the probability that the voltage regulator

   (i) fails during your ownership? and
  (ii) will exceed its life for 8 years.

8.76. Suppose that it is known that the number of items produced in a factory during a week is a random variable with mean 50.

   (i) What can be said about the probability that this week's production will exceed 75 items?

(ii) If the variance of a week's production is known to equal 25, then what can said about the probability that this week's production will be between 40 and 60?

8.77. From past experience, a professor knows that the test score of a student taking his final examination is a random variable with mean 75.

(i) Give an upper bound to the probability that a student's test score will exceed 85.
(ii) Probability that a student's test score will be between 65 and 85, if the variance of the score is 36.
(iii) Find a such that $P\{|X - 75| < a\} \geq 0.90$.

8.78. Data from Indian Meteorological Department (IMD) indicate that the yearly precipitation in New Delhi is normal with mean 12.08 inches and a standard deviation of 3.1 inches.

(i) Find the probability that the total precipitation during the next 2 years will exceed 25 inches.
(ii) Find the probability that the next year's precipitation will exceed that of the following year by more than 3 inches, assuming the precipitation totals for the next 2 years are independent.

8.79. The following are the center thickness measurements for 100 randomly chosen spherical mirrors (mm). It is assumed that the target center thickness is 6.8 mm and specification limits are ±0.2 mm around the target.

6.904	6.732	6.610	6.808	6.795	6.802	6.826	6.787	6.665	6.765
6.844	6.706	6.799	6.714	6.767	6.827	6.962	6.702	6.829	6.769
6.767	6.833	6.978	6.713	6.827	6.644	6.853	6.635	6.917	6.761
6.830	6.764	6.832	6.778	6.767	6.711	6.682	6.726	6.891	6.769
6.784	6.922	6.671	6.762	6.977	6.751	6.617	6.822	6.843	6.758
6.785	6.814	6.764	6.773	6.797	6.908	6.715	6.831	6.942	6.781
6.740	6.884	6.820	6.855	6.848	6.797	6.835	6.798	6.832	6.730
6.914	6.829	6.782	6.740	6.999	6.859	6.797	6.802	6.811	6.869
6.910	6.792	6.807	6.870	6.730	6.843	6.773	6.817	6.903	6.767
6.768	6.754	6.668	6.871	6.804	6.776	6.716	6.898	6.858	6.672

(i) Perform normality checks on the data.
(ii) Using the sample standard deviation s as an estimate for σ, estimate C_p, C_{pk}, and P_p, P_{pk} and interpret the results.
(iii) Find the short-term sigma level of the process

8.80. A manufacturer of a wide variety of seals, gaskets, and O-rings used in a number of industries, such as automotive, chemical processing, oil refining, medical, and aerospace. The O-rings are classified by two important characteristics: cross-sectional width and inside diameter. The particular O-rings is specified to have a cross-sectional width of 0.275 inches with an inside diameter of 4.725 inches. The lower and upper specifications for the inside diameter are 4.692 and 4.758 inches. The company has established strict quality control procedures for all its O-ring processes. The following table presents the inside diameter measurements of O-ring for 30 subgroups of size 5.

Subgroups	X_1	X_2	X_3	X_4	X_5
1	4.7227	4.7274	4.7189	4.7201	4.7201
2	4.7204	4.7133	4.7111	4.7335	4.7279
3	4.7129	4.7219	4.7222	4.7384	4.7235
4	4.7275	4.7302	4.7430	4.7183	4.7221
5	4.7346	4.7314	4.7355	4.7237	4.7293
6	4.7228	4.7197	4.7383	4.7188	4.7114
7	4.7235	4.7289	4.7265	4.7290	4.7214
8	4.7253	4.7197	4.7313	4.7312	4.7197
9	4.7293	4.7449	4.7285	4.7369	4.7324
10	4.7268	4.7306	4.7381	4.7279	4.7259
11	4.7213	4.7340	4.7077	4.7353	4.7381
12	4.7224	4.7356	4.7304	4.7166	4.7213
13	4.7357	4.7241	4.7267	4.7288	4.7363
14	4.7229	4.7160	4.7274	4.7295	4.7221
15	4.7145	4.7231	4.7349	4.7288	4.7253
16	4.7311	4.7262	4.7278	4.7322	4.7304
17	4.7096	4.7229	4.7329	4.7201	4.7299
18	4.7231	4.7262	4.7296	4.7289	4.7167
19	4.7386	4.7366	4.7288	4.7321	4.7183
20	4.7251	4.7400	4.7191	4.7162	4.7288
21	4.7297	4.7125	4.7137	4.7262	4.7206
22	4.7223	4.7120	4.7294	4.7372	4.7200
23	4.7242	4.7282	4.7362	4.7401	4.7343
24	4.7233	4.7259	4.7243	4.7275	4.7360
25	4.7236	4.7243	4.7296	4.7225	4.7368
26	4.7319	4.7255	4.7251	4.7283	4.7251
27	4.7100	4.7234	4.7156	4.7233	4.7370
28	4.7275	4.7161	4.7281	4.7256	4.7263
29	4.7212	4.7198	4.7334	4.7225	4.7241
30	4.7263	4.7379	4.7124	4.7308	4.7200

(i)   Check whether the characteristic under study is according to normal,
(ii) Carry out a process capability study and interpret the results, and
(iii) Find the short-term sigma level of the process

References

1. Abramowitx, M., Stegun, I.A.: Handbook of Mathematical functions. Applied Mathematics Series 55. National Bureau of Standards, Washington (1968)
2. Bovee, C.L., Thill, J.V.: Marketing. McGraw-Hill, New York (1992)
3. Casella, G., Berger, R.L.: Statistical Inference. Thompson Learning, Duxburry (2002)
4. Finch, B.J.: A new way to listen to the customer. Qual. Prog. **30**, 73–76 (1997)
5. Gryna, F.M., Chua, R.C.H., Defeo, J.A.: Juran's Quality Planning & Analysis for Enterprise Quality, 5th edn. Tata McGraw-Hill, New Delhi (2007)
6. Hutson, F.G.: Cause, effect, efficiency & soft systems models. J. Oper. Res. Soc. **44**(4), 333–344 (1992)
7. Ishikawa, K: Guide to Quality Control. Asian Productivity Organization (1976). ISBN 92-833-1036-5
8. Kubiak, T.M., Benbow, D.W.: The Certified Six Sigma Black Belt Handbook, 2nd edn. Dorling Kindersley (India) Pvt. Ltd., New Delhi (2010)
9. Meeker, W.Q., Escobar, L.A.: Statistical methods for reliability data. Wiley, New York (1998)
10. Montgomery, D.C.: Introduction to Statistical Quality Control. Wiley, India (2003)
11. Montgomery, D.C., Runger, G.C.: Applied Statistics and Probability for Engineers. Wiley, New Delhi (2011)
12. Muir, A.: Lean Six Sigma way. McGraw-Hill, New York (2006)
13. Muralidharan, K.: On Sample size calculation. J. Math. Interdisc. Sci. **3**(1) (2014)
14. Muralidharan, K., Syamsundar, A.: Statistical Methods for Quality, Reliability and Maintainability. PHI, New Delhi (2012)
15. Murthy, D.N.P., Xie, M., Jiang, R.: Weibull Models. Wiley Inter Science, New Jersey (2004)
16. Oakland, J.S.: Statistical Process Control. Elsevier, New Delhi (2005)
17. Pande, P.S., Newuman, R.P., Cavanagh, R.R.: The Six Sigma way. Tata McGraw-Hill, New Delhi (2003)
18. Rinne, H.: The Weibull Distribution: A Handbook. Taylor and Francis Group, LLC, New York (2009)
19. Ross, S.M.: Introduction to Probability and Statistics for Engineers and Scientists. Academic Press, Elsevier, Amsterdam (2009)

Chapter 9
Analyze Phase

Analyze phase tasks and deliverables

- Validate gaps in requirements versus current metrics with vital causes
- Quantify opportunity to close gaps
- Prioritize root causes and identify the most contributing one
- Establish a proper process model in terms of $y = f(x)$
- Suggest possible relationships and associations between control variables
- Identify the most contributing KPIVs and KPOVs through statistical analysis
- Test the significance of various KPIVs on KPOVs
- Conduct a revised root cause analysis to identify the vital causes

9.1 Process Mapping

In every Six Sigma improvement project, understanding the process is essential. The process map and flowchart is, therefore, very common in the measure phase. It is also used in the analyze phase for identifying improvement potential compared to similar processes and in the control phase to institutionalize the changes made to the process. Flowcharts can vary tremendously in terms of complexity, ranging from the simplest to very advanced charts. When improving variation, a simple flowchart is often applied in the measure phase to map the Xs (input variables) and Y (output variable) of the process or product to be improved.

For quality systems, it is advantageous to represent system structure and relationships using flowcharts. A flowchart provides a picture of the steps that are needed to understand a process. Flowcharts are widely used in industry and have become a key tool in the development of information systems, quality management systems, and employee handbooks. The main value of the flowchart resides in the identification and mapping of activities in processes, so that the main flows of products and information are visualized and made known to everyone. Figure 9.1 depicts such a process where information flow is established through supplier and

© Springer India 2015
K. Muralidharan, *Six Sigma for Organizational Excellence*,
DOI 10.1007/978-81-322-2325-2_9

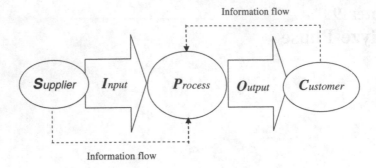

Fig. 9.1 Information flow through a process

customer; whereas the natural flow of information is happening through S–I–P–O–C in a phased manner.

The information flow helps one to understand the process flow through SIPOC, to tackle cycle time problems, and to identify opportunities to reduce process costs. An alternative (or supplement) to a detailed process flowchart is a high-level *process map* that shows only a few major process steps as activity symbols. For each of these symbols, key process input variables (KPIVs) to the activity are listed on one side of the symbol, while key process output variables (KPOVs) to the activity are listed on the other side of the symbol. Note that a KPIV can be a CTQ_X and a KPOV can be a CTQ_Y. The best way of identifying causes and effects of a problem is through the principle of brainstorming.

9.2 Brainstorming

Brainstorming is a simple but effective technique for generating many ideas among a group of people within a short span of time to solve a given problem [30, 34]. It also helps to get to the root causes of a problem through the joint efforts of the people associated with problem solving. Brainstorming is necessary:

- To find a solution to a specific problem
- To suggest counter measures when the target has not been achieved
- To select a theme/topic
- To find causes of a specific problem.

The *principles of brainstorming* involve the following procedures:

- The uniqueness of each participant's knowledge is tapped to develop new insights
- Deferment of evaluation develops the appropriate psychologically safe climate for ideation
- Ideas of one participant tend to trigger off ideas in the brains of other group members

- Free association encourages fruitful ideation
- The pressure of time-bound sessions in a non-threatening atmosphere is conducive to a high productivity of ideas.

The *steps to brainstorming* are

- Choose the topic
- Each member, in rotation, is asked for ideas
- Each member offers only one idea per turn, regardless of how many he/she has in mind
- This continues until all the ideas have been exhausted
- Ideas are to be recorded and displayed on transparent sheets.

An *effective brainstorming*:

- Give everyone a chance to speak
- Do not criticize ideas
- Record all ideas
- Expand ideas and use them to come up with more ideas
- Allow free expression in any language that the person is most comfortable with in giving the idea.

Some *key points* that need special attention are as follows:

- Acquaint participants with the technique of brainstorming
- Obtain the commitment of the management to encourage the development and implementation of worthwhile ideas
- Set up appropriate criteria to evaluate the ideas considered worthwhile
- Screen all generated ideas
- Keep participants informed of the final choice of the ideas and the actions initiated.

The *benefits of brainstorming* can be summarized as

- Individual can generate limited ideas, and the group produces more ideas
- Ideas are improved upon by members
- The presence of others increases creativity
- Pooling of ideas and resources is made possible by coming together as a group

Thus, after a successful brainstorming, we are now left with plenty of ideas and inputs, which will help us to identify the root causes of the problem. For organizing potential causes, we can use cause-and-effect diagram (see Sect. 8.9.3), FMEA, tree diagram, or an affinity diagram. Next, we will discuss the FMEA techniques followed by various statistical tools for analyzing the same. At the end of this chapter, we will discuss the root cause analysis and its practical significance in Six Sigma projects.

9.3 Failure Modes and Effects Analysis

Failure modes and effects analysis (FMEA) is a set of guidelines, a process, and a form of identifying and prioritizing potential failures and problems in order to facilitate process improvement [1, 2, 21, 23]. These guidelines help a manager, or an improvement team, or a process owner to focus on the energy and resources of prevention and to monitor the activities of a process where they are most likely to pay off. The FMEA method has many applications in a Six Sigma environment in terms of looking for problems not only in work processes and improvements but also in data collection activities, VOC efforts, and procedures.

There are two types of FMEA: one is design FMEA and the other is process FMEA. Design FMEA applications mainly include component, sub-system, and main system. Process FMEA applications include assembly machines, workstations, gauges, procurement, training of operators, and tests (Table 9.1).

Table 9.1 An FMEA chart

The major benefits derived from a properly implemented FMEA are as follows:

1. It provides a documented method for selecting a design with a high probability of successful operation and safety.
2. A documented uniform method of assessing potential failure mechanisms, failure modes, and their impact on system operation, resulting in a list of failure modes ranked according to the seriousness of their system impact and likelihood of occurrence.
3. Early identification of single failure points (SFPs) and system interface problems, which may be critical to a mission's success and/or safety. They also provide a method of verifying that switching between redundant elements is not jeopardized by postulated single failures.
4. An effective method for evaluating the effect of proposed changes in the design and/or operational procedures on the mission's success and safety.
5. A basis for in-flight troubleshooting procedures and for locating performance monitoring and fault-detection devices.
6. A criteria for early planning of tests.

The FMEA should be updated whenever:

- A new cycle begins (new product/process)
- Changes are made to the operating conditions
- A change is made in the design
- New regulations are instituted
- Customer feedback indicates a problem (Table 9.2).

Table 9.2 Rating definitions

Rating	Severity	Occurrence	Detection
High 10	Hazardous without warning	Very high and almost inevitable	Cannot detect or detection with very low probability
	Loss of primary function	High repeated failures	Remote or low chance of detection
	Loss of secondary function	Moderate failures	Low detection probability
	Minor defect	Occasional failures	Moderate detection probability
Low 1	No effect	Failure unlikely	Almost certain detection

9.3.1 Design FMEA

Within a design FMEA, manufacturing and/or process engineering input is important to ensure that the process will produce according to design specifications. A team should consider including knowledgeable representation from design, test, reliability, materials, service, and manufacturing/process organizations. When beginning a design FMEA, the responsible design engineer compiles documents that provide insight into the design intent. Design intent is expressed as a list of what the design is expected to do.

The design FMEA has the following components as described below:

- *Header information*: Documents the system/sub-system/component and supplies other information about when the FMEA was created and by whom.
- *Item/function*: Contains the name and number of the analyzed item. Includes a concise, exact, and easy-to-understand explanation of the function of the item task.
- *Potential failure mode*: Describes the ways in which a design could fail to perform its intended function.
- *Potential effect of failure*: Contains the effects of the failure mode on the function from an internal or external customer point of view.
- *Severity*: Assesses the seriousness of the effect of the potential failure mode to the next component, sub-system, or system, if it should occur. Estimation is typically based on a 1–10 scale where 10 is the most serious, 5 is low, and 0 is no effect.
- *Classification*: Includes optional information such as critical characteristics that may require additional process controls.
- *Potential cause of failure*: Indicates a design weakness that causes the potential failure mode.
- *Occurrence*: Estimates the likelihood that a specific cause will occur. Estimation is usually based on a 1–10 scale where 10 is very high (failure is almost inevitable), 5 is low, and 1 is remote (failure is unlikely).
- *Current design controls*: Lists activities such as design verification tests, design reviews, DOEs, and tolerance analysis that ensure occurrence criteria.
- *Detection*: Assessment of the ability of the current design control to detect the subsequent failure mode. Assessment is based on a 1–10 scale where 10 is absolute uncertainty (there is no control), 5 is moderate (moderate chance that the design control will detect a potential cause), 1 is almost certain (design control will almost certainly detect a potential cause).
- *Risk priority number (RPN)*: Product of severity, occurrence, and detection rankings. The ranking of RPN prioritizes design concerns.
- *Recommended action*: Intent of this entry is to institute actions.
- *Responsibility for recommended action*: Documents the organization and individual responsibility for recommended action.
- *Actions taken*: Describes implementation action and effective date.

- *Resulting RPN*: Contains the re-calculated RPN resulting from corrective actions that affected previous severity, occurrence, and detection rankings. Blanks indicate no action.

9.3.2 Process FMEA

For a process FMEA, design engineering input is important to ensure appropriate focus on the important design needs. A team should consider including knowledgeable representation from design, manufacturing/process, quality, reliability, tooling, and operators.

The advantages of process FMEA are that they

- Improve the quality, reliability, and safety of a product/process
- Improve company image and competitiveness
- Increase user satisfaction
- Reduce system development time and cost
- Collect information to reduce future failures and capture engineering knowledge
- Reduce the potential for warranty concerns
- Early identification and elimination of potential failure modes
- Emphasize problem prevention
- Minimize late changes and associated cost
- Act as catalyst for teamwork and idea exchange between functions
- Reduce the possibility of same the kind of failure in future
- Reduce impact on the company profit margin
- Improve production yield.

Thus, an FMEA provides a structured way to identify all possible modes of failure in a process. That is, it first identifies all the ways in which the process can fail to do what it's supposed to do. Then, the effect analysis determines the ramifications of each failure on other activities and results. FMEA also ranks and prioritizes the causes and effects of failures, and it can be applied to a product or service as well as to a process.

Having identified the causes and effects, the next step is to identify those variables that are really critical for ensuring the quality of the process. For this, we carry out a detailed analysis through statistical methods. From next section onward, we will study some of the important statistical tools and their implementation procedures, to identify the CTQ_Xs and CTQ_Ys identified through the FMEA techniques. This chapter ends with the discussion of root cause analysis, where we suggest the recommendations for each analysis tool for further investigations and critical analysis. Chapter 10 discusses some other advanced statistical tools for identifying and improving the CTQ_Xs, which will create a platform for controlling and sustaining the improvement process.

9.4 Histogram and Normality

Histogram summarizes data from a process and graphically presents the frequency distribution in bar form. It helps to answer the questions: Is the process capable of meeting customer requirements? and Do processed CTQ_X's best describes CTQ_Y's? A *histogram* consists of a set of rectangles having bases on a horizontal axis (*x*-axis), with centers at the class marks and lengths equal to the class interval sizes and areas proportional to the class frequencies (*y*-axis). If class intervals all have an equal size, the heights of the rectangles are proportional to the class frequencies and it is then customary to take the heights numerically equal to the class frequencies. If the class intervals do not have an equal size, these heights must be adjusted.

A *frequency polygon* is a line graph of the class frequency plotted against the *class mark* (midpoint of the class interval). It can be obtained by connecting the midpoints of the tops of the rectangles in the histogram. If the midpoints of the rectangles in a histogram are connected using a free hand draw, then we get a frequency curve. A histogram with normal curve is presented in Fig. 9.2.

9.4.1 Probability Plotting

Probability plots are an important tool for analyzing process data and are particularly used to

- Assess the adequacy of a particular distribution model (*Goodness-of-fit test*)
- Provide graphical estimates of probabilities and distribution quantiles
- Display the results of a parametric maximum likelihood fit along with the data.

The idea of probability plotting is like this: if x_i is the *i*th order statistic of a sample from a distribution $F(x)$, then the points $\left(F^{-1}\left(\frac{i}{n}\right), x_{(i)}\right), i = 1, 2, \ldots, n$ will be scattered around a straight line with slope 1 passing through the origin. In general, if the true CDF is $F\left(\frac{x-\mu}{\sigma}\right)$ having its *p*th fractile as $x_p = \mu + \sigma F^{-1}(p)$, then the points

Fig. 9.2 Histogram with normal curve

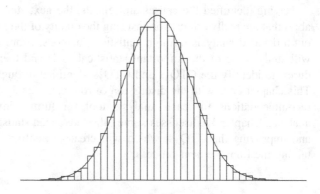

$\left(F^{-1}\left(\frac{i}{n}\right), x_{(i)}\right), i = 1, 2, \ldots, n$ will be scattered around a straight line with intercept μ and slope σ.

Since the expected value of $F\left(x_{(i)}\right)$ is equal to $i/(n+1)$, the plotting position is placed at $p_i = i/(n+1)$ corresponding to $x_{(i)}$. The other plotting position may be $p_i = (i - 0.5)/n$. For details, one may see [5, 6, 24]. The estimates of intercept and slope further help to estimate the parameters involved in a probability model graphically. In the absence of any other estimators, one can use these for modeling and inferences. The graphical procedure also helps to assess the suitability of a model and hence is used as a goodness-of-fit procedure. Specifically, the normal probability plot is very important for capability studies and performance studies in quality control, which is discussed below.

9.4.2 Normal Probability Plot

The most frequently used probability plot is the normal plot. As mentioned earlier, this plot helps to assess analytically the reasonableness of particular distributional model. The quantile function for normal distribution is

$$x_p = \mu + \sigma\Phi^{-1}(p), \tag{9.1}$$

where $\Phi^{-1}(p)$ is the pth quantile of the standard normal distribution. Then, the graph of $x_{(i)}$ versus $\Phi^{-1}\left(\frac{i}{n}\right), i = 1, 2, \ldots, n-1$ will give a straight-line plot. Since $\Phi^{-1}\left(\frac{n}{n}\right) = \infty$, it is better to consider the coordinate $\Phi^{-1}\left(\frac{i}{n+1}\right)$ or $\Phi^{-1}\left(\frac{i-3/8}{n+1/4}\right)$ (see [41]). The slope of the line on the observation versus quantile scale is $(1/\sigma)$. Any normal CDF plots as a straight line with a positive slope.

Example 9.1 From a manufacturing process, a sample of 120 piston rings is taken and their diameters are observed as given in the data below:

Ring diameter (in mm)	No. of piston rings
73.980–73.985	6
73.985–73.990	10
73.990–73.995	19
73.995–74.000	23
74.000–74.005	22
74.005–74.010	22
74.010–74.015	13
74.015–74.020	5

Prepare a normal probability plot.

Solution To check for normality, we construct the cumulative frequency table as follows:

Ring diameter $(x_{(i)})$	No. of piston rings	Cumulative frequency (CF)	CF/(N + 1) * 100
73.9825	6	6	4.958678
73.9875	10	16	13.22314
73.9925	19	35	28.92562
73.9975	23	58	47.93388
74.0025	22	80	66.1157
74.0075	22	102	84.29752
74.0125	13	115	95.04132
74.0175	5	120	99.17355

Then, the graph of $x_{(i)}$ versus $\Phi^{-1}\left(\frac{i}{n}\right) = \frac{CF}{N+1} \times 100$ is shown in Fig. 9.3.

Conclusion. Since all the points falls on a straight line, it is possible to conclude that the ring diameter follows normal.

A normal probability plot can also be developed by creating the corresponding normal scores as follows:

- Order the data from smallest to largest
- Find mean and standard deviation of the data
- Compute $Z_i = \frac{x_i - \bar{x}}{\sigma}, i = 1, 2, \ldots, n$
- Plot the points and assess the linearity

Example 9.2 An important quality characteristic in a bronze casting is the percentage of copper in the casting. Suppose that copper determinations are made on samples of three castings as shown in Table 9.3.

Fig. 9.3 Normal probability plot for piston rings

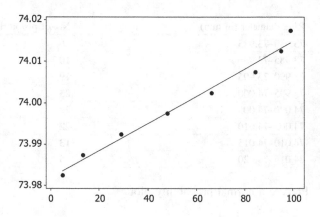

Table 9.3 Data on copper determinations	Sup group	x_1	x_2	x_3
	1	87.15	86.97	86.69
	2	87.07	87.20	86.95
	3	86.58	87.03	86.54
	4	87.62	86.89	87.23
	5	86.93	87.34	86.81
	6	86.58	86.75	86.55
	7	87.43	87.06	86.78
	8	87.54	87.29	87.45
	9	87.16	86.71	86.87
	10	87.25	87.38	87.23
	11	87.31	86.52	86.27
	12	86.60	86.62	87.49
	13	87.22	87.51	87.31
	14	86.85	87.99	87.28
	15	87.29	86.35	87.13
	16	86.85	86.90	87.49
	17	87.23	86.81	86.95
	18	86.90	86.01	86.52
	19	87.18	86.96	86.97
	20	87.07	87.06	86.99
	21	87.13	87.88	86.56
	22	86.77	87.34	86.90
	23	86.68	87.74	87.30
	24	87.52	87.72	86.94
	25	86.96	87.45	87.05

Perform normality check for the data.

Solution The probability plots in MINITAB are shown in Fig. 9.4. Note that the p-value in each case is larger than 0.05. Hence, the assumption of normality is accepted in all the three cases.

9.5 Parameter Estimation

The process of drawing conclusions about the nature of some system on the basis of data subject to random variation is called *statistical inference*. According to Cox [10], Freedman [16], Casella and Berger [9], and Young and Smith [40], a statistical inference begins with a search of a suitable estimate of the parameter of interest. Estimates can be viewed as the descriptive summary of the data set. These estimates are used to make inferences about the underlying probability model.

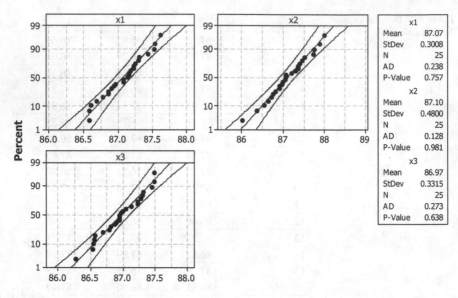

Fig. 9.4 Normal probability plot for copper determinants

There are two approaches to estimation and several methods for each of them. They are *point estimation* and *interval estimation*. In point estimation, a numerical value for the parameter θ is calculated. In interval estimation, a region is determined in such a way that this region covers the true vector θ with a specified and predetermined probability $1 - \alpha$, called *level of confidence*. This region is an interval, hence the name interval estimation.

9.5.1 Point Estimation

Any statistic used to estimate the value of an unknown parameter θ is called an *estimator*. The observed value of the estimator is called the *estimate*. Let $X_1, X_2, \ldots, X_n$ be the sample variates and x_i be the realization of X_i in a given sample. Then, $T_n = T(X_1, X_2, \ldots, X_n)$ is called an estimator and is a random variable, and the numerical value obtained through the realization say $\hat{\theta} = t(x_1, x_2, \ldots, x_n)$ is called an estimate.

There are many methods available for estimation. One of the popular methods of estimation is the maximum likelihood method of estimation, as it satisfies a number of properties of an estimator.

Maximum likelihood estimation. Suppose that the random variables $X_1, X_2, \ldots, X_n$, whose joint distribution, $f(x_1, x_2, \ldots, x_n)$ is assumed known, except for an unknown parameter θ, are to be observed. The problem of interest is to use the observed values to estimate θ. Because θ is assumed unknown, we write $f(.)$ as a function of θ. Then, the maximum likelihood estimate (MLE) $\hat{\theta}$ is defined to be that value of θ maximizing $f(x_1, x_2, \ldots, x_n | \theta)$ where $x_1, x_2, \ldots, x_n$ are the observed values. The function $f(x_1, x_2, \ldots, x_n | \theta)$ is often referred to as the *likelihood function* of θ.

9.5.1.1 Maximum Likelihood Estimation in Exponential Population

Suppose $X_1, X_2, \ldots, X_n$ are independent and identically distributed exponential random variables, each having the same unknown mean θ. In this case, the joint density function is given by

$$f(x_1, x_2, \ldots, x_n) = f_{X_1}(x_1) f_{X_2}(x_2) \ldots f_{X_n}(x_n)$$

$$= \frac{1}{\theta} e^{-x_1/\theta} \frac{1}{\theta} e^{-x_2/\theta} \ldots \frac{1}{\theta} e^{-x_n/\theta} \qquad (9.2)$$

$$= \frac{1}{\theta^n} \exp\left\{ -\sum_{i=1}^{n} x_i / \theta \right\}$$

For convenience of maximization, we consider the log-likelihood function as

$$\log L(\theta | \text{data}) = -n \log \theta - \frac{1}{\theta} \sum_{i=1}^{n} x_i$$

Then, the ML estimate of θ is obtained by solving $\frac{\partial \log L}{\partial \theta} = 0$. And the estimator is given by

$$\hat{\theta} = \frac{1}{n} \sum_{i=1}^{n} x_i \qquad (9.3)$$

Note that $\frac{\partial^2 \log L}{\partial \theta^2} < 0$. Therefore, $\hat{\theta} = \frac{1}{n} \sum_{i=1}^{n} x_i$ is the MLE of θ.

9.5.1.2 Maximum Likelihood Estimation in Normal Population

Suppose $X_1, X_2, \ldots, X_n$ are independent and identically distributed normal random variables, each having the same unknown mean μ and unknown standard deviation σ. The joint density function is given by

$$f(x_1, x_2, \ldots, x_n | \mu, \sigma) = f_{X_1}(x_1) f_{X_2}(x_2) \ldots f_{X_n}(x_n)$$

$$= \left(\frac{1}{2\pi} \right)^{n/2} \frac{1}{\sigma^n} \exp\left\{ -\frac{1}{2\sigma^2} \sum_{i=1}^{n} (x_i - \mu)^2 \right\} \qquad (9.4)$$

The log-likelihood function is

$$\log f(x_1, x_2, \ldots, x_n | \mu, \sigma) = -\frac{n}{2} \log(2\pi) - n \log \sigma - \frac{1}{2\sigma^2} \sum_{i=1}^{n} (x_i - \mu)^2$$

Then, the ML estimate of μ and σ are the solution of the likelihood equations:

$$\frac{\partial \log f(x_1, x_2, \ldots, x_n | \mu, \sigma)}{\partial \mu} = 0 \Rightarrow \frac{1}{\sigma^2} \sum_{i=1}^{n} (x_i - \mu) = 0$$

and

$$\frac{\partial \log f(x_1, x_2, \ldots, x_n | \mu, \sigma)}{\partial \sigma} = 0 \Rightarrow -\frac{n}{\sigma} + \frac{1}{\sigma^3} \sum_{i=1}^{n} (x_i - \mu)^2 = 0$$

Solving the above equations, we get the estimates of μ and σ

$$\hat{\mu} = \frac{1}{n} \sum_{i=1}^{n} x_i = \bar{x} \tag{9.5}$$

and

$$\hat{\sigma} = \left[\frac{1}{n} \sum_{i=1}^{n} (x_i - \hat{\mu})^2 \right]^{1/2}$$
$$= \left[\frac{1}{n} \sum_{i=1}^{n} (x_i - \bar{x})^2 \right]^{1/2} \tag{9.6}$$

It should be noted that the MLE of the standard deviation σ differs from the sample standard deviation

$$s = \left[\frac{1}{n-1} \sum_{i=1}^{n} (x_i - \bar{x})^2 \right]^{1/2}$$

However, for large n, these two estimators of σ are approximately equal.

Example 9.3 Consider the following observations from normal distributions. Obtain the ML estimates of the parameters.

42.5770	37.8522	46.2559	52.3097	47.0930
52.5798	46.2865	53.9004	42.9649	41.2775
63.8564	42.0441	42.9401	55.105	54.6081
58.3437	47.3574	44.6553	53.2667	43.9971

Solution The estimates are obtained as

$$\hat{\mu} = \frac{1}{n} \sum_{i=1}^{n} x_i = 48.4635$$

and

$$\hat{\sigma} = \left[\frac{1}{n} \sum_{i=1}^{n} (x_i - \hat{\mu})^2 \right]^{1/2}$$

$$= \left[\frac{1}{n} \sum_{i=1}^{n} (x_i - 48.4635)^2 \right]^{1/2}$$

$$= 6.5146$$

9.5.2 Confidence Interval Estimation

A confidence interval is a range of plausible values for the population parameter. The confidence interval plays a very important role in parameter estimation. Instead of a point estimate, an interval estimate provides a range of values of the parameter of interest. The width of the confidence interval is sometimes very important to estimate the precision of the estimate. Narrow intervals are desirable, since they give a narrow range of values likely to contain the parameter of interest, meaning that our knowledge about the parameter value is precise. In Fig. 9.5, a typical graph of confidence interval is shown. The graph shows 100 simulated 95 % level confidence intervals for samples of size 5 from a population with mean $\mu = 3$. Intervals marked with bold end points did not capture $\mu = 3$.

A confidence interval depends on a specified confidence level. In order to suggest a confidence interval, we need to have a lower and upper limit for the estimate. The larger the interval, the greater is our confidence that the interval does

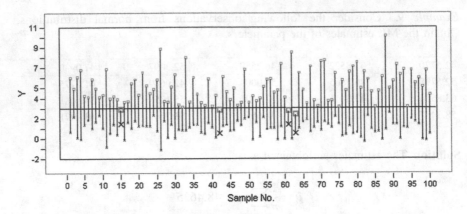

Fig. 9.5 95 % confidence interval (*Source* Six Sigma MBB notes, ISI Bangalore)

contain the true population mean. It is to be noted that the true population mean is a constant and is not a variable. Therefore, the interval that we specify will depend on the sample mean. For example, if the sample mean is 40 and the standard error of the mean is 5, we may specify our interval estimate as (35, 45). That is, the interval spans one standard error of the mean on either side of the sample mean. On the other hand, if we specify the interval estimate as (30, 50), i.e., spanning two standard error of the mean on either side of the sample mean, we are more confident that the latter interval contains the true population mean as compared to the former. However, if the confidence level is raised too high, the corresponding interval may become too wide to be of any practical use.

The confidence level, therefore, may be defined as the probability that the interval estimate will contain the true value of the population parameter that is being estimated. If we say that a 95 % confidence interval for the population mean is obtained by spanning 1.96 times the standard error of the mean of either side of the sample mean, we mean that we take a large number of samples of size n, say 1000 and obtain the interval estimates from each of these 1000 samples and then 95 % of these interval estimates would contain the true population mean. The 95 % confidence interval for population mean μ is given by $\bar{x} \pm 1.96 \frac{\sigma}{\sqrt{n}}$. Similarly, one can construct confidence interval for any given confidence level. To do so, recall that z_α is such that

$$P(|Z| > z_\alpha) = \alpha \tag{9.7}$$

where Z is a standard normal random variable. This implies that for any α

$$P(-z_{\alpha/2} < Z < z_{\alpha/2}) = 1 - \alpha \tag{9.8}$$

As a result, we see that

$$P\left(-z_{\alpha/2} < \sqrt{n}\frac{(\bar{x} - \mu)}{\sigma} < z_{\alpha/2}\right) = 1 - \alpha$$

or

$$P\left(-z_{\alpha/2}\frac{\sigma}{\sqrt{n}} < \bar{x} - \mu < \frac{\sigma}{\sqrt{n}}z_{\alpha/2}\right) = 1 - \alpha$$

or

$$P\left(\bar{x} - z_{\alpha/2}\frac{\sigma}{\sqrt{n}} < \mu < \bar{x} + \frac{\sigma}{\sqrt{n}}z_{\alpha/2}\right) = 1 - \alpha \tag{9.9}$$

The above probability is shown in the Fig. 9.6.

When the variance is unknown, the confidence interval for a normal mean is constructed using the t-distribution. In that case, the $100(1 - \alpha)$ % confidence interval for mean is

$$P(\bar{x} - t_{\alpha/2,n-1}\frac{s}{\sqrt{n}} < \mu < \bar{x} + \frac{s}{\sqrt{n}}t_{\alpha/2,n-1}) = 1 - \alpha \tag{9.10}$$

which is shown graphically in Fig. 9.7.

Confidence interval plays a very important role in managerial decision making. When point estimates are not sensible for statistical conclusions, it is better to use confidence interval, as it provides a range of values of a parameter of interest with a given confidence. Even for computing sample size for a particular sampling procedure (see Chap. 6 for details), we use confidential interval technique. In various tables, below, we discuss the confidence intervals associated with various statistical test procedures (Table 9.4).

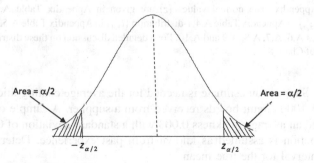

Fig. 9.6 $P(-z_{\alpha/2} < Z < z_{\alpha/2}) = 1 - \alpha$

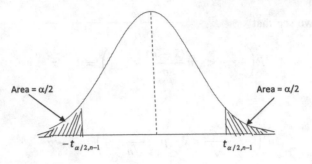

Fig. 9.7 $P(-t_{\alpha/2} < T_{n-1} < t_{\alpha/2,n-1}) = 1 - \alpha$

Table 9.4 $100(1 - \alpha)\%$ confidence intervals for mean

Test name	Assumption	Confidence interval
z-test for a population mean	σ^2 known	$\bar{x} \pm z_{\alpha/2} \frac{\sigma}{\sqrt{n}}$
z-test for two population means	$\sigma_1^2 = \sigma_2^2 = \sigma^2$	$(\bar{x}_1 - \bar{x}_2) \pm z_{\alpha/2}\sigma\sqrt{\frac{1}{n_1} + \frac{1}{n_2}}$
z-test for two population means	$\sigma_1^2 \neq \sigma_2^2$	$(\bar{x}_1 - \bar{x}_2) \pm z_{\alpha/2}\sqrt{\frac{\sigma_1^2}{n_1} + \frac{\sigma_2^2}{n_2}}$
t-test for a population mean	σ^2 unknown and small samples	$\bar{x} \pm t_{\alpha/2,n-1} \frac{s}{\sqrt{n}}, s = \sqrt{\frac{\sum (x_i - \bar{x})^2}{n-1}}$
t-test for two population means	Variances unknown and equal	$(\bar{x}_1 - \bar{x}_2) \pm t_{\alpha/2,n_1+n_2-2}s\sqrt{\frac{1}{n_1} + \frac{1}{n_2}}$, $s^2 = \frac{(n_1-1)s_1^2 + (n_2-1)s_2^2}{n_1+n_2-2}$
t-test for two population means	Variances unknown and unequal	$(\bar{x}_1 - \bar{x}_2) \pm t_{\alpha/2,v}\sqrt{\frac{s_1^2}{n_1} + \frac{s_2^2}{n_2}}$, $v = \frac{\left(\frac{s_1^2}{n_1} + \frac{s_2^2}{n_2}\right)^2}{\frac{(s_1^2/n_1)^2}{n_1-1} + \frac{(s_2^2/n_2)^2}{n_2-1}}$
t-test for two population means (paired t-test)	Observations are in pairs $(d_i = x_i - y_i, i = 1, 2, \ldots, n)$	$\bar{d} \pm t_{\alpha/2,n-1}\frac{s_d}{\sqrt{n}}$, $s_d = \sqrt{\frac{1}{n-1}\sum (d_i - \bar{d})^2}$

Note At various places, we will be using the table values of normal, t, chi-square, and F-distributions, as per their suitability and relevance. These values are very well tabulated and are presented in various tables in the Appendix. The normal values (z) are given in Appendix Table A.3, chi-square distribution $(\chi^2_{\alpha,n})$ in Appendix Table A.4, t-distribution $(t_{\alpha,n})$ in Appendix Table A.5, and $F_{\alpha(m,n)}$ in Appendix Tables A.6, A.7, A.8, A.9 and A.10. For a detailed discussion on these distributions, refer to Sect. 8.12.5 of Chap. 8

Example 9.4 Suppose an estimate is needed for the average coating thickness for a population of 1000 circuit boards received from a supplier. A sample of 36 circuit boards yielded an average thickness 0.003 with a standard deviation of 0.0005. The standard deviation is assumed as known from past experience. Determine 95 % confidence interval for the true mean.

Solution We have

$$n = 36, \bar{x} = 0.003, \sigma = 0.0005 \text{ and } \alpha = 0.05. \text{ For } \alpha = 0.05, z_{\alpha/2} = 1.96,$$

Thus, the 95 % confidence intervals for mean is

$$\bar{x} \pm z_{\alpha/2} \frac{\sigma}{\sqrt{n}} = (0.00284, 0.00316).$$

Example 9.5 Let X be the time in seconds between phone calls received at a call center company. Fifteen observations of X are

$$82, 42, 185, 66, 384, 27, 334, 545, 650, 127, 35, 45, 285, 133, 120$$

Assuming that X has an exponential distribution with mean θ obtain unbiased estimate of θ. Also obtain 95 % confidence interval for θ.

Solution Here, $\sum x_i = 3060$; therefore,

$$\hat{\theta} = 204$$

Then, the 95 % CI for θ is

$$\left(\frac{2 \times \sum x_i}{\chi^2_{\alpha/2,2n}}, \frac{2 \times \sum x_i}{\chi^2_{1-\alpha/2,2n}} \right) = \left(\frac{2 \times 3060}{46.9772}, \frac{2 \times 3060}{16.7908} \right)$$
$$= (130.27, 364.49)$$

Example 9.6 A dairy refuses to accept milk which gives, by a standard test, more than 5000 colonies of bacteria per milliliter. A shipment of 100 cans of milk is tested, using a random sample of ten 1-ml samples. The following data are obtained:

$$5370, 4890, 5100, 4500, 5260, 5150, 4900, 4760, 4700 \text{ and } 4870$$

Compute 95 % confidence interval for the average bacteria count.

Solution Here, variance is unknown; therefore, we use t-test for constructing confidence interval. The 95 % confidence interval for mean is $\bar{x} \pm t_{\alpha/2,n-1} \frac{s}{\sqrt{n}}$, where

$$\bar{x} = 4950 \quad \text{and} \quad s = \sqrt{\frac{\sum (x_i - \bar{x})^2}{n-1}} = 268.45$$

Therefore, the 95 % CI for μ is (4743.649, 5156.351) (Table 9.5).

Table 9.5 $100(1 - \alpha)\%$ confidence interval for proportion

Test name	Assumption	Confidence interval
z-test for a given proportion p_0	$n \geq 30$	$\hat{p} \pm z_{\alpha/2}\sqrt{\frac{\hat{p}(1-\hat{p})}{n}}, \hat{p} = \frac{X}{n}$
z-test for equality of two proportions	Different sample sizes $n \geq 30$	$(\hat{p}_1 - \hat{p}_2) \pm z_{\alpha/2}\sqrt{\frac{\hat{p}_1(1-\hat{p}_1)}{n_1} + \frac{\hat{p}_2(1-\hat{p}_2)}{n_2}},$ $\hat{p}_1 = \frac{X_1}{n_1}; \hat{p}_2 = \frac{X_2}{n_2}$

Example 9.7 In the Example 9.4 discussed above, assume that the sampling process yields four defectives. Determine 95 % confidence interval for the true population.

Solution We have, for $\alpha = 0.05$, $z_{\alpha/2} = 1.96$ and $\hat{p} = \frac{4}{36} = 0.1111$.

Thus, the 95 % confidence intervals for mean are

$$\hat{p} \pm z_{\alpha/2}\sqrt{\frac{\hat{p}(1-\hat{p})}{n}} \text{ are } (0.00844, 0.21376).$$

Example 9.8 An investigation into the performance of two machines, in a factory manufacturing large numbers of bobbins, gives the following result:

	No. of bobbins examined	No. of bobbins found defective
Machine 1	375	17
Machine 2	450	22

Obtain 90 % confidence limit for difference in the performance of the two machines.

Solution Here,

$$n_1 = 375, n_2 = 450, \hat{p}_1 = \frac{17}{375}, \hat{p}_2 = \frac{22}{450}$$

and for $\alpha = 0.10$, $z_{\alpha/2} = 1.645$ (Table 9.6).

Then, the 90 % CI for difference in performance of the machines are

$$(\hat{p}_1 - \hat{p}_2) \pm z_{\alpha/2}\sqrt{\frac{\hat{p}_1(1-\hat{p}_1)}{n_1} + \frac{\hat{p}_2(1-\hat{p}_2)}{n_2}} = (-0.01839, 0.020774)$$

Example 9.9 In the Example 9.4 discussed above, assume that variance is not known and the sampling process yields $s = 0.1$. Determine 95 % confidence interval for the true variance.

Table 9.6 $100(1 - \alpha)\%$ confidence intervals for variance

Test name	Assumption	Confidence interval
χ^2 test between a sample variance and an assumed population variance σ_0^2	Sample is drawn from a normal population	$\frac{(n-1)s^2}{\chi_{\alpha/2,n-1}^2} \leq \sigma^2 \leq \frac{(n-1)s^2}{\chi_{1-\alpha/2,n-1}^2}$
χ^2 test between a population variance and an assumed variance σ_0^2	Sample is drawn from a normal population	$\frac{(n-1)s^2}{\chi_{\alpha/2,n-1}^2} \leq \sigma^2 \leq \frac{(n-1)s^2}{\chi_{1-\alpha/2,n-1}^2}$
F-test for two population variances	Both the samples are drawn from a normal population, μ known	$\frac{s_1^2}{s_2^2} f_{1-\alpha/2,n_2,n_1} \leq \frac{\sigma_1^2}{\sigma_2^2} \leq \frac{s_1^2}{s_2^2} f_{\alpha/2,n_2,n_1}$
F-test for two population variances	Both the samples are drawn from a normal population, μ unknown	$\frac{s_1^2}{s_2^2} f_{1-\alpha/2,n_2-1,n_1-1} \leq \frac{\sigma_1^2}{\sigma_2^2} \leq \frac{s_1^2}{s_2^2} f_{\alpha/2,n_2-1,n_1-1}$

Solution We have $n = 36, \bar{x} = 0.003, s = 0.1$ and $\alpha = 0.05$. For $\alpha = 0.05$, $\chi_{\alpha/2,n-1}^2 = 53.203$ and $\chi_{1-\alpha/2,n-1}^2 = 20.569$

Thus, the 95 % confidence intervals for mean are

$$\frac{(n-1)s^2}{\chi_{\alpha/2,n-1}^2} \leq \sigma^2 \leq \frac{(n-1)s^2}{\chi_{1-\alpha/2,n-1}^2} = (0.00658, 0.01702).$$

9.6 Testing of Hypothesis

A *statistical hypothesis test* is a method of making decisions using experimental data. It is an assumption about a population parameter. This assumption may or may not be true. The best way to determine whether a statistical hypothesis is true would be to examine the entire population. Since that is often impractical, researchers typically examine a random sample from the population. If sample data are not consistent with the statistical hypothesis, the hypothesis is rejected. In statistics, a result is called *statistically significant* if it is unlikely to have occurred by chance. The phrase "*test of significance*" was coined by Ronald Fisher. Critical tests of this kind may be called tests of significance, and when such tests are available, we may discover whether a second sample is or is not significantly different from the first.

Hypothesis testing is sometimes called *confirmatory data analysis*, in contrast to *exploratory data analysis*. In frequency probability, these decisions are almost always made using null-hypothesis tests. One use of hypothesis testing is deciding whether experimental results contain enough information to cast doubt on conventional wisdom. Statistical hypothesis testing is a key technique of frequentist statistical inference and is widely used, but also much criticized. Other approaches to reaching a decision based on data are available via decision theory and optimal decisions. Note that in accepting a given hypothesis, we are not actually claiming that it is true, but rather we are saying that the resulting data appear to be consistent with it.

There are two types of statistical hypotheses

- *Null hypothesis*. The null hypothesis, denoted by H_0, is usually the hypothesis that sample observations result purely from chance
- *Alternative hypothesis*. The alternative hypothesis, denoted by H_1 or H_a, is the hypothesis that sample observations are influenced by some non-random cause

For example, suppose we wanted to determine whether a coin is fair and balanced. A null hypothesis might be that half the flips would result in heads and half in tails. The alternative hypothesis might be that the number of heads and tails would be very different. Symbolically, these hypotheses would be expressed as $H_0: p = 0.5$ versus $H_1: p \neq 0.5$. Suppose we flipped the coin 50 times, resulting in 40 heads and 10 tails. Given this result, we would be inclined to reject the null hypothesis. We would conclude, based on the evidence, that the coin was probably not fair and balanced.

Decision errors. Two types of errors can result from a hypothesis test. They are as follows

- *Type I error*. A Type I error occurs when the researcher rejects a null hypothesis when it is true. The probability of committing a Type I error is called the *significance level*. This probability is also called *alpha* and is often denoted by α
- *Type II error*. A Type II error occurs when the researcher fails to reject a null hypothesis that is false. The probability of committing a Type II error is called *beta* and is often denoted by β. The probability of *not* committing a Type II error is called the *power* of the test

Decision rules. The analysis plan includes decision rules for rejecting the null hypothesis. In practice, statisticians describe these decision rules in two ways—with reference to a p-value and with reference to a region of acceptance.

- *p-value*. The strength of evidence in support of a null hypothesis is measured by the *p-value*. Suppose the test statistic is equal to S. The p-value is the probability of observing a test statistic as extreme as S, assuming the null hypothesis is true. If the p-value is less than the significance level, we reject the null hypothesis
- *Region of acceptance*. The *region of acceptance* is a range of values. If the test statistic falls within the region of acceptance, the null hypothesis is not rejected. The region of acceptance is defined so that the chance of making a Type I error is equal to the significance level

The set of values outside the region of acceptance is called the *region of rejection* or *critical region*. The critical region of a hypothesis test is the set of all outcomes which, if they occur, will lead us to decide that there is a difference and hence, the null hypothesis is rejected. In such cases, we say that the hypothesis has been rejected at the α level of significance.

Steps involved in a hypothesis test. Statisticians follow a formal process to determine whether to reject a null hypothesis, based on sample data. This process, called *hypothesis testing*, consists of five steps.

1. State the hypotheses. This involves stating the null and alternative hypotheses. The hypotheses are stated in such a way that they are mutually exclusive. That is, if one is true, the other must be false.
2. Specify a suitable level of significance α.
3. Formulate an analysis plan. The analysis plan describes how to use sample data to evaluate the null hypothesis. The evaluation often focuses on a single test statistic.
4. Analyze sample data. Find the value of the test statistic (mean score, proportion, t-score, z-score, chi-square statistics, etc.) described in the analysis plan.
5. Interpret results. Apply the decision rules described in the analysis plan. If the value of the test statistic is unlikely, based on the null hypothesis, reject the null hypothesis.

Most of the statistical softwares use the p-value approach for rejection of test statistics. A p-value is a measure of how much evidence you have against the null hypothesis. The smaller the p-value, the more evidence you have. One may combine the p-value with the significance level to make a decision on a given test of hypothesis. In such a case, if the p-value is less than some threshold (usually 0.05, sometimes a bit larger like 0.10 or a bit smaller like 0.01), then you reject the null hypothesis.

9.6.1 Parametric Tests

Note that the basic assumption of all parametric tests is normal. Suppose now that in order to test a specific null hypothesis H_0, a population sample of size n, say $X_1, X_2, \ldots, X_n$, is to be observed. Based on these values, we must decide whether or not to accept H_0. A test for H_0 can be specified by defining a region C in n-dimensional space with the provision that the hypothesis is to be rejected if the random sample $X_1, X_2, \ldots, X_n$ turns out to lie in C and is to be accepted otherwise. The region C is called the *critical region* or *rejection region*. That is, the statistical test determined by the critical region C is the one such that

$$\text{accepts } H_0 \quad \text{if } (X_1, X_2, \ldots, X_n) \notin C \tag{9.11}$$

and

$$\text{reject } H_0 \quad \text{if } (X_1, X_2, \ldots, X_n) \in C \tag{9.12}$$

For instance, we are interested in testing the null hypothesis

$$H_0 \colon \mu = \mu_0$$

against the alternative hypothesis

$$H_1 \colon \mu \neq \mu_0$$

where μ_0 is some specified constant.

Since $\bar{x} = \frac{1}{n} \sum_{i=1}^{n} x_i$ is a reasonable estimator of μ, it seems reasonable to accept H_0 if $\bar{x}$ is not too far from μ_0. That is, the critical region of the test would be of the form

$$C = \{(X_1, X_2, \ldots, X_n) : |\bar{x} - \mu_0| > c\} \tag{9.13}$$

where c is such that

$$P\{|\bar{x} - \mu_0| > c\} = \alpha$$

That is,

$$P\left\{ \left| \frac{\bar{x} - \mu_0}{\sigma/\sqrt{n}} \right| > c \frac{\sqrt{n}}{\sigma} \right\} = \alpha$$

or, equivalently,

$$2P\left\{ Z > c \frac{\sqrt{n}}{\sigma} \right\} = \alpha$$

or

$$P\left\{ Z > c \frac{\sqrt{n}}{\sigma} \right\} = \alpha/2$$

where Z is a standard normal random variable. Also recall that

$$P\{Z > z_{\alpha/2}\} = \alpha/2$$

Thus,

$$c\frac{\sqrt{n}}{\sigma} = z_{\alpha/2}$$

or

$$c = z_{\alpha/2}\frac{\sigma}{\sqrt{n}}$$

Thus, the significance level α test is to

$$\text{reject } H_0 \quad \text{if } \frac{\sqrt{n}}{\sigma}|\bar{x} - \mu_0| > z_{\alpha/2} \tag{9.14}$$

and

$$\text{accept } H_0 \quad \text{if } \frac{\sqrt{n}}{\sigma}|\bar{x} - \mu_0| \le z_{\alpha/2} \tag{9.15}$$

Figure 9.8 shows the critical region corresponds to the test of H_0.

Recall that the critical region depends on the choice of the significance level. Also the correct level of significance to be used in a given situation depends on the individual circumstances involved in that situation. For instance, if rejecting a null hypothesis H_0 would result in large costs that would thus be lost if H_0 were indeed true, then we might elect to be quite conservative and so choose a significance level of 0.05 or 0.01. Also, if we initially feel strongly that H_0 was correct, then we would set a very low significance level. In practice, the significance level is often not set in advance, but rather the data are looked at to determine the resultant p-value. Sometimes, this critical significance level is much larger than what we use, and so the null hypothesis can be readily accepted. At other times, the p-value is so small that it is clear that the hypothesis should be rejected. Hence, it is interesting to compute the p-value for a better conclusion.

In various tables below, we summarize the tests for means, proportion, and variances with their corresponding critical regions. Table 9.7 comprehends the hypothesis tests for means.

Example 9.10 A manufacturer supplies the rear axles for mail trucks. These axles must be able to withstand 80,000 pounds per square inch in stress tests, but an excessively strong axle raises production costs significantly. Past experience indicates that the standard deviation of the strength of its axles is 4000 pounds per square inch. The manufacturer selects a sample of 100 axles from production, tests them, and finds that the mean stress capacity of the sample is 79,600 pounds per square inch. If the axle manufacturer uses a significance level of 0.05 in testing, will the axles meet his stress requirements?

Table 9.7 Hypothesis tests for means

Test name	Assumptions	Test statistic	H_0	H_1	Rejection criteria
z-test for a population mean	σ^2 known	$z = \dfrac{\bar{x}-\mu_0}{\sigma/\sqrt{n}}$	$\mu = \mu_0$	$\mu < \mu_0$ $\mu > \mu_0$ $\mu \neq \mu_0$	$z < -z_\alpha$ $z > z_\alpha$ $|z| \geq z_{\alpha/2}$
z-test for two population means	$\sigma_1^2 = \sigma_2^2 = \sigma^2$	$z = \dfrac{(\bar{x}_1-\bar{x}_2)-\delta}{\sigma\sqrt{\frac{1}{n_1}+\frac{1}{n_2}}}$	$\mu_1 - \mu_2 = \delta$	$\mu_1 - \mu_2 < \delta$ $\mu_1 - \mu_2 > \delta$ $\mu_1 - \mu_2 \neq \delta$	$z < -z_\alpha$ $z > z_\alpha$ $|z| \geq z_{\alpha/2}$
z-test for two population means	$\sigma_1^2 \neq \sigma_2^2$	$z = \dfrac{(\bar{x}_1-\bar{x}_2)-\delta}{\sqrt{\frac{s_1^2}{n_1}+\frac{s_2^2}{n_2}}}$	$\mu_1 - \mu_2 = \delta$	$\mu_1 - \mu_2 < \delta$ $\mu_1 - \mu_2 > \delta$ $\mu_1 - \mu_2 \neq \delta$	$z < -z_\alpha$ $z > z_\alpha$ $|z| \geq z_{\alpha/2}$
t-test for a population mean	σ^2 unknown	$t = \dfrac{\bar{x}-\mu_0}{s/\sqrt{n}}$	$\mu = \mu_0$	$\mu < \mu_0$ $\mu > \mu_0$ $\mu \neq \mu_0$	$t < -t_{\alpha,n-1}$ $t > t_{\alpha,n-1}$ $|t| \geq t_{\alpha/2,n-1}$
t-test for two population means	Variances unknown and equal	$t = \dfrac{(\bar{x}_1-\bar{x}_2)-\delta}{s\sqrt{\frac{1}{n_1}+\frac{1}{n_2}}}$, $s^2 = \dfrac{(n_1-1)s_1^2 + (n_2-1)s_2^2}{n_1+n_2-2}$	$\mu_1 - \mu_2 = \delta$	$\mu_1 - \mu_2 < \delta$ $\mu_1 - \mu_2 > \delta$ $\mu_1 - \mu_2 \neq \delta$	$t < -t_{\alpha,n_1+n_2-2}$ $t > t_{\alpha,n_1+n_2-2}$ $|t| \geq t_{\alpha/2,n_1+n_2-2}$
t-test for two population means	Variances unknown and unequal	$t = \dfrac{(\bar{x}_1-\bar{x}_2)-\delta}{\sqrt{\frac{s_1^2}{n_1}+\frac{s_2^2}{n_2}}}$	$\mu_1 - \mu_2 = \delta$	$\mu_1 - \mu_2 < \delta$ $\mu_1 - \mu_2 > \delta$ $\mu_1 - \mu_2 \neq \delta$	$t < -t_{\alpha,v}$ [a] $t > t_{\alpha,v}$ $|t| \geq t_{\alpha/2,v}$
t-test for two population means (paired t-test)	Observations are in pairs	$t = \dfrac{(\bar{x}_1-\bar{x}_2)-\delta}{s_d/\sqrt{n}}$, $s_d = \sqrt{\dfrac{1}{n-1}\sum(d_i-\bar{d})^2}$	$\mu_d = \delta$	$\mu_d < \delta$ $\mu_d > \delta$ $\mu_d \neq \delta$	$t < -t_{\alpha,n-1}$ $t > t_{\alpha,n-1}$ $|t| \geq t_{\alpha/2,n-1}$

[a] $v = \dfrac{\left(\frac{s_1^2}{n_1}+\frac{s_2^2}{n_2}\right)^2}{\frac{(s_1^2/n_1)^2}{n_1-1}+\frac{(s_2^2/n_2)^2}{n_2-1}}$

Solution As per the requirements, we carry out a test for mean using z-test as follows:

1. H_0: $\mu = 80,000$ versus H_1: $\mu \neq 80,000$
2. Significance level, $\alpha = 0.05$
3. Test statistics (since σ is known here)

$$z = \frac{\bar{x} - \mu_0}{\sigma/\sqrt{n}},$$

where $n = 100, \bar{x} = 79,600, \mu_0 = 80,000$ and $\sigma = 4000$. On substitution, $z = -1$.
4. For $\alpha = 0.05, z_{\alpha/2} = 1.96$. Since $|z| < z_{\alpha/2}$, we do not reject the hypothesis (see also Fig. 9.8 for graphical presentation of the critical region).
5. Since the hypothesis is not rejected, we conclude that there is no significant difference between the hypothesized mean of 80,000 and the observed mean of the sample axles. Therefore, the manufacturer should not reject the production run as meeting the stress requirements (Fig 9.9).

A 95 % confidence interval for μ is (78816, 80384). The p-value can be computed as

$$p\text{-value} = P(|z| > 1) = 2P(z > 1) = 0.3173$$

Fig. 9.8 Critical region for a two-sided hypothesis

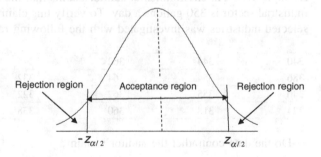

Fig. 9.9 Test statistic and critical region

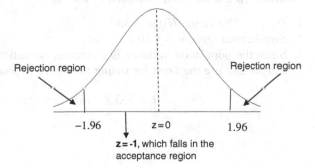

Example 9.11 Consider a test of $H_0: \mu \leq 100$ versus $H_0: \mu > 100$. Suppose that a sample of size 20 has a sample mean $\bar{x} = 105$. Determine the *p*-value of this outcome if the population standard deviation is known to equal (a) 5; (b) 10; and (c) 15.

Solution Given $n = 20, \bar{x} = 105, \mu_0 = 100$.

To compute the *p*-value, first we compute the test statistic value using

$$z = \frac{\bar{x} - \mu_0}{\sigma/\sqrt{n}}$$

(a) when $\sigma = 5$, $z = \frac{105-100}{5/\sqrt{20}} = 4.4721$, and hence,

$$p\text{-value} = P(Z > 4.4721) = 3.8727 \times 10^{-6}$$

(b) when $\sigma = 10$, $z = \frac{105-100}{10/\sqrt{20}} = 2.2361$, and hence,

$$p\text{-value} = P(Z > 2.2361) = 0.0127$$

(c) when $\sigma = 15$, $z = \frac{105-100}{15/\sqrt{20}} = 1.4907$, and hence,

$$p\text{-value} = P(Z > 1.4907) = 0.0680$$

Example 9.12 An environmental auditor claims that the average water used in an industrial sector is 350 gallons a day. To verify this claim, a study of 20 randomly selected industries was investigated with the following results:

340	344	362	375	356
386	354	364	332	402
340	355	362	322	372
324	318	360	338	370

Do the data contradict the auditor's claim?

Solution To test the claim, we need to test

1. $H_0: \mu = 350$ versus $H_1: \mu \neq 350$
2. Significance level, $\alpha = 0.10$
3. Since the population variance is unknown, we estimate the same through the sample and use the *t*-test for testing the above hypothesis. Thus, we obtain

$$\bar{x} = 353.8 \quad \text{and} \quad s = 21.8478$$

Hence, the test statistics is

$$t = \frac{\bar{x} - \mu_0}{s/\sqrt{n}} = 0.7778,$$

4. Now, for $\alpha = 0.10$, $t_{0.05,19} = 1.730$. Since $|t| < t_{\alpha/2,n-1}$, we accept the hypothesis.
5. Since the hypothesis is accepted, we conclude that the auditor's claim is not contradicted.

The p-value is obtained as

$$p\text{-value} = P(|t| > 0.7778) = 2P(t > 0.7778) = 0.4462,$$

indicating that the null hypothesis would be accepted at any reasonable significance level and thus that the data are not inconsistent with the claim of the auditor.

Example 9.13 An experiment is conducted to compare the precision of two brands of mercury detectors in measuring mercury concentration in the air. During the noon of one day in a downtown city area, 7 measurements of the mercury concentration are made with brand A instrument and 6 measurements are made with brand B instrument. The measurements per cubic meter of air are as follows:

Brand A	0.95	0.82	0.78	0.96	0.71	0.86	0.99
Brand B	0.89	0.91	0.94	0.91	0.90	0.89	

Do the data provide strong evidence that brand B measures mercury concentration in the air more precisely?

Solution Since the population variances are unknown and the sample sizes are very less, we use t-test for two sample mean tests.

1. $H_0: \mu_1 = \mu_2$ versus $H_1: \mu_1 < \mu_2$
2. Significance level, $\alpha = 0.05$
3. Test statistics

$$t_{cal} = \frac{(\bar{x}_1 - \bar{x}_2) - \delta}{s\sqrt{\frac{1}{n_1} + \frac{1}{n_2}}},$$

where $n_1 = 7, n_2 = 6, \delta = 0, \bar{x} = 0.867143, s_1 = 0.104198, \bar{y} = 0.90667, s_1 = 0.018619$ and

$$s^2 = \frac{(n_1 - 1)s_1^2 + (n_2 - 1)s_2^2}{n_1 + n_2 - 2} = 0.077972$$

On substitution, $t_{cal} = -0.9111$.

4. For $\alpha = 0.05$, $t_{0.05,11} = 1.795885$. Since $t_{cal} > -t_{0.05,11}$, we accept the null hypothesis.
5. Since the hypothesis is accepted, we do not have any strong evidence to say that brand B measures mercury concentration in the air more precisely

The *p*-value of this test is obtained as

$$p\text{-value} = P(t_{cal} < -0.9111) = P(t_{cal} > 0.9111) = 0.1908$$

The hypothesis tests for proportions and variances are given in Tables 9.8 and 9.9, respectively.

Example 9.13 Historical data indicate that 4 % of the components produced by a certain company are defective. A particularly acrimonious labor dispute has recently been resolved, and the management is curious about whether it will result in any change in this figure of 4 %. If a random sample of 500 items indicated 16 defectives, is this significant evidence, at the 5 % level of significance, to conclude that a change has occurred?

Table 9.8 Hypothesis tests for proportions

Test name	Assumption	Test statistic	H_0	H_1	Rejection criteria
z-test for a given proportion p_0	$n \geq 30$	$z = \dfrac{p-p_0}{\sqrt{\frac{p_0(1-p_0)}{n}}}, p = \frac{x}{n}$	$p = p_0$	$p < p_0$ $p > p_0$ $p \neq p_0$	$z < -z_\alpha$ $z > z_\alpha$ $\lvert z \rvert \geq z_{\alpha/2}$
z-test for equality of two proportions	Different sample size $n \geq 30$	$z = \dfrac{(p_1 - p_2)}{\sqrt{P(1-P)\left(\frac{1}{n_1}+\frac{1}{n_2}\right)}}$ $P = \dfrac{n_1 p_1 + n_2 p_2}{n_1 + n_2}$	$p_1 = p_2$	$p_1 < p_2$ $p_1 > p_2$ $p_1 \neq p_2$	$z < -z_\alpha$ $z > z_\alpha$ $\lvert z \rvert \geq z_{\alpha/2}$

Table 9.9 Hypothesis tests for variances

Test name	Assumption	Test statistic	H_0	H_1	Rejection criteria
χ^2 test between a sample variance and an assumed population variance σ_0^2	Sample is drawn from a normal population	$\chi^2 = \dfrac{(n-1)s^2}{\sigma_0^2}$	$\sigma^2 = \sigma_0^2$	$\sigma^2 < \sigma_0^2$ $\sigma^2 > \sigma_0^2$ $\sigma^2 \neq \sigma_0^2$	$\chi^2 < \chi^2_{1-\alpha,n-1}$ $\chi^2 > \chi^2_{\alpha,n-1}$ $\chi^2 > \chi^2_{\alpha/2,n-1}$ or $\chi^2 < \chi^2_{1-\alpha/2,n-1}$
F-test for two population variances	Both the samples are drawn from a normal population	$F = \dfrac{s_1^2}{s_2^2}$	$\sigma_1^2 = \sigma_2^2$	$\sigma_1^2 < \sigma_2^2$ $\sigma_1^2 > \sigma_2^2$ $\sigma_1^2 \neq \sigma_2^2$	$F < f_{1-\alpha,n_1-1,n_2-1}$ $F > f_{\alpha,n_1-1,n_2-1}$ $F < f_{1-\alpha,n_1-1,n_2-1}$ or $F > f_{\alpha,n_1-1,n_2-1}$

Solution To be able to conclude that a change has occurred, the data need to be strong enough to reject the null hypothesis, for which, we test the following:

1. $H_0: p = 0.04$ versus $H_0: p \neq 0.04$
2. Significance level, $\alpha = 0.05$
3. Test statistics

$$z = \frac{p - p_0}{\sqrt{\frac{p_0(1-p_0)}{n}}}$$

where $n = 500, p = \frac{16}{500} = 0.032$. On substitution, we get $z = -0.91287$
4. For $\alpha = 0.05, z_{\alpha/2} = 1.96$. Since $|z| < z_{\alpha/2}$, we do not reject the hypothesis.
5. Since the hypothesis is accepted, there is no strong evidence to reject the null hypothesis. That is, there is no change in the manufacturing the process.

The p-value of this test is obtained as

$$p\text{-value} = P(|z| > 0.91287) = 2P(z > 0.91287) = 0.3613$$

Example 9.14 A coal-fired power plant is considering two different systems for pollution abatement. The first system has reduced the emission of pollutants to acceptable levels 68 % of the time, as determined from 200 air samples. The second more expensive system has reduced the emission of pollutants to acceptable levels 76 % of the time, as determined from 250 air samples. If the expensive system is significantly more effective than the inexpensive system in reducing pollutants to acceptable levels, then the management of the power plant will install the expensive system. Which system will be installed if management uses a significance level of 0.10 in making its decision?

Solution As per the requirements, we carry out a test for equality of proportion as follows:

1. $H_0: p_1 = p_2$ versus $H_0: p_1 < p_2$
2. Significance level, $\alpha = 0.10$
3. Test statistics

$$z = \frac{(p_1 - p_2)}{\sqrt{P(1 - P)\left(\frac{1}{n_1} + \frac{1}{n_2}\right)}},$$

where $n_1 = 200, p_1 = 0.68, n_2 = 250, p_2 = 0.76$. On substitution, we get

$$P = \frac{n_1 p_1 + n_2 p_2}{n_1 + n_2} = 0.724 \quad \text{and} \quad z = -1.89$$

4. For $\alpha = 0.10$, $z_\alpha = 1.645$. Since $z < -z_\alpha$, we reject the hypothesis.
5. Since the hypothesis is rejected, we conclude that there is some significant difference between the two systems. Therefore, it is recommended to install the more expensive system in reducing pollutants to acceptable levels.

The p-value of this test is obtained as

$$p\text{-value} = P(z < -1.89) = P(z > 1.89) = 0.0587$$

Example 9.15 A project was undertaken to reduce the variance of the coating on circuit boards from 0.0005. A lot of 1000 circuit boards were received for inspection. After randomly sampling 36 boards, it is observed that the mean thickness is 0.003 and variance is 0.00025. Has the variance been reduced?

Solution As per the requirements, we carry out a test for variance as follows:

1. H_0: $\sigma^2 = \sigma_0^2$ versus H_0: $\sigma^2 < \sigma_0^2$
2. Significance level, $\alpha = 0.05$
3. Test statistics

$$\chi^2 = \frac{(n-1)s^2}{\sigma_0^2},$$

where $n = 36, s^2 = 0.00025, \sigma_0^2 = 0.0005$. On substitution, we get

$$\chi^2 = \frac{35 \times 0.00025}{0.0005} = 17.5$$

4. For $\alpha = 0.05$, $\chi_{0.95,35}^2 = 22.465$. Since $\chi^2 < \chi_{1-\alpha,n-1}^2$, we reject the hypothesis.
5. Since the hypothesis is rejected, we conclude that there has been a reduction in the variance of the coating thickness.

The p-value of this test is obtained as

$$p\text{-value} = P(\chi^2 < 17.5) = 0.0059$$

Example 9.16 Two samples are drawn from two normal populations. From the following data test whether the two samples have the same variance at 5 % level:

| Sample-1: | 60 | 65 | 71 | 74 | 76 | 82 | 85 | 87 | | |
| Sample-2: | 61 | 66 | 67 | 85 | 78 | 63 | 85 | 86 | 88 | 91 |

Solution We use F-test to test the sample variances as follows:

1. $H_0: \sigma_1^2 = \sigma_2^2$ versus $H_1: \sigma_1^2 \neq \sigma_2^2$
2. Significance level, $\alpha = 0.05$
3. Test statistics

$$F = \frac{s_1^2}{s_2^2},$$

where $s_1^2 = 90.857, s_2^2 = 133.33$. On substitution, we get $F = 1.467$

4. For $\alpha = 0.05$, $F_{0.05,(9,7)} = 3.68$. Since $F < f_{1-\alpha, n_1-1, n_2-1}$, we accept the hypothesis.
5. Since the hypothesis is accepted, we conclude that two samples have the same variances.

The p-value of this test is obtained as

$$p\text{-value} = P(F < 1.467) = 0.3137$$

9.6.2 Nonparametric Tests

Here, we develop some hypothesis tests in situations where the data come from a probability distribution whose underlying form is not specified. That is, it will not be assumed that the underlying distribution is normal, or exponential, or any other given type. Because no particular parametric form for the distribution is assumed, such tests are called *nonparametric* or *distribution-free* tests.

Most of the confidence interval procedures and hypothesis testing problems discussed previously are based on the assumption that the underlying distribution is normal, and hence, the inference procedure is called *parametric methods*. The strength of a nonparametric test resides in the fact that it can be applied without any assumption to the form of the underlying distribution. If there is justification for assuming a particular parametric form, such as normal, then the relevant parametric test should be employed. Nonparametric methods are generally applied to small sample data and do not make use of the actual observations. They are often based on ranks and positional values. Hence, under no circumstances, a nonparametric test will be more efficient than a parametric test [18, 27].

Some of the useful and easy to apply nonparametric tests are discussed below. These tests are sometimes viewed as alternatives to the tests based on normal, t, and F-tests.

9.6.2.1 Test for Randomness

A basic assumption in much of statistical analysis is that a set of data constitutes a random sample from some population. However, it is sometimes the case that the data are not generated by a truly random process but by one that may follow a trend or a type of cyclical pattern. Hence, a test for randomness using *runs test* (H_0: data are random) is carried out as follows.

Let us suppose that each of the data values is either a 0 or a 1. That is, we shall assume that each data value can be dichotomized as being either a *success* or a *failure*. Let $X_1, X_2, \ldots, X_N$ denote the set of data. Any consecutive sequence of either 1s or 0s is called a *run*. For instance,

$$1\,0\,0\,1\,1\,1\,0\,1\,1\,0\,0\,0\,0\,1\,1\,0\,1\,0\,1\,1\,1\,1\,0\,0\,0\,1$$

contains 13 runs: 7 runs of 1s and 6 runs of 0s. Suppose the data set contains n 1s and m 0s, such that $n + m = N$. Let R denote the number of runs. If H_0 is true, then $X_1, X_2, \ldots, X_N$ would be equally likely to be any of the $\frac{N!}{n!m!}$ permutations of 1s and 0s. Hence, the probability mass function of R under H_0 is

$$P_{H_0}(R = 2k) = 2 \frac{\binom{m-1}{k-1}\binom{n-1}{k-1}}{\binom{m+n}{n}} \qquad (9.16)$$

and

$$P_{H_0}(R = 2k + 1) = \frac{\binom{m-1}{k-1}\binom{n-1}{k} + \binom{m-1}{k}\binom{n-1}{k-1}}{\binom{m+n}{n}} \qquad (9.17)$$

It can be shown that when n and m are both large and H_0 is true, R will have approximately a normal distribution with mean and standard deviation given by

$$\mu_R = \frac{2mn}{m+n} + 1 \qquad (9.18)$$

and

$$\sigma_R = \sqrt{\frac{2mn(2mn - m - n)}{(m+n)^2(m+n-1)}} \qquad (9.19)$$

Therefore, for an observed number of runs r and for large n and m,

$$P_{H_0}(R \le r) = P_{H_0}\left\{ \frac{R - \mu_R}{\sigma_R} \le \frac{r - \mu_R}{\sigma_R} \right\}$$

$$= P_{H_0}\left\{ Z \le \frac{r - \mu_R}{\sigma_R} \right\}$$

$$= \Phi\left(\frac{r - \mu_R}{\sigma_R} \right)$$

and similarly,

$$P_{H_0}(R \ge r) = 1 - \Phi\left(\frac{r - \mu_R}{\sigma_R} \right)$$

and the p-value of the test is

$$p\text{-value} = 2 \min\left\{ \Phi\left(\frac{r - \mu_R}{\sigma_R} \right), 1 - \Phi\left(\frac{r - \mu_R}{\sigma_R} \right) \right\} \tag{9.20}$$

Example 9.17 Suppose that a sequence of sixty 1s and sixty 0s resulted in 75 runs. Test whether the sequence is random.

Solution Since the sequence is large, we use normal approximation to test the randomness, as follows:

1. H_0: data are random versus H_1: data are not random.
2. $\alpha = 5\%$.
3. Test statistics

$$z = \frac{r - \mu_R}{\sigma_R}$$

where $n = m = 60$, $r = 75$, $\mu_R = 61$ and $\sigma_R = 5.454$. Hence, $z = 2.5669$.
4. For $\alpha = 5\%$, $z_{\alpha/2} = 1.96$. Since $|z| > z_{\alpha/2}$, we reject the null hypothesis.
5. Since the hypothesis is rejected, we conclude that the sequence is not random.

The p-value of the test is

$$p\text{-value} = 2 \min\{\Phi(2.567), 1 - \Phi(2.567)\}$$
$$= 0.0102$$

If the number of runs was equal to 70 instead of 75, then the test statistics value would have been $z = 1.650$. In this situation, for $\alpha = 5\%$, the null hypothesis would have been accepted. That is, randomness would have been concluded in the data set with a p-value

$$p\text{-value} = 2\min\{\Phi(1.65), 1 - \Phi(1.65)\}$$
$$= 0.0990$$

9.6.2.2 Sign Test

Consider a random variable X having a continuous distribution $F(x)$, and let $\tilde{\mu}$ be the median of the distribution. The sign test is used to test the median $\tilde{\mu}$ of a continuous population. That is to test

$$H_0: \tilde{\mu} = \tilde{\mu}_0 \tag{9.21}$$

versus

$$H_1: \tilde{\mu} \neq \tilde{\mu}_0$$

This hypothesis can easily be tested by noting that each of the observations will, independently, be less than $\tilde{\mu}_0$ with the probability $p = F(\tilde{\mu}_0)$. If we define

$$I_i = \begin{cases} 1 & \text{if } X_i - \tilde{\mu}_0 > 0 \\ 0 & \text{otherwise} \end{cases} \tag{9.22}$$

then $I_1, I_2, \ldots, I_n$ are independent Bernoulli random variables with parameter $p = F(\tilde{\mu}_0)$. If the null hypothesis is true, then $P(X_i - \tilde{\mu}_0 < 0) = P(X_i - \tilde{\mu}_0 > 0) = 0.5$. An appropriate test is that of

$$R^+ = \sum_{i=1}^{n} I_i \tag{9.23}$$

the number of positive differences. Therefore, to test the null hypothesis, we are really testing that the number of positive signs is a value of a binomial random variable with parameter $p = 0.5$. The p-value corresponding to this test for the observed number of positive signs r^+ is such that

$$p\text{-value} = \begin{cases} 2P(R^+ \leq r^+), & \text{if } r^+ < n/2 \\ 2P(R^+ \geq r^+), & \text{if } r^+ > n/2 \end{cases} \tag{9.24}$$

If the p-value is less than some preselected level α, the test is rejected; otherwise, it is accepted.

For a one-sided hypothesis of the form $H_0: \tilde{\mu} = \tilde{\mu}_0$ versus $H_1: \tilde{\mu} > \tilde{\mu}_0$, the hypothesis is rejected, if the observed number of positive signsobserved is sufficiently large; that is, if p-value $= P(R^+ \geq r^+)$ is less than some preselected level α, the test is rejected; otherwise, it is accepted. Similarly, for a left-tailed alternative hypothesis, the hypothesis is rejected if p-value $= P(R^+ \leq r^+)$ is less than some preselected level α.

For a large sample, under H_0, the test can be approximated to a normal approximation with mean $\mu = n \times p = 0.5n$ and variance $\sigma^2 = n \times p \times (1 - p) = 0.25n$. Hence, a large sample approximation of the test statistics is given by

$$Z = \frac{R^+ - 0.5n}{\sqrt{0.25n}} \tag{9.25}$$

The statistic Z may be then compared with the Z_α values for specified significance level α.

Example 9.18 The following are the measurements of the breaking strength of a certain kind of 2-in. cotton ribbons in pounds

| 163 | 165 | 166 | 161 | 171 | 158 | 151 | 162 | 163 | 139 | 172 | 165 | 148 | 169 |

Test the hypothesis that the median breaking strength is 160 pounds against the alternative that it is greater than 160 pounds at 5 % level of significance.

Solution Here, $n = 14$ and $\tilde{\mu}_0 = 160$. Since the data do not follow normal distribution (see Fig. 9.10), we use a nonparametric test for this data.

It is required to test

1. H_0: $\tilde{\mu} = 160$ against H_1: $\tilde{\mu} > 160$
2. $\alpha = 5\%$

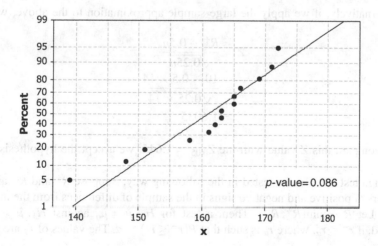

Fig. 9.10 Probability plot of the data

3. Test statistics calculation

x_i	163	165	166	161	171	158	151	162	163	139	172	165	148	169
$x_i - 160$	3	5	6	1	11	-2	-9	2	3	-21	12	5	-12	9
l_i	1	1	1	1	1	0	0	1	1	0	1	1	0	1

Now,

$$R^+ = \sum_{i=1}^{n} I_i = 10$$

4. The hypothesis is rejected, if the number of positive signs observed is sufficiently large

$$p\text{-value} = P(R^+ \geq 10)$$

$$= \sum_{r=10}^{14} \binom{14}{r} (0.5)^r (0.5)^{14-r}$$

$$= 1 - \sum_{r=0}^{9} \binom{14}{r} (0.5)^{14}$$

$$= 1 - 0.91022$$

$$= 0.08978 \text{ (From Appendix Table A.1)}$$

5. Since p-value $= 0.08978$, which is larger than 0.05, the hypothesis is accepted.

Therefore, we conclude that the median breaking strength is greater than 160 pounds.

Alternatively, if we apply the large sample approximation to the above, we get

$$Z = \frac{R^+ - 0.5n}{\sqrt{0.25n}}$$

$$= \frac{10 - 0.5 \times 14}{\sqrt{0.25 \times 14}}$$

$$= 1.6035$$

If we compare this Z-value with the $Z_{0.05} = 1.645$, we accept the hypothesis.

A sign test can also be tested in the following way: Suppose R^+ and R^- are the numbers of positive and negative signs in the sample of differences from the median value. Let $R = \min(R^+, R^-)$. Then, a test for $H_0: \tilde{\mu} = \tilde{\mu}_0$ against $H_1: \tilde{\mu} > \tilde{\mu}_0$ is rejected, if $r^- \leq r_\alpha$, where r_α is such that $P(r^- \leq r_\alpha) = \alpha$. The values of r_α are given in Appendix Table A.14. If the alternative hypothesis is $H_1: \tilde{\mu} > \tilde{\mu}_0$, then the test $H_0: \tilde{\mu} = \tilde{\mu}_0$ is rejected if $r^+ \geq r_\alpha$, where r_α is such that $P(r^+ \leq r_\alpha) = \alpha$. For a two-sided alternative hypothesis, $H_1: \tilde{\mu} \neq \tilde{\mu}_0$, the test is rejected if $r \geq r_\alpha$, where r_α is such that $P(r \leq r_\alpha) = \alpha$.

For the example discussed above, $R = \min(R^+, R^-) = \min(10, 4) = 4$, and the $r_{0.05} = 3$ (for a one-sided test). Since the alternative hypothesis under consideration is $H_1: \tilde{\mu} > 160$, the hypothesis $H_0: \tilde{\mu} = 160$ is rejected if $r^- \leq r_\alpha$. Here, $r^- = 4$ and $r_{0.05} = 3$ (from Appendix Table A.14). Since $r^- > r_\alpha$, the test is accepted. This conclusion is in tune with the conclusion derived through the p-value comparison done previously.

A sign test can also be used for testing two samples by comparing their common medians. This is equivalent to comparing two means of samples using t-test. A sign test in this case compares only the signs of differences of observations. We illustrate this using an example below.

Example 9.19 In an elementary school, 15 pairs of 1st grade children were formed on the basis of similarity of intelligence and family background. One child from each pair was taught to read by method-I and the other child by method-II. After a period of training, the children were given reading tests with the following results.

M-I	65	68	70	65	64	62	73	75	72	78	64	73	79	80	67
M-II	63	68	68	60	65	60	72	75	73	70	66	70	77	78	65

Use sign test to test whether the two methods are equivalent or not.

Solution The normal probability plot of M-I and M-II is shown in Fig. 9.11. Although, both the data sets are normal, we use the sign test procedure for comparing the two distributions here.

1. $H_0: \tilde{\mu}_1 = \tilde{\mu}_2$ against $H_0: \tilde{\mu}_1 \neq \tilde{\mu}_2$
2. $\alpha = 5\%$
3. Test statistics calculation

X_i	65	68	70	65	64	62	73	75	72	78	64	73	77	80	67
Y_i	63	69	68	60	65	60	72	76	73	70	66	70	79	78	65
I_i	1	0	1	1	0	1	1	0	0	1	0	1	0	1	1

Now,

$$R^+ = \sum_{i=1}^{n} I_i = 9$$

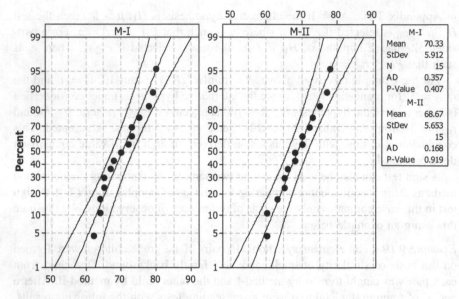

Fig. 9.11 Normal probability plot of M-I and M-II

4. The hypothesis is rejected, if observed number of positive signs is sufficiently large.

$$p\text{-value} = P(R^+ \geq 9)$$

$$= \sum_{r=9}^{15} \binom{15}{r} (0.5)^r (0.5)^{15-r}$$

$$= 1 - \sum_{r=0}^{8} \binom{15}{r} (0.5)^{15}$$

$$= 1 - 0.69638$$

$$= 0.30362 \text{ (From Appendix Table A.1)}$$

5. Since p-value = 0.30362, which is larger than 0.05, the hypothesis is accepted. Therefore, we conclude that the two methods are same.

If we use the large sample approximation, then the value of the standardized statistic is $Z = \frac{R^+ - 0.5n}{\sqrt{0.25n}} = 0.7746$, which is much less than the table value $Z_{0.05} = 1.96$, and therefore, the null hypothesis of equality of median is accepted, concluding that the two methods are same.

An identical conclusion can be drawn, if we use a two-sample t-test (most suitable in this case), the computed value of the t-test is 0.79 (assuming equal variances) with a p-value = 0.437, suggesting a strong support for the null hypothesis.

The exact test based on the positive and negative ranks suggests a different conclusion in this case. Note that $R^- = \sum_{i=1}^{n} I_i = 6$; therefore, $R = \min(R^+, R^-) = 6$. The table value of the test is $r_{0.05} = 3$ (for a two-sided test from Appendix Table A.14). Since the alternative hypothesis under consideration is $H_1: \tilde{\mu}_1 \neq \tilde{\mu}_2$, the hypothesis $H_0: \tilde{\mu}_1 = \tilde{\mu}_2$ is rejected if $r \geq r_\alpha$. Since $r = 6$ is larger than $r_{0.05} = 3$, the test is rejected. Therefore, we conclude that the medians of the two samples are not the same. Hence, the two methods may be different.

9.6.2.3 Signed-Rank Test

Consider a random variable X having a continuous distribution $F(x)$, and let μ_0 be the median of the distribution. The signed-rank test is to test whether the distribution is symmetric to the median μ_0. That is to test

$$H_0: \mu = \mu_0 \tag{9.26}$$

against

$$H_1: \mu \neq \mu_0$$

Let $Y_i = X_i - \mu_0, i = 1, 2, \ldots, n$ and rank the absolute values $|Y_1|, |Y_2|, \ldots, |Y_n|$. Let W^+ be the sum of positive ranks and W^- be the sum of negative ranks. If $W = \min(W^+, W^-)$, then a test for $H_0: \mu = \mu_0$ against $H_1: \mu \neq \mu_0$ is rejected if $w \leq w_\alpha$, where w is the observed values of W. Further, w_α is such that $P(w \leq w_\alpha) = \alpha$ under H_0. For a one-sided alternative of the form $H_1: \mu > \mu_0$, the H_0 is rejected if $w^- \leq w_\alpha$, and if the alternative of the form $H_1: \mu < \mu_0$, the H_0 is rejected if $w^+ \leq w_\alpha$. Appendix Table A.15 provides the values of w_α for various combination of n and α. The test is due to Wilcoxon and hence is also known as Wilcoxon signed-rank test.

To find the distribution under H_0, we set $j = 1, 2, \ldots, n$,

$$l_j = \begin{cases} 1 & \text{if the } j\text{th smallest value comes from a data} \\ & \text{value that is smaller than } \mu_0 \\ 0 & \text{otherwise} \end{cases} \tag{9.27}$$

Then the signed-rank test statistic is written as

$$W^* = \sum_{j=1}^{n} j l_j \tag{9.28}$$

The mean and variance of the test statistic under H_0 is obtained as

$$E[W^*] = \mu_{W^*} = E\left[\sum_{j=1}^{n} jI_j\right]$$

$$= \sum_{j=1}^{n} \frac{j}{2} = \frac{n(n+1)}{4} \tag{9.29}$$

and

$$\text{Var}[W^*] = \sigma_{W^*}^2 = \text{Var}\left[\sum_{j=1}^{n} jI_j\right]$$

$$= \sum_{j=1}^{n} j^2 \text{Var}(I_j) \tag{9.30}$$

$$= \sum_{j=1}^{n} \frac{j^2}{4} = \frac{n(n+1)(2n+1)}{24}$$

Hence, for large samples, one may use the approximation

$$z = \frac{W^* - \frac{n(n+1)}{4}}{\sqrt{\frac{n(n+1)(2n+1)}{24}}} \tag{9.31}$$

z is now a standard normal distribution and can be compared with Z_α values for a specified significance level α.

Example 9.20 Consider the observations: 4.2, 1.8, 5.3, 1.7, 2.4, and 3.8. Test whether the data are symmetric about the median $\mu_0 = 2$.

Solution Since the sequence is small, we proceed as follows:

1. $H_0: \mu_0 = 2$ versus $H_1: \mu_0 \neq 2$
2. $\alpha = 5\%$
3. Test statistics calculation

x_i	$x_i - \mu_0$	$|Y_i|$	Rank $(|Y_i|)$	I_j
4.2	2.2	2.2	5	1
1.8	−0.2	0.2	1	0
5.3	3.3	3.3	6	1
1.7	−0.3	0.3	2	0
2.4	0.4	0.4	3	1
3.8	1.8	1.8	4	1

Therefore,

$$w^+ = 18, \quad w^- = 3, \quad \text{and} \quad w = \min(w^+, w^-) = 3$$

4. For $\alpha = 5\%$, $w_{0.05} = 1$. $w \le w_\alpha$. Since $w \ge w_\alpha$, we accept the null hypothesis.
5. Since the hypothesis is accepted, we conclude that the data are symmetric about the median.

Alternatively, using the large sample approximation, we get

$$z = \frac{W^* - \mu_{W^*}}{\sigma_{W^*}},$$

where $W^* = \sum_{j=1}^{n} jI_j = 18$, $\mu_{W^*} = 5$, and $\sigma_{W^*} = 4.7696$. Hence, $z = 1.5724$. For $\alpha = 5\%$, $z_{\alpha/2} = 1.96$ for a two-sided alternative hypothesis. Since $|z| < z_{\alpha/2}$, we accept the null hypothesis and conclude that the data are symmetric about the median. The p-value of the test is

$$p\text{-value} = 2P(|z| > 0.7303)$$
$$= 0.4652$$

9.6.2.4 Mann–Whitney–Wilcoxon Test

Suppose that one is considering two different methods for producing items having measurable characteristics with an interest in determining whether the two methods result in statistically identical items. To solve this problem, let $X_1, X_2, \ldots, X_m$ denote a sample of measurable values of m items produced by method 1, and $Y_1, Y_2, \ldots, Y_n$ be the corresponding sample of measurable values of n items produced by method 2, respectively. Let F and G, both assumed to be continuous, denote the distribution functions of the two samples, respectively, then the hypothesis we wish to test is $H_0: F = G$ or, equivalently, $H_0: \mu_1 = \mu_2$.

To test the above hypothesis, define

$$Z_{ij} = \begin{cases} 1, & X_i < Y_j \\ 0, & X_i > Y_j \end{cases}, \quad i = 1, 2, \ldots, m; j = 1, 2, \ldots, n \tag{9.32}$$

and write

$$U = \sum_{i=1}^{m} \sum_{j=1}^{n} Z_{ij} \tag{9.33}$$

Note that $\sum_{j=1}^{n} Z_{ij}$ is the number of Y_j's that are larger than X_i's, and hence, U is the number of values of $X_1, X_2, \ldots, X_m$ that are smaller than each of $Y_1, Y_2, \ldots, Y_n$. The test statistic U is called the Mann–Whitney U-test. Then, to test $H_0: F = G$ against $H_1: F > G$, we reject H_0 if $U \geq u_\alpha$, where u_α is such that $P(U \geq u_\alpha) \leq \alpha$. For practical convenience, U is defined for small sample groups of observations. The critical values corresponding to this test are given in Appendix A.16 for some selected values of m, n, and α.

When the null hypothesis is true and so $F = G$, it follows that all $m + n$ data come from the same distribution, and hence, all $(m + n)!$ possible rankings of the values $X_1, X_2, \ldots, X_n$ and $Y_1, Y_2, \ldots, Y_m$ are equally likely. Therefore, under H_0, the mean and variance of U is given by

$$E_{H_0}[U] = \mu_U = \frac{nm}{2}$$

and

$$\mathrm{Var}_{H_0}[U] = \sigma_U^2 = \frac{mn(n + m + 1)}{12}$$

Hence,

$$z = \frac{U - \frac{nm}{2}}{\sqrt{\frac{mn(n+m+1)}{12}}} \quad \sim \quad N(0, 1) \tag{9.34}$$

Example 9.21 An experiment is designed to compare two treatments against corrosion yielded the following data in pieces of wire subjected to the two treatments:

Treatment-1	59.4	72.1	68.0	66.2	58.5	
Treatment-2	65.2	67.1	69.4	78.2	74.0	80.3

The data represent the maximum depth of pits in units of one thousandth of an inch. Do the data indicate any significant difference in distribution?

Solution Here, $m = 5$ and $n = 6$.

1. $H_0: F = G$ versus $H_0: F > G$
2. $\alpha = 5\%$
3. Test statistics

$$U = \sum_{i=1}^{m} \sum_{j=1}^{n} Z_{ij}$$

$$= 2 + 3 + 4 + 5 + 5 + 5 = 24$$

4. From Appendix Table A.17, the critical value u_α for $m = 5$, $n = 6$, and $\alpha = 0.01$ is 27. Since $U < u_\alpha$, the test is accepted
5. Since the test is accepted, we conclude that there is no significant difference between the two treatment combinations.

We now use the large sample approximation to see whether the same conclusion persists for the test. For large samples, we have

$$z = \frac{U - \frac{nm}{2}}{\sqrt{\frac{mn(n+m+1)}{12}}}$$

$$= \frac{24 - 15}{\sqrt{30}} = 1.6431$$

For $\alpha = 0.01$, $z_\alpha = 2.33$. Since $z < z_\alpha$, we accept the null hypothesis and conclude that the samples are coming from same population. The p-value of the test is

$$p\text{-value} = P(z > 1.6431)$$
$$= 0.0502$$

9.6.2.5 Kruskal–Wallis Test

The U-test discussed above is a test for deciding whether or not two samples come from the same population. A generalization of this for k samples is provided by the Kruskal–Wallis H-test, or briefly the H-test.

Suppose that we have k samples of sizes $n_1, n_2, \ldots, n_k$, with the total size of all the samples taken together given by $n = n_1 + n_2 + \cdots + n_k$. Further, suppose that the data from all the samples taken together are ranked and that the sum of the ranks of each samples are $R_1, R_2, \ldots, R_k$, respectively. If we define the statistic

$$H = \frac{12}{n(n+1)} \sum_{i=1}^{k} \frac{R_i^2}{n_i} - 3(n+1) \tag{9.35}$$

then under H_0, the sampling distribution of H follows a chi-square distribution with $k - 1$ degrees of freedom, provided that the samples sizes are at least 5. In case there are too many ties among the observations in the sample data, the value of H given in (9.35) is smaller than it should be. In such cases, we use the corrected version of H statistics denoted by H_c, by dividing the value given in statistics (9.35) by the correction factor

$$1 - \frac{\sum (T^3 - T)}{n^3 - n}$$

where T is the number of ties corresponding to each observation, and where the sum is taken over by all the tied observations.

The H-test provides a nonparametric method equivalent to the analysis of variance (ANOVA) for one-way classification or one-factor experiments.

Example 9.22 A company wishes to purchase one out of five different machines: A, B, C, D, and E. In an experiment designed to determine whether there is a performance differences among the machines, five experienced operators were asked to work on the machines for an equal number of times. The data in terms of number of units produced by each machine are as follows:

A	68	72	77	42	53
B	72	53	63	53	48
C	60	82	64	75	72
D	48	61	57	64	50
E	64	65	70	68	53

Test the hypothesis that there is no significant difference between the machines at 0.05 level of significance.

Solution Here, $n_1 = n_2 = n_3 = n_4 = n_5 = 5$. By arranging all the values in ascending order of magnitude, the data will be like this:

$42, 48, 48, 50, 53, 53, 53, 53, 57, 60, 61, 63, 64, 64, 64, 65, 68, 68, 70, 72, 72, 72, 75, 77, 82.$

The ranks correspond to each machine and their totals are shown in Table 9.10.

Table 9.10 Table of ranks

Machines	Ranks					Rank total
A	17.5	21	24	1	6.5	70
B	21	6.5	12	6.5	2.5	48.5
C	10	25	14	23	21	93
D	2.5	11	9	14	4	40.5
E	14	16	19	17.5	6.5	73

Hence, to test $H_0: \mu_1 = \mu_2 = \mu_3 = \mu_4 = \mu_5$, we use the test statistic

$$H = \frac{12}{n(n+1)} \sum_{i=1}^{k} \frac{R_i^2}{n_i} - 3(n+1)$$

$$= \frac{12}{25(25+1)} \left[\frac{(70)^2}{5} + \frac{(48.5)^2}{5} + \frac{(93)^2}{5} + \frac{(40.5)^2}{5} + \frac{(73)^2}{5} \right] - 3(26)$$

$$= 6.44$$

For $k - 1 = 4$, the table value for 0.05 significance level is $\chi^2_{(0.05)4} = 9.49$, which is larger than the calculated value. Hence, we cannot reject the null hypothesis. Hence, it is accepted that there is no significant difference between the machines.

9.6.3 Goodness-of-Fit Tests

In this section, we are going to answer a very important question regarding sampled data, suppose a data set is available to us. How to decide a good probability model for the data set is a very pertinent question that arises in most cases. When one has some indication of the distribution of a population by probabilistic reasoning or otherwise, it is often possible to fit such theoretical distributions (also called *model* or *expected* distributions) to frequency distributions obtained from a sample of the population. The method generally uses the mean and standard deviation of the sample to estimate the mean and standard deviation of the population. To test, how good the fit is going to be? We use the chi-square test for *goodness of fit*.

9.6.3.1 Goodness-of-Fit Tests When Parameters Are Specified

Suppose that n independent random variables $Y_1, Y_2, \ldots, Y_n$ each taking on one of the values 1, 2, ..., k are to be observed and we are interested in testing the null hypothesis that $\{p_i, i = 1, 2, \ldots, k\}$ is the probability mass function of the variable Y_j. That is, we want to test the hypothesis

$$H_0: P[Y = i] = p_i, \quad i = 1, 2, \ldots, k \tag{9.36}$$

versus

$$H_1: P[Y = i] \neq p_i, \quad \text{for some } i = 1, 2, \ldots, k$$

To test the above hypothesis, we proceed as follows:

Let $X_i, i = 1, 2, \ldots, k$ denote the number of the Y_j's that are equal to i. Then, each Y_j will independently be equal to i with the probability $P[Y = i]$. Therefore, it follows that when H_0 is true, X_i is binomial with parameters n and p_i. Hence, $E(X_i) = np_i$ and so $(X_i - np_i)^2$ will be an indication as to how likely it appears that p_i indeed equals the probability that $Y = i$. When this is large, then it is an indication that H_0 is not correct. This reasoning helps us to consider the test statistic

$$\chi^2 = \sum_{i=1}^{k} \frac{(X_i - np_i)^2}{np_i}$$
$$= \sum_{i=1}^{k} \frac{(O_i - E_i)^2}{E_i} \tag{9.37}$$

where $O_i = X_i$ and $E_i = np_i$. The test is rejected when χ^2 is large. To determine the critical region for a specified significance level α, we compute the probability

$$P_{H_o}(\chi^2 \geq c) = \alpha$$

For large k, the value of c is equal to $\chi^2_{\alpha, k-1}$. The p-value of such a test is

$$p\text{-value} = P_{H_0}(\chi^2_{k-1} \geq \chi^2) \tag{9.38}$$

Example 9.23 Suppose the TCP company requires that a fresher seeking placement in the company be interviewed by three of their executives. Each executive gives the candidate either a positive or negative rating. The table below summarizes the interview results of 100 candidates:

Positive ranking from three interviewers	Number of candidates receiving each of these rankings
0	18
1	47
2	24
3	11

The director of the company thinks that a candidate scoring 40 % or more may be placed for the job. Does the above information support his claim? (Assume $\alpha = 0.05$).

Solution Let

Y: the positive ranking of the interviewers. Then $Y \sim BD(3, 0.40)$.

Then, to test H_0: $p = 0.40$ versus H_1: $p \neq 0.40$, we proceed as follows:

$$P(X = 0) = \binom{3}{0}(0.4)^0(0.6)^3 = 0.216$$

$$P(X = 1) = \binom{3}{1}(0.4)^1(0.6)^2 = 0.4320$$

$$P(X = 2) = \binom{3}{2}(0.4)^2(0.6)^1 = 0.2880$$

$$P(X = 3) = \binom{3}{3}(0.4)^3(0.6)^0 = 0.0640$$

The calculation of the test statistics is shown in the following table:

Y	O_i	p_i	$E_i = np_i$	$(O_i - E_i)^2$	$(O_i - E_i)^2 / E_i$
0	18	0.2160	21.6	12.96	0.6000
1	47	0.4320	43.2	14.44	0.3343
2	24	0.2880	28.8	23.04	0.8000
3	11	0.0640	6.4	21.16	3.3063
					5.0406

Therefore,

$$\chi^2 = \sum_{i=1}^{k} \frac{(O_i - E_i)^2}{E_i} = 5.0406$$

For $k = 4$, $\chi^2_{0.05,3} = 7.815$. Since $\chi^2 < \chi^2_{0.05,3}$, we do not reject H_0. That is, the director of the company can go ahead with the placement, if a candidate scores 40 % or more at a significance level of $\alpha = 0.05$. The p-value of the test is obtained as

$$P-\text{value} = P(\chi^2_{(3)} > 5.0406) = 0.1688.$$

Example 9.24 Suppose the weekly number of accidents over a 30-week period is as follows:

8	0	0	1	3	4	0	2	12	5
1	8	0	2	0	1	9	3	4	5
3	3	4	7	4	0	1	2	1	2

Test the hypothesis that the number of accidents in a week follows a Poisson distribution.

Solution Let

Y: the number of accidents in a week. Then $Y \sim PD(\lambda)$

Then, to test

H_0: the data follow a Poisson distribution versus
H_0: the data do not follow a Poisson distribution

we first estimate the parameter

$$\hat{\lambda} = \frac{95}{30} = 3.16667. \text{ Hence } Y \sim PD(3.16667)$$

For computational simplicity, we classify the number of accidents conveniently into a five groups and compute the expected frequencies as follows:

Y	O_i	p_i	$E_i = np_i$
0	6	0.04214	1.2642
1	5	0.13346	4.0038
2–3	8	0.43434	13.030
4–5	6	0.28841	8.6523
>5	5	0.10164	3.0492

Note that the expected frequencies of the first and last classes are less than 5. This is adjusted with the classes close to these, and the revised calculation table is shown below:

O_i	p_i	$E_i = np_i$	$(O_i - E_i)^2$	$(O_i - E_i)^2 / E_i$
11	0.13346	5.268	32.85582	6.23686864
8	0.43434	13.030	25.3009	1.94174213
11	0.28841	11.7015	0.492102	0.04205463
				8.2206654

Therefore,

$$\chi^2 = \sum_{i=1}^{k} \frac{(O_i - E_i)^2}{E_i} = 8.2206654$$

Since the two classes are combined, the degrees of freedom now become 2 (=5 − 2 − 1), and therefore, $\chi^2_{0.05,2} = 5.9915$. Since $\chi^2 > \chi^2_{0.05,2}$, we reject H_0. That is, the accident data follow a Poisson distribution. The p-value of this test is obtained as

$$p\text{-value} = P(\chi^2_{(2)} > 8.2206654) = 0.0164.$$

9.6.3.2 Tests of Independence in Contingency Tables

Consider that a population is classified into two distinct attributes or categories. Suppose there are r possible values of the X-attributes and s possible values of the Y-attributes. Then, Table 9.11 is called an $r \times s$ contingency table.
Let

$$P_{ij} = P\{X = i, Y = j\} \quad \text{for } i = 1, 2, \ldots, r, \quad j = 1, 2, \ldots, s$$

which represents the probability that a randomly chosen member of the population will have X-attributes i and Y-attributes j. Suppose

$$p_i = P\{X = i\} = \sum_{j=1}^{s} P_{ij}, \quad i = 1, 2, \ldots, r$$

and

$$q_j = P\{Y = j\} = \sum_{i=1}^{r} P_{ij}, \quad j = 1, 2, \ldots, s$$

are, respectively, the marginal probabilities of X and Y. Then, we are interested in testing the hypothesis that the two attributes are independent. That is,

Table 9.11 A contingency table	X	Y			
		1	2	$\ldots$	s
	1	O_{11}	O_{12}	$\ldots$	O_{1s}
	2	O_{21}	O_{22}	$\ldots$	O_{2s}
	$\ldots$	$\ldots$	$\ldots$	$\ldots$	$\ldots$
	r	O_{r1}	O_{r2}	$\ldots$	O_{rs}

$$H_0: P_{ij} = p_i q_j, \quad \text{for all} \quad i = 1, 2, \ldots, r, \quad j = 1, 2, \ldots, s \tag{9.39}$$

against the alternative

$$H_0: P_{ij} \neq p_i q_j, \quad \text{for some} \quad i = 1, 2, \ldots, r, \quad j = 1, 2, \ldots, s$$

To test the above hypothesis, we first need to estimate the quantities p_i and q_j under H_0. Suppose that n members of the population have been sampled, with the result that n_{ij} of them have been simultaneously observed for the attribute X and Y. Note that $\sum_{ij} n_{ij} = n_,$, and if $n_i = \sum_j n_{ij}$ and $n_j = \sum_i n_{ij}$, then the estimates of p_i and q_j are, respectively, obtained as

$$\hat{p}_i = \frac{n_i}{n}, \quad i = 1, 2, \ldots, r$$

and

$$\hat{q}_j = \frac{n_j}{n}, \quad j = 1, 2, \ldots, s$$

Then, under H_0, the test statistic is obtained as

$$
\begin{aligned}
\chi^2 &= \sum_{i=1}^{r} \sum_{j=1}^{s} \frac{\left(n_{ij} - n\hat{p}_i\hat{q}_j\right)^2}{n\hat{p}_i\hat{q}_j} \\
&= \sum_{i=1}^{r} \sum_{j=1}^{s} \frac{\left(n_{ij} - E_{ij}\right)^2}{E_{ij}}
\end{aligned}
\tag{9.40}
$$

where $E_{ij} = n\hat{p}_i\hat{q}_j = \frac{n_i n_j}{n}$. When H_0 is true, (9.40) follows a $\chi^2_{\alpha,(r-1)(s-1)}$. Hence, the null hypothesis is rejected if $\chi^2 \geq \chi^2_{\alpha,(r-1)(s-1)}$; otherwise it is accepted. Let us look at an example to illustrate this test procedure.

Example 9.25 A company operates four machines on three shifts daily. The table below presents the data on machine breakdowns reported during a six-month period:

Shifts	Machines				Total
	A	B	C	D	
1	10	12	6	7	35
2	10	24	9	10	53
3	13	20	7	10	50
Total	33	56	22	27	138

The problem is to check whether the machines and shifts have any influence on breakdowns.

Solution Here, we are interested in testing whether the machine causing the breakdown and the shift in which the breakdown occurred are independent. That is,

H_0: the shift and machines are independent

H_1: the shift and machines are not independent

The expected frequencies of each cell are calculated using the formula $E_{ij} = \frac{n_i n_j}{n}$, for each $i = 1, 2, \ldots, r$ and $j = 1, 2, \ldots, s$. They are shown in the brackets of each cell.

Shifts	Machines				Total
	A	B	C	D	
1	10 (8)	12 (14)	6 (6)	7 (7)	35
2	10 (13)	24 (22)	9 (8)	10 (10)	53
3	13 (12)	20 (20)	7 (8)	19 (10)	50
Total	33	56	22	27	138

Hence,

$$\chi^2 = \sum_{i=1}^{r} \sum_{j=1}^{s} \frac{\left(n_{ij} - E_{ij}\right)^2}{E_{ij}}$$

$$= \frac{(10-8)^2}{8} + \frac{(12-14)^2}{14} + \cdots + \frac{(19-10)^2}{10}$$

$$= 10.0932$$

Now, for $\alpha = 0.05$ and $(r-1)(s-1) = 6$ degrees of freedom, $\chi^2_{0.05,6} = 12.592$. Since $\chi^2 < \chi^2_{\alpha,(r-1)(s-1)}$, the hypothesis is accepted. The p-value associated with this test is

$$p\text{-value} = P(\chi^2_{(6)} \geq 10.0932) = 0.1208$$

9.7 Modeling Relationship Between Variables

A model is a formal framework for representing the basic features of a complex system by a few central relationships. They simplify the reality through variables of the model. For problem-solving and problem-analyzing process, we need a model. A mathematical model best explains the relationship between independent and dependent variables through the functional form $y = f(x)$, which is the crux of Six Sigma projects. They broadly define the essential features of a physical system or process in mathematical terms. Charts, graphs, and diagrams are used for understanding the pattern of a model and often help to establish associations and correlations between variables. Below, we discuss some important tools of model development through diagrams and charts.

9.7.1 Scatter Diagram and Correlations Study

A scatter diagram is a graph (see Fig. 9.12) that helps to visualize the relationship between two variables. It can be used to check whether one variable is related to another variable and is an effective way to communicate the relationship you find. The scatter plots may be used

- To study and identify possible relationships between the changes observed in two different sets of variables
- To understand the relationships between variables
- To discover whether two variables are related
- To find out if changes in one variable are associated with changes in the other

Scatter plots are also used to test for a cause-and-effect relationship. While interpreting such a relationship, one has to take proper care. Even for correlated data, the actual cause of variation may not be reflected in the diagram. Also the absence of *correlation* does not mean there is no *causation*.

If $x_1, x_2, \ldots, x_n$ and $y_1, y_2, \ldots, y_n$ are two pairs of observations, then the correlation between x and y is obtained as

$$r = \frac{\sum_{i=1}^{n} (x_i - \bar{x})(y_i - \bar{y})}{\sqrt{\sum_{i=1}^{n} (x_i - \bar{x})^2} \sqrt{\sum_{i=1}^{n} (y_i - \bar{y})^2}} \qquad (9.41)$$

Fig. 9.12 Scatter diagram of marks against anxiety scores

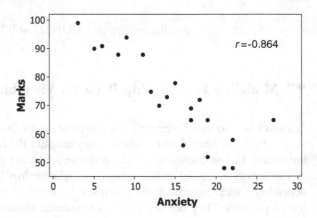

A value of $r = 1$ indicates perfect positive correlation and $r = -1$ indicates perfect negative correlation. A value of $r = 0$ implies the variables are independent and have no association.

Instead of actual observations, if the observations are given in terms of ranks, then the rank correlation can be found from the following formula

$$\rho = 1 - \frac{6 \sum_{i=1}^{n} d_i^2}{n(n^2 - 1)} \tag{9.42}$$

where d_i is the difference between the ranks of X and Y.

Example 9.26 Following are the anxiety scores before a major training program and the marks in mathematics in that test for a group of 10th standard students:

Anxiety	12	16	6	22	17	14	8	27	18	19
Marks	75	56	91	48	69	73	88	65	72	65
Anxiety	11	9	21	3	15	17	22	19	13	5
Marks	88	94	48	99	78	65	58	52	70	90

(a) Construct a scatter plot and describe its nature.
(b) Compute Karl Pearson's correlation coefficient between the two variables and interpret it.

Solution

(a) The scatter plot is presented in Fig. 9.10, which clearly shows a negative association.
(b) The Karl Pearson's correlation coefficient between the two variables is -0.864, which says that the anxiety score is highly negatively associated with the performance. That is, the anxiety can adversely affect the performance in the examination.

Example 9.27 Calculate spearman's coefficient of rank correlation for

X	36	56	20	65	42	33	44	50	15	60
Y	50	35	70	25	58	75	60	45	80	38

Solution Since the actual observations are given, we need to find the corresponding ranks. The calculations are shown below:

X	Y	R_1	R_2	d^2
36	50	4	5	1
56	35	8	2	36
20	70	2	8	36
65	25	10	1	81
42	58	5	6	1
33	75	3	9	36
44	60	6	7	1
50	45	7	4	9
15	80	1	10	81
60	38	9	3	36
				318

Hence,

$$\rho = 1 - \frac{6\sum_{i=1}^{n} d_i^2}{n(n^2 - 1)}$$
$$= 1 - \frac{6 \times 318}{10(10^2 - 1)}$$
$$= -0.9272$$

9.7.1.1 Correlation Versus Causation

Correlation is a measure of the strength of association between two quantitative variables. It is a directed quantity, while causation is non-directed. Many causes can lead to correlations, and the converse may not be true in general. If one action causes another, then they are most certainly correlated. But just because two things occur together does not mean that one caused the other, even if it seems to make sense. The idea that correlation and causation are connected is certainly true; where there is causation, there is likely to be correlation. Indeed, correlation is used when inferring causation; the important point is that such inferences are not always correct because there are other possibilities. Statisticians suggest that the shortest true statement that can be made about causality and correlation is one of the following:

- "Empirically observed covariation is a necessary but not sufficient condition for causality."
- "Correlation is not causation but it sure is a hint."

Technically speaking, correlation measures the degree of linearity between two variables assumed to be completely independent of each other. Correlated occurrences may be due to a common cause. There are several reasons why common sense conclusions about cause and effect might be wrong. For example, the fact that red hair is correlated with blue eyes stems from a common genetic specification which codes for both. A correlation may also be observed when there is causality behind it, for example, it is well established that cigarette smoking not only correlates with lung cancer, but actually causes it. But in order to establish cause, we would have to rule out the possibility that smokers are more likely to live in urban areas, where there is more pollution or any other possible explanation for the observed correlation.

An action or occurrence can cause another (such as smoking causes lung cancer), or it can correlate with another (such as smoking is correlated with alcoholism). Unfortunately, our intuition can lead us astray when it comes to distinguishing between causality and correlation. For example, eating breakfast has long been correlated with success in school for elementary school children. It would be easy to conclude that eating breakfast causes students to be better learners. It turns out, however, that those who do not eat breakfast are also more likely to be absent or tardy and it is absenteeism that is playing a significant role in their poor performance.

Many studies are actually designed to test a correlation, but are suggestive of "reasons" for the correlation. It is possible to show that "kids who watch cartoons are more likely to have eating disorders," a correlation between cartoon watching and eating disorders, but then, they incorrectly conclude that watching cartoons gives kids eating disorders. In many cases, it seems obvious that one action causes another. However, there are also many cases when it is not so clear. In the case of cartoon-watching anorexics, we can neither exclude nor embrace the hypothesis that the television is a cause of the problem. Therefore, a convincing argument for causality is difficult to make in these cases.

In general, it is extremely difficult to establish causality between two correlated events or observances. In contrast, there are many statistical tools to establish a statistically significant correlation. The most effective way of establishing causation is through a controlled study. In a controlled study, two groups of people who are comparable in almost every way are given two different sets of experiences (such as one group watching soap operas and the other game shows), and the outcome is compared. If the two groups have substantially different outcomes, then the different experiences may have caused the different outcome.

Obviously, there are many limitations for controlled studies: For example, if we want to know whether cigarette smoking really causes lung cancer, then it would be problematic to take two comparable groups and make one smoke while denying cigarettes to the other in order to see the effect. Ethically, it is a wrong experiment to perform. This is why epidemiological (or observational) studies are so important. These are studies in which large groups of people are followed over time, and their behavior and outcome are also observed. In these studies, it is extremely difficult (though sometimes still possible) to tease out cause and effect versus a mere correlation.

Fig. 9.13 Scatter diagram
with possible causation

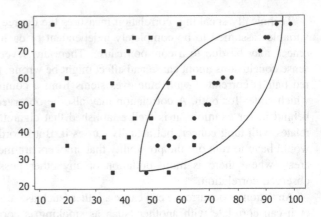

Figure 9.13 depicts a scatter plot showing some negative correlation between two variables. A close look at the points within the two lines covered in (20, 30) shows there is no correlation. So one needs to understand the real causes behind relationship mechanisms using close monitoring of observations to understand the correlation and causation. A "lurking variable" can also affect the relationship between variables, for which one should use stratification and classification techniques to understand the phenomena. Typically, one can only establish correlation, unless the effects are extremely notable and there is no reasonable explanation that challenges causality. When the stakes are high, people are much more likely to jump to causal conclusions. Most often, correlation can be a hint for causation.

Before contemplating the causation effect, one should use some confirmatory analysis to see the effect of correlation. A possible rejection of tests of hypothesis about population correlation (ρ) equal to zero is the answer to it. If correlation value of zero is rejected, then one can conclude that the variables are independent.

In order to test the independence of the variables, that is to test $H_0: \rho = 0$ versus $H_1: \rho \neq 0$, we use an approximate t-test as given below:

$$t = \frac{r}{\sqrt{1 - r^2}} \sqrt{n - 2} \qquad (9.43)$$

which follows a t-distribution with $n - 2$ degrees of freedom. For large values of t, we reject the null hypothesis of independence.

Example 9.28 An auto manufacturing company wanted to investigate how the price of one of its car model depreciates with age. The research department at the company took a sample of eight cars of this model and collected the following information regarding the ages (in years) and prices (in hundreds of dollars) of these cars.

Age	8	3	6	9	2	5	6	3
Price	18	94	50	21	145	42	36	99

Test the hypothesis that age does not have any dependence on the price for $\alpha = 0.01$

Solution Here, we test

1. $H_0: \rho = 0$ versus $H_1: \rho \neq 0$
2. $\alpha = 0.01$
3. Test statistics

$$t = \frac{r}{\sqrt{1 - r^2}}\sqrt{n - 2} = \frac{-0.923}{\sqrt{1 - 0.923^2}}\sqrt{6} = -5.8755$$

4. If $|t| \geq t_{\alpha/2, n-2}$, we reject H_0. Here, $t_{\alpha/2, n-2} = t_{0.005, 6} = 3.7074$. Hence, the test is rejected.
5. Since the test is rejected, it is concluded that there is dependence between the variables. That is, depreciation will be low as age increases.

9.7.2 Regression Analysis

Having identified the inputs and outputs, the next important concern is to establish possible relationships between them. Many engineering and scientific applications generally look for functional relations between variables. This is possible through the study of regression analysis. For instance, in a chemical process, we might be interested in the relationship between the output of the process, the temperature at which it occurs, and the amount of catalyst employed. Knowledge of such a relationship would enable the engineers to predict the output for various values of the temperature and the amount of catalyst.

In many situations, there is a single *response* (output) variable y, also called the *dependent* variable, which depends on the value of a set of *input*, also called *independent* variables $x_1, x_2, \ldots, x_k$. This is expressed as $y = f(x_1, x_2, \ldots, x_k)$. The simplest type of relationship between the Y and the input variable is a linear relationship, which is expressed as

$$y = \beta_0 + \beta_1 x_1 + \beta_2 x_2 + \cdots + \beta_k x_k \qquad (9.44)$$

where $\beta_0, \beta_1, \ldots, \beta_k$ are some constants. If this relationship holds for the knowledge of $\beta_0, \beta_1, \ldots, \beta_k$, then it would be possible to exactly predict the response for any set of input values. However, in practice, such precision is almost never attainable, and the most that one can expect is that (0.1) would be valid, subject to some random error. With this the model, (9.44) becomes

$$y = \beta_0 + \beta_1 x_1 + \beta_2 x_2 + \cdots + \beta_k x_k + \varepsilon \qquad (9.45)$$

where ε is assumed to have zero mean and constant variance. Another way of expressing (9.44) is in terms of expected response given inputs $x = (x_1, x_2, \ldots, x_k)$ as $E(Y|x)$. Thus,

$$E(Y|x) = y = \beta_0 + \beta_1 x_1 + \beta_2 x_2 + \cdots + \beta_k x_k \qquad (9.46)$$

Equation (9.46) is called a linear regression equation, as it describes the regression of Y on the set of independent variables $x_1, x_2, \ldots, x_k$. The quantities $\beta_0, \beta_1, \ldots, \beta_k$ are called the regression coefficients (or parameters) and must usually be estimated from the given data. A regression equation containing a single independent variable, that is, for which $k = 1$, is called a *simple regression* equation, whereas one containing many independent variables is called a *multiple regression* equation.

Thus, a simple linear regression model supposes a linear relationship between the mean response and the value of a single independent variable. It can be expressed as

$$y = \beta_0 + \beta_1 x + \varepsilon \qquad (9.47)$$

where x is the value of the independent variable, and ε is assumed to have mean zero and constant variance. In regression theory, the line of best fit is also called the regression equation or regression line or a prediction equation (or line) or fitted line or simply a model.

Example 9.29 The following data indicate the relationship between x, the specific gravity of a wood sample, and y, its maximum crushing strength in compression parallel to the grain:

x	0.41	0.46	0.44	0.47	0.42	0.39	0.41	0.44	0.43	0.44
y	1850	2620	2340	2690	2160	1760	2500	2750	2730	3120

Some of the practical questions one may be interested in are as follows: (i) Is a linear relationship reasonable? (ii) What is the exact linear relationship between variables? (iii) What is the correlation between variables? and (iv) What is the predicted maximum crushing strength of a wood sample whose specific gravity is 0.43? We will answer these questions one by one below: A plot of y_i versus x_i called a *scatter diagram* is given in Fig. 9.14.

To construct the actual relationship (or a mathematical mode) and using the same for estimation and prediction, we need to understand the mathematical theory of curve fitting as discussed below.

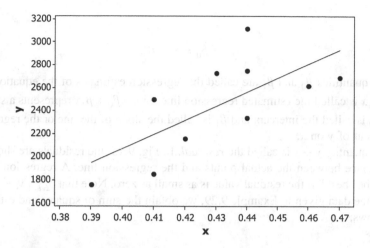

Fig. 9.14 Scatter plot with regression line

9.7.2.1 Least Square Estimators of the Regression Parameters

To determine the estimators of β_0 and β_1, we use the principle of least squares (PLS). If A and B are the estimators of β_0 and β_1, respectively, then the sum of squared differences between the estimated responses and the actual response values —call it error sum of squares (ESS)—are given by

$$\text{ESS} = \sum_{i=1}^{n} (y_i - \hat{y}_i)^2 = \sum_{i=1}^{n} (y_i - A - Bx_i)^2 \tag{9.48}$$

Then, the method of PLS chooses as estimators of β_0 and β_1 the values of A and B that minimizes the ESS. Thus, minimizing ESS with respect to A and B, we get the following normal equations:

$$\frac{\partial \text{ESS}}{\partial A} = -2 \sum_{i=1}^{n} (y_i - A - Bx_i) = 0 \Rightarrow \sum_{i=1}^{n} y_i = nA + B \sum_{i=1}^{n} x_i \tag{9.49}$$

$$\frac{\partial \text{ESS}}{\partial B} = -2 \sum_{i=1}^{n} (y_i - A - Bx_i)x_i = 0 \Rightarrow \sum_{i=1}^{n} x_i y_i = A \sum_{i=1}^{n} x_i + B \sum_{i=1}^{n} x_i^2 \tag{9.50}$$

If we denote $\bar{y} = \frac{1}{n}\sum_{i=1}^{n} y_i$ and $\bar{x} = \frac{1}{n}\sum_{i=1}^{n} x_i$, then the estimators can be simplified as follows:

$$B = \hat{\beta}_1 = \frac{\sum_{i=1}^{n} x_i y_i - n\bar{x}\bar{y}}{\sum_{i=1}^{n} x_i^2 - n\bar{x}^2} \tag{9.51}$$

and

$$A = \hat{\beta}_0 = \bar{y} - B\bar{x} \qquad (9.52)$$

The quantities $\hat{\beta}_0$ and $\hat{\beta}_1$ are called the regression estimates of the equation, and $\hat{\beta}_0 + \hat{\beta}_1 x$ is called the estimated regression line. Since $\hat{\beta}_0 + \hat{\beta}_1 x$ represents a straight line, $\hat{\beta}_0$ is called the intercept and $\hat{\beta}_1$ is called the slope of the line or the regression coefficient of y on x.

The quantity $y - \hat{y}$ is called the *residual*. In Fig. 9.15, the residuals are shown as the distance between the actual points and the regression line. A regression line is said to be "best," if the residual value is as small as zero. Note that $\sum_{i=1}^{n} y_i - \hat{y} = 0$.

For the data given in Example 9.29, we obtain the sum of squares and estimates as follows:

Thus,

$$\sum_{i=1}^{n} x_i = 4.31, \sum_{i=1}^{n} y_i = 24{,}520, \sum_{i=1}^{n} x_i^2 = 1.8629, \sum_{i=1}^{n} y_i^2 = 61{,}761{,}600 \quad \text{and}$$

$$\sum_{i=1}^{n} x_i y_i = 10{,}632.9$$

Hence,

$$\bar{x} = 0.431, \quad \text{and} \quad \bar{y} = 2452$$

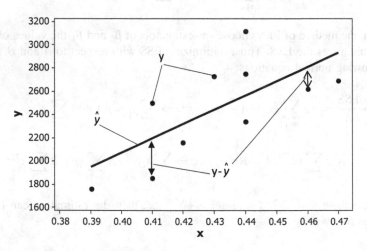

Fig. 9.15 Regression line and residual

$$\hat{\beta}_1 = \frac{\sum_{i=1}^{n} x_i y_i - n\overline{xy}}{\sum_{i=1}^{n} x_i^2 - n\overline{x}^2}$$

$$= \frac{1372.8 - 10 \times 0.431 \times 2452}{1.8629 - 10(0.431^2)}$$

$$= 12245.75$$

and

$$\hat{\beta}_0 = \overline{y} - \hat{\beta}_1 \overline{x}$$

$$= 2452 - 12245.75 \times 0.431$$

$$= -2825.92$$

Hence, the regression line of y on x is obtained as

$$y = -2825.92 + 12245.75x \qquad (9.53)$$

Equation (9.53) is called the regression equation or estimating equation (or prediction equation) of crushing strength on specific gravity. Therefore, the maximum crushing strength of a wood sample whose specific gravity 0.43 is 2439.753 [upon substituting $x = 0.43$ in Eq. (9.53)]. Proceeding in a similar way, one can construct a regression line of x on y as

$$x = -0.334 + 0.000039y \qquad (9.54)$$

Equation (9.54) is the prediction model for predicting the value of the specific gravity for a given crushing strength.

There are a number of relationships existing between the covariance, regression coefficient, and correlation coefficient. For example, if the regression coefficient of y on x, say β_{yx} is

$$\hat{\beta}_1 = \beta_{yx} = \frac{\sum_{i=1}^{n} x_i y_i - n\overline{xy}}{\sum_{i=1}^{n} x_i^2 - n\overline{x}^2}$$

$$= \frac{\text{Cov}(X, Y)}{\sigma_x^2} = \frac{r\sigma_y}{\sigma_x} \qquad (9.55)$$

Similarly, the regression coefficient of x on y, say β_{xy} is

$$\beta_{xy} = \frac{\sum_{i=1}^{n} x_i y_i - n\overline{xy}}{\sum_{i=1}^{n} y_i^2 - n\overline{y}^2}$$

$$= \frac{\text{Cov}(X, Y)}{\sigma_y^2} = \frac{r\sigma_x}{\sigma_y} \qquad (9.56)$$

Thus,

$$r = \pm\sqrt{\beta_{xy}\beta_{yx}} \tag{9.57}$$

For the example discussed above, since both the regression coefficients are positive, the correlation coefficient is obtained as

$$r = \sqrt{12245.75 \times 0.000039}$$
$$= 0.69$$

The regression coefficients play a very important role in understanding the behavior of the associations and the influence of the independent variable on the dependent variable. The presence of the regression coefficient in the equation makes the inference more meaningful than the absence of it. Therefore, the inferential study associated with the regression coefficient is essential for decision making. Especially, the test for $H_0: \beta_1 = 0$ against $H_1: \beta_1 \neq 0$ is an important test of hypothesis for a linear regression line.

9.7.2.2 Statistical Inference on Regression Coefficients

In order to propose any inference based on the parameters, we first need to make some distributional assumptions on the random errors specified in the model. The usual approach is to assume that the random errors are independent normal random variables having mean 0 and variance σ^2. Thus, if y_i is the response corresponding to the input values x_i, then $y_i \sim N(0, \sigma^2)$. Note that the variance of the random component does not depend on the input value but rather is a constant. This value σ^2 is not assumed to be known but rather must be estimated from the data.
If we denote

$$S_{xx} = \sum_{i=1}^{n}(x_i - \bar{x})^2 = \sum_{i=1}^{n}x_i^2 - n\bar{x}^2 \tag{9.58}$$

$$S_{yy} = \sum_{i=1}^{n}(y_i - \bar{y})^2 = \sum_{i=1}^{n}y_i^2 - n\bar{y}^2 \tag{9.59}$$

$$S_{xy} = \sum_{i=1}^{n}(x_i - \bar{x})(y_i - \bar{y}) = \sum_{i=1}^{n}x_iy_i - n\bar{x}\bar{y} \tag{9.60}$$

and

$$\text{ESS} = \frac{S_{xx}S_{yy} - S_{xy}^2}{S_{xx}} \tag{9.61}$$

then the least square estimates can be expressed as

$$\hat{\beta}_1 = \frac{S_{xy}}{S_{xx}} \tag{9.62}$$

and

$$\hat{\beta}_0 = \bar{y} - B\bar{x} \tag{9.63}$$

Further, the distributions of $\hat{\beta}_0$ and $\hat{\beta}_1$ are, respectively, given as

$$\hat{\beta}_0 \sim N\left(\beta_0, \frac{\sigma^2 \sum_{i=1}^{n} x_i^2}{n S_{xx}}\right) \tag{9.64}$$

and

$$\hat{\beta}_1 \sim N\left(\beta_1, \frac{\sigma^2}{S_{xx}}\right) \tag{9.65}$$

As mentioned earlier, a test for $H_0: \beta_1 = 0$ against $H_1: \beta_1 \neq 0$ is important due to the fact that it is equivalent to state that the mean response does not depend on the input, or, equivalently, that there is no regression effect on the input variable, we use the test statistic

$$\frac{\hat{\beta}_1 - \beta_{H_0}}{\sqrt{\frac{\sigma^2}{S_{xx}}}} = \sqrt{S_{xx}} \frac{\hat{\beta}_1 - \beta_{H_0}}{\sigma} \sim N(0, 1)$$

and is independent of $\frac{\text{ESS}}{\sigma^2} \sim \chi_{(n-2)}^2$. Hence, from the definition of a t-random variable, it follows that

$$\sqrt{\frac{(n-2)S_{xx}}{\text{ESS}}}\left(\hat{\beta}_1 - \beta_{H_0}\right) \sim t_{n-2} \tag{9.66}$$

If H_0 is true, then

$$\sqrt{\frac{(n-2)S_{xx}}{\text{ESS}}}\hat{\beta}_1 \sim t_{n-2}$$

Hence, a significance level α test for H_0: $\beta_1 = 0$ is to be rejected if

$$\sqrt{\frac{(n-2)S_{xx}}{\text{ESS}}}|\hat{\beta}_1| > t_{(\alpha/2,n-2)} \qquad (9.67)$$

Otherwise, the test is accepted. The p-value of the test is obtained as

$$p\text{-value} = P\left(|t_{\alpha/2,n-2}| \geq \sqrt{\frac{(n-2)S_{xx}}{\text{ESS}}}|\hat{\beta}_1|\right) \qquad (9.68)$$

Interpreting p-values in regression analysis

- If p-value ≥ 0.05

  - Do not reject the H_0
  - There is not enough evidence to say there is a statistically significant slope
  - If there is a true slope, either the variation is too large or the sample size is too small to detect it

- If p-value < 0.05

  - Reject the H_0, conclude that H_1 is true
  - There is a statistically significant slope
  - Evaluate importance of the relationship and investigate further for causes

For the example under consideration, suppose we have to test H_0: $\beta_1 = 0$ against H_1: $\beta_1 \neq 0$. On analyzing Table 9.12, we get

Table 9.12 Regression analysis table

	x	y	X^2	Y^2	xy
	0.41	1850	0.1681	3,422,500	758.5
	0.46	2620	0.2116	6,864,400	1205.2
	0.44	2340	0.1936	5,475,600	1029.6
	0.47	2690	0.2209	7,236,100	1264.3
	0.42	2160	0.1764	4,665,600	907.2
	0.39	1760	0.1521	3,097,600	686.4
	0.41	2500	0.1681	6,250,000	1025
	0.44	2750	0.1936	7,562,500	1210
	0.43	2730	0.1849	7,452,900	1173.9
	0.44	3120	0.1936	9,734,400	1372.8
Total	4.31	24520	1.8629	61,761,600	10632.9

$$S_{xx} = \sum_{i=1}^{n} x_i^2 - n\bar{x}^2$$

$$= 1.8629 - 10(0.431)^2$$

$$= 0.00529$$

$$S_{yy} = \sum_{i=1}^{n} y_i^2 - n\bar{y}^2$$

$$= 617,61600 - 10(2452)^2$$

$$= 163,8560$$

$$S_{xy} = \sum_{i=1}^{n} x_i y_i - n\overline{xy}$$

$$= 10632.9 - 10(0.431)(2452)$$

$$= 64.78$$

and

$$\text{ESS} = \frac{S_{xx}S_{yy} - S_{xy}^2}{S_{xx}}$$

$$= \frac{(0.00529)(1,638,560) - 64.78^2}{0.00529}$$

$$= 84,5280.5$$

Hence,

$$\sqrt{\frac{(n-2)S_{xx}}{\text{ESS}}}|\hat{\beta}_1| = \sqrt{\frac{(10-2)(0.00529)}{845,280.5}} \times 12,245.75$$

$$= 2.740045$$

Now, the test $H_0: \beta_1 = 0$ is rejected if $\sqrt{\frac{(n-2)S_{xx}}{\text{ESS}}}|\hat{\beta}_1| > t_{(\alpha/2,n-2)}$; otherwise, the test is accepted. From Appendix Table A.5, we get $t_{(\alpha/2,n-2)} = t_{0.025,8} = 2.306$. Since the calculated value is larger than $t_{0.025,8} = 2.306$, the test is rejected.

The p-value of the test is obtained as

$$p\text{-value} = P\left(|t_{n-2}| \geq \sqrt{\frac{(n-2)S_{xx}}{\text{ESS}}}|\hat{\beta}_1|\right)$$

$$= P(|t_{n-2}| \geq 2.740045)$$

$$= 0.02544$$

According to the MINITAB output given below, the *p*-value of the regression coefficient is 0.025. All other estimates are almost similar to the manually obtained estimates.

MINITAB Regression Analysis: y versus x

```
The regression equation is
y = - 2826 + 12246 x


Predictor   Coef  SE Coef      T      P
Constant   -2826     1929  -1.46  0.181
x          12246     4469   2.74  0.025


S = 325.054   R-Sq = 48.4%   R-Sq(adj) = 42.0%


Analysis of Variance

Source          DF       SS       MS     F      P
Regression       1   793279   793279  7.51  0.025
Residual Error   8   845281   105660
Total            9  1638560
```

The suitability of a regression model can also be assessed through the value of adjusted R^2, R^2_{adj}, where

$$R^2_{adj} = 1 - \frac{\text{ESS}/(n-p)}{S_{yy}/(n-1)} \tag{9.69}$$

The model that maximizes R^2_{adj} is considered to be a good candidate for the best regression equation. Since $S_{yy}/(n-1)$ is a constant, the model that maximizes the R^2_{adj} value also minimizes the mean square error (See [13, 28]). This assessment is most suitable when there are more than two explanatory variables in the model. In the above example, ESS = 845,280.5 and $S_{yy} = 1,638,560$, and hence,

$$R^2_{adj} = 1 - \frac{845,280.5/8}{1,638,560/9}$$

$$= 0.4196$$

which according to MINITAB output is 0.42, that is, 42 %.

9.7.2.3 The Coefficient of Determination and the Sample Correlation Coefficient

In one of the previous sections, we have discussed the method of establishing the extent of association between response and input values through the correlation coefficient. Suppose we wanted to measure the amount of variation in the set of response values $y_1, y_2, \ldots, y_n$ corresponding to the set of input values $x_1, x_2, \ldots, x_n$. The variation in the values of y_i arises from two factors. First, because the input values x_i are different, the response variables y_i all have different mean values, which will result in some variation in their values. Second, the variation also arises from the fact that even when the differences in the input values are taken into account, each of the response variables y_i has variance σ^2 and thus will not exactly equal the predicted value at its input x_i.

In order to understand how much of the variation in the values of the response variables is due to the different input values, let us recall that the quantity

$$\text{ESS} = \frac{S_{xx}S_{yy} - S_{xy}^2}{S_{xx}}$$

measures the remaining amount of variation in the response values after the different input values have been taken into account. Then, $S_{yy} - \text{ESS}$ represents the amount of variation in the response variables that is *explained* by the different input values, and so the quantity R^2 defined by

$$R^2 = \frac{S_{yy} - \text{ESS}}{S_{yy}}$$
$$= 1 - \frac{\text{ESS}}{S_{yy}} \tag{9.70}$$

represents the proportion of the variation in the response variables that is explained by the different input values. Here, R^2 is called the *coefficient of determination*. The value of R^2 will be between 0 and 1. A value of R^2 near 1 indicates that most of the variation of the response data is explained by the different input values, whereas a value of R^2 near 0 indicates that little of the variation is explained by the different input values. If $R^2 = 0.833$, it means that almost 83 % of the variation in y variable can be explained by the x variable. About 17 % of the variation is unexplained. Under these circumstances, we can be reasonably comfortable with a prediction for the y variable.

The value of R^2 is often used as an indicator of how well the regression model fits the data, with a value near 1 indicating a good fit and one near 0 indicating a poor fit. In other words, if the regression model is able to explain most of the variation in the response data, then it is considered to fit the data well.

Recall once again the sample correlation coefficient as

$$r = \frac{\sum_{i=1}^{n} (x_i - \bar{x})(y_i - \bar{y})}{\sqrt{\sum_{i=1}^{n} (x_i - \bar{x})^2} \sqrt{\sum_{i=1}^{n} (y_i - \bar{y})^2}}$$

$$= \frac{S_{xy}}{\sqrt{S_{xx}S_{yy}}}$$

Therefore,

$$r^2 = \frac{S_{xy}^2}{S_{xx}S_{yy}}$$

$$= \frac{S_{xx}S_{yy} - S_{xx}\text{ESS}}{S_{xx}S_{yy}}$$

$$= 1 - \frac{\text{ESS}}{S_{yy}}$$

$$= R^2$$

Hence,

$$|r| = \sqrt{R^2} \tag{9.71}$$

That is, the sample correlation coefficient is the square root of the coefficient of determination. The sign of r is the same as that of the regression coefficient, B. This representation of r gives another meaning of the sample coefficient r. For example, if $r = 0.8$, it implies that a simple linear regression model for these data explains 64 % (since $R^2 = 0.64$) of the variation in the response values. That is, 64 % of the variation in the response values is explained by the different input values.

9.7.2.4 Multiple Linear Regressions

A method that describes the statistical relationship between a response and two or more predictors is called multiple linear regressions. For example, a relationship of weight can be studied on height and age. The response variable (Y) here is weight, and the predictor variables (X's) are height and age. Regression often uses the method of least squares, which determines the equation for the straight line that minimizes the sum of the squared vertical distances between the data points and the line. A multiple linear regression equation in k independent variables can be described as

$$y_i = \beta_0 + \beta_1 x_{i1} + \beta_2 x_{i2} + \cdots + \beta_k x_{ik} + \varepsilon_i$$

The above equations can be suitably written in matrix form as

$$y = X\beta + \varepsilon \qquad (9.72)$$

As discussed in Sect. 9.7.2.1, the parameters in the above model are also estimated through the PLS. The $(k + 1)$ normal equations can be obtained as

$$\sum_{i=1}^{n} y_i = n\beta_0 + \beta_1 \sum_{i=1}^{n} x_{i1} + \beta_2 \sum_{i=1}^{n} x_{i2} + \cdots + \beta_k \sum_{i=1}^{n} x_{ik}$$

$$\sum_{i=1}^{n} x_{i1} y_i = \beta_0 \sum_{i=1}^{n} x_{i1} + \beta_1 \sum_{i=1}^{n} x_{i1}^2 + \beta_2 \sum_{i=1}^{n} x_{i1} x_{i2} + \cdots + \beta_k \sum_{i=1}^{n} x_{i1} x_{ik}$$

$$\cdots \qquad \cdots \qquad \cdots \qquad \cdots \qquad \cdots$$

$$\sum_{i=1}^{n} x_{ik} y_i = \beta_0 \sum_{i=1}^{n} x_{ik} + \beta_1 \sum_{i=1}^{n} x_{i1} x_{ik} + \beta_2 \sum_{i=1}^{n} x_{ik} x_{i2} + \cdots + \beta_k \sum_{i=1}^{n} x_{ik}^2$$

The matrix equivalent of the above equations is

$$X'Y = X'X\hat{\beta} \qquad (9.73)$$

Thus, the least square estimate of β is

$$\hat{\beta} = (X'X)^{-1} X'Y \qquad (9.74)$$

Hence, the estimated regression line is

$$\hat{Y} = X\hat{\beta}$$

and the residual is $e = y - \hat{y}$. If we define

$$\text{ESS} = \sum_{i=1}^{n} (y_i - \hat{y}_i)^2 = \sum_{i=1}^{n} e_i^2 = ee'$$

then an unbiased estimate of σ^2 is obtained as

$$\hat{\sigma}^2 = \frac{\text{ESS}}{n - k - 1} \qquad (9.75)$$

and the covariance matrix of $\hat{\beta}$ is

$$\mathrm{Cov}(\hat{\beta}) = \sigma^2 (X'X)^{-1} \qquad (9.76)$$

Note that Eq. (9.72) involves many parameters, where some of them may be significant and some of them may not be so. In order to decide the influence of explanatory variables, one needs to carry out tests of hypotheses of each parameter associated with the variables in the model. So the variable selection in this process may be a tedious task, and hence, model building becomes a redundant activity. Several criteria may be used for evaluating and comparing regression models obtained. The use of R^2 and R^2_{adj} values also help to decide the best models, most of the time. The other criteria for choosing best regression equations are described below.

Stepwise regression. Stepwise regression removes and adds variables to the regression model for the purpose of identifying a useful subset of the predictors. MINITAB provides three commonly used procedures:

- Standard stepwise regression (adds and removes variables), forward selection (adds variables), and backward elimination (removes variables)
- It begins by selecting the single independent variable (entire set of predictors), that is, the "best" predictor which maximizes R^2. Then, it adds (eliminates) variables in sequential manner, in order of importance and at each step it increases R^2
- When you choose the stepwise method, you can enter a starting set of predictor variables in predictors in the initial model. These variables are removed if their p-values are greater than α to enter value. If you want keep variables in the model regardless of their p-values, enter them in predictors to be included in every model in the main dialog box

Best subsets regression. Best subsets regression identifies the best-fitting regression models that can be constructed with the predictor variables you specify. Best subsets regression is an efficient way to identify models that achieve your goals with as few predictors as possible. Subset models may actually estimate the regression coefficients and predict future responses with smaller variance than the full model using all predictors.

MINITAB software examines all possible subsets of the predictors, beginning with all models containing one predictor, and then, all models containing two predictors, and so on. By default, MINITAB displays the two best models for each number of predictors. For example, suppose you conduct a best subsets regression with three predictors. MINITAB will report the best and second best one-predictor models, followed by the best and second best two-predictor models, followed by the full model containing all three predictors.

- It generates regression models using the maximum R^2 criterion by first examining all one-predictor regression models and then selecting the two-predictor models giving the largest R^2. It examines all two-predictor models, selects the two models with the largest R^2, and displays information on these two models. This process continues until the model contains all predictors.

- $C_p = \dfrac{\text{ESS}_p}{\text{MSE}_m} - (n - 2p)$ where ESS_p is error sum of squares for the best model with "p" parameters, and MSE_m is the mean square error for the model with all "m" predictors.

- We look for models where C_p is small and is also close to p, the number of parameters in the model.

Dealing with multicollinearity. One of the assumptions of model accuracy is that X's are not correlated. But this may not be true always. Variance inflation factor (VIF) detects multicollinearity or correlation among predictors. The VIF measures how much the variance of an estimated regression coefficient increases if your predictors are correlated. VIF = 1 indicates no relation among predictors; VIF > 1 indicates that the predictors are correlated; VIF > 5 indicates that the regression coefficients are poorly estimated. In such cases, one should consider the options to break up the multicollinearity by collecting additional data, deleting predictors, using different predictors, or an alternative to least square regression, etc.

A matrix plot is a two-dimensional matrix of individual plots. Matrix plots are good for, among other things, seeing the two-variable relationships among a number of variables all at once. This helps to identify the meaningful relationships with a single graph.

Example 9.30 The time to failure (y) of a machine component is related to the operating voltage (x_1), the motor speed in revolutions per minute (x_2), and the operating temperature (x_3). A designed experiment is run in the R&D laboratory, and the following data are obtained:

y	x_1	x_2	x_3
2145	110	750	140
2155	110	850	180
2220	110	1000	140
2225	110	1100	180
2260	120	750	140
2266	120	850	180
2334	120	1000	140
2340	130	1000	180
2212	115	840	150
2180	115	880	150

Analyze the data by fitting a multiple regression model to the data

Solution Here, we fit the multiple linear regression models as

$$y = \beta_0 + \beta_1 x_1 + \beta_2 x_2 + \beta_3 x_3 + \varepsilon$$

where

$$\hat{\beta} = (X'X)^{-1}X'y$$

$$= \begin{bmatrix} 40.2168 & -0.280855 & -0.0051922 & -0.0180646 \\ -0.2809 & 0.002663 & -0.0000008 & -0.0001727 \\ -0.0052 & -0.000001 & 0.0000091 & -0.0000185 \\ -0.0181 & -0.000173 & -0.0000185 & 0.0003465 \end{bmatrix} \begin{bmatrix} 22337 \\ 2594430 \\ 20179580 \\ 3530540 \end{bmatrix}$$

$$= \begin{bmatrix} 1109 \\ 8.64 \\ 0.261 \\ -0.711 \end{bmatrix} = \begin{bmatrix} \hat{\beta}_0 \\ \hat{\beta}_1 \\ \hat{\beta}_2 \\ \hat{\beta}_3 \end{bmatrix}$$

Thus, the regression equation is

$$\hat{y} = 1109 + 8.64 x_1 + 0.261 x_2 - 0.711 x_3$$

The MINITAB output of the regression analysis is given below:

MINITAB Regression Analysis: y versus x1, x2, x3

```
The regression equation is
y = 1109 + 8.64 x1 + 0.261 x2 - 0.711 x3


Predictor      Coef   SE Coef       T      P   VIF
Constant     1108.7     175.9    6.30  0.001
x1            8.639     1.431    6.04  0.001   1.0
x2          0.26077   0.08361    3.12  0.021   1.1
x3          -0.7114    0.5162   -1.38  0.217   1.2


S = 27.7328    R-Sq = 88.7%    R-Sq(adj) = 83.0%

PRESS = 14409.9    R-Sq(pred) = 64.66%


Analysis of Variance

Source          DF      SS      MS       F      P
Regression       3   36159   12053   15.67  0.003
Residual Error   6    4615     769
Total            9   40774
```

Analysis interpretation The R^2 value indicates that the predictors explain 88.7 % of the variance in failure time. The adjusted R^2 is 83 %, which accounts for the number of predictors in the model. Both values indicate that the model fits the data well. The predicted R^2 value is 64.66 %, which is not close to the R^2 and adjusted R^2 values. The model does not appear to be a good fit for prediction. Possibly, the presence of the variable x_3, which is not significant at $\alpha = 0.05$, may be causing the model to be unfit for prediction. The quantity "PRESS" stands for "prediction error sum of squares" and is a measure of how well the model for one experiment is likely to predict in a new experiment. Small values of PRESS are desirable. The PRESS and predicted R^2 are related in the following way:

$$R^2_{predicted} = 1 - \frac{PRESS}{S_{yy}} = 1 - \frac{14,409.9}{40,774} = 0.6466$$

The *p*-value in the analysis of variance Table (0.003) shows that the model estimated by the regression procedure is significant at an α-level of 0.05. This indicates that at least one coefficient is different from zero. The *p*-values for the estimated coefficients of x_1 and x_2 are both less than 0.05, indicating that they are significantly related to the failure time of the component. Also, the VIF in each case is close to 1, and hence, the multicollinearity is not very serious.

Interpretation of graphs The normal probability plot given in Fig. 9.16 shows that the residuals are approximately linear and consistent with normal distribution. Although, the lower extreme point could be an outlier, its overall effect will be not

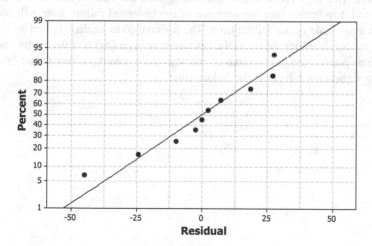

Fig. 9.16 Normal probability plot of residuals

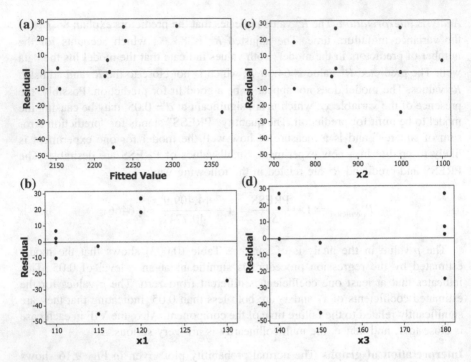

Fig. 9.17   Residual plots

significant. The Fig. 9.17a is the residual against the fitted value of the time to failure, and it indicates that the variance of the observed failure time will increase with the magnitude of the failure time. The residual plots against x_1 and x_2 indicate that the failure time may go up as the operating voltage (x_1) and the motor speed in revolutions per minute (x_2) increases, whereas the operating temperature (x_3) may not have significant effect on the failure time.

MINITAB Regression Analysis: y versus x1, x2

```
The regression equation is
y = 1072 + 8.28 x1 + 0.223 x2


Predictor      Coef    SE Coef      T      P   VIF
Constant     1071.6      184.6   5.80  0.001
x1            8.285      1.495   5.54  0.001   1.0
x2          0.22289    0.08387   2.66  0.033   1.0


S = 29.4601    R-Sq = 85.1%    R-Sq(adj) = 80.8%

PRESS = 16558.6    R-Sq(pred) = 59.39%


Analysis of Variance

Source           DF       SS      MS      F      P
Regression        2    34699   17349  19.99  0.001
Residual Error    7     6075     868
Total             9    40774
```

Both the descriptive and graphical analyses support the exclusion of the variable x_3 from the model. Thus, a model with only x_1 and x_2 may be more appropriate for predicting the lifetime of the component.

MINITAB Best Subsets Regression: y versus x1, x2, x3

```
Response is y

                               Mallows            x x x
Vars  R-Sq  R-Sq(adj)          C-p        S      1 2 3
  1   70.1       66.3          9.9   39.058      X
  1   19.8        9.7         36.5   63.949        X
  2   85.1       80.8          3.9   29.460      X X
  2   70.3       61.9         11.7   41.571      X   X
  3   88.7       83.0          4.0   27.733      X X X
```

Analysis interpretation We look for models where C_p is small and is also close to p (= 4), the number of parameters in the model. Here, C_p is small when x_1 and x_2 are included in the model. The C_p value is also low for all the three variables included in the model. Hence, a regression model with x_1 and x_2 or a model with all variables included will yield the expected result in this example.

Checking a regression model
Regression analysis does not end once the regression model is fit. You should examine residual plots and other diagnostic statistics to determine whether your

Table 9.13 Characteristics of a good regression model

Characteristics of a model	Possible solutions	Tools to verify
Linear relationship between response and predictors	• Add higher order term to model • Transform variables	• Residuals versus variables plot
Residuals are independent of (not correlated with) one another	• Add new predictor • Use time series analysis • Add lag variable	• Durbin–Watson statistic • Residuals versus order plot
Residuals have constant variance	• Transform variables • Use weighted least squares	• Residuals versus fit plot
Residuals are normally distributed	• Transform variables • Check for outliers	• Histogram of residuals • Residuals versus fit plot • Normality test

model is adequate and the assumptions of regression have been met. If your model is inadequate, it will not correctly represent your data. For example,

- The standard errors of the coefficients may be biased, leading to incorrect t-values and p-values
- Coefficients may have the wrong sign
- The model may be overly influenced by one or two points
- VIF may be too large
- p-values are significant, but low R^2 value

In Table 9.13, we present some of the characteristics of a good regression model and its possible solutions.

For any other problem, one may need to check to see whether the data are entered correctly, especially observations identified as unusual, then try to determine the cause of the problem, etc., and then proceed. A suitable transformation also helps in constructing good models in such situations.

9.7.3 Nonlinear Regression

In situations where the functional relationship between the response y and the independent variable x cannot be adequately approximated by a linear relationship, we must try for nonlinear relationships such as polynomial, exponential, logarithmic, or any other suitable relationship. Specifically, if the form of the relationship can be determined, it is possible sometimes, to transform it into a linear form, by a change of variables. For instance, if the intensity of a life component is related to time t by the functional form

$$h(t) = \exp(\alpha + \beta t) \tag{9.77}$$

On taking logarithm, this can be expressed as

$$\log h(t) = \alpha + \beta t$$

If we let $y = \log h(t)$, then the above model can be expressed as

$$y = \alpha + \beta t + \varepsilon$$

The parameters can be now estimated by the usual method of least square approach. Similarly, for any kind of nonlinear relationship, one can get back to linearity through suitable transformation of the variables and proceed with the estimation of the parameters involved in the model. In the next two subsections, we discuss the fitting of nonlinear models for a given situation.

9.7.3.1 Polynomial Regression

A general polynomial regression between two variables can be expressed as

$$y = \beta_0 + \beta_1 x + \beta_2 x^2 + \cdots + \beta_k x^k + \varepsilon \tag{9.78}$$

where $\beta_0, \beta_1, \beta_2, \ldots, \beta_k$ are regression coefficients that would have to be estimated through the model. If the data set consists of n pairs $(x_i, y_i), i = 1, 2, \ldots, n$, then the least square estimators of $\beta_0, \beta_1, \beta_2, \ldots, \beta_k$—call them $B_0, B_1, B_2, \ldots, B_k$—are those values which minimize

$$\sum_{i=1}^{n} (y_i - B_0 - B_1 x_i - \cdots - B_k x_i^k)^2$$

The normal equations corresponding with the above minimization will lead to the following $(k + 1)$ equations:

$$\sum_{i=1}^{n} y_i = nB_0 + B_1 \sum_{i=1}^{n} x_i + B_2 \sum_{i=1}^{n} x_i^2 \cdots - B_k \sum_{i=1}^{n} x_i^k$$

$$\sum_{i=1}^{n} x_i y_i = B_0 \sum_{i=1}^{n} x_i + B_1 \sum_{i=1}^{n} x_i^2 + B_2 \sum_{i=1}^{n} x_i^3 \cdots - B_k \sum_{i=1}^{n} x_i^{k+1} \tag{9.79}$$

$$- - - \qquad - - - \qquad - - -$$

$$\sum_{i=1}^{n} x_i^r y_i = B_0 \sum_{i=1}^{n} x_i^r + B_1 \sum_{i=1}^{n} x_i^{r+1} + B_2 \sum_{i=1}^{n} x_i^{r+2} \cdots - B_k \sum_{i=1}^{n} x_i^{2k}$$

In fitting a polynomial to a set of data points, it is often possible to determine the necessary degree of the polynomial by a study of the scatter diagram. Therefore, one should try to fit lower degree polynomial to start with and understand the necessity of higher degree polynomials. Unlike linear regression, it is extremely risky to use a polynomial fit to predict the value of a response at an input level x_0 that is far away from the input levels $x_i, i = 1, 2, \ldots, n$ used in finding the polynomial fit.

Example 9.31 Fit a polynomial regression to the following data:

x	1	2	3	4	5	6	7	8	9	10
y	20.6	30.8	55	71.4	97.3	131.8	156.3	197.3	238.7	291.7

Solution Here, $n = 10$, and the sums and sum of squares for fitting various polynomials are obtained from the table below.

x	y	x^2	x^3	x^4	xy	x^2y
1	20.6	1	1	1	20.6	20.6
2	30.8	4	8	16	61.6	123.2
3	55	9	27	81	165	495
4	71.4	16	64	256	285.6	1142.4
5	97.3	25	125	625	486.5	2432.5
6	131.8	36	216	1296	790.8	4744.8
7	156.3	49	343	2401	1094.1	7658.7
8	197.3	64	512	4096	1578.4	12,627.2
9	238.7	81	729	6561	2148.3	19,334.7
10	291.7	100	1000	10,000	2917	29,170
55	1290.9	385	3025	25,333	9547.9	77749.1

Therefore,

$$\sum_{i=1}^{10} x_i = 55, \sum_{i=1}^{10} x_i^2 = 385, \sum_{i=1}^{10} x_i^3 = 3025, \sum_{i=1}^{10} x_i^4 = 25{,}333$$

$$\sum_{i=1}^{10} y_i = 1291.1, \sum_{i=1}^{10} x_i y_i = 9549.3, \sum_{i=1}^{10} x_i^2 y_i = 77{,}758.9$$

Using the normal equations given in (9.79), we get the following three normal equations for a second degree polynomial:

$$10B_0 + 55B_1 + 385B_2 = 1291.1$$
$$55B_0 + 385B_1 + 3025B_2 = 9549.3$$
$$385B_0 + 3025B_1 + 25{,}333B_2 = 77{,}758.9$$

On solving, the least square equations, we get

$$B_0 = 12.593, B_1 = 6.326, B_2 = 2.123$$

Thus, the estimated quadratic regression equation is

$$y = 12.593 + 6.326x + 2.123x^2$$

Similarly, a cubic plot for the above data has the equation

$$y = 5.157 + 12.94x + 0.685x^2 + 0.08726x^3$$

And the linear regression equation fitted to the data is

$$y = -34.1 + 29.7x$$

The fitted line plots are shown in Fig. 9.18. Note that the quadratic and cubic plots are almost same.

The MINITAB output is given below. Note that the R^2 is very high, and hence, the model fit is very good. Also the p-value is 0.0, which shows that the regression coefficients are highly significant.

Fig. 9.18 Fitted line plots

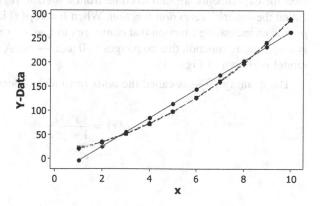

MINITAB Polynomial Regression Analysis: y versus x

```
The regression equation is
y = 12.64 + 6.297 x + 2.125 x**2


S = 3.69652    R-Sq = 99.9%   R-Sq(adj) = 99.8%


Analysis of Variance

Source       DF       SS       MS       F       P
Regression    2   75020.1  37510.1  2745.12  0.000
Error         7      95.6     13.7
Total         9   75115.8


Sequential Analysis of Variance

Source       DF       SS       F       P
Linear        1   72635.9  234.32  0.000
Quadratic     1    2384.3  174.49  0.000
```

9.7.3.2 Logistic Regression Models for Binary Output Data

Often in Six Sigma projects, one may observe the output data in terms of binary outputs such as (0, 1), (success, failure), (positive, negative), (non-defective, defective), (observed, not-observed), (true, false), (yes, no), (present, absent), and (confirmed, non-confirmed). If we suppose that the experiments can be performed at various levels, and that an experiment performed at level x will result in success with a probability $p(x)$, $-\infty < x < \infty$. If $p(x)$ is of the form

$$p(x) = \frac{e^{a+bx}}{1 + e^{a+bx}} \tag{9.80}$$

then the experiments are said to come from a *logistic regression model* and $p(x)$ is called the *logistic regression function*. When $b = 0$, $p(x)$ is a constant; if $b > 0$, then $p(x)$ is an increasing function that converges to 1 as $x \to \infty$, and if $b < 0$, then $p(x)$ is a decreasing function that converges to 0 as $x \to \infty$. A typical logistic regression model is shown in Fig. 9.19.

The quantity $\frac{p(x)}{1-p(x)}$ is called the *odds ratio* and is denoted as $o(x)$. That is,

$$o(x) = \frac{p(x)}{1 - p(x)}$$
$$= e^{a+bx} \tag{9.81}$$

Fig. 9.19 Logistic regression

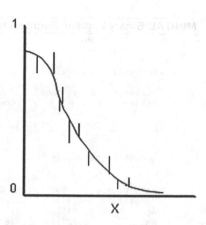

Thus, when $b > 0$, the odds increase exponentially in the input level x; when $b < 0$, the odds decrease exponentially in the input level x. The log odds (*logit*) are a linear function:

$$\log[o(x)] = a + bx \tag{9.82}$$

Another model related to this is the *probit* model, where $p(x)$ is equal to the probability that a standard normal random variable is less than $(a + bx)$. That is,

$$p(x) = \Phi(a + bx) = \frac{1}{\sqrt{2\pi} \int\limits_{-\infty}^{a+bx} e^{-y^2/2} dy} \tag{9.83}$$

Example 9.33 The data in Table 9.14 represent 50 diabetic patient's BMI measure collected through a clinical trial. Fit a binary logistic regression and suggest the dependency of BMI on gender.

Solution Here, we model gender (response variable) on age and BMI measurement. Since gender is a binary variable, we use binary logistic regression. The MINITAB output of the same is shown below.

MINITAB Binary Logistic Regression: Gender_2 versus Age, BMI

```
Link Function: Logit


Response Information

Variable  Value  Count
Gender    M        28    (Event)
          F        22
          Total    50


Logistic Regression Table

                                               Odds     95% CI
Predictor      Coef      SE Coef       Z      P  Ratio  Lower  Upper
Constant    -5.40883    2.18020    -2.48  0.013
Age          0.0071136  0.0183123    0.39  0.698  1.01   0.97   1.04
BMI          0.227152   0.0767057    2.96  0.003  1.26   1.08   1.46


Log-Likelihood = -28.175
Test that all slopes are zero: G = 12.244, DF = 2, P-Value = 0.002


Goodness-of-Fit Tests

Method            Chi-Square  DF      P
Pearson            46.3791    47   0.498
Deviance           56.3491    47   0.165
Hosmer-Lemeshow    13.7359     8   0.089


Measures of Association:
(Between the Response Variable and Predicted Probabilities)

Pairs        Number  Percent  Summary Measures
Concordant     466    75.6    Somers' D                 0.52
Discordant     148    24.0    Goodman-Kruskal Gamma     0.52
Ties             2     0.3    Kendall's Tau-a           0.26
Total          616   100.0
```

Table 9.14 Data on diabetic patients

Gender	Age	BMI	Gender	Age	BMI
M	30	23.92	M	38	24.54
F	40	19.71	F	44	24.74
F	55	14.20	M	45	20.07
M	65	27.92	M	85	20.90
M	63	18.75	M	15	23.32
M	26	24.89	F	25	17.96
F	28	27.06	F	35	14.53
F	59	24.03	M	65	16.05
F	63	21.34	F	58	24.92
F	57	19.13	M	48	33.73
M	24	25.04	M	42	29.90
M	45	35.21	M	52	27.99
M	44	34.72	M	56	27.24

(continued)

Table 9.14 (continued)

Gender	Age	BMI	Gender	Age	BMI
F	34	20.93	F	57	17.96
F	33	19.15	M	62	17.11
M	29	29.40	M	86	20.55
F	58	19.47	M	16	31.98
F	45	20.32	F	85	17.31
F	56	21.14	M	29	21.34
F	49	28.67	F	27	17.01
M	24	31.89	M	43	20.66
M	26	32.46	F	27	25.86
M	29	29.67	M	65	27.24
M	35	24.09	M	68	29.44
F	37	23.71	F	74	24.45

Analysis interpretation There are 28 male and 22 female patients under study. The logistic table indicates that the regression coefficient of BMI is significant ($p < 0.05$) and hence, it is not zero. The odds ratio for BMI is slightly more than 1, indicating that the BMI has some dependency on gender. The variable age does not show any significance on gender. The log-likelihood indicates that the regression coefficient of one of the factor is not zero (p-value = 0.004). All goodness-of-fit tests have p-value larger than 0.05, and hence, the model is adequate for the given data. Two of the summary measures show more than 50 % association between the response and predicted probabilities. Figure 9.20 presents the delta chi-Square versus probability analysis of the data, which shows that one observation has high delta chi-square value and it corresponds to the observation 20 (Female aged 49 having BMI 28.67). Possibly, the middle-aged woman may be obese.

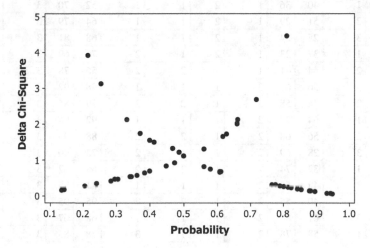

Fig. 9.20 Delta chi-square versus probability

Example 9.34 Table 9.15 presents the data on 60 fliers along with their performance rating for different airlines. The following codes are used for data collection:

Gender: 1 = Male, 2 = Female
Airline: 1 = Jet, 2 = Spice, 3 = Indigo
Class: 1 = Economy, 2 = Business
Flier type: 1 = Frequent, 2 = Occasional, 3 = First timer

Table 9.15 Data on airline versus flier type and rankings

Gender	F_type	R-1	R-2	Airline	class	Gender	F_type	R-1	R-2	Airline	class
1	1	40	36	1	1	1	1	52	65	2	2
1	1	28	28	1	1	1	1	70	80	2	2
2	2	36	30	1	1	1	3	73	79	2	2
1	3	32	28	1	1	1	1	72	88	2	2
1	1	60	40	1	1	1	2	73	89	2	2
1	1	12	14	1	1	1	1	71	72	2	2
1	1	32	26	1	1	2	2	55	58	2	2
1	2	36	30	1	1	1	1	68	67	2	2
1	3	44	38	1	1	2	1	81	85	2	2
2	2	36	35	1	1	2	1	78	80	2	2
2	2	40	42	1	2	1	2	92	95	3	1
1	2	68	49	1	2	1	1	56	60	3	1
2	1	20	24	1	2	1	1	64	70	3	1
1	1	33	35	1	2	1	1	72	78	3	1
1	1	65	40	1	2	2	1	48	65	3	1
1	1	40	36	1	2	1	1	52	70	3	1
1	2	51	29	1	2	1	1	64	79	3	1
1	3	25	24	1	2	1	1	68	81	3	1
2	1	37	23	1	2	1	2	76	69	3	1
1	2	44	41	1	2	2	2	56	78	3	1
1	3	56	67	2	1	1	1	88	92	3	2
1	2	48	58	2	1	2	2	79	85	3	2
1	1	64	78	2	1	1	1	92	94	3	2
1	2	56	68	2	1	2	2	88	93	3	2
2	3	28	69	2	1	1	2	73	90	3	2
2	1	32	74	2	1	2	2	68	67	3	2
2	3	42	55	2	1	1	1	81	85	3	2
1	3	40	55	2	1	2	2	95	95	3	2
1	3	61	80	2	1	1	1	68	67	3	2
1	1	58	78	2	1	1	3	78	83	3	2

Carry out a nominal logistic regression to model the airline type.

Solution Since the variable airline is a nominal type, we carry out a nominal logistic regression given the other variables. The MINITAB output of the same is presented below:

MINITAB Nominal Logistic Regression: F_type versus Gender, Rating-1, ...

```
Response Information

Variable  Value  Count
Airline   3       20    (Reference Event)
          2       20
          1       20
          Total   60
```

```
Factor Information

Factor   Levels  Values
Gender      2     1, 2
```

Logistic Regression Table

Predictor	Coef	SE Coef	Z	P	Odds Ratio	95% CI Lower	Upper
Logit 1: (2/3)							
Constant	3.47172	2.78021	1.25	0.212			
Gender-2	-0.405095	0.846254	-0.48	0.632	0.67	0.13	3.50
Rating-1	-0.0777531	0.0431199	-1.80	0.071	0.93	0.85	1.01
Rating-2	0.0143592	0.0527084	0.27	0.785	1.01	0.91	1.12
flier_type	0.434364	0.503451	0.86	0.388	1.54	0.58	4.14
Logit 2: (1/3)							
Constant	110.111	14531.4	0.01	0.994			
Gender-2	-1.61167	4556.57	-0.00	1.000	0.20	0.00	*
Rating-1	0.648952	276.176	0.00	0.998	1.91	0.00	*
Rating-2	-2.69989	281.963	-0.01	0.992	0.07	0.00	*
flier_type	-1.51026	4515.07	-0.00	1.000	0.22	0.00	*

```
Log-Likelihood = -22.865
Test that all slopes are zero: G = 86.103, DF = 8, P-Value = 0.000
```

Goodness-of-Fit Tests

Method	Chi-Square	DF	P
Pearson	35.8823	106	1.000
Deviance	42.9576	106	1.000

Analysis interpretation There are 20 travelers each of Jet, Spice, and Indigo. Since the reference event is "Indigo" flight, the output is given as *logit* functions of "Spice and Indigo" and "Jet and Indigo". The high p-values for the Logit-1 function indicate that there is sufficient evidence to conclude that Gender-2 ($p = 0.632$) or flier type ($p = 0.388$) affected the choice of the flight Spice, over Indigo. Also, the coefficient -0.40509 for Gender-2 (= female) with an odds ratio 0.67 indicates that Female fliers are less likely to prefer Spice over Indigo. In Logit-2 function, the coefficient -1.61167 for Gender-2 (= female) with an odds ratio 0.2 indicates that

Female fliers are more likely to prefer Jet over the Indigo airline. Similarly, the coefficient −1.51026 for flier type with an odds ratio 0.22 indicates that the airline Jet does matter for frequent and other types of fliers.

The low *p*-value of the log-likelihood test indicates that there is sufficient evidence to conclude that at least one coefficient of regression is nonzero. Also, both goodness-of-fit tests have a *p*-value much higher than 0.05, indicating that the model fits the data adequately.

Example 9.35 For the data in Example 9.34, we now carry out an ordinal logistic regression to model the airline on the two ratings. The MINITAB output of the same is presented below:

MINITAB Ordinal Logistic Regression: Airline versus Rating-1, Rating-2

Link Function: Logit

Response Information

Variable	Value	Count
Airline	1	20
	2	20
	3	20
	Total	60

Logistic Regression Table

Predictor	Coef	SE Coef	Z	P	Odds Ratio	95% CI Lower	Upper
Const(1)	7.20041	1.57142	4.58	0.000			
Const(2)	10.9536	2.08090	5.26	0.000			
Rating-1	-0.0052650	0.0286377	-0.18	0.854	0.99	0.94	1.05
Rating-2	-0.138464	0.0326776	-4.24	0.000	0.87	0.82	0.93

Log-Likelihood = -32.849
Test that all slopes are zero: G = 66.135, DF = 2, P-Value = 0.000

Goodness-of-Fit Tests

Method	Chi-Square	DF	P
Pearson	61.5002	106	1.000
Deviance	59.1064	106	1.000

Measures of Association:
(Between the Response Variable and Predicted Probabilities)

Pairs	Number	Percent	Summary Measures	
Concordant	1072	89.3	Somers' D	0.79
Discordant	124	10.3	Goodman-Kruskal Gamma	0.79
Ties	4	0.3	Kendall's Tau-a	0.54
Total	1200	100.0		

Analysis interpretation There are 20 travelers each of Jet, Spice, and Indigo. The high p-value of Rating-1 and low p-value of Rating-2 indicate that the ratings are not uniform (or significant) for airline type. The Rating-2 is highly significant ($p = 0.000$), which indicates that the experience of travel on a particular airline changes the rating preference. This is further established by conducting a paired t-test on the two rankings and we obtained the t-value $= -2.85$, with a p-value $= 0.006$. The low p-value of the log-likelihood ratio test indicates that one of the regression coefficients is not zero. Both the goodness-of-fit tests have a high p-value indicating the suitability of the model. Similarly, the higher value of summary measures also indicates the better predictability of the model.

9.8 Analysis of Variance

Consider this situation: A manufacturing company is considering purchasing, in large quantity, one out of four different computer packages designed to model a new programming task. A group of executives in this company has claimed that these packages are basically interchangeable, in that the one chosen will have little effect on the final competence of its user. To test this hypothesis, the company has decided to choose 60 of its executives and divide them into four groups of size 15. Each member in group i will then be given the software package i, $i = 1, 2, 3, 4$, to learn the new task. When all the executives complete their study, a comprehensive examination will be taken. The company wants to use the results of this examination to determine whether the computer software packages are really interchangeable or not.

In order to conclude that the software packages are indeed interchangeable, the average test scores in all the groups should be similar, and to conclude that the packages are essentially different, there should be a large variation among these average test scores. Similarly, what could be concluded, if the members of the first group score significantly higher than those of the other groups? or, is software package 1 superior to any other software package? Are the executives in group 1 better programmers? To be able to reach such a conclusion, the method of division of the 60 executives into four groups should be done completely in random. So, let us suppose that the division of the executives was indeed random. It is now probably reasonable to suppose that the test score of a given individual should be approximately a normal random variable having parameters that depend on the package from which he was taught. Also, it is probably reasonable to suppose that whereas the average test score of an academician will depend on the software package he/she was exposed to, the variability in the test scores will result from the inherent variation of 60 different people and not from the particular package used.

Thus, if we let $X_{ij}, i = 1, 2, 3, 4, j = 1, 2, \ldots, 15$ denote the test score of the jth executive in group i, a reasonable model might be able to suppose that the X_{ij} are independent random variables with X_{ij} having a normal distribution with unknown mean μ_i and unknown variance σ^2. The hypothesis that the software packages are interchangeable is then equivalent to the hypothesis $H_0: \mu_1 = \mu_2 = \mu_3 = \mu_4$. The technique used to test such a hypothesis for a multitude of parameters relating to population means is known as the *analysis of variance* (ANOVA). If the test of hypothesis is for two populations, then one can use z-test or t-test as per the practical significance.

9.8.1 One-Way Classification or One-Factor Experiments

In a one-way classification, measurements (or observations) are obtained for r independent groups of samples (or treatments), where each treatment repeats s times. Thus, we have Table 9.16. In the above example, $r = 4$ and $s = 15$. Thus, $X_{ij}, i = 1, 2, \ldots, r, j = 1, 2, \ldots, s$ denote the measurement in the ith row and jth column.

Model of ANOVA In Table 9.14, each column is assumed to be a random sample of size s from the population for that particular treatment. If μ_i is the mean of the ith treatment, then the observations X_{ij} will differ from the population mean μ_i by a *chance error* (or *random error*), denoted by ε_{ij}. Thus,

$$X_{ij} = \mu_i + \varepsilon_{ij}$$

These errors are assumed to be normally distributed with an unknown mean 0 and variance σ^2. If μ is the mean of the population for all treatments and if we let $\alpha_i = \mu_i - \mu$, so that $\mu_i = \mu + \alpha_i$, then the model becomes

$$X_{ij} = \mu + \alpha_i + \varepsilon_{ij} \tag{9.84}$$

where $\sum_i \alpha_i = 0$. Then, the null hypothesis that all treatment means are equal is given by

$$H_0: \alpha_1 = \alpha_2 = \cdots = \alpha_r = 0$$

Table 9.16 One-way classification

Treatment-1	Treatment-2	...	Treatment-r
X_{11}	X_{21}	...	X_{r1}
X_{12}	X_{22}	...	X_{r2}
...	...	...	...
X_{1s}	X_{2s}	...	X_{rs}

or, equivalently

$$H_0: \mu_1 = \mu_2 = \cdots = \mu_r = \mu$$

If H_0 is true, the treatment populations will all have the same normal distribution. In such a case, there is just one treatment population. That is, there is no significant difference between the treatments. To test the above hypothesis, we use F-test for equal means, and the test statistic is given as

$$F = \frac{\hat{S}_B^2}{\hat{S}_W^2}, \quad \text{which has } F\text{-distribution} \quad \text{with } r - 1 \text{ and } r(s - 1) \text{ degrees of freedom.}$$

where $\hat{S}_B^2$ is the mean square variation between treatments, and $\hat{S}_W^2$ is the mean square variation within treatments. The hypothesis of equal means is rejected if $F > F_{\alpha,(r-1,r(s-1))}$.

ANOVA table To prepare the ANOVA table, we use the following notations:
  Let

$$T = \sum_{i=1}^{r} \sum_{j=1}^{s} X_{ij}, \quad \overline{X} = \frac{T}{rs}, \quad T_{i.} = \sum_{j=1}^{s} X_{ij} \quad \text{and} \quad \overline{X}_{i.} = \frac{T_{i.}}{s}$$

Then,

$$\text{total variation} = V = \sum_{i=1}^{r} \sum_{j=1}^{s} (X_{ij} - \overline{X})^2 \qquad (9.85)$$

$$= \sum_{i=1}^{r} \sum_{j=1}^{s} X_{ij}^2 - \frac{T^2}{rs}$$

Since,

$$X_{ij} - \overline{X} = (X_{ij} - \overline{X}_{i.}) + (\overline{X}_{i.} - \overline{X})$$

therefore,

$$\sum_{i=1}^{r} \sum_{j=1}^{s} (X_{ij} - \overline{X})^2 = \sum_{i=1}^{r} \sum_{j=1}^{s} (X_{ij} - \overline{X}_{i.})^2 + \sum_{i=1}^{r} \sum_{j=1}^{s} (\overline{X}_{i.} - \overline{X})^2$$

$$= \sum_{i=1}^{r} \sum_{j=1}^{s} (X_{ij} - \overline{X}_{i.})^2 + s \sum_{i=1}^{r} (\overline{X}_{i.} - \overline{X})^2$$

that is,

$$\text{total variation} = \text{variation within treatments} + \text{variation between treatments}$$

$$V = V_W + V_B \tag{9.86}$$

where

$$V_W = \sum_{i=1}^{r} \sum_{j=1}^{s} (X_{ij} - \overline{X}_{i.})^2 \tag{9.87}$$

and

$$V_B = s \sum_{i=1}^{r} (\overline{X}_{i.} - \overline{X})^2$$

$$= \frac{1}{s} \sum_{i=1}^{r} T_{i.}^2 - \frac{T^2}{rs} \tag{9.88}$$

The one-way analysis of variance for equal number of observations is presented in Table 9.17. In the table, the mean sum of squares is obtained by dividing the sum of squares by their corresponding degrees of freedom.

In the ANOVA presented above, every treatment has equal number of observations. That is, r treatments have s number of observations. Suppose each treatment has unequal number of observations, say $n_1, n_2, \ldots, n_r, \left(\sum_{i=1}^{r} n_i = n \right)$, then the above results are modified as follows:

$$V = \sum_{i=1}^{r} \sum_{j=1}^{n_i} (X_{ij} - \overline{X})^2$$

$$= \sum_{i=1}^{r} \sum_{j=1}^{n_i} X_{ij}^2 - \frac{T^2}{n} \tag{9.89}$$

Table 9.17 One-way ANOVA for equal number of observations

Source of variation	Degrees of freedom	Sum of squares	Mean sum of squares	F ratio
Between treatments	$r - 1$	$V_B = s \sum_{i=1}^{r} (\overline{X}_{i.} - \overline{X})^2$	$\hat{S}_B^2 = \frac{V_B}{r-1}$	$\hat{S}_B^2 / \hat{S}_W^2$
Within treatments	$r(s-1)$	$V_W = \sum_{i=1}^{r} \sum_{j=1}^{s} (X_{ij} - \overline{X}_{i.})^2$	$\hat{S}_W^2 = \frac{V_W}{r(s-1)}$	
Total	$r(s-1)$	$V = \sum_{i=1}^{r} \sum_{j=1}^{s} (X_{ij} - \overline{X})^2$		

$$V_B = \sum_{i=1}^{r} n_i (\overline{X}_{i.} - \overline{X})^2$$

$$= \sum_{i=1}^{r} \frac{T_{i.}^2}{n_i} - \frac{T^2}{n} \qquad (9.90)$$

and

$$V_W = V - V_B \qquad (9.91)$$

Accordingly, the ANOVA table for unequal number of observations is presented in Table 9.18.

Example 9.36 A company wishes to purchase one out of five machines A, B, C, D, or E. In an experiment designed to test whether there is a difference in the machines performance, each of the five different operators collected the following data equal number of times. The table below shows the number of units produced per machine.

A	B	C	D	E
68	72	60	48	64
72	53	82	61	65
77	63	64	57	70
42	53	75	64	68
53	48	72	50	53

Test the hypothesis that there is no significant difference among the machines at 5 % significance level.

Solution Here, $r = s = 5$, and we want to test

$H_0: \mu_A = \mu_B = \mu_C = \mu_D = \mu_E$ versus

$H_1:$ at least one of the machine performances is not the same.

The box plot of each treatment is shown in Fig. 9.21. As such, there is no influence of outliers or inliers in the data set.

Table 9.18 Analysis of variance table for unequal number of observations

Source of variation	Degrees of freedom	Sum of squares	Mean sum of squares	F ratio
Between treatments	$r - 1$	$V_B = \sum_{i=1}^{s} n_i (\overline{X}_{i.} - \overline{X})^2$	$\hat{S}_B^2 = \frac{V_B}{r-1}$	$\hat{S}_B^2 / \hat{S}_W^2$
Within treatments	$n - r$	$V_W = \sum_{i=1}^{r} \sum_{j=1}^{n_i} (X_{ij} - \overline{X}_{i.})^2$	$\hat{S}_W^2 = \frac{V_W}{n-r}$	
Total	$n - 1$	$V = \sum_{i=1}^{r} \sum_{j=1}^{s} (X_{ij} - \overline{X})^2$		

Fig. 9.21 Box plot on
machine performance

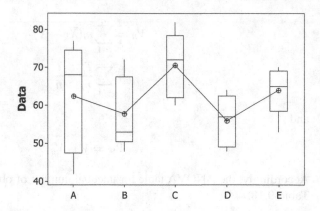

The complete analysis is shown below:

	A	B	C	D	E	Total
	68	72	60	48	64	312
	72	53	82	61	65	333
	77	63	64	57	70	331
	42	53	75	64	68	302
	53	48	72	50	53	276
$T_{i.}$	312	289	353	280	320	1554
$T_{i.}^2$	97,344	83,521	124,609	78,400	102,400	486,274

Now,

$$T = \sum_{i=1}^{r} \sum_{j=1}^{s} X_{ij} = 1554$$

$$\overline{X} = \frac{T}{rs} = \frac{1554}{25} = 62.16$$

$$V = \sum_{i=1}^{r} \sum_{j=1}^{s} X_{ij}^2 - \frac{T^2}{n} = 99,138 - \frac{1554^2}{25} = 2541.36$$

$$V_B = \frac{1}{s} \sum_{i=1}^{r} T_{i.}^2 - \frac{T^2}{rs} = \frac{486,274}{5} - \frac{1554^2}{25} = 658.16$$

and

$$V_W = V - V_B = 1883.2$$

The analysis of variance of the machine performance is given in Table 9.19.

From Appendix Table A.7, the F-table value for 4 and 20 degrees of freedom for
0.05 level of significance is $F_{0.05,(4,20)} = 2.87$. Since $F < F_{0.05,(4,20)}$, we cannot

Table 9.19 Analysis of variance for machine performance

Source of variation	Degrees of freedom	Sum of squares	Mean sum of squares	F ratio
Between treatments	$r - 1 = 4$	$V_B = 658.16$	$\hat{S}_B^2 = 164.54$	$\hat{S}_B^2/\hat{S}_W^2 = 1.7475$
Within treatments	$r(s - 1) = 20$	$V_W = 1883.2$	$\hat{S}_W^2 = 94.16$	
Total	$r(s - 1) = 24$	$V = 2541.36$		

reject H_0 and conclude that the machine performance is the same throughout. The p-value of the test is

$$p\text{-value} = P(F_{(4,20)} \geq 1.7475) = 0.179$$

MINITAB One-way ANOVA

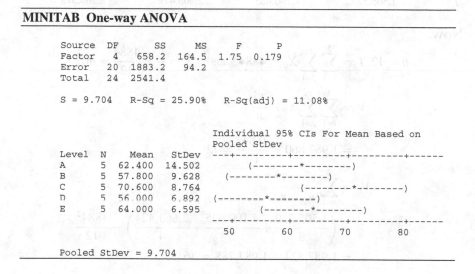

```
Source   DF     SS      MS     F      P
Factor    4    658.2  164.5  1.75  0.179
Error    20   1883.2   94.2
Total    24   2541.4

S = 9.704    R-Sq = 25.90%    R-Sq(adj) = 11.08%


                                 Individual 95% CIs For Mean Based on
                                 Pooled StDev
Level  N    Mean    StDev   ---+---------+---------+---------+------
A      5   62.400  14.502                (--------*--------)
B      5   57.800   9.628      (--------*--------)
C      5   70.600   8.764                   (--------*--------)
D      5   56.000   6.892   (--------*--------)
E      5   64.000   6.595            (--------*--------)
                            ---+---------+---------+---------+------
                              50        60        70        80


Pooled StDev = 9.704
```

Example 9.37 The data below present the lifetime in hours of samples from three different types of television tubes manufactured by a company.

Sample 1	407	411	409		
Sample 2	404	406	408	405	402
Sample 3	410	408	406	408	

Decide whether there is any difference among the three types of tubes at 0.01 level of significance.

Solution Here, the sample sizes are unequal and $r = 3, n_1 = 3, n_2 = 5, n_3 = 4$, and $n = 12$. We are interested in testing

$$H_0: \mu_1 = \mu_2 = \mu_3 \text{ versus}$$

H_1: the lifetime of at least one of the tube is not the same.

The complete analysis is shown below:

	Sample 1	Sample 2	Sample 3	Total
	407	404	410	1221
	411	406	408	1225
	409	408	406	1223
		405	408	813
		402		402
$T_{i.}$	1227	2025	1632	4884
$T_{i.}^2$	1,505,529	4,100,625	2,663,424	8,269,578

Now,

$$n = 12, T = \sum_{i=1}^{3} \sum_{j=1}^{n_i} X_{ij} = 4884 \quad \text{and} \quad \overline{X} = \frac{T}{n} = \frac{4884}{12} = 407$$

$$V = \sum_{i=1}^{3} \sum_{j=1}^{n_i} X_{ij}^2 - \frac{T^2}{n}$$

$$= 1,987,860 - \frac{4884^2}{12} = 72$$

$$V_B = \sum_{i=1}^{3} \frac{T_{i.}^2}{n_i} - \frac{T^2}{n}$$

$$= \left(\frac{1,505,529}{3} + \frac{4,100,625}{5} + \frac{2,663,424}{4} \right) - \frac{4884^2}{12}$$

$$= 1,987,824 - 1,987,788 = 36$$

and

$$V_W = V - V_B = 36$$

The analysis of variance for the tube lifetime is shown in Table 9.20.

From Appendix Table A.6, the F-table value for 2 and 9 degrees of freedom for 0.01 level of significance is $F_{0.05,(2,9)} = 4.26$. Since $F > F_{0.05,(2,9)}$, we reject H_0 and conclude that the lifetime of the machines is not the same. The p-value of the test is

$$p\text{-value} = P(F_{(2,9)} \geq 4.5) = 0.044$$

Table 9.20 ANOVA for tube lifetime

Source of variation	Degrees of freedom	Sum of squares	Mean sum of squares	F ratio
Between treatments	$r - 1 = 2$	$V_B = 36$	$\hat{S}_B^2 = 18$	$\hat{S}_B^2 / \hat{S}_W^2 = 4.5$
Within treatments	$n - r = 9$	$V_W = 36$	$\hat{S}_W^2 = 4$	
Total	$n - 1 = 11$	$V = 72$		

MINITAB One-way ANOVA for life times

```
Source  DF    SS     MS    F      P
Factor  2   36.00  18.00  4.50  0.044
Error   9   36.00   4.00
Total  11   72.00

S = 2    R-Sq = 50.00%    R-Sq(adj) = 38.89%

                                  Individual 95% CIs For Mean Based on
                                  Pooled StDev
Level      N    Mean   StDev   --------+---------+---------+---------+-
Sample 1   3  409.00   2.00                        (---------*---------)
Sample 2   5  405.00   2.24    (-------*-------)
Sample 3   4  408.00   1.63                  (--------*--------)
                               --------+---------+---------+---------+-
                                  405.0     407.5     410.0     412.5

Pooled StDev = 2.00
```

9.8.2 Two-Way Classification or Two-Factor Experiments

The two-way classification is just a generalization of the one-way classification, where we assume that there is one experimental value corresponding to each treatment and block. Thus, two experimental variations due to treatments and blocks will be identified in this model. Let X_{ij} be the observation corresponding to ith treatment and jth block. Thus, an $r \times s$ two-factor experiment can be represented as in Table 9.21.

The model of a two-way factor experiment can be written as

$$X_{ij} = \mu + \alpha_i + \beta_j + \varepsilon_{ij} \tag{9.92}$$

where $\sum_i \alpha_i = 0$ and $\sum_j \beta_j = 0$. Here, μ is the population grand mean, α_i is the treatment effect, β_j is the effect due to blocks, and ε_{ij} is the effect due to random

Table 9.21 Two-way classification

Treatments	Blocks				
	1	2	...	s	Total
1	X_{11}	X_{12}	...	X_{1s}	$X_{1.}$
2	X_{21}	X_{22}	...	X_{2s}	$X_{2.}$
...	...	...	...	...	...
r	X_{r1}	X_{r2}	...	X_{rs}	$X_{r.}$
Total	$X_{.1}$	$X_{.2}$	...	$X_{.s}$	T

(chance or error) variation. We assume that ε_{ij} is normally distributed with mean 0 and variance σ^2, so that X_{ij} is normally distributed with mean μ and variance σ^2. Here, we are interested in testing two hypotheses namely,

(i) All treatment (row) means are equal: That is, $H_0^{(1)}: \alpha_1 = \alpha_2 = \cdots = \alpha_r = 0$ and

(ii) All block (column) means are equal: That is, $H_0^{(2)}: \beta_1 = \beta_2 = \cdots = \beta_s = 0$

For testing the above hypothesis, we use a two-way ANOVA table. For the computation of the sum of squares and variances associated with treatments and blocks, we use the notations: $T = \sum_{i=1}^{r}\sum_{j=1}^{s} X_{ij}$, $\overline{X} = \frac{T}{rs}$, $\overline{X}_{i.} = \frac{1}{s}\sum_{j=1}^{s} X_{ij}$, $\overline{X}_{j} = \frac{1}{r}\sum_{i=1}^{r} X_{ij}$ and proceed as before as done for one-way factor analysis. The analysis of variance table is presented in Table 9.22.

The critical regions for the hypotheses (1) and (2) shown above are given as follows:

(1) Reject $H_0^{(1)}$ if $\hat{S}_T^2/\hat{S}_E^2 \geq F_{\alpha,(r-1,(r-1)(s-1))}$ and

(2) Reject $H_0^{(2)}$ if $\hat{S}_B^2/\hat{S}_E^2 \geq F_{\alpha,(s-1,(r-1)(s-1))}$

In the two-way classification discussed above, we have considered only one replication for each treatment and block. Instead, if we replicate the experiment for t times, then there will be an interaction effect due to treatment and blocks. In such a situation, the two-way model is modified as

Table 9.22 Two-way ANOVA table

Source of variation	Degrees of freedom	Sum of squares	Mean sum of squares	F ratio
Between treatments	$r-1$	$V_T = s\sum_{i=1}^{r} (\overline{X}_{i.} - \overline{X})^2$	$\hat{S}_T^2 = \frac{V_T}{r-1}$	$\hat{S}_T^2/\hat{S}_E^2$
Between blocks	$s-1$	$V_B = r\sum_{j=1}^{s} (\overline{X}_{j} - \overline{X})^2$	$\hat{S}_B^2 = \frac{V_B}{s-1}$	$\hat{S}_B^2/\hat{S}_E^2$
Error or Residual	$(r-1)(s-1)$	$V_e = V - V_T - V_B$	$\hat{S}_E^2 = \frac{V_e}{(r-1)(s-1)}$	
Total	$rs-1$	$V = \sum_{i=1}^{r}\sum_{j=1}^{s} (X_{ij} - \overline{X})^2$		

$$X_{ijk} = \mu + \alpha_i + \beta_j + \gamma_{ij} + \varepsilon_{ijk} \tag{9.93}$$

where the quantity γ_{ij} is the interaction effect, such that $\sum_i \gamma_{ij} = 0$ and $\sum_j \gamma_{ij} = 0$. With the interaction present in the model, there are three hypotheses to test in the model. They are

(i) All treatment (row) means are equal: That is, $H_0^{(1)}: \alpha_1 = \alpha_2 = \cdots = \alpha_r = 0$

(ii) All block (column) means are equal: That is, $H_0^{(2)}: \beta_1 = \beta_2 = \cdots = \beta_s = 0$, and

(iii) There is no interaction between treatments and blocks: That is $H_0^{(3)}: \gamma_{ij} = 0, \forall i, j$

The modified analysis of variance table with the interaction effect is shown in Table 9.23.

The critical regions for the hypotheses (1), (2), and (3) discussed above are given as follows:

(1) Reject $H_0^{(1)}$ if $\hat{S}_T^2 / \hat{S}_E^2 \geq F_{\alpha,(r-1,rs(t-1))}$

(2) Reject $H_0^{(2)}$ if $\hat{S}_B^2 / \hat{S}_E^2 \geq F_{\alpha,(s-1,rs(t-1))}$ and

(3) Reject $H_0^{(3)}$ if $\hat{S}_I^2 / \hat{S}_E^2 \geq F_{\alpha,((r-1)(s-1),rs(t-1))}$

Example 9.38 Here, we will recall the data discussed in Example 9.36 where we consider the block effect as the operator effect (or column effect) along with the treatment effect (row effect) and then treat it as a two-way classification model. Table 9.24 represents the two-way classified data with necessary computations.

Table 9.23 Two-way ANOVA table with interaction

Source of variation	Degrees of freedom	Sum of squares	Mean sum of squares	F ratio
Between treatments	$r - 1$	$V_T = st \sum_{i=1}^{r} (\overline{X}_{i..} - \overline{X})^2$	$\hat{S}_T^2 = \frac{V_T}{r-1}$	$\hat{S}_T^2 / \hat{S}_E^2$
Between blocks	$s - 1$	$V_B = rt \sum_{j=1}^{s} (\overline{X}_{.j.} - \overline{X})^2$	$\hat{S}_B^2 = \frac{V_B}{s-1}$	$\hat{S}_B^2 / \hat{S}_E^2$
Interaction	$(r - 1)$ $(s - 1)$	$V_I = t \sum_{j=1}^{s} (\overline{X}_{ij.} - \overline{X}_{i..} - \overline{X}_{.j.} - \overline{X})^2$	$\hat{S}_I^2 = \frac{V_I}{(r-1)(s-1)}$	$\hat{S}_I^2 / \hat{S}_E^2$
Error or Residual	$rs(t - 1)$	$V_E = V - V_T - V_B - V_I$	$\hat{S}_E^2 = \frac{V_W}{r(s-1)}$	
Total	$rs(t - 1)$	$V V - \sum_{i=1}^{r} \sum_{j=1}^{s} \sum_{k=1}^{t} (X_{ijk} - \overline{X})^2$		

Table 9.24 Data and analysis table

Treatments	O-1	O-2	O-3	O-4	O-5	$T_{i.}$	$T_{i.}^2$
A	68	72	77	42	53	312	97,344
B	72	53	63	53	48	289	83,521
C	60	82	64	75	72	353	124,609
D	48	61	57	64	50	280	78,400
E	64	65	70	68	53	320	102,400
T_j	312	333	331	302	276	1554	486,274
T_j^2	97,344	110,889	109,561	91,204	76,176	485,174	

Now,

$$T = \sum_{i=1}^{r}\sum_{j=1}^{s} X_{ij} = 1554$$

$$\overline{X} = \frac{T}{rs} = \frac{1554}{25} = 62.16$$

$$V = \sum_{i=1}^{r}\sum_{j=1}^{n_i} X_{ij}^2 - \frac{T^2}{rs} = 99{,}138 - \frac{1554^2}{25} = 2541.36$$

$$V_T = \frac{1}{s}\sum_{i=1}^{r} T_{i.}^2 - \frac{T^2}{rs} = \frac{486{,}274}{5} - \frac{1554^2}{25} = 658.16$$

$$V_B = \frac{1}{r}\sum_{j=1}^{s} T_j^2 - \frac{T^2}{rs} = \frac{485{,}174}{5} - \frac{1554^2}{25} = 438.16$$

and

$$V_W = V - V_T - V_B = 1445.04$$

The analysis of variance is shown in Table 9.23.

From Appendix Table A.8, the F-table value for 4 and 16 degrees of freedom for 0.05 level of significance is $F_{0.05,(4,16)} = 3.0069$. Since $F < F_{0.05,(4,16)}$, in both the cases, we cannot reject the hypothesis and hence conclude that the machine's performance and operators are same throughout. The p-value of the test for treatment effect is p-value $= P(F_{(4,16)} \geq 1.822) = 0.1739$ and for block effect is $P(F_{(4,16)} \geq 1.213) = 0.3438$ (Table 9.25).

Table 9.25 Analysis of variance table for machine performance

Source of variation	Degrees of freedom	Sum of squares	Mean sum of squares	F ratio
Between treatments	4	$V_T = 658.16$	$\hat{S}_T^2 = 164.54$	$\hat{S}_T^2/\hat{S}_E^2 = 1.822$
Between blocks	4	$V_B = 438.16$	$\hat{S}_B^2 = 109.54$	$\hat{S}_B^2/\hat{S}_E^2 = 1.213$
Error	16	$V_e = 1445.04$	$\hat{S}_E^2 = 90.315$	
Total	24	$V = 2541.36$		

Fig. 9.22 Individual value plot of yield versus machine and operator

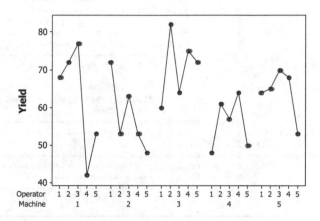

To Analyze the data in MINITAB, we reorganize the above data as shown below. The individual value plot (Fig. 9.22) and analysis output are shown below:

Row	Operator	Machine	Yield	Row	Operator	Machine	Yield
1	1	1	68	14	3	4	57
2	1	2	72	15	3	5	70
3	1	3	60	16	4	1	42
4	1	4	48	17	4	2	53
5	1	5	64	18	4	3	75
6	2	1	72	19	4	4	64
7	2	2	53	20	4	5	68
8	2	3	82	21	5	1	53
9	2	4	61	22	5	2	48
10	2	5	65	23	5	3	72
11	3	1	77	24	5	4	50
12	3	2	63	25	5	5	53
13	3	3	64				

MINITAB Two-way ANOVA: Yield versus Machine, Block

```
Source    DF       SS        MS        F       P
Machine    4    658.16    164.540    1.82    0.174
Operator   4    438.16    109.540    1.21    0.344
Error     16   1445.04     90.315
Total     24   2541.36

S = 9.503    R-Sq = 43.14%    R-Sq(adj) = 14.71%

              Individual 95% CIs For Mean Based on
              Pooled StDev
Machine  Mean    ---+---------+---------+---------+------
1        62.4          (--------*--------)
2        57.8       (--------*--------)
3        70.6                  (--------*---------)
4        56.0    (--------*--------)
5        64.0             (--------*--------)
                 ---+---------+---------+---------+------
                   50        60        70        80


              Individual 95% CIs For Mean Based on
              Pooled StDev
Block  Mean    --+---------+---------+---------+-------
1      62.4           (----------*----------)
2      66.6             (----------*-----------)
3      66.2            (-----------*----------)
4      60.4        (----------*-----------)
5      55.2    (----------*----------)
               --+---------+---------+---------+-------
                48.0      56.0      64.0      72.0
```

Example 9.39 (Two-way classification with replication). Table 9.26 gives the number of articles produced by four different operators working on two different types of machine, I and II, on different days of the week. Determine at the 0.05 level of significance whether there are any differences (a) between the operators and (b) between the machines.

Table 9.26 Machine versus operator

Operator	Machine I					Machine II				
	Mon.	Tue.	Wed.	Thu.	Fri.	Mon.	Tue.	Wed.	Thu.	Fri.
A	15	18	17	20	12	14	16	18	17	15
B	12	16	14	18	11	11	15	12	16	12
C	14	17	18	16	13	12	14	16	14	11
D	19	16	21	23	18	17	15	18	20	17

Solution Here, we carry out a two-factor ANOVA with replication, where factor I will be treated as the operator and factor II will be the machine type. Accordingly, we rearrange the data for analysis in MINITAB as shown below:

Row	Operator	Machine	Articles	Row	Operator	Machine	Articles
1	1	1	15	21	3	1	14
2	1	1	18	22	3	1	17
3	1	1	17	23	3	1	18
4	1	1	20	24	3	1	16
5	1	1	12	25	3	1	13
6	1	2	14	26	3	2	12
7	1	2	16	27	3	2	14
8	1	2	18	28	3	2	16
9	1	2	17	29	3	2	14
10	1	2	15	30	3	2	11
11	2	1	12	31	4	1	19
12	2	1	16	32	4	1	16
13	2	1	14	33	4	1	21
14	2	1	18	34	4	1	23
15	2	1	11	35	4	1	18
16	2	2	11	36	4	2	17
17	2	2	15	37	4	2	15

(continued)

MINITAB Two-way ANOVA: Articles versus Operator, Machine

```
Source        DF    SS       MS        F      P
Operator       3  129.8   43.2667   7.98   0.000
Machine        1   19.6   19.6000   3.61   0.066
Interaction    3    5.4    1.8000   0.33   0.802
Error         32  173.6    5.4250
Total         39  328.4

S = 2.329   R-Sq = 47.14%   R-Sq(adj) = 35.57%


                   Individual 95% CIs For Mean Based on
                   Pooled StDev
Operator   Mean    ---------+---------+---------+---------+
1          16.2                   (-------*-------)
2          13.7    (------*-------)
3          14.5       (-------*------)
4          18.4                         (-------*-------)
                   ---------+---------+---------+---------+
                        14.0      16.0      18.0      20.0

                   Individual 95% CIs For Mean Based on
                   Pooled StDev
Machine    Mean    -+---------+---------+---------+--------
1          16.4                 (----------*----------)
2          15.0    (----------*----------)
                   -+---------+---------+---------+--------
                     14.0      15.0      16.0      17.0
```

Row	Operator	Machine	Articles	Row	Operator	Machine	Articles
18	2	2	12	38	4	2	18
19	2	2	16	39	4	2	20
20	2	2	12	40	4	2	17

The main effect and interaction plots are shown in Figs. 9.23 and 9.24, respectively. Figure 9.24 does not show any interaction between the operator and the machine (see also the ANOVA table, the p-value is not significant at 0.05 level). The operators 1 and 2 are consistent in producing the articles, whereas operator 3

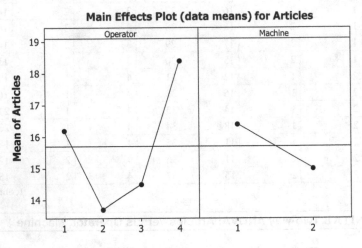

Fig. 9.23 Main effect plot

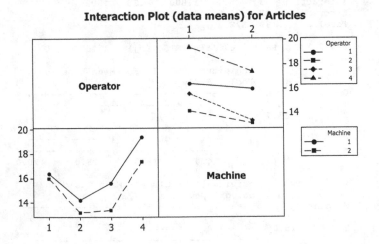

Fig. 9.24 Interaction plot

and 4 are not consistent. Operator 1 is more or less close to the target, and operator 4 is above the average of producing the article. However, the operators 2 and 3 are below the average of the production. Although machine types are significant, the interaction effects of operators are consistent.

9.8.3 Three-Way Classification

We now consider a three-way classification of data which often arises as a combination of column, row, and treatments of a process. The model of a three-way factor experiment can be written as

$$X_{ijk} = \mu + \alpha_i + \beta_j + \gamma_k + \varepsilon_{ij} \begin{cases} i = 1, 2, \ldots, s \\ j = 1, 2, \ldots, s \\ k = 1, 2, \ldots, s \end{cases} \tag{9.94}$$

where $\sum_i \alpha_i = 0$, $\sum_j \beta_j = 0$ and $\sum_k \gamma_k = 0$. Here, μ is the population grand mean, α_i denotes the row effects, β_j is the effect due to the column, γ_k is the effect due to kth treatment, and ε_{ij} is the effect due to random (chance or error) variation. We assume that ε_{ij} is normally distributed with mean 0 and variance σ^2, so that X_{ij} is normally distributed with mean μ and variance σ^2. Since the arrangement of observations is according to a square, the setup can also be called a Latin square arrangement of observations, and the design associated with such a design is called Latin square design (LSD).

Here, we are interested in testing three hypotheses, namely

(i) All row means are equal: That is, $H_0^{(1)}: \alpha_1 = \alpha_2 = \cdots = \alpha_s = 0$

(ii) All column means are equal: That is, $H_0^{(2)}: \beta_1 = \beta_2 = \cdots = \beta_s = 0$, and

(iii) All treatment means are equal: That is, $H_0^{(3)}: \gamma_1 = \gamma_2 = \cdots = \gamma_s = 0$

The analysis of variance table for three-way classification is presented in Table 9.27.

Table 9.27 Three-way ANOVA table with interaction

Source of variation	Degrees of freedom	Sum of squares	Mean sum of squares	F ratio
Rows	$s - 1$	$V_R = \frac{1}{s}\sum_{i=1}^{s} x_{i\ldots}^2 - \frac{T^2}{s^2}$	$\hat{S}_R^2 = \frac{V_R}{s-1}$	$\hat{S}_R^2/\hat{S}_E^2$
Columns	$s - 1$	$V_C = \frac{1}{s}\sum_{j=1}^{s} x_{.j.}^2 - \frac{T^2}{s^2}$	$\hat{S}_C^2 = \frac{V_C}{s-1}$	$\hat{S}_C^2/\hat{S}_E^2$
Treatments	$(s - 1)$	$V_T = \frac{1}{s}\sum_{k=1}^{s} x_{..k}^2 - \frac{T^2}{s^2}$	$S_T^2 = \frac{V_T}{(s-1)}$	$\hat{S}_T^2/\hat{S}_E^2$
Error	$(s - 1)$ $(s - 2)$	$V_E = V - V_R - V_C - V_T$	$\hat{S}_E^2 = \frac{V_E}{(s-1)(s-2)}$	
Total	$s^2 - 1$	$V = \sum_{i=1}^{s}\sum_{j=1}^{s}\sum_{k=1}^{s} X_{ijk}^2 - \frac{T^2}{s^2}$		

Table 9.28 Observations of a chemical process

Observations				Total
A (18)	C (21)	D (25)	B (11)	75
D (22)	B (12)	A (15)	C (19)	68
B (15)	A (20)	C (23)	D (24)	82
C (22)	D (21)	B (10)	A (17)	70
Total	77	74	73	71	295

Example 9.40 An engineer wishes to test the effects of four different treatments (A, B, C, and D) on the yield of a chemical process. In order to eliminate the sources of error due to variability in the process, he uses a Latin square arrangement of observations along with the totals as shown in Table 9.28. Perform an analysis of variance to determine whether there is any difference between the treatments at 0.05 and 0.01 levels of significance?

Solution Here, $s = 4$. $T = 295$. The treatment totals are

$$A = 18 + 20 + 15 + 17 = 70$$
$$B = 15 + 12 + 10 + 11 = 48$$
$$C = 22 + 21 + 23 + 19 = 85$$
$$D = 22 + 21 + 25 + 24 = 92$$

Hence,

$$V_R = \frac{1}{s}\sum_{i=1}^{s} x_{i..}^2 - \frac{T^2}{s^2} = \frac{1}{4}(75^2 + 68^2 + 82^2 + 70^2) - \left(\frac{295}{4}\right)^2 = 29.19$$

$$V_C = \frac{1}{s}\sum_{j=1}^{s} x_{.j.}^2 - \frac{T^2}{s^2} = \frac{1}{4}(77^2 + 74^2 + 73^2 + 71^2) - \left(\frac{295}{4}\right)^2 = 4.69$$

$$V_T = \frac{1}{s}\sum_{k=1}^{s} x_{..k}^2 - \frac{T^2}{s^2} = \frac{1}{4}(70^2 + 48^2 + 85^2 + 92^2) - \left(\frac{295}{4}\right)^2 = 284.19$$

and

$$V = \sum_{i=1}^{s}\sum_{j=1}^{s}\sum_{k=1}^{s} x_{ijk}^2 - \frac{T^2}{s^2} = (18^2 + 21^2 + \cdots + 17^2) - \left(\frac{295}{4}\right)^2 = 329.94$$

The analysis of variance of the chemical process is shown in Table 9.29.

Since $F_{0.95,(3,6)} = 4.76$, we reject at the 0.05 level of significance that the row means are equal, whereas the hypothesis of column mean is accepted and we conclude that there is no significant difference in columns. Similarly, the hypothesis of treatment effect is also rejected showing high significance between treatments.

Table 9.29 Three-way ANOVA table with interaction

Source of variation	Degrees of freedom	Sum of squares	Mean sum of squares	F ratio
Rows	3	29.19	9.73	4.92
Columns	3	4.69	1.563	0.79
Treatments	3	284.19	94.73	47.9
Error	6	11.87	1.978	
Total	15	329.94		

Now as $F_{0.99,(3,6)} = 9.78$, both row and column effects become non-significant. The MINITAB data structure (Table 9.30) and analysis are presented below:

MINITAB: General Linear Model: Yield versus Rows, Columns, Treatment

---

```
Factor      Type   Levels  Values
Rows        fixed    4     1, 2, 3, 4
Columns     fixed    4     1, 2, 3, 4
Treatment   fixed    4     1, 2, 3, 4


Analysis of Variance for Yield, using Adjusted SS for Tests

Source      DF    Seq SS    Adj SS   Adj MS      F      P
Rows         3   29.188    29.188    9.729    4.92  0.047
Columns      3    4.688     4.687    1.562    0.79  0.542
Treatment    3  284.188   284.188   94.729   47.86  0.000
Error        6   11.875    11.875    1.979
Total       15  329.938


S = 1.40683   R-Sq = 96.40%   R-Sq(adj) = 91.00%


Unusual Observations for Yield

Obs     Yield       Fit   SE Fit   Residual   St Resid
  3   25.0000   23.1250   1.1122     1.8750       2.18 R
```

The residual plots and main effect plots are, respectively, given in Figs. 9.25 and 9.26. As such, there are no unusual patterns in the residual plots. The main effect plots clearly show that the treatment effect is highly significant.

The problems discussed above are resultant of industrial, engineering, and laboratory experimentations done under close monitoring of the process. In many situations, the experimentation involves the activities of finding suitable treatment combinations, factor selection, and level settings as the case may be. Hence, experiment must be planned carefully to understand the process, and this is achieved through design of experiments (DOE) techniques. The *experimental designs* are generally used as confirmatory analysis or as a method for improvement processes. In Chap. 10, we will take up these issues in detail for improving the process activities.

Table 9.30 Data structure

Rows	Columns	Treatment	Yield
1	1	1	18
1	2	3	21
1	3	4	25
1	4	2	11
2	1	4	22
2	2	2	12
2	3	1	15
2	4	3	19
3	1	2	15
3	2	1	20
3	3	3	23
3	4	4	24
4	1	3	22
4	2	4	21
4	3	2	10
4	4	1	17

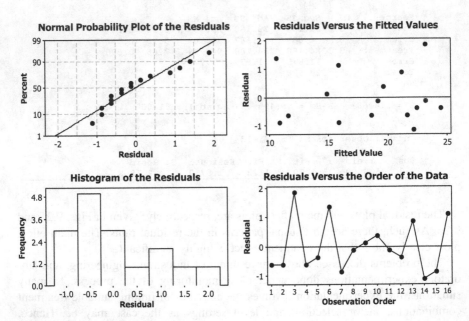

Fig. 9.25 Residual plots for yield

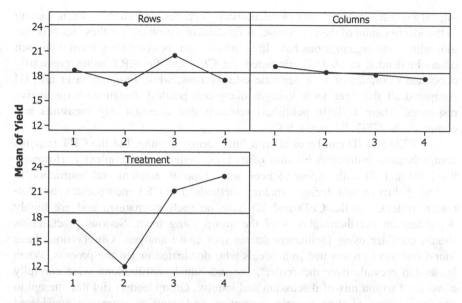

Fig. 9.26 Main effect plot (fitted means) for yield

9.9 Root Cause Analysis

To solve a problem, one must first recognize and understand what is causing the problem. According to Wilson et al. [39], a root cause is the most basic reason for an undesirable condition or problem. If the real cause of the problem is not identified, then one is merely addressing the symptoms and the problem will continue to exist. For this reason, identifying and eliminating root causes of problems is of utmost importance [14, 37]. Root cause analysis is the process of identifying causal factors using a structured approach with techniques designed to provide a focus for identifying and resolving problems.

The commonly used root cause analysis tools are the cause-and-effect diagram (CED) (see Sect. 8.9.3), Pareto analysis (see Sect. 8.9.1), FMEA (see Sect. 9.3), FTA (see Sect. 9.9.1), 5-Why's tools (see Sect. 9.9.2), etc. Some other related tools are the interrelationship diagram (ID) and the current reality tree (CRT). A separate treatment on these methods is discussed in Anderson and Fagerhaug [3], Arcaro [4], Brassard [7], Brassard and Ritter [8], Cox and Spencer [11], Dettmer [12], Lepore and Cohen [22], Moran et al. [29], Robson [33], Scheinkopf [35], Smith [36], and references contained therein.

Ishikawa [20] advocated the CED as a tool for breaking down potential causes into more detailed categories, so that they can be organized into and related as factors that help identify the root causes. In contrast, Mizuno [25] supported the ID as a tool to quantify the relationships between factors and thereby classify potential causal issues or drivers. Finally, Goldratt [18] championed the CRT as a tool to find logical inter-dependent chains of relationships between undesirable effects leading to the identification of the core cause. A fundamental problem for these tools is that individuals and organizations have little information for comparing them with each other. Fredendall et al. [17] compared the CED and the CRT using previously published examples of their separate effectiveness, while Pasquarella et al. [31] compared all the three tools using a one-group posttest design with qualitative responses. There is little published research that quantitatively measures and compares the CED, ID, and CRT.

The CED and ID can be used with little formal training, but the CRT requires comprehensive instruction because of its logic system and complexity. However, the CED and ID both appear to need some type of supplemental instruction in critical evaluation and decision making methods. The CRT incorporates the evaluation system, but the CED and ID have no such mechanism and are highly dependent on the thoroughness of the group using them. Serious practitioners should consider using facilitators during root cause analysis. Observations have found that most groups had individuals who dominated or led the process. When leadership prevails over the project, the individual contributions were carefully considered with a mix of discussion and inquiry. Group leaders did not attempt to convince others of their superior expertise, and conflicts were not considered threatening. In contrast, groups that were dominated did not encourage discussion, and differences of opinion were viewed as disruptive. An experienced facilitator could encourage group members to raise difficult and potentially conflicting viewpoints, so the best ideas would emerge. These tools can be used to their greatest potential with repeated practice. Like other developed skills, the more the groups use the tools, the better they become with them.

9.9.1 Fault Tree Analysis

A FTA provides a graphical means of evaluating the relationships between different parts of the system. FTA is a tree diagram (see Fig. 9.27, accessed through http://en. wikipedia.org/wiki/files/fault_tree.png) that shows failures and/or defects in increasing levels of detail. It helps to narrow the root cause and focus on preventions. These models incorporate predictions based on parts-count failure rates taken from historical data. While the predictions are often not accurate in an absolute sense, they are valuable to assess relative differences in design alternatives.

FTA was originally developed in 1962 at Bell Laboratories by H.A. Watson, under a US Air Force Ballistics Systems Division contract to evaluate the Minuteman I Intercontinental Ballistic Missile (ICBM) Launch Control System (see [15, 32, 38]) for more details. FTA attempts to model and analyze failure processes of engineering and biological systems. FTA is basically composed of logic diagrams that display the state of the system and is constructed using graphical design techniques. Originally, engineers were responsible for the development of FTA, as a deep knowledge of the system under analysis is required.

Fig. 9.27 A fault tree
diagram

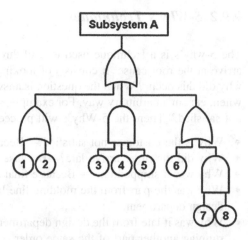

In the diagram, the failure of components is represented by circles. The failure can happen due to one or more components in a system or due to failure of all components at a time, where components can be arranged in a parallel setup, or in a series setup. FTA is a prominent method of presenting information. The compelling message of FTA presentations is that the entire spectrum of failures has been investigated. The deficiency with FTA presentations is that the presented data may not have adequate analysis and understanding. FTA of failures should prompt inquiry into the source data. The participation and communication from product reliability experts are crucial in this process.

FTA is a logical tool that assists in uncovering of potential root causes of defects or equivalent failures. The output of FTA is the result of various levels of contributing factors or potential causes for failure. FTA uses AND and OR function gates to illustrate symptoms of failure down to real root causes.

Steps to FTA:

1. Write down failure as an output for the top or gate (Level 1)
2. Determine what input (symptoms) could be considered a contributing element (Level 2)
3. Continue breaking down the failure with additional gate levels (Level 3, 4, or more)
4. Always ask: Can this gate output be true with any input (or function), or does all input need to be present for a true gate output (and function)?
5. Finalize and date the diagram

Application of FTA

- To allow a backward approach to systematically identify the potential causes of failures
- To provide an overview of interrelationships between causes and failures
- To break down failure indications into more detailed input branches

9.9.2 5-Why's Techniques

The 5-why's is a technique used to drill through the layers of cause and effect to arrive at the root cause. It consists of looking at an undesirable result and asking, why did this occur? When the question is answered, the next question asked is how, when, etc., in an arbitrary way. For example, suppose the problem is "A customer is not satisfied." Then, the 5-Why's will proceed this way:

- Why is the customer not satisfied?—Because the order arrived late.
- Why did the order arrive late?—Because it was shipped late.
- Why was it shipped late?—Because final assembly was not completed on time.
- Why was the part from the molding line late?—Because it arrived late from the design department.
- Why was it late from the design department?—Because the design machine was running another part of the same order.

Most of the time, somebody will be responsible for the things to happen. The above approach helps one to identify those 5-W's (who, when, where, what, why), which answer most of the above questions and by which, one can very easily narrow down the real cause and effect of an action.

9.10 Readying for the Improve Phase

All the work of defining, measuring, and analyzing process problems pays off in the improve phase, if your team and organization handle it well. The project personnel associated with each phase should complete their responsibility in coordination with the improvement team. Lack of creativity, incomplete assignments, failure to think out-of-the-box solutions, careless identification of KPIV's, haphazard implementation, organizational resistance, etc., will affect the Six Sigma project success. Before moving to the improvement stage, some of the areas that need special attention are as follows:

- What possible actions or ideas will help us address the root cause of the problem and achieve the required goal?
- Which of these ideas form workable potential solutions?
- Which solution will most likely achieve our goal with the least cost and disruption?
- How do we test our chosen solution to ensure its effectiveness and then implement it permanently?

A project leader at this point should be confined to those causes and effects, where the improvement is likely to be certain and viable for implementation. Sporadic use of technical tools and dependency on management ethics may kill the benefits of Six Sigma, when improvements are not attached to the customer's requirements.

9.11 Analyze Checklists

Before moving on to the improve phase of DMAIC, check whether the following are in place:

- Examine the process and ensure that the bottlenecks are identified; disconnections and redundancies are properly addressed
- Conduct a value and cycle time analysis, locating areas where time and resources are devoted to tasks not critical to the customer
- Analyze data about the process and its performance to help stratify the problem, understand reasons for variation, and identify potential root causes
- Evaluate whether the project should focus on process design or redesign as opposed to process improvement and confine decision to the project charter
- Ensure key workings of the process, so that a new process to meet the needs of the customer efficiently and effectively is in place
- Develop root cause hypothesis to explain the problem being solved
- Investigate and verify the root cause hypothesis to fix "vital few" variables

9.12 Relevance for Managers

The core of any Six Sigma project is its sound analysis and its caliber to deliver the specified objective. Application of statistical methods and adoption of tools and techniques will enable and empower managers/administrators for better utilization of resources, enhanced productivity and quality control in manufacturing, automobiles, or services sector equally. Besides engineering knowledge and management skills, the analysis skill is also necessary for any Six Sigma project. One can move from simple to hard-core statistical tools and methods for setting standards and performances. This is the essence of this chapter.

Although explorative data analysis (EDA) is used for finding relationship between variables by the use of graphical techniques (box plot, scatter diagram, Pareto charts, probability plots, etc.), it is also necessary to acquire knowledge on factors impacting performances and variations with reference to market expectations and customer requirements. They can be effectively studied through inferential statistics (point estimation, confidence estimation, analysis of variance, multivariate analysis, etc.), projection and trend analysis (correlation and regression studies etc.), and finally through the confirmatory analysis (testing of hypothesis, DOE, and validating causes). The necessity for establishing the functional form of a Six Sigma project, that is $y - f(x)$, is fulfilled in this chapter.

Knowledge of statistical methods alone does not work many a times. One should also have the supporting tools to substantiate the causes and effects, which is fulfilled through some of the effective management tools. They include some of the frequently used techniques such as brainstorming, process mapping, failure modes and effect analysis, etc. Since the objective of analysis phase is to identify the inputs (x's) impacting the process output (y's), the professionals should be very clear about the appropriate use of statistical and management tools at this stage. Only then the success of any Six Sigma project can be ascertained. It is cautioned that over use of any of these tools can derail the project at any point of time. All these issues are discussed in this chapter.

Exercises

9.1. Discuss various steps involved in brainstorming. State the advantages of carrying out such an exercise.

9.2. What is FMEA? Discuss various types of FMEA and their benefits.

9.3. What is risk priority number? How is it used for identifying the vital few from critical many?

9.4. What is normal probability plot? State the steps involved in the construction of normal probability plot.

9.5. Discuss various steps involved in maximum likelihood estimation.

9.6. Let X be a random variable with the following probability distribution:

$$f(x) = \begin{cases} (\theta + 1)x^{\theta}, & 0 \leq x \leq 1 \\ 0, & \text{otherwise} \end{cases}$$

Find the ML estimator of θ, based on a random sample of size n. Also get the point estimate of θ for the samples: 0.13, 0.54, 0.63, 0.44, 0.12, 0.78, 0.59, and 0.39.

9.7. Let $X \sim N(\mu, \sigma^2)$ equal the length in (centimeters) of certain species of fish when caught in the spring. A random sample of $n = 15$ observation of X is

15.2	5.1	18.0	8.7	16.5	9.8	6.8	
12.0	17.8	25.4	19.2	15.8	23.0	13.1	19.0

Find the estimate of μ and σ^2.

9.8. Twenty persons with normal hearing were tested for reaction time to an auditory stimulus, and $\overline{X}$ were found to be 0.145 s. If the standard deviation of reaction time is known to be 0.025 s, then 99 % confidence limits on the mean reaction time for normal hearing people to this stimulus is obtained. How large the sample is, if we wish to be at least 99 % confidence that the error in estimating the true value of the mean is less than 0.01?

9.9. Fifteen bars of steel produced by process-I have mean breaking strength 62.2 with standard deviation 8.7, while 12 bars produced by process-II have mean breaking strength 57.5 with standard deviation 11.6. There is not enough ground to say that the population standard deviations are equal. Obtain 95 % confidence limit for difference between two populations mean.

9.10. The following data are the lives in hours of 2 batches of electric lamps. Find 95 % confidence limit for difference between the batches in respect of average length of life. Assume population variances are unknown and equal.
Batch 1   1505, 1556, 1801, 1629, 1644, 1607, 1825, 1748
Batch 2   1799, 1618, 1604, 1655, 1708, 1675, 1728

9.11. In a random sample of 85 automobile engine crankshaft bearings, 10 have a surface finish that is rougher than the specifications allow. Obtain a 95 % confidence interval for the proportion of roughness specifications. How large the sample if we wish to be at least 95 % confidence that the error in estimating the true value of p is less than 0.02?

9.12 A rivet is to be inserted into a hole. A random sample of $n = 15$ parts is selected, and the hole diameter is measured. The sample standard deviation of the hole diameter measurements is $s = 0.008$ mm. Construct a 99 % confidence bound for σ^2.

9.13 The percentage of titanium in an alloy used in aerospace castings is measured in randomly selected parts. The sample standard deviation is $s = 4.8$ mg. Calculate a 95 % two-sided confidence interval for σ.

9.14. Let X is the time in seconds between phone calls at Domino's pizza company. Fifteen observations of X are as follows: 82, 42, 185, 66, 384, 27, 334, 545, 650, 127, 35, 45, 285, 133, 120. Assuming that X has an exponential distribution with mean θ, obtain unbiased estimate of θ. Also obtain 95 % confidence interval for θ.

9.15. The fraction of defective integrated circuits produced in a photolithography process is being studied. A random sample of 350 circuits is tested, revealing 15 defectives. Calculate 98 % confidence interval on the fraction of defective circuits produced by this particular tool.

9.16. Distinguish between

    (i)   Null and alternative hypothesis
    (ii) Parametric and nonparametric tests
    (iii) Type I and Type II error
    (iv) Significance level and confidence coefficient

9.17. A single observation of a random variable having an exponential distribution is used to test the null hypothesis that the mean of the distribution is $\theta = 2$ against the alternative that it is $\theta = 5$. The null hypothesis is accepted if the observed value of the random variable is less than 3. Find the probability of Type I error and Type II error.

9.18. A single observation of a random variable having a geometric distribution is used to test the null hypothesis $p = 1/3$ against the alternative hypothesis

$p = 1/2$. The null hypothesis is rejected if $X \geq 2$. Find the probability of type I error and power of the test.

9.19. From a population with Poisson distribution with parameter θ, the test for the hypothesis H_0: $\theta = 2$ against H_1: $\theta = 1$ is rejected if $X \leq 3$. Find size and power of the test.

9.20. Discuss various steps involved in testing of hypothesis.

9.21. Scores on a standard IQ test are normally distributed with standard deviation of 15. In the whole population, the mean of this distribution is 100. A company has a selection test of its own, which it claims will pick out people whose IQ is better than average. Twenty of its employee, selected using this test, is given the standard IQ test also and on this, their mean score is 104.2. Examine the company's claim.

9.22. A research team is willing to assume that systolic blood pressure in a certain population of males is approximately normally distributed with standard deviation of 16. A simple random sample of 64 males form the population had a mean systolic blood pressure reading of 133. At the 0.05 level of significance, do these data provide sufficient evidence for us to conclude that the population mean is greater than 130?

9.23. Analysis of a random sample consisting of $m = 20$ specimens of cold-rolled steel to determine yield strengths resulted in a sample average strength of $\bar{x} = 29.8$ ksi. A second random sample of $n = 25$ two-side galvanized steel specimens gave a sample average strength of $\bar{y} = 34.7$ ksi. Assuming that two yield strength distributions are normal with $\sigma_1 = 4.0$ and $\sigma_1 = 5.0$ ksi, does the data indicate that the corresponding true average yield strengths μ_1 and μ_2 are different? Use $= \alpha = 0.01$.

9.24. Two brand of truck tyres are being compared by a trucking firm. A random sample of 50 tyres of brand X (group-1) had an average life of 45,000 miles, while the average life of random sample of 40 of brand Y tyres (group-2) was 46,500 miles. Assuming that population SD was 2,000 miles for both brands of truck tyres. Is there any significant difference in quality at 0.01 level? Construct 95 % confidence interval for the difference between the mean lifetimes of the two brands of truck tyres.

9.25. In a sample of 400 parts manufactured by a factory, the number of defective parts was found to be 30. The company, however, claimed that only 5 % of their product is defective. Is the claim tenable?

9.26. An investigation of the performance of two methods, in a factory manufacturing large numbers of bobbins, gives the following results:

	Sample of bobbins examined	Fraction of bobbins found defective
Methods 1	300	0.20
Methods 2	250	0.32

Test whether there is any significant difference in the performance of the two methods.

9.27. Consider the case of a pharmaceutical manufacturing company testing two new compounds intended to reduce blood pressure levels. The compounds are administered to two different sets of laboratory animals. In group one, 71 of 100 animals tested respond to drug 1 with lower blood-pressure levels. In group two, 58 of 90 animals tested respond to drug 2 with lower blood-pressure levels. The company wants to test at the 0.05 levels whether there is a difference between the effectiveness of these two drugs.

9.28. A machine is used to fill containers with a liquid product. Fill volume can be assumed to be normally distributed. A random sample of 10 containers is selected and the net contents are

$$120.03 \quad 120.04 \quad 120.05 \quad 11.96 \quad 120.05 \quad 120.01 \quad 120.02 \quad 11.98$$
$$120.02 \quad 11.99$$

The manufacturer wishes to be sure that the mean net contents exceed 12.00. What conclusions can be drawn from these data at 5 % level of significance?

9.29. A researcher examined the admissions to a mental health clinic's emergency room on days when the moon was full. For the 12 days with full moons from August 1971 through July 1972, the number of people admitted was as follows:

$$5 \quad 13 \quad 14 \quad 12 \quad 6 \quad 9 \quad 13 \quad 16 \quad 25 \quad 13 \quad 14 \quad 20$$

Assume that the above data are normally distributed. It was observed that the average number of admissions on other days is 11.2. Test the hypothesis that on full moon, average number of admissions is higher than on other days.

9.30. Let μ_1 and μ_2 denote true average tread lives for two competing brands of radial tyres. Test: $H_0: \mu_1 - \mu_2 = 0$ against $H_1: \mu_1 - \mu_2 \neq 0$ at level 0.05 using the following data:

$$m = 45, \bar{x} = 42,500, s_1 = 2200, n = 45, \bar{y} = 40,400, s_2 = 1900.$$

9.31. Tensile strength tests were carried out on two different grades of wire rod, resulting in accompanying data:

Grade	Sample size	Sample mean (kg/mm^2)	Sample sd.
AISI 1064	$m = 129$	$\bar{x} = 107.6$	$s_1 = 1.3$
SISI 1078	$n = 129$	$\bar{y} = 123.6$	$s_2 = 2.0$

Does the data provide compelling evidence for concluding that true average strength for the 1078 grade exceeds that for 1064 grade by more than 10 kg/mm^2. Use $\alpha = 0.05$.

9.32. Two experiments, A and B, take repeated measurements on the length of a copper wire. On the basis of the data obtained given below, test whether B's measurements are more accurate then A's at 1 % level of significance (It may be supposed that the readings taken by both are unbiased).

A's measurements (in mm)		B's measurements (in mm)	
12.47	12.44	12.06	12.34
11.90	12.13	12.23	12.46
12.77	11.86	12.46	12.39
11.96	12.25	11.98	
12.78	12.29	12.22	

9.33. In a sample of 8 observations, the sum of the squared deviations of items from the mean was 94.5. In another sample of 10 observations, the value was 101.7. Test whether the difference is significant at 5 % level.

9.34. Over a period of 6 consecutive days, the opening prices (in rupees) of two well-known stocks were observed and recorded as follows:

Stock #1	390.9	397.0	417.7	389.6	414.2	422.6
Stock #2	423.3	391.6	421.0	409.2	464.7	450.2

Test whether the variabilities in the opening prices of the two stocks are the same at 10 % level.

9.35. A machine that automatically controls the amount of ribbon on a tape was recently been installed. This machine will be judged to be effective if the standard deviation of the amount of ribbon on a tape is less than 0.15 cm. If a sample of 20 tapes yields a sample variance of $s^2 = 0.025$ cm^2, are we justified in concluding that the machine is ineffective?

9.36. Weights in kg of 10 students are as follows: 38, 40, 45, 53, 47, 43, 55, 48, 52, and 49. Can we say variance of the distribution of weights of all students from which the sample of students was drawn is equal to 20 kg^2?

9.37. A manufacturer of breakfast cereal uses a machine to insert randomly one of two types of toys in each box. The company wants randomness so that every child in the neighborhood does not get the same toy. Testers choose samples of 60 successive boxes to see whether the machine is properly mixing the two types of toys. Using the symbols A and B to represent the two types of toys, a tester reported that one such batch looked like this:

B, A, B, B, B, A, A, A, B, B, A, B, B, B, B, B, A,
A, A, A, B, A, B, A, A, B, B, B, A, A, B, A, A, A,
A, B, B, A, B, B, A, A, A, A, B, B, A, B, B, B, B,
A, A, B, B, A, B, A, A, B, B.

9.38. The following are 15 measurements of the octane rating of a certain kind of gasoline:

| 97.5 | 95.2 | 97.3 | 96 | 96.8 | 100.3 | 98.2 | 98.5 | 94.9 |
| 94.9 | 97.4 | 95.3 | 93.2 | 99.1 | 96.1 | 97.6 | 98.2 | 98.5 |

Use the signed-rank test at the 0.05 level of significance to test whether or not the mean octane rating of the given kind of gasoline is 98.5.

9.39. From the data given below, test the hypothesis that the median length of a variety of rice is 13 mm against the alterative it is not at 5 % level of significance using (i) sign test and (ii) Wilcoxon signed-rank test:

| 9.62 | 14.46 | 10.19 | 15.02 | 15.92 | 10.27 |
| 14.16 | 12.26 | 13.55 | 11.13 | 13.00 | 16.26 |

9.40. The following are measurement of the breaking strength of a certain kind of 2-in. cotton ribbon in pounds:

| 163 | 165 | 166 | 161 | 171 | 158 | 151 |
| 162 | 163 | 139 | 172 | 165 | 148 | 169 |

Using (i) sign test and (ii) Wilcoxon signed-rank test to test the hypothesis that the median breaking strength is 160 pounds against alternative that it is greater than 160 pounds at 5 % level of significance.

9.41. Following are four sets of eight measurements each of the smoothness of a certain type of paper, obtained in four different laboratories:

Laboratory	Smoothness measurements							
A	38.7	41.5	43.8	44.5	45.5	46.0	47.7	58.0
B	39.2	39.3	39.7	41.4	41.8	42.9	43.3	45.8
C	34.0	35.0	39.0	40.0	43.0	43.0	44.0	45.0
D	34.0	34.8	34.8	35.4	37.2	37.8	41.2	42.8

Assume that the measurements from the different laboratories are equally variable and test the hypothesis of no systematic differences among the laboratories. Use $\alpha = 0.05$.

9.42. State the practical relevance of goodness-of-fit tests.

9.43. The following data show the result of throwing 12 dice 4096 times, a throw of 4, 5, or 6 being called a success (x).

x	0	1	2	3	4	5	6	7	8	9	10	11	12
f	0	7	60	198	430	731	948	847	536	257	71	11	0

Fit an appropriate distribution and test goodness of fit.

9.44. A set of 8 coins was tossed 256 times and the frequencies of throws observed were as follows:

No. of heads	0	1	2	3	4	5	6	7	8
Freq. of throws	2	6	24	63	64	50	36	10	1

Fit an appropriate distribution and test goodness of fit (coin is not unbiased).

9.45. In a 72-h period on long holiday weekend, there were a total of 306 fatal automobile accidents. The data are as follows:

No. of fatal accidents/h	0	1	2	3	4	5	6	7	8
No. of hours	0	4	10	15	13	12	6	5	7

Fit an appropriate distribution and test goodness of fit.

9.46. Fit a Poisson distribution to the following data with respect to the no. of red blood corpuscles (x) per cell.

x	0	1	2	3	4	5
No. of cells	142	156	69	27	5	1

9.47. The following table gives the distance covered by disc thrown by 180 students of same age. The distance travelled by disc is assumed to be exponentially distributed. Test the goodness of fit for the following data (use $\alpha = 2\,\%$).

Distance in (m)	10–25	25–40	40–55	55–70	70–85
No. of students	6	64	94	14	2

9.48. Given the following data:

2.483757	0.68226	−0.66346	1.157925	−0.83395
0.098728	0.046796	0.181101	1.769654	0.890532
0.88203	0.318287	1.412159	0.067944	0.73047
0.79096	2.276302	0.600866	1.452486	0.7941
0.81216	1.9146	1.727667	0.6284	0.04634
0.08444	0.03164	−0.81461	0.417698	0.1145
0.202217	1.909366	−0.09212	0.434873	0.06059
1.69556	0.366229	0.407952	0.63167	0.66537
0.45837	0.28793	0.703662	0.035159	1.744934
0.03003	0.91154	-0.90242	0.026739	0.37434

Use goodness-of-fit test to decide a suitable distribution

9.49. Consider the data given below

x	1	2	3	4	5	6	7	8	9
y	9	8	10	12	11	13	14	16	15

(i) Plot a scatter diagram and hence interpret it
(ii) Fit a regression line of y on x and x on y
(iii) Predict the value of y for the given values of x when x = 10 and 20
(iv) Test the significance of regression coefficient.

9.50. The following results are available with a linear regression study:
Regression equations: $6y = 5x + 90$ and $15x = 8y + 130$; $\sigma_x^2 = 4$
Find $\bar{x}, \bar{y}, \sigma_y$ and r_{xy}.

9.51. The following are the data on molar ratio and intrinsic viscosity of an experimental process:

Ratio	1.0	0.9	0.8	0.7	0.6	0.5	0.4	0.3
Viscosity	0.45	0.20	0.34	0.58	0.70	0.57	0.55	0.44

(i) Plot a scatter diagram. Does a linear relationship reasonable?
(ii) Estimate the regression coefficients
(iii) Predict the molar ratio if viscosity is 0.43.

9.52. The following calculations have been made for closing prices (x) of 12 stocks at the Bombay Stock Exchange on a certain day along with the volume of sales (in '000) of shares (y): $\Sigma x = 580$, $\Sigma y = 370$, $\Sigma xy = 11,494$, and $\Sigma x^2 = 41,658$. Find the line of regression of y on x and estimate the volume of sales of a share when closing price of a stock is 19.

9.53. Distinguish between the coefficient of determination and the sample correlation coefficient.

9.54. The following are sample data provided by a moving company on the weights of size shipments, the distances they were moved, and the damage that was incurred.

Weight (1000 lb)	4.0	3.0	1.6	1.2	3.4	4.8
Distance (1000 miles)	1.5	2.2	1.0	2.8	0.8	1.6
Damage (Dollars)	160	112	69	90	123	186

(i) Identify dependent and independent variables.
(ii) Fit an appropriate equation to estimate the damage when a shipment weighing 2400 pounds is moved 1200 miles.
(iii) Calculate the partial and multiple correlation coefficient and interpret them.
(iv) Compute coefficient of determination of above regression model and interpret it.

(v) Compute coefficient of determination when influence of distance is removed from the regression model and comment on it.

(vi) Compute coefficient of determination when influence of weight is removed from the regression model and comment on it.

9.55. The following table gives the 15 artificial data of three explanatory variables and a response variable.

Observation	X_1	X_2	X_3	Y
1	10.1	19.6	28.3	9.7
2	9.5	20.5	28.9	10.1
3	10.7	20.2	31.0	10.3
4	9.9	21.5	31.7	10.0
5	10.3	21.1	31.1	10.0
6	10.8	20.4	29.2	10.0
7	10.5	20.9	29.1	10.8
8	9.9	19.6	28.8	10.3
9	9.7	20.7	31.0	9.6
10	9.3	19.7	30.3	9.9
11	12.0	26.0	34.0	0.7
12	11.0	34.0	34.0	0.1
13	3.1	2.2	0.3	0.6
14	2.3	1.6	2.0	0.0
15	0.8	2.9	1.6	0.1

(i) Construct the scatter diagrams of Y versus X_1, Y versus X_2, and Y versus X_3, respectively, and interpret them.

(ii) Suggest an appropriate regression model with necessary assumptions.

(iii) Compute the predicted values using the fitted equation obtained in (ii).

(iv) Obtain an unbiased estimate of residual variance.

(v) Construct an analysis of variance table and test at 5 % level of significance the hypothesis that variables X_1, X_2, and X_3 are linearly related to Y.

(vi) Compute the coefficient of determination (R^2) and interpret it. Compute the adjusted R^2.

(vii) Test whether variable X_2 can be dropped from the model.

(viii) Test whether variables X_2 and X_3 can be dropped from the model.

(ix) Compute the estimated variance and covariance of the sample regression coefficients.

(x) Perform individual tests of the significance of the regression coefficients.

(xi) Compute the individual 95 % confidence intervals for the regression coefficients in the model described in (ii).

(xii) Construct and interpret the graphs of

    a. Residual versus the predicted values.

    b. Residual against each explanatory variable.

(x) For an individual in a study with $X_1 = 10.5$, $X_2 = 20.0$, and $X_3 = 29.5$, compute the standard error of

    a. The estimate of the population mean.

    b. The prediction of Y for a new individual.

9.56. A consumer agency wants to check whether the mean lives of four brands of outer batteries, which sell for nearly the same price, are the same. The agency randomly selected a few batteries of each brand and tested them, and the following table gives the lives of these batteries in thousands of hours.

Brand A	Brand B	Brand C	Brand D
74	53	57	56
78	51	81	51
51	47	77	49
56	59	68	43
65		71	

Identify the design and carry out the following:

(i) Test at 5 % level of significance, the mean lifetime of each of these four brands of batteries is the same?

(ii) Test at 5 % level of significance, the mean lifetime of Brand B and Brand D of batteries are same?

9.57. Three technicians used each of four thermometers to measure the melting point of hydroquinone. The recoded melting points (in °C) are shown below:

Thermometer	Technician		
	1	2	3
1	174.0	173.0	173.5
2	173.0	172.0	173.0
3	171.5	171.0	173.0
4	173.5	171.0	172.5

Identify the design, specify the model, and analyze it.

9.58. Set up ANOVA for the following information relating to archers. Each one shoots three arrows at each of three distances, 30, 40, and 50 yards. Scores for every archer at each distance are given below:

Archers	30 yards	40 yards	50 yards
1	70	51	27
	65	56	44
	90	69	30
2	92	65	50
	100	63	40
	104	75	30
3	93	68	56
	94	93	55
	123	98	77
4	118	116	76
	141	108	70
	98	110	67

At 5 % level of significance, test the hypothesis that

(i)   There is no difference due to the three distances
(ii) There is no difference due to archers
(iii) The archers and the distances do not interact

9.59. A three-way classified data along with their yields are given as follows:

B (220)	F (98)	D (149)	A (92)	E (282)	C (160)
A (74)	E (238)	B (163)	C (228)	F (48)	D (168)
D (188)	C (279)	F (118)	E (278)	B (176)	A (133)
E (295)	B (242)	A (640)	D (104)	C (213)	F (163)
C (187)	D (90)	E (242)	F (96)	A (66)	B (188)
F (90)	A (124)	C (195)	B (109)	D (79)	E (211)

The six different types of treatment effects (A, B, C, D, E, and F) are found to have effect on row-wise and column-wise. Stating clearly the model, analyze the data to find if there is any difference among the treatments.

9.60. Discuss the practical importance of carrying out root cause analysis. What are the tools used in this analysis?

9.61. What is fault tree analysis? How are they applied?

9.62. Discuss briefly the 5-Why's technique.

References

1. AIAG.: Potential Failure Mode and Effect Analysis. Automotive Industry Action Group, (1993)
2. AIAG.: Potential Failure Mode and Effect Analysis, 4th edn. Automotive Industry Action Group, (2008) ISBN: 9781605341361
3. Anderson, B., Fagerhaug, T.: Root Cause Analysis: Simplified Tools and Techniques. ASQ Quality Press, Milwaukee (2000)
4. Arcaro, J.S.: TQM Facilitator's Guide. St. Lucie Press, Boca Raton (1997)
5. Barnett, V.: Convenient Plotting Positions for The Normal Distribution. Marcel Dekker, New York (1976)
6. Blom, G.: Statistical estimates and transformed beta variables. Wiley, New York (1958)
7. Brassard, M.: The Memory Jogger Plus+: Featuring the Seven Management and Planning Tools. Goal/QPC, Salem (1996)
8. Brassard, M., Ritter, D.: The Memory Jogger II: A Pocket Guide of Tools for Continuous Improvement and Effective Planning. GOAL/QPC, Salem (1994)
9. Casella, G., Berger, R.L.: Statistical Inference. Duxbury Press, Pacific Grove (2001)
10. Cox, D. R.: Principles of Statistical Inference. CUP, Cambridge (2006)
11. Cox III, J. F., Spencer, M. S.:The Constraints Management Handbook. St. Lucie Press, Boca Raton (1998)
12. Dettmer, H.W.: Goldratt's Theory of Constraints. ASQC Press, Milwaukee (1997)
13. Draper, N., Smith, H.: Applied Regression Analysis, 3rd edn. Wiley, New York (1998)
14. Dew, J.R.: In search of the root cause. Qual. Prog. 24(3), 97–107 (1991)
15. Ericson, C.: Fault tree analysis—a history. In: Proceedings of the 17th International Systems Safety Conference (1999). Retrieved 2010-01-17
16. Freedman, D. A.: Statistical Models: Theory and Practice (revised edition), Cambridge University Press, Cambridge (2009)
17. Fredendall, L.D., Patterson, J.W., Lenhartz, C., Mitchell, B.C.: What should be changed? Qual. Prog. 35(1), 50–59 (2002)
18. Goldratt, E.M.: It's not luck. North River Press, Great Barrington (1994)
19. Hollander, M., Wolfe, D.A.: Non-parametric statistical methods. Marcel Dekker, New York (2003)
20. Ishikawa, K.: Guide to Quality Control, 2nd edn. Asian Productivity Organization, Tokyo (1982)
21. Kmenta, S., Ishii, K.: Scenario-based failure modes and effects analysis using expected cost. J. Mech. Des. 126(6), 1027 (2004)
22. Lepore, D., Cohen, O.: Deming and Goldratt: The Theory of Constraints and the System of Profound Knowledge. North River Press, Great Barrington (1999)
23. Marvin, R., Arnljot, H.: System Reliability Theory, Models, Statistical Methods, and Applications. Wiley Series in probability and statistics, 2nd edn., p. 88 (2004)
24. Meeker, W.Q., Escobar, L.A.: Statistical Methods for Reliability Data. Wiley, New York (1998)
25. Mizuno, S. (Ed.): Management for Quality Improvement: The Seven New QC Tools. Productivity Press, Cambridge (Original work published in 1979) (1988)
26. Muralidharan, K., Syamsunder, A.: Statistical Methods for Quality, Reliability and Maintainability. PHI India Ltd, New Delhi (2012)
27. Montgomery, D.C., Runger, G.C.: Applied Statistics and Probability for Engineers, 5th edn. Wiley, India (2011)
28. Montgomery, D.C., Peck, E.A., Vining, G.G.: Introduction to Linear Regression Analysis, 4th edn. Wiley, New York (2006)
29. Moran, J.W., Talbot, R.P., Benson, R.M.: A Guide to Graphical Problem-Solving Processes. ASQC Quality Press, Milwaukee (1990)

30. Osborn, A.F.: Applied Imagination: Principles and Procedures of Creative Problem Solving, 3rd edn (revised). Charles Scribner's Sons, New York (1963)
31. Pasquarella, M., Mitchell, B., Suerken, K.: A comparison on thinking processes and total quality management tools. In: 1997 APICS Constraints Management Proceedings: Make Common Sense a Common Practice, Denver, CO, APICS, Falls Church, pp. 59–65 (1997)
32. Rechard, R.P.: Historical relationship between performance assessment for radioactive waste disposal and other types of risk assessment in the United States. Risk Anal. **19**(5), 763–807 (1999). doi:10.1023/A:1007058325258. SAND99-1147 J. Retrieved 2010-01-22
33. Robson, M.: Problem Solving in Groups, 2nd edn. Gower, Brookfield (1993)
34. Santanen, E., Briggs, R.O., de Vreede, G.-J.: Causal Relationships in Creative Problem Solving: Comparing Facilitation Interventions for Ideation. J. Manag. Inf. Syst. **20**(4), 167–198 (2004)
35. Scheinkopf, L.J.: Thinking for a Change: Putting the TOC Thinking Processes to Use. St. Lucie Press, Boca Raton (1999)
36. Smith, D.: The Measurement Nightmare: How the Theory of Constraints can Resolve Conflicting Strategies, Policies, and Measures. St. Lucie Press, Boca Raton (2000)
37. Sproull, B.: Process Problem Solving: A Guide for Maintenance and Operations Teams. Productivity Press, Portland (2001)
38. Vesely, W. et al.: Fault Tree Handbook with Aerospace Applications. National Aeronautics and Space Administration (2002). Retrieved 2010-01-17
39. Wilson, P.F., Dell, L.D., Anderson, G.F.: Root Cause Analysis: A Tool for Total Quality Management. ASQC Quality Press, Milwaukee (1993)
40. Young, G.A., Smith, R.L.: Essentials of Statistical Inference. CUP, Cambridge (2005)
41. Zacks, S.: Introduction to Reliability Analysis. Springer, New York (1992)

Chapter 10
Improve Phase

Improve phase tasks and deliverables

- Suggest an improvement implementation plan
- Develop potential improvements or solutions for root causes
- Develop evaluation criteria

    - Prioritize solution options for each root cause
    - Examine solutions with a short-term and long-term approach
    - Weigh the costs and benefits of "quick-hit" versus more difficult solution options

- Select and implement the improved process and metric
- Measure results and conduct a designed experiment
- Validate solutions for improvement using statistical analysis
- Evaluate whether improvements meet targets
- Evaluate for risk
- Document the follow-up steps
- Exit review

The stage is now set for suggesting incremental changes and solutions to eliminate or reduce defects, costs, or cycle time without changing the basic structure of the process. All root causes will be subjected to testing and validation using strong statistical analysis. The findings will be supported by management inputs and the Six Sigma personals experience in dealing with the set targets and goals. We now discuss some of the incremental tools required to achieve the deliverables below.

10.1 Balanced Scorecard (BSC)

The concept of a balanced scorecard became popular following research studies published in the Harvard Business Review articles of Kaplan and Norton and ultimately led to the 1996 publication of the standard business book on the subject, titled

© Springer India 2015
K. Muralidharan, *Six Sigma for Organizational Excellence*,
DOI 10.1007/978-81-322-2325-2_10

The Balanced Scorecard [11]. The authors define the balanced scorecard (BSC) as "organized around four distinct performance perspectives—financial, customer, internal, and innovation and learning. The name reflects the balance provided between short-term and long-term objectives, between financial and non-financial measures, between lagging and leading indicators, and between external and internal performance perspectives." As data are collected at various points throughout the organization, the need to summarize many measures—so that the top-level leadership can gain an effective idea of what is happening in the company—becomes critical. One of the most popular and useful tools we can use to reach that high-level view is the BSC. The BSC is a flexible tool for selecting and displaying "key performance indicator" measures about the business in an easy-to-read format. Many organizations not involved in Six Sigma, including many government agencies, are using the BSC to establish common performance measures and keep a closer eye on the business.

The key performance indicators (KPIs) are both financial and non-financial metrics that reflect an organization's key business drivers (KBDs) also known as critical success factors (CSFs). The KPIs are defined in terms of the SMART acronym as follows:

- *Specific*—KPIs should be laser-focused and process-based
- *Measurable*—KPIs must be quantitative and easily determined
- *Achievable*—KPIs must be set with regard to benchmarking levels, yet remain obtainable
- *Relevant*—KPIs should be linked to the organization's strategies, goals, and objectives
- *Time bound*—KPI levels should reflect a specific period of time, and they should never be open-ended. A time bound KPI can improve employee focus

As a part of improvement activity, the KPIs should address:

- Performance standards for each key output, as defined by the customers
- Observable and (if possible) measurable service standards for key interfaces with customers
- Analysis of performance and service standards based on their relative importance to customers and customer segments and their impact on business strategy
- Analyzing and prioritizing requirements as per the business strategy

Note that a great deal of effort in an SS project is for improving the value-adding activities to the project. Lean thinking focuses on non-value-added activities—those activities that occur in every enterprise but do not add value for the customer. Below, we discuss some of the tools for identifying and eliminating or reducing waste of all kinds.

10.2 Kaizen Events

Kaizen continuous improvement is the foundation of Japanese management. The concept was introduced by Masaoki Imai of Japan and has received worldwide fame and acceptability in improving ongoing products and processes [33]. The Kaizen philosophy assumes that our way of life—be it our working life, our social life, or our home life—deserves to be constantly improved. It involves everyone, including both managers and workers for ongoing improvement. The concept has helped Japanese companies to generate a process-oriented way of thinking and develop strategies that assure continuous improvement involving people at all levels of the organizational hierarchy. The message of Kaizen strategy is that not a day should go in vain without any improvement being made somewhere in the company.

Kaizen is an umbrella concept covering those "uniquely Japanese" practices such as customer orientation, QC circles, suggestion system, automation, discipline in the workplace, total productivity maintenance (TPM), just-in-time, zero defects, small-group activities, cooperative labor management relations, and new product development. According to Kaizen, management has two major components: maintenance and improvement. Maintenance refers to activities directed toward maintaining current technological, managerial, and operating standards, and improvement refers to those directed toward improving current standards. Under its maintenance functions, management performs its assigned tasks so that everybody in the company can follow the established standard operating procedure (SOP). This means that management must first establish policies, rules, directives, and procedures for all major operations and then see to it that everybody follows SOP. If people are able to follow the standard but they do not, management must introduce discipline. If people are unable to follow the standard, management must either provide training, or review or revise the standard so that people can follow it.

Top Management

- Be determined to introduce Kaizen as a corporate strategy
- Provide support and direction for Kaizen by allocating resources
- Establish policy for Kaizen and cross-functional goals
- Realize Kaizen goals through policy deployment and audits
- Build systems, procedures, and structures conducive to Kaizen

Middle Management

- Deploy and implement Kaizen goals as directed by top management
- Use Kaizen in functional capabilities
- Establish, maintain, and upgrade standards
- Make employees Kaizen conscious through intensive training programs
- Help employees develop skills and tools for problem solving

Supervisors

- Use Kaizen in functional roles
- Formulate plans for Kaizen and provide guidance to workers
- Improve communication with workers and sustain high morale
- Support small-group activities and the individual suggestion system
- Introduce discipline in the workplace
- Provide Kaizen suggestions

Workers

- Engage in Kaizen through the suggestion system and small-group activities
- Practice discipline in the workshop
- Engage in continuous self-development to become better problem solvers
- Enhance skills and job performance expertise with cross-education

Thus, maintenance and improvement of the standards is one of the main objectives of Japanese management. Now, improvement can be broken down between Kaizen and innovation. Kaizen signifies small improvements made in the status quo as a result of ongoing efforts. Innovation involves a drastic improvement in the status quo as a result of a large investment in new technology and/or equipment.

Lee et al. [13] believe that Kaizen can be very effective in industrial technology curricula and could not only benefit our graduates in future employment, but also improve the understanding of this technique in manufacturing settings across the USA. As cooperation between industries and educational institutions increases, this type of instruction could also further reduce the gap between educational practice and industrial needs.

Kaizen can be best achieved through the 5S approach, which is considered to be one of the waste elimination tools and is described below.

10.3 5S Implementation

A planned, systematic follow-up of activities becomes an important feature of any quality management process that helps to continuously emphasize the organizational commitment to quality improvement. Hence, a comprehensive approach to quality management is essential. This is offered by the 5S approach. The 5S's stand for five Japanese words: Seiri (structurize), Seiton (systemize), Seiso (sanitize), Seiketsu (standardize), and Shitsuke (self-discipline). These practices are useful not only in improving their physical environment, but also in improving their thinking processes.

- *Seiri* is about separating the things which are necessary for the job from those that are not and keeping the number of the former as low as possible and at a convenient location.

- *Seiton* means to put things in order. Things must be kept in order so that they are ready for use when needed. This can be achieved by deciding where things belong, analyze the status quo, putting things back where they belong, etc.
- *Seiso* means keep the workplace clean. Everybody in the organization is responsible for this.
- *Seiketsu* means continually and repeatedly maintaining neatness and cleanliness in the organization.
- *Shitsuke* means instilling the ability of doing things in way they are supposed to be done.

The 5S system is often a starting place for implementing Lean operations, so it is important that it be done properly and periodically. Recall that, shitsuke, the fifth "S", essentially calls for performing the first four S's on an ongoing basis. The success of the 5S practices depends not only on the management but also on the people who are part of the entire quality process. It is the commitment and the driving force which enables a good practice of neatness, cleanliness, and standardization in an organization. That is why Kaizen advocates the practice of 5S in a workplace.

10.4 The 3*M*'s Technique

The three M's are considered to be a part of Kaizen activities used to reduce waste and to put a break on frequent occurrences of waste and inconsistencies in the process. The three interconnected Toyota Production System (TPS) waste components are *Muda* (waste), *Muri* (overburden), and *Mura* (unevenness).

Muda are no value-added, wasteful activities that cause long cycle time and excess inventory of a production process. There are seven kinds of Muda:

- Over-production—Having more (output, supplies, etc) on hand than is needed to meet customer demand
- Waiting—Process stops when an input, decision, etc is not available
- Transportation or conveyance—Movement of people, information, documents, etc
- Processing—Performing unnecessary processing that the client does not want
- Defects—Checking inputs, outputs or intermediate steps confirming with customer (internal or external)
- Inventory—Maintaining a demand/supply that is no longer used
- Motion—Unnecessary movement of inventories, files, transports

Muda is the result of Mura. Mura (unevenness) in production will result in excess inventory and overburdening of production schedules. When different transactional projects are handled simultaneously, then the consequences of changing priorities of transactions while they are in progress and having to correct defects in transactions by reworking them can cause unevenness in the process. These behaviors create enough problems in the system that specialized expediters may be responsible for hand carrying emergency orders through the system. Even though these particular orders are handled quickly, it is done at the expense of every other transaction in the system.

Muri is nothing but overburdening the people, machinery, or processes that result in quality or safety problems. Failing to accommodate variation in customer demands can result in swings of over- and underproduction. The emphasis for production in times of heavy demand can be made at the expense of overtime, canceled training, or system upgrades. These factors can cause stress to the workers. The result can be impatience with customers, canceled transactions, or increase in errors and Muda.

Through the systematic implementation of 5S, it is possible to minimize the effect of Muda, Mura, and Muri. While these techniques work quite well, they should best be applied to the areas where you have identified the problems during the measure and analyze phases of your project.

10.5 Kanban

A *Kanban* is a visual signal for over-production. It eliminates wastes in a controlled way. A Kanban is a system that signals the need to replenish stock or materials or to produce more of an item. This philosophy also helps the manufacturers to reduce waste in their process. This principle works on a "pull" system as opposite to the traditional "push" system. In a pull system, the customer order process withdraws the needed items from a supermarket, and the supplying process produces to replenish what was withdrawn.

Kanban system need not be elaborate, sophisticated, or even computerized to be effective. A system is best controlled when material and information flow into and out of the process in a smooth and rational manner. If process inputs arrive before they are needed, unnecessary confusion, inventory, and costs generally occur. If process outputs are not synchronized with downstream processes, the result often is delays, disappointed customers, and associated costs. A Kanban system may be used to simplify and improve resupply procedures.

When a business process has a perfect one piece flow, each business process will generate a "pull" signal to the upstream business process to supply one article for processing. This is extremely difficult to achieve, so a small batch of material is usually "pulled" instead of a single article. This unit, called a *Kanban*, is not the same as the size of the pitch unit. When the intermediate supply of an article decreases to a predetermined, minimum amount (safety stock), a Kanban signal is sent upstream to generate a batch of intermediate material. The downstream process can keep using the safety stock until the next batch arrives. The calculation of these parameters depends on some economic parameters, the size of the batch, stock-out-costs, replenishment cycle time, etc. Variation in inventory delivery cycle time, discounts for large batch forward buying, and changes in consumption rate complicate the calculation. When the inventory of these intermediate articles increases, the ability to economically make quick design changes will be impossible. The effect of variation in customer demand can be smoothed out in manufacturing processes by allowing changes in the level of inventory. Using a properly designed deterministic inventory model, one can find the optimum level of cycle time which reduces the overall variation in the process.

10.6 Design of Experiments

One of the best and statistically proved improvement tools is the design of experiments (DOE).

The DOE is the process of planning experiments so that appropriate data will be collected, minimum number of experiments will be performed to acquire the necessary technical information, and suitable statistical methods will be used to analyze the collected data. A DOE is carried out by researchers or engineers in all fields of study to compare the effects of several conditions or to discover something new. If an experiment is to be performed most efficiently, then a scientific approach to planning must be considered. The statistical approach to experimental design is necessary if we wish to draw meaningful conclusions from the data. Thus, there are two aspects to any experimental design: the design of experiment and the statistical analysis of the collected data. They are closely related, since the method of statistical analysis depends on the design employed.

As discussed earlier at many places in this book, a Six Sigma project emphasizes the relationship between input and output in the form $y = f(x)$, where y represents the response variable of importance for the customers and x represents input variables which are called factors in DOE. The question is which of the factors are important to reach good values on the response variable and how to determine the levels of the factors. The DOE plays a major role in many engineering activities. For instance, DOE is used for

1. Improving the performance of a manufacturing process. The optimal values of process variables can be economically determined by application of DOE.
2. The development of new processes. The application of DOE methods early in process development can result in reduced development time, reduced variability of target requirements, and enhanced process yields.
3. Screening important factors.
4. Engineering design activities such as evaluation of material alternations, comparison of basic design configurations, and selection of design parameters so that the product is robust to a wide variety of field conditions.
5. Empirical model building to determine the functional relationship between x and y.

According to Montgomery [19], some applications of experimental designs in engineering design include the following:

- Evaluation and comparison of basic design configurations
- Evaluation of material alternatives
- Selection of design parameters so that the product will work well under a wide variety of field conditions, so that the product is robust
- Determination of key product design parameters that impact product performance.

Now consider some examples:

Example 10.1 The yield of a particular chemical process is being studied. The two important variables are thought to be pressure and temperature. Two levels of each factor are selected and are run for three replicates each. The data on yield are as follows:

Temperature (°C)	Pressure (psi)	
	100	200
150	90.4	90.7
	90.2	90.6
	90.1	90.5
250	90.3	90.6
	90.5	90.8
	90.7	90.9

Here, y = yield and the factors are temperature (x_1) and pressure (x_2). In this problem, the engineers want to answer the following questions:

1. What effects do x_1 and x_2 have on y?
2. What effects do the 250 °C of x_1 and 100 psig of x_2 have on y?
3. Is there a choice of x_2 that would give uniform yield regardless of x_1?
4. Is there a choice of x_2 that would give uniform y regardless of x_1?

We will answer some of these questions after developing the idea of DOE below. Now consider another example:

Example 10.2 Firing time of explosive switches depends on three factors:

A　the metal used
B　the amount of primary initiator and
C　the packing pressure of the explosive

Two levels of each factor are considered for manufacturing the switches. They include two metal compositions for factor A, say $a_0(=\text{soft})$ and $a_1(=\text{hard})$; two specific amount of initiators for factor B, say $b_0(=10 \text{ mg})$ and $b_1(=15 \text{ mg})$; and two packing pressure for factor C, say $c_0(=12{,}000 \text{ psi})$ and $c_1(=20{,}000 \text{ psi})$. For each of these eight combinations, two switches are manufactured and tested.

The order of manufacture and testing of switches were completely randomized. The data on time in seconds are given below:

$a_0b_0c_0$	$a_0b_0c_1$	$a_0b_1c_0$	$a_0b_1c_1$	$a_1b_0c_0$	$a_1b_0c_1$	$a_1b_1c_0$	$a_1b_1c_1$
40	35	29	27	32	25	16	11
39	30	34	28	30	26	12	13

In this problem, the engineers may be interested in answering the following questions:

1. What effects does each factor have on the firing time of the switch?
2. What effects does the hard metal have on the firing time of the switch?
3. How does the amount of 10 mg primary initiator and 2000 psi packing pressure affect the variation in the manufacturing?
4. How does the metal type jointly affect the packing pressure and amount of initiator?
5. Is there a choice of metal that would reduce the firing time regardless of the packing pressure?
6. Is there a choice of metal that would go with a particular choice of initiator?
7. Is there a choice of combination of metal, amount of primary initiator, and the packing pressure that would reduce the manufacturing time?

In both the examples above, the engineers are concerned with either a particular level of the settings or the optimum combination of variables that will affect the quality and overall improvement of the process. In such situations, a designed experiment supported by a good data analysis and interpretation is essential. This is achieved through the use of DOE. In the first example, there are two factors (temperature and factors) used at two levels each, and hence, the design is called a 2^2 factorial design. In the second case, there are three factors (A, B, and C) used at two levels each, and hence, the design is called a 2^3 factorial design. Thus, a design having m factors each used at n levels will be called an n^m factorial design.

To understand the concept of DOE better, we will first provide a few fundamental definitions:

- An *experiment* is a device to obtain answers to some scientific query. An experiment can be either comparative or absolute. In comparative experiments, we compare the effects of two or more factors on some population characteristics, for example, comparison of different treatment combinations, varieties of crops, different diets or medicines in a dietary experiment. In an absolute experiment, we determine the absolute value of some population characteristics, for example, the IQ of a group of students.
- An *experimental unit* is the smallest entity receiving a particular treatment that yields a value of the response variable.
- A *treatment* is the specific setting or combination of factor levels for an experimental unit.

- A *factor* is an independent variable or assignable cause that may affect the responses and of which different levels are included in the experiment. Factors are also known as explanatory variables, predictor variables, or input variables.
- The *response variable* is the output variable that shows the observed results or value of an experimental treatment. It is sometimes known as the dependent variable. There may be multiple response variables in an experimental design.
- The *observed value* is a particular value of a response variable determined as a result of a test or measurement.
- A *level* is the setting or assignment of a factor at a specific value.
- An *effect* is the relationship between a factor and a response variable. Note that there can be many factors and response variables. They include main effect, dispersion effect, and interaction effect.
- *Noise factor* is an independent variable that is difficult or too expensive to control as part of standard experimental conditions. Hence, they are also called random factors.
- *Experimental error* is the variation that occurs in the response variable beyond that accounted for by the factors, blocks, or other assignable sources.
- An *experimental run* is a single performance of the experiment for a specific set of treatment combinations.

10.6.1 Principles of Experimentation

The tool, DOE, was developed in the 1920s by the British scientist Sir Ronald A. Fisher (1890–1962) as a tool in agricultural research. The first industrial application was performed in order to examine factors leading to improved barley growth for the Dublin brewery. After its original introduction to the brewery industry, factorial design, a class of design in DOE, began to be applied in industries such as agriculture, cotton, wool, and chemistry. Box (1919–), an American scientist, and Taguchi (1924–2012), a Japanese scientist, have contributed significantly to the usage of DOE where variation and design are the central considerations. Large manufacturing industries in Japan, Europe, and the USA have applied DOE from the 1970s. However, DOE remained a specialist tool, and it was with Six Sigma that DOE was first brought to the attention of top management as a powerful tool to achieve cost savings and income growth through improvements in variation, cycle time, yield, and design. DOE is now an integral part of the Six Sigma training program.

For the validity of statistical analysis and enhancing the precision of the experiments, it is important to identify and critically examine the following principles of experimentations:

- *Replication*. Replication means the repetition of the same treatment in a number of plots, the purpose being to get an estimate of the error variance of the experiment. Each repetition of the experiment is called a replicate. Replication increases the precision of the estimates of the effects in an experiment.

- *Randomization*. Randomization is used to assign treatments to experimental units so that each unit has an equal chance of being assigned a particular treatment, thereby minimizing the effect of variation from uncontrolled noise factors. A completely randomized design (CRD) is one in which the treatments are assigned at random to the full set of experimental units. No blocks are involved in a CRD.
- *Blocking or local control*. A block is a collection of experimental units more homogeneous than the full set of experimental units. Blocks are usually selected to allow for special causes, in addition to those introduced as factors to be studied. Blocking refers to the method of including blocks in an experiment in order to broaden the applicability of the conclusions or minimize the impact of selected assignable causes. Randomization of the experiment is restricted and occurs within blocks. The block effect is the resulting effect from a block in an experimental design. Existence of a block effect generally means that the method of blocking was appropriate and that an assignable cause has been found. A design that uses blocking and randomization is the randomized block design (RBD).
- *Order*. Refers to the sequence in which the runs of an experiment will be conducted. Generally, two types of runs are observed in an experiment. They are standard order and run order. The standard order shows what the order of the runs in an experiment would be if the experiment was done in Yate's order. Run order shows what the order of the runs in an experiment would be if the experiment was run in random order. Random order works to spread the effects of noise variables.
- *Confounding*. Confounding or *aliasing* is a design technique for arranging a complete factorial experiment in blocks, where the block size is smaller than the number of treatment combinations in one replicate. Confounding occurs when factors or interactions are not distinguishable from one another. There are two types of confounding: *total confounding* and *partial confounding*. If an effect is of little or no interest, this effect may be confounded with the incomplete block differences in all the replicates. It is then called total confounding for the relevant effect. If an effect is confounded with incomplete block differences in one or some replicates, another effect is confounded in another or some other replicates and so on, and then, the effects are partially confounded with block differences.
- *Resolution*. In the context of experimental design, resolution refers to the level of confounding in a fractional factorial design. Resolutions are numbered with roman numerals. The common resolutions are as follows:

  - Resolution III: where no main effects are confounded with another main effect and main effects are confounded with two-factor interactions
  - Resolution IV: where no main effects are confounded with another main effect, and no main effects are confounded with two-factor interactions, main effects are confounded with three-factor interactions, and two-factor interactions may be confounded with other two-factor interactions

– Resolution III: where no main effects are confounded with another main effect, and no main effects are confounded with two-factor interactions, main effects are confounded with four-factor interactions, no two-factor interactions are confounded with other two-factor interactions, and two-factor interactions may be confounded with other three-factor interactions

Many designs are classified according to the above characteristics. In the next section, we discuss some commonly used designs in statistical literature.

10.6.2 Classification of Design of Experiments

There are many different types of DOE. They may be classified according to the allocation of factor combinations and the degree of randomization of experiments.

- *Factorial design*: This is a design for investigating all possible treatment combinations which are formed from the factors under consideration. The order in which possible treatment combinations are selected is completely random. Single-factor, two-factor and three-factor factorial designs belong to this class, as do 2^k (k factors at two levels) and 3^k (k factors at three levels) factorial designs.
- *Fractional factorial design*: This is a design for investigating a fraction of all possible treatment combinations which are formed from the factors under investigation. Designs using tables of orthogonal arrays, Plackett–Burman designs, and Latin square designs (LSDs) are fractional factorial designs. This type of design is used when the cost of the experiment is high and the experiment is time-consuming.
- *Randomized complete block design, split-plot design, and nested design*: All possible treatment combinations are tested in these designs, but some form of restriction is imposed on randomization. For instance, a design in which each block contains all possible treatments, and the only randomization of treatments is within the blocks, is called the randomized complete block design.
- *Incomplete block design*: If every treatment is not present in every block in a randomized complete block design, it is an incomplete block design. This design is used when we may not be able to run all the treatments in each block because of shortage of experimental apparatus or inadequate facilities.
- *Response surface design and mixture design*: This is a design where the objective is to explore a regression model to find a functional relationship between the response variable and the factors involved, and to find the optimal conditions of the factors. Central composite designs, rotatable designs, simplex designs, mixture designs, and evolutionary operation (EVOP) designs belong to this class. Mixture designs are used for experiments in which the various components are mixed in proportions constrained to sum to unity.
- *Robust design*: Taguchi [30] developed the foundations of robust design, which are often called parameter design and tolerance design. The concept of robust

design is used to find a set of conditions for design variables which are robust to noise and to achieve the smallest variation in a product's function about a desired target value. Tables of orthogonal arrays are extensively used for robust design. For references related to robust design, see Taguchi [30], Park [23], Logothetis and Wynn [16].

- *Balanced design*: In a balanced design, all treatment combinations have the same number of observations. If replication in a design exists, it would be balanced only if the replication was consistent across all the treatment combinations.

There are many designs which can be used in combinations of both. For instance, balanced incomplete block design (BIBD), partially incomplete block design (PBIBD), etc., designed using various combinations of designs. A randomized (complete) block design (RCBD) is a special case of two-way analysis, and LSD is a special case of three-way classification carried out in Chap. 9. Below, we study the factorial design concepts in detail.

10.6.3 General Two-Factor Factorial Design

A general case of two-factor factorial design can be explained as follows: Let y_{ijk} be the observed response when factor A is at the ith level ($i = 1, 2, \ldots, r$) and factor B is at the jth level ($j = 1, 2, \ldots, s$) for the kth replicate ($k = 1, 2, \ldots, t$). Thus, if the observations are all random, then the total number of observations ($=rst$) will be arranged as per Table 10.1.

The above observations can be described in a model as

$$y_{ijk} = \mu + \alpha_i + \beta_j + (\alpha\beta)_{ij} + \varepsilon_{ijk}, \quad \begin{cases} i = 1, 2, \ldots, r \\ j = 1, 2, \ldots s \\ k = 1, 2, \ldots, t \end{cases} \quad (10.1)$$

where μ is the overall mean effect, α_i is the effect of ith level of row factor A, β_j is the effect of jth level of column factor B, and ε_{ijk} is the effect due to random (chance or error) variation. Both the factors are assumed to be fixed and $\sum_i \alpha_i = 0$ and $\sum_j \beta_j = 0$. We assume that ε_{ij} is normally distributed with mean 0 and variance σ^2. Here, we are interested in testing three hypotheses, namely

Table 10.1 A general two-factor factorial design

		Factor—B			
		1	2	...	s
Factor—A	1	$y_{111}, y_{112}, \ldots, y_{11t}$	$y_{121}, y_{122}, \ldots, y_{12t}$	...	$y_{1s1}, y_{1s2}, \ldots, y_{1st}$
	2	$y_{211}, y_{212}, \ldots, y_{21t}$	$y_{221}, y_{222}, \ldots, y_{22t}$	...	$y_{2s1}, y_{2s2}, \ldots, y_{2st}$
	...	...	...	...	...
	r	$y_{r11}, y_{r12}, \ldots, y_{r1t}$	$y_{r21}, y_{r22}, \ldots, y_{r2t}$	...	$y_{rs1}, y_{rs2}, \ldots, y_{rst}$

(i)   All row treatment effects are equal: That is, $H_0^{(1)}$: $\alpha_1 = \alpha_2 = \cdots = \alpha_r = 0$

(ii)   All column treatment effects are equal: That is, $H_0^{(2)}$: $\beta_1 = \beta_2 = \cdots = \beta_s = 0$, and

(iii)   All row and column treatment interaction effects are equal: That is, $H_0^{(3)}$: $(\alpha\beta)_{ij} = 0$ for all i and j.

If we denote $y_{i..}$ as the total of all observations under the ith level of factor A, $y_{.j.}$ as the total of all observations under the jth level of factor B, $y_{ij.}$ as the total of all observations under the ijth cell of factor A and B, and $y_{...}$ as the total of all observations. Then, the sum of squares in the ANOVA Table 10.2 can be computed as follows:

$$SS_T = \sum_{i=1}^{r}\sum_{j=1}^{s}\sum_{k=1}^{t} y_{ijk}^2 - \frac{y_{...}^2}{rst} \tag{10.2}$$

$$SS_A = \frac{1}{st}\sum_{i=1}^{r} y_{i..}^2 - \frac{y_{...}^2}{rst} \tag{10.3}$$

$$SS_B = \frac{1}{rt}\sum_{j=1}^{s} y_{.j.}^2 - \frac{y_{...}^2}{rst} \tag{10.4}$$

$$SS_{AB} = \frac{1}{t}\sum_{i=1}^{r}\sum_{j=1}^{s} y_{ij.}^2 - \frac{y_{...}^2}{rst} - SS_A - SS_B \tag{10.5}$$

and

$$SS_E = SS_T - SS_A - SS_B - SS_{AB} \tag{10.6}$$

For large values of F ratio, we reject the hypothesis. We now study two particular cases of two-factor factorial design and perform their analysis.

Table 10.2 Two-factor factorial experiment analysis of variance table

Source of variation	Degrees of freedom	Sum of squares	Mean sum of squares	F ratio
A treatments	$r-1$	SS_A	$MS_A = \frac{SS_A}{r-1}$	$\frac{MS_A}{MS_E}$
B treatments	$s-1$	SS_B	$MS_B = \frac{SS_B}{s-1}$	$\frac{MS_B}{MS_E}$
AB interaction	$(r-1)(s-1)$	SS_{AB}	$MS_{AB} = \frac{SS_{AB}}{(r-1)(s-1)}$	$\frac{MS_{AB}}{MS_E}$
Error or residual	$rs(t-1)$	SS_E	$MS_E = \frac{SS_E}{rs(t-1)}$	
Total	$rst-1$	SS_T		

10.6.4 2^2 Factorial Design

This is a particular case of two-factor factorial experiment discussed in Table 10.1, where $r = s = 2$, that is, the factors A and B are used at two levels each and the experiment is replicated for t times. For such a design, we have $2^2 xt = 4t$ observations. Specifically, if $t = 4$, we will have 16 observations. For the Example discussed in 10.1 (yield of a particular chemical process), we have $r = s = 2$ and $t = 3$, and hence, there are 12 observations. Let the levels of factors (A = Temperature and B = Pressure) may be arbitrarily called "low" and "high," and the natural order of the treatment combinations of 2^2 design be (1), a, b, ab. Then, the complete experiment with all treatment combinations along with the replications is described as follows:

Factor			Treatment combinations		Replicates			Total
A	B	AB	Description	Labels	1	2	3	
−	−	+	A low, B low	(1)	90.4	90.2	90.1	270.7
+	−	−	A high, B low	a	90.3	90.5	90.7	271.5
−	+	−	A low, B high	b	90.7	90.6	90.5	271.8
+	+	+	A high, B high	ab	90.6	90.8	90.9	272.3

The average *main effects* of A and B can be computed as follows:

$$A = \frac{1}{2t}[ab + a - b - (1)] \tag{10.7}$$

$$B = \frac{1}{2t}[ab - a + b - (1)] \tag{10.8}$$

and the average *interaction effect* AB is obtained as follows:

$$AB = \frac{1}{2t}[ab - a - b + (1)] \tag{10.9}$$

For the Example discussed in 10.1, we estimate the average main effects for factors as follows:

$$A = \frac{1}{2(3)}[272.3 + 271.5 - 271.8 - 270.7]$$

$$= 0.216667$$

$$B = \frac{1}{2(3)}[272.3 - 271.5 + 271.8 - 270.7]$$

$$= 0.316667$$

and

$$AB = \frac{1}{2(3)} [272.3 - 271.5 - 271.8 + 270.7]$$
$$= -0.05$$

Here, both the main effects A and B are positive and significant. The interaction effect AB is negative and is very small. This suggests that increasing temperature from 150 to 250 °C and pressure from 100 to 200 psi may improve the yield.

Using the sum of squares for A, B, and AB, we can compute the *contrast* for A, B, and AB as follows:

$$\text{Contrast}_A = ab + a - b - (1)$$
$$\text{Contrast}_B = ab - a + b - (1)$$

and

$$\text{Contrast}_{AB} = ab - a - b + (1)$$

Also, these three contrasts are *orthogonal*. These contrasts are also called the *total effect* of A, B, and AB, respectively. The contrasts help us to get the sum of squares as follows:

$$SS_A = \frac{(\text{contrast}_A)^2}{2^2 t} = \frac{1}{4t} [ab + a - b - (1)]^2 \tag{10.10}$$

$$SS_B = \frac{(\text{contrast}_B)^2}{2^2 t} = \frac{1}{4t} [ab - a + b - (1)]^2 \tag{10.11}$$

and

$$SS_{AB} = \frac{(\text{contrast}_{AB})^2}{2^2 t} = \frac{1}{4t} [ab - a - b + (1)]^2 \tag{10.12}$$

For the example under investigation, we obtain the sum of squares as follows:

$$SS_A = \frac{1}{4(3)} [272.3 + 271.5 - 271.8 - 270.7]^2 = 0.140833$$

$$SS_B = \frac{1}{4(3)} [272.3 - 271.5 + 271.8 - 270.7]^2 = 0.300833$$

and

$$SS_{AB} = \frac{1}{4(3)} [272.3 - 271.5 - 271.8 + 270.7]^2 = 0.0075$$

The SS_A, SS_B, and SS_{AB} can also be computed using the formula discussed in (10.2–10.6) above. The total sum of squares is obtained as follows:

$$SS_T = \sum_{i=1}^{r} \sum_{j=1}^{s} \sum_{k=1}^{t} y_{ijk}^2 - \frac{y_{...}^2}{rst}$$

$$= 90.4^2 + 90.3^2 + \cdots + 90.9^2 - \frac{1086.3^2}{12}$$

$$= 0.6425$$

Hence,

$$SS_E = 0.193$$

The complete ANOVA table is presented in Table 10.3.

The table value of F-distribution gives $F_{0.05(1,8)} = 5.32$, for a significance level of 5 %. Since the F ratio for both A and B is larger than the table value, the hypothesis is rejected. That is, the temperature and pressure levels are significant. Since the F ratio for AB is smaller than the table value, it is concluded that there is no interaction effect between temperature and pressure.

The above experiment can be suitably expressed in a regression model as follows: Let x_1 be the *coded value* of the amount of temperature and let x_2 be the *coded value* of the amount of pressure. Then, the yield of a particular chemical process is expressed in a regression model as follows:

$$y = \beta_0 + \beta_1 x_1 + \beta_2 x_2 + \varepsilon$$

where β_0 is the average of all observations, and β_1 and β_2 are the corresponding factor effect estimates. One can also add interaction effect in the model, provided they are significant.

Table 10.3 Analysis of variance table for the chemical process data

Source of variation	Degrees of freedom	Sum of squares	Mean sum of squares	F ratio
A	1	0.140833	0.14083	5.8276
B	1	0.300833	0.30083	12.4483
AB	1	0.0075	0.00750	0.31035
Error	8	0.193	0.02417	
Total	11	0.6425		

The MINITAB output of the above analysis is presented below: Table 10.4 shows the full factorial design with their run orders.

The normal probability plot of the effects is shown in Fig. 10.1. The plot clearly shows that the effects A and B are significant.

Table 10.4 2^2 Full factorial design for chemical process with three replicates

Std order	Run order	Center Pt	Blocks	Temperature	Pressure	Yield
12	1	1	1	250	200	90.6
1	2	1	1	150	100	90.4
3	3	1	1	150	200	90.7
8	4	1	1	250	200	90.8
4	5	1	1	250	200	90.9
6	6	1	1	250	100	90.3
5	7	1	1	150	100	90.2
2	8	1	1	250	100	90.5
10	9	1	1	250	100	90.7
11	10	1	1	150	200	90.6
7	11	1	1	150	200	90.5
9	12	1	1	150	100	90.1

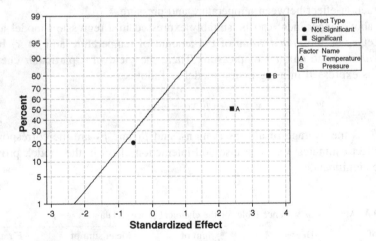

Fig. 10.1 Normal probability plot of the standardized effects

MINITAB Full Factorial Design

```
Factors:    2    Base Design:           2, 4
Runs:      12    Replicates:            3
Blocks:     1    Center pts (total):    0
```

Factorial Fit: Yield versus Temperature, Pressure

Estimated Effects and Coefficients for Yield (coded units)

Term	Effect	Coef	SE Coef	T	P
Constant		90.5250	0.04488	2017.21	0.000
Temperature	0.2167	0.1083	0.04488	2.41	0.042
Pressure	0.3167	0.1583	0.04488	3.53	0.008
Temperature*Pressure	-0.0500	-0.0250	0.04488	-0.56	0.593

S = 0.155456   R-Sq = 69.91%   R-Sq(adj) = 58.63%

Analysis of Variance for Yield (coded units)

Source	DF	Seq SS	Adj SS	Adj MS	F	P
Main Effects	2	0.441667	0.441667	0.220833	9.14	0.009
2-Way Interactions	1	0.007500	0.007500	0.007500	0.31	0.593
Residual Error	8	0.193333	0.193333	0.024167		
Pure Error	8	0.193333	0.193333	0.024167		
Total	11	0.642500				

Estimated Coefficients for Yield using data in uncoded units

Term	Coef
Constant	89.3167
Temperature	0.00366667
Pressure	0.00516667
Temperature*Pressure	-1.00000E-05

Figure 10.2 shows the Pareto chart for standardized effects. It is apparent that the factor B explains maximum variation followed by the factor A. This is supported by the main effect plots given in Fig. 10.3. The interaction plot is shown in Fig. 10.4.

Fig. 10.2 Pareto chart of the standardized effects

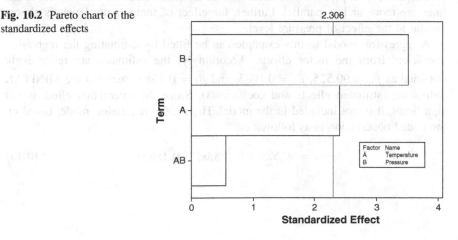

Fig. 10.3 Main effects plot
(data means) for yield

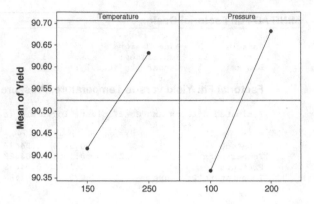

Fig. 10.4 Interaction plot
(data means) for yield

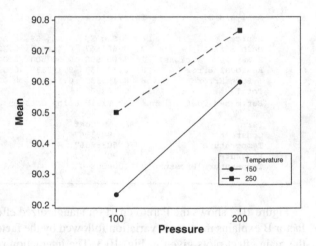

The interaction plot shows that there is no interaction between the factors as the lines are more or less parallel. Further, the effect of temperature levels is almost similar to the effect of pressure level.

A regression model to this example can be fitted by estimating the regression coefficient from the factor effects. Accordingly, the estimates are respectively obtained as $\hat{\beta}_0 = 90.525$, $\hat{\beta}_1 = 0.1083$, and $\hat{\beta}_2 = 0.1583$ (see also the MINITAB output on estimated effects and coefficients). Since the interaction effect is not significant, it is not included in the model. Hence, the regression model based on the coded observation is as follows:

$$y = 90.525 + 0.1083x_1 + 0.1583x_2 \qquad (10.13)$$

The equivalent uncoded model based on the original data can be obtained as follows:

$$y = 89.3167 + 0.003666667x_1 + 0.00516667x_2 \qquad (10.14)$$

The estimated value of the response based on the coded regression model (10.13) is given in Table 10.5. The residual analysis is presented in Fig. 10.5.

The residual analysis does not show any kind of unusual observations. The residuals are normal, and the run orders are also random. The response surface plot and the contour plot of response against pressure and temperature are presented in Fig. 10.6a, b, respectively. Both the surface plot and contour plot show that the yield increases as temperature and pressure increase.

Table 10.5 Estimated responses and residuals

A	B	Replicates			$\hat{y}$	Residuals of each replicates		
		y_1	y_2	y_3		$y_1 - \hat{y}$	$y_2 - \hat{y}$	$y_3 - \hat{y}$
−	−	90.4	90.2	90.1	90.2584	0.1416	−0.0584	−0.1584
+	−	90.3	90.5	90.7	90.475	−0.175	0.025	0.225
−	+	90.7	90.6	90.5	90.575	0.125	0.025	−0.075
+	+	90.6	90.8	90.9	90.7916	−0.1916	0.0084	0.1084

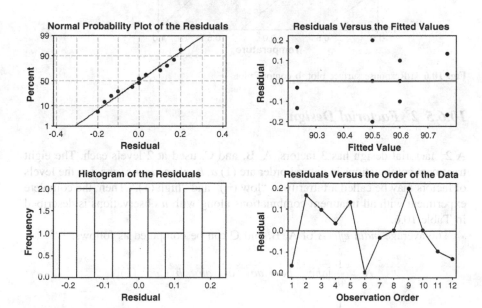

Fig. 10.5 Residual analysis of yield

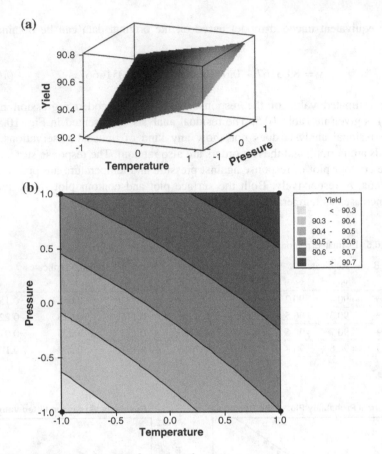

Fig. 10.6 a Response surface plot, **b** contour plot

10.6.5 2^3 Factorial Design

A 2^3 factorial design has 3 factors, A, B, and C, used at 2 levels each. The eight treatment combinations in natural order are (1) a, b, ab, c, ac, bc, abc. Let the levels of factors may be called arbitrarily as "low (−)" and "high (+)." Then, the complete experiment with all treatment combinations along with n observations is described in Table 10.6.

The average *main effects* of A, B, and C can be computed as follows:

$$A = \frac{1}{4n}[abc - bc + ac - c + ab - b + a - (1)]$$

$$B = \frac{1}{4n}[abc + bc - ac - c + ab + b - a - (1)]$$

Table 10.6 2^3 Factorial design layout

Run	A	B	C	Treatment combinations	1	2	...	n	Total
1	−	−	−	(1)	...	...	...	...	...
2	+	−	−	a	...	...	...	...	...
3	−	+	−	b	...	...	...	...	...
4	+	+	−	ab	...	...	...	...	...
5	−	−	+	c	...	...	...	...	...
6	+	−	+	ac	...	...	...	...	...
7	−	+	+	bc	...	...	...	...	...
8	+	+	+	abc	...	...	...	...	...

$$C = \frac{1}{4n}[abc + bc + ac + c - ab - b - a - (1)]$$

and the average *interaction effect* AB, AC, BC, and ABC is obtained as follows:

$$AB = \frac{1}{4n}[abc - bc - ac + c + ab - b - a + (1)]$$

$$AC = \frac{1}{4n}[abc - bc + ac - c - ab + b - a + (1)]$$

$$BC = \frac{1}{4n}[abc + bc - ac - c - ab - b + a + (1)]$$

$$ABC = \frac{1}{4n}[abc - bc - ac + c - ab + b + a - (1)]$$

Using the sum of squares for A, B, and AB, we can compute the *contrast* for A, B, and AB as follows:

$$\text{Contrast}_A = abc - bc + ac - c + ab - b + a - (1)$$

$$\text{Contrast}_B = abc + bc - ac - c + ab + b - a - (1)$$

and

$$\text{Contrast}_C = abc + bc + ac + c - ab - b - a - (1)$$

Also, these three contrasts are *orthogonal*. These contrasts are also called the *total effect* of A, B, and AB, respectively. The contrasts help us to get the sum of squares as follows:

$$SS_A = \frac{(\text{contrast}_A)^2}{2^3 n}$$
$$= \frac{1}{8n}[abc - bc + ac - c + ab - b + a - (1)]^2$$

$$SS_B = \frac{(\text{contrast}_B)^2}{2^3 n}$$
$$= \frac{1}{8n}[abc + bc - ac - c + ab + b - a - (1)]^2$$

and

$$SS_C = \frac{(\text{contrast}_C)^2}{2^3 n}$$
$$= \frac{1}{8n}[abc + bc + ac + c - ab - b - a - (1)]^2$$

Similarly, other sum of squares can also be obtained (Table 10.7).

We now design a 2^3 factorial design for the Example described in 10.2 (firing time of explosive switches). In Table 10.8, the same data are presented in a designed experiment having three factors each of two level with two replicates.

This is an example for 2^3 factorial design with $r = 3$, $s = 2$, and $t = 2$, where factor A is the metal used, B is the amount of primary initiator, and C is the packing pressure of the explosive. Table 10.9 shows the 2^3 factorial data collection sheet along with the average response variable firing time in seconds.

The main effect plot for the yield (firing time) is presented in Fig. 10.7. Factors A (the metal used) and B (the amount of primary initiator) impact the firing time significantly, whereas factor C (the packing pressure of the explosive) may not impact too much.

Table 10.7 A 2^3 factorial design showing interaction effects

Treatment combinations	A	B	C	AB	AC	BC	ABC
(1)	−	−	−	+	+	+	−
a	+	−	−	−	−	+	+
b	−	+	−	−	+	−	+
ab	+	+	−	+	−	−	−
c	−	−	+	+	−	−	+
ac	+	−	+	−	+	−	−
bc	−	+	+	−	−	+	−
abc	+	+	+	+	+	+	+

Table 10.8 Data on firing time of explosive switches

	Factor A			
	a_0(=soft)		a_1(=hard)	
	Factor B		Factor B	
Factor C	b_0(=10 mg)	b_1(=15 mg)	b_0(=10 mg)	b_1(=15 mg)
c_0(=12,000 psi)	40	29	32	16
	39	34	30	12
c_1(=20,000 psi)	35	27	25	11
	30	28	26	13

Table 10.9 A 2^3 factorial design along with the responses

Std order	Run order	Center Pt	Blocks	A	B	C	Firing time
11	1	1	1	Soft	15	12,000	34
10	2	1	1	Hard	10	12,000	32
14	3	1	1	Hard	10	20,000	25
7	4	1	1	Soft	15	20,000	27
8	5	1	1	Hard	15	20,000	11
2	6	1	1	Hard	10	12,000	30
16	7	1	1	Hard	15	20,000	13
3	8	1	1	Soft	15	12,000	29
1	9	1	1	Soft	10	12,000	40
13	10	1	1	Soft	10	20,000	35
4	11	1	1	Hard	15	12,000	16
12	12	1	1	Hard	15	12,000	12
6	13	1	1	Hard	10	20,000	26
5	14	1	1	Soft	10	20,000	30
9	15	1	1	Soft	10	12,000	39
15	16	1	1	Soft	15	20,000	28

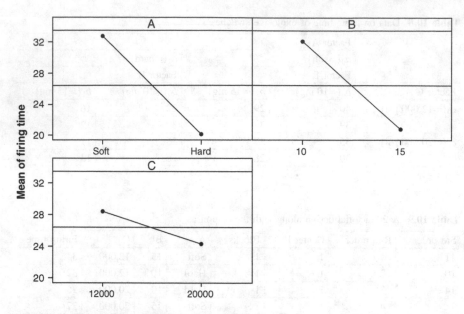

Fig. 10.7 Main effects plot (data means) for firing time

The Pareto chart of the standardized effects is shown in Fig. 10.8. It is very clear that the factor A followed by factor B contributes variation largely. The analysis of variance in the MINITAB output shows that factors A, B, C, and AB are highly significant as their p value is less than 0.05.

Fig. 10.8 Pareto chart of the standardized effects

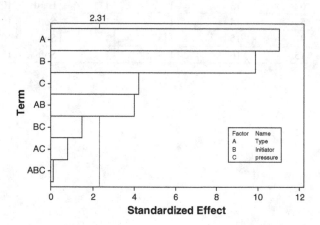

MINITAB Full Factorial Design

```
Factors:   3   Base Design:          3, 8
Runs:     16   Replicates:            2
Blocks:    1   Center pts (total):    0
```

Estimated Effects and Coefficients for firing time (coded units)

Term	Effect	Coef	SE Coef	T	P
Constant		26.688	0.5484	48.66	0.000
Type	-12.125	-6.062	0.5484	-11.05	0.000
Initiator	-10.875	-5.438	0.5484	-9.91	0.000
pressure	-4.625	-2.313	0.5484	-4.22	0.003
Type*Initiator	-4.375	-2.188	0.5484	-3.99	0.004
Type*pressure	0.875	0.437	0.5484	0.80	0.448
Initiator*pressure	1.625	0.813	0.5484	1.48	0.177
Type*Initiator*pressure	0.125	0.063	0.5484	0.11	0.912

S = 2.19374   R-Sq = 96.98%   R-Sq(adj) = 94.34%

Analysis of Variance for firing time (coded units)

Source	DF	Seq SS	Adj SS	Adj MS	F	P
Main Effects	3	1146.69	1146.69	382.229	79.42	0.000
2-Way Interactions	3	90.19	90.19	30.063	6.25	0.017
3-Way Interactions	1	0.06	0.06	0.063	0.01	0.912
Residual Error	8	38.50	38.50	4.813		
Pure Error	8	38.50	38.50	4.813		
Total	15	1275.44				

Estimated Coefficients for firing time using data in uncoded units

Term	Coef
Constant	79.3750
Type	4.3750
Initiator	-3.47500
pressure	-0.00159375
Type*Initiator	-0.975000
Type*pressure	0.000031250
Initiator*pressure	8.12500E-05
Type*Initiator*pressure	6.25000E-06

Figure 10.9 presents the complete interaction plots of factors according to their levels. This plot also supports the evidence of interaction between the factors, type of the metal used, and initiator. The pressure does not show any dependency on the type of metal used and the initiator. Also, there is absolutely no higher interactions existing between the variables.

A regression model to this example can be fitted by estimating the regression coefficient from the factor effects. We first refer to the MINITAB-coded output on estimated effects and coefficients. Since A, B, C, and AB are significant, we include those variables in the model. Accordingly, the regression model based on *coded* observation is as follows:

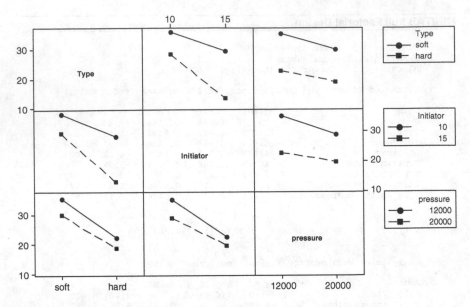

Fig. 10.9 Interaction plot (data means) for firing time

$$\hat{y} = 26.688 - 6.0625x_1 - 5.438x_2 - 2.313x_3 - 2.188x_{12} \quad (10.15)$$

The equivalent *uncoded* model based on the original data can be obtained as follows:

$$\hat{y} = 79.375 + 4.375x_1 - 3.475x_2 - 0.00159375x_3 - 0.975x_{12} \quad (10.16)$$

The prediction based on the coded observation is worked out as follows: Suppose $x_1 = -1$, $x_2 = -1$, and $x_1 = +1$, then the predicted value of y is as follows:

$$y = 26.688 - 6.0625(-1) - 5.438(-1) - 2.313(+1) - 2.188(-1)(-1)$$
$$= 33.6875.$$

The actual value of y at this setting is 35. Similarly, when $x_1 = +1$, $x_2 = -1$, and $x_1 = +1$, then the predicted value of y is as follows:

$$y = 26.688 - 6.0625(+1) - 5.438(-1) - 2.313(+1) - 2.188(+1)(-1)$$
$$= 25.9385,$$

and the actual value of y at this setting is 25.

We now consider an example for 2^4 factorial experiment and carry out the analysis based on full factorial, for transformed variable and a fractional replicate plan experiments.

Table 10.10 Data on surface roughness

Run	A	B	C	D	Surface roughness
1	−	−	−	−	0.00340
2	+	−	−	−	0.00362
3	−	+	−	−	0.00301
4	+	+	−	−	0.00182
5	−	−	+	−	0.00280
6	+	−	+	−	0.00290
7	−	+	+	−	0.00252
8	+	+	+	−	0.00160
9	−	−	−	+	0.00336
10	+	−	−	+	0.00344
11	−	+	−	+	0.00308
12	+	+	−	+	0.00184
13	−	−	+	+	0.00269
14	+	−	+	+	0.00284
15	−	+	+	+	0.00253
16	+	+	+	+	0.00163

Example 10.3 An engineer has performed an experiment to study the effect of four factors on the surface roughness of a machined part. The factors (and their levels) are A = tool angle (12°, 15°), B = cutting fluid viscosity (300, 400), C = feed rate (10 in/min, 15 in/min), and D = cutting fluid cooler used (no, yes). The data from this experiment (with the factors coded to the usual −1, +1 levels) are shown in Table 10.10.

We try to answer the following questions:

1. Estimate the factor effects. Plot the effect estimates on a normal probability plot and select a tentative model
2. Fit the model identified in part (1) and analyze the residuals. Is there any indication of model inadequacy?
3. Repeat the analysis from parts (1) and (2) using 1/y as the response variable. Is there any indication that the transformation has been useful?
4. Fit the model in terms of coded variables that can be used to predict the surface roughness. Convert this prediction equation into a model in the natural variables.
5. Carry out a half-replicate plan of this experiment and conclude.

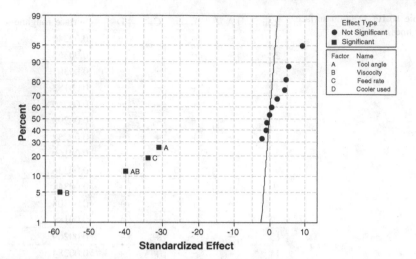

Fig. 10.10 Normal probability plot of the standardized effects

Solution

1. 2^4 Fractional factorial experiment (Table 10.10)

MINITAB Factorial Fit: Roughness versus Tool angle, Viscosity, Feed rate, Cooler used

Estimated Effects and Coefficients for y (coded units)

Term	Effect	Coef	SE Coef	T	P
Constant		0.002693	0.000007	359.00	0.002
Tool angle	-0.000462	-0.000231	0.000007	-30.83	0.021
Viscosity	-0.000877	-0.000439	0.000007	-58.50	0.011
Feed rate	-0.000507	-0.000254	0.000007	-33.83	0.019
Cooler used	-0.000032	-0.000016	0.000007	-2.17	0.275
Tool angle*Viscosity	-0.000600	-0.000300	0.000007	-40.00	0.016
Tool angle*Feed rate	0.000070	0.000035	0.000007	4.67	0.134
Tool angle*Cooler used	-0.000015	-0.000007	0.000008	-1.00	0.500
Viscosity*Feed rate	0.000140	0.000070	0.000008	9.33	0.068
Viscosity*Cooler used	0.000065	0.000033	0.000007	4.33	0.144
Feed rate*Cooler used	0.000000	0.000000	0.000008	0.00	1.000
Tool angle*Viscosity*Feed rate	0.000083	0.000041	0.000007	5.50	0.114
Tool angle*Viscosity*Cooler used	0.000008	0.000004	0.000008	0.50	0.705
Tool angle*Feed rate*Cooler used	0.000033	0.000016	0.000008	2.17	0.275
Viscosity*Feed rate*Cooler used	-0.000013	-0.000006	0.000008	-0.83	0.558

The estimated factor effects of A, B, C, and AB are significant as their p value is less than 0.05. The factor BC also is significant roughly at 0.06 levels. This is in agreement with the normal probability plot of standardized effects (Fig. 10.10) where significant factors are shown in dark square boxes. Hence, a good model can be formulated using the factors A, B, C, and AB.

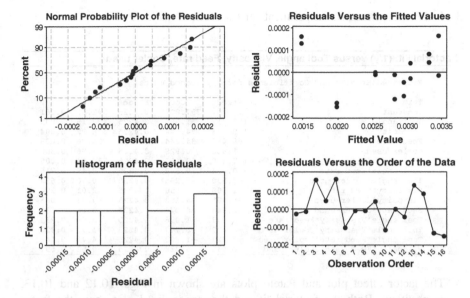

Fig. 10.11 Residual plots for estimated model

2. Analysis of residuals for fitted model. Here, the residual analysis is carried out using the significant effects identified through the factorial experiment done in question (1) above. The residual plot presented in Fig. 10.11 does not show any kind of unusual patterns, and hence, the model is adequate.

Factorial Fit: y versus Tool angle, Viscosity, Feed rate

```
Estimated Effects and Coefficients for y (coded units)

Term                     Effect      Coef   SE Coef       T      P
Constant                          0.002693  0.000030   91.25  0.000
Tool angle            -0.000462 -0.000231  0.000030   -7.84  0.000
Viscosity             -0.000877 -0.000439  0.000030  -14.87  0.000
Feed rate             -0.000507 -0.000254  0.000030   -8.60  0.000
Tool angle*Viscosity  -0.000600 -0.000300  0.000030  -10.17  0.000


S = 0.000118023   R-Sq = 97.66%   R-Sq(adj) = 96.81%


Analysis of Variance for y (coded units)

Source              DF    Seq SS     Adj SS     Adj MS       F      P
Main Effects         3  0.00000497 0.00000497 0.00000166  118.83  0.000
2-Way Interactions   1  0.00000144 0.00000144 0.00000144  103.38  0.000
Residual Error      11  0.00000015 0.00000015 0.00000001
  Lack of Fit        3  0.00000013 0.00000013 0.00000004   11.93  0.003
  Pure Error         8  0.00000003 0.00000003 0.00000000
Total               15  0.00000656
```

3. 2^4 Fractional factorial experiment for transformed variable ($1/y$)

Factorial Fit: (1/y) versus Tool angle, Viscosity, Feed rate, Cooler used

---

```
Estimated Effects and Coefficients for 1/y (coded units)

Term                                    Effect    Coef  SE Coef        T      P
Constant                                        397.807  0.4225   941.62  0.001
Tool angle                             103.212  51.606  0.4225   122.15  0.005
Viscosity                              149.488  74.744  0.4225   176.92  0.004
Feed rate                               68.488  34.244  0.4225    81.06  0.008
Cooler used                              1.656   0.828  0.4225     1.96  0.300
Tool angle*Viscosity                   117.398  58.699  0.4225   138.94  0.005
Tool angle*Feed rate                     0.404   0.202  0.4225     0.48  0.716
Tool angle*Cooler used                  -0.590  -0.295  0.4225    -0.70  0.612
Viscosity*Feed rate                      1.700   0.850  0.4225     2.01  0.294
Viscosity*Cooler used                   -8.305  -4.152  0.4225    -9.83  0.065
Feed rate*Cooler used                    0.548   0.274  0.4225     0.65  0.634
Tool angle*Viscosity*Feed rate           2.193   1.096  0.4225     2.59  0.234
Tool angle*Viscosity*Cooler used        -1.499  -0.749  0.4225    -1.77  0.327
Tool angle*Feed rate*Cooler used        -3.723  -1.862  0.4225    -4.41  0.142
Viscosity*Feed rate*Cooler used         -0.435  -0.218  0.4225    -0.52  0.697
```

The factor effect plot and Pareto plots are shown in Figs. 10.12 and 10.13, respectively. Both the factorial fit and the standardized plot show the factor A, B, C, and AB as significant. Hence, there is no gain in experimenting it with a transformed variable.

4. From all the analysis table and the plots of factor effects, it is possible to conclude a model with factors A, B, C, and AB as they are significant at 0.05 level. Hence, a regression model in coded form is as follows:

$$\hat{y} = 397.807 + 51.606x_1 + 74.744x_2 + 34.244x_3 + 58.699x_{12}$$

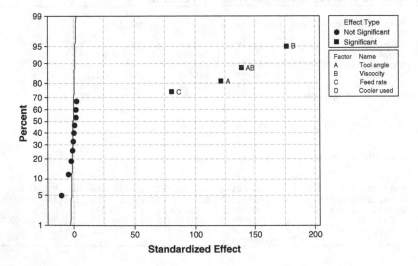

Fig. 10.12 Normal probability plot of the standardized effects for (1/roughness)

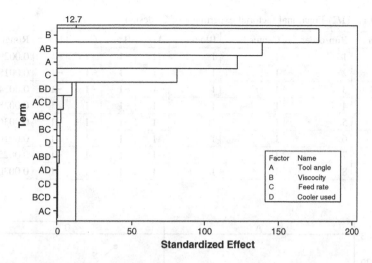

Fig. 10.13 Pareto chart of the standardized effects

5. 2^{4-1} Fractional factorial experiment (Table 10.11; Fig. 10.14)

MINITAB Factorial Fit: roughness versus A, B, C

```
Estimated Effects and Coefficients for yield (coded units)

Term          Effect       Coef   SE Coef       T       P
Constant               0.002709  0.000049   55.56   0.011
A          -0.000448  -0.000224  0.000049   -4.59   0.137
B          -0.000943  -0.000471  0.000049   -9.67   0.066
C          -0.000508  -0.000254  0.000049   -5.21   0.121
A*B        -0.000608  -0.000304  0.000049   -6.23   0.101
A*C         0.000037   0.000019  0.000049    0.38   0.766
B*C         0.000153   0.000076  0.000049    1.56   0.362


S = 0.000137886   R-Sq = 99.46%   R-Sq(adj) = 96.20%


Analysis of Variance for yield (coded units)

Source              DF      Seq SS      Adj SS      Adj MS       F       P
Main Effects         3  0.00000269  0.00000269  0.00000090   47.20   0.106
2-Way Interactions   3  0.00000079  0.00000079  0.00000026   13.81   0.195
Residual Error       1  0.00000002  0.00000002  0.00000002
Total                7  0.00000350
```

Here, the factorial fit analysis and the main effect plots support only one factor, that is, factor B as significant, whereas the other factors are not significant. Even interaction effects are not very serious. To understand better, one may need to carry out the experiment with different runs and standard orders.

Table 10.11 (1/2)-Fractional factorial experiment of 2^4 design

Std order	Run order	Center Pt	Blocks	A	B	C	D	Roughness
6	1	1	1	1	−1	1	−1	0.00290
4	2	1	1	1	1	−1	−1	0.00182
1	3	1	1	−1	−1	−1	−1	0.00340
5	4	1	1	−1	−1	1	1	0.00280
2	5	1	1	1	−1	−1	1	0.00362
8	6	1	1	1	1	1	1	0.00160
7	7	1	1	−1	1	1	−1	0.00252
3	8	1	1	−1	1	−1	1	0.00301

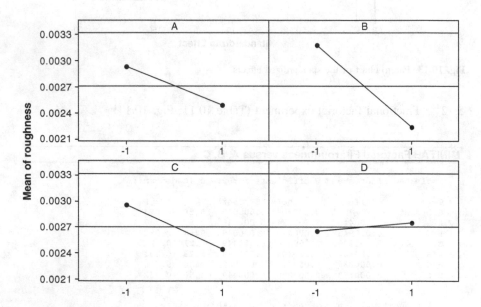

Fig. 10.14 Main effects plot (data means) for roughness

10.7 Robust Designs

The most basic product feature is performance, i.e., the output—the picture clarity of a camera, the alignment of wheels, the color density of a television set, the turning radius of an automobile, etc. To create such an output, engineers use engineering principles to combine inputs of materials, parts, components, assemblies, settings, etc. For each input, the engineer identifies parameters and specifies numerical values to achieve the required output of the final product. For each parameter, the specifications state a target (or nominal) value and a tolerance range around the target. The process is called "parameter and tolerance design."

In selecting these target values, it is useful to set values so that the performance of the product in the field is not affected by variability in manufacturing or field conditions. Such a design is called *robust design*. Robust designs provide optimum performance simultaneously with variation in manufacturing and filed conditions.

Thus, the situation where the design of a product/process produces consistent, high-level performance, despite being subjected to a wide range of changing customer and manufacturing conditions, is called the state of *robustness*. Robustness is synonymous to high-quality function. This means that the goal of an engineer is to search for and achieve "robust" designs for products and processes. It is only through achieving this state engineers can enhance their technological capabilities and hence improve corporate profit. Genichi Taguchi was the first person to develop a method for determining the optimum target values of product and process parameters that minimize variation while keeping the mean on target (see [19, 30–32]). This is facilitated through a three-phase design strategy, namely system design, parameter design, and tolerance design.

The *system design* phase requires the technical knowledge and experience of engineers and scientists to visualize the creation of a product or process that provides the function which the customer is expecting. In this phase, innovation, ingenuity, creativity, and knowledge of new technology are critical for the conceptualization of a new design. This is the phase where concept selection is determined and the search for the best available technology to produce the desired function is intended.

The *parametric design* phase is the most important step in off-line quality engineering, and it is here the engineer can obtain technological capability to improve the quality of function, while maintaining or reducing costs. This capability will lead to the production of high-quality, low-cost product that satisfies the voice of customer (VOC). In parametric design, the goal is to determine "how" to synthesize product or process parameter values that render robust function suitably designed around noise rather than control or eliminate it. The strategy of this phase is to begin experimentation on the design utilizing low-cost, low-grade materials and components. In this manner, when robustness is achieved, it is found despite the fact that low-cost alternatives were utilized thus reducing costs.

The *tolerance design* phase brings further improvements to the optimized design and focuses on the trade-off between quality and cost. In other words, tolerance design evaluates the cost implications behind controlling noise. Hence, tolerance design can be thought of as the "rational" manipulation of tolerances, materials, and components. The key word here is rational, because the engineer should only consider tightening tolerances or upgrading materials and components for factors which are found to have had a significant impact on quality through parameter design experimentation. Although both parameter design and tolerance design are considered optimization phases, the attitudes are substantially different. In parameter design, the strategy is to reduce the effect of noise (variability) without increasing costs, whereas in tolerance design, the strategy is to remove the noise itself by increasing cost.

10.7.1 Robust Parameter Design

The goal of parameter design is to determine how to synthesize levels for product or process parameters to achieve robustness at least cost. In the Taguchi approach, the DOE is employed to investigate control factors in the presence of noise factors. *Control factors* are factors that can be controlled during the design of a product or process. During design, one wants to set their values for optimum product performance and minimum variation. *Noise factors* are factors that cannot be easily controlled (e.g., ambient temperature, variation, and deterioration of materials), but one wants to minimize them. Taguchi recommends using fractional factorial experiments and orthogonal designs to identify the critical control factors and set their levels to minimize variation.

The inclusion of these two types of factors is required in order that their interaction be evaluated. It is through this evaluation that the state of robustness is achieved. The measure used to evaluate this interaction, with regard to robustness, is the signal-to-noise (S/N) ratio. This ratio is a function of the two sample statistics, namely mean and variance. One of the first steps in parameter design study is to identify the types of quality characteristic to be measured, in order to improve performance of function. They are as follows:

- Nominal-the-best—Dimension, viscosity, roughness, etc.
- Smaller-the-better—Wear, shrinkage, deterioration, etc.
- Larger-the-better—Strength, lifetime, efficiency, etc.
- Classified attributes—Appearance, taste types (good or bad), grades (first, second etc.)
- Dynamic characteristics—Speed, acceleration, transmission rate, etc.

A nominal-the-best characteristic is a characteristic which has a specified target value and a tolerance given around the ideal value. A smaller-the-better characteristic has its ideal value zero, and larger-the-better characteristic has its value infinity. Classified attribute characteristics are ones which are not amenable to a continuous scale. Therefore, subjective judgment is used to classify quality on some discretely graded scale. Finally, dynamic characteristics are those where the output from the system is expected to change as the input to the system changes.

Dynamic characteristics best express the engineering intent of the product or process function. Therefore, Dr. Taguchi strongly recommends that all product systems and manufacturing processes be treated as dynamic systems, in order to achieve greatest optimization and benefits. Although few practitioners throughout the world, including Japan, currently utilize the "dynamic" approach for robust designs, the concept is rapidly becoming widely recognized as the most efficient and effective means of developing products and processes. In fact, Dr. Taguchi has often stated that "the adoption and continued utilization of the dynamic approach represents the path that virtually all world-class organizations will take to establish themselves as leaders in their industries, irrespective of their geographical location" (Six Sigma MBB lecture notes, ISI Bangalore). Below, we discuss an engineering problem to illustrate the parameter design strategy, where we use the larger-the-better characteristic of the process.

Example 10.4 We use the data discussed in Example 10.3 above, with an additional measurement on roughness. The control factors and their levels are as follows:

Factor code	Factor descriptions	Levels
A	Tool angle	12°, 15°
B	Cutting fluid viscosity	300, 400
C	Feed rate	10, 15 in/min
D	Cutting fluid cooler used	No, yes

Here, we consider the noise factors as the roughness measurement. The layout out of the experiment is L_8 orthogonal array, with one noise factor at two levels (inner and outer). Thus, we have the experimental layout as shown in Table 10.12.

The S/N ratio for larger-the-better characteristic of the process is computed using the quantity.

$$S/N = 10x \log_{10}\left[\text{sum}\left(\frac{1}{y^2}\right)/n\right] \tag{10.17}$$

As stated earlier, here the focus is on evaluating the interaction between control and noise factors. To determine which factors impact the variability and which factors have the ability to adjust the average, we use the S/N ratio computed using the formula (10.17). The S/N ratio response table is completed by calculating the average S/N ratio for each level of each factor. For example, the average S/N ratio for factor A at level -1 (i.e., at 12° angle) is as follows:

Table 10.12 L_8 orthogonal array and S/N ratio

A	B	C	D	Roughness		$\bar{y}$	s	S/N ratio
				Inner (y_1)	Outer (y_2)			
12	300	10	No	0.00340	0.00323	0.003315	0.00012	49.5989
12	300	15	Yes	0.00269	0.00292	0.002805	0.000163	51.06325
12	400	10	Yes	0.00308	0.00291	0.002995	0.00012	50.48256
12	400	15	No	0.00252	0.00269	0.002605	0.00012	51.69772
15	300	10	Yes	0.00344	0.00315	0.003295	0.000205	49.66813
15	300	15	No	0.00290	0.00322	0.00306	0.000226	50.32121
15	400	10	No	0.00182	0.00230	0.00206	0.000339	53.89991
15	400	15	Yes	0.00163	0.00162	0.001625	7.07E−06	55.78306

$$A_{-1} = \frac{49.5989 + 51.06325 + 50.48256 + 51.69772}{4}$$

$$= 50.71061$$

and the average S/N ratio for factor A at level +1 (i.e. at 15° angle) is as follows:

$$A_{+1} = \frac{49.66813 + 50.32121 + 53.89991 + 55.78306}{4}$$

$$= 52.41808$$

Similarly, the S/N ratio for other factors can also be computed. Table 10.13 presents the response table for S/N ratio and absolute difference of factor levels for all the factors. The difference (Δ) represents the relative significance of each factor that has on maximizing robustness. Also, the preferred level of each factor is identified. Thus, the factors as per their relative importance (based on $|\Delta|$) are B, A, C, and D, and the factor levels as per their significance are B_{+1} and A_{+1}.

Similarly, in Table 10.14, we present the response table for mean for various levels of factors A, B, C, and D. This table reveals that factors B and A have the strongest impact on adjusting or tuning the average to the target value. The factor levels that impact the process average are B_{-1} and A_{-1}. See also the lower panel of the MINITAB output shown below. Note that all the response table values match with the numerical calculations.

Table 10.13 Response table for S/N ratio

Level	A	B	C	D
−1	50.71061	50.16287	50.91237	51.37943
+1	52.41808	52.96581	52.21631	51.74925
Δ = Difference	−1.70747	−2.80294	−1.30393	−0.36981
$|\Delta|$	1.70747	2.80294	1.30393	0.36981
Rank	2	1	3	4

Table 10.14 Response table for average

Level	A	B	C	D
−1	0.00293	0.003119	0.002916	0.00276
+1	0.00251	0.002321	0.002524	0.00268
Δ = Difference	0.00042	0.000798	0.000393	0.00008
$|\Delta|$	0.00042	0.000798	0.000393	0.00008
Rank	2	1	3	4

Notice that, factors C and D are relatively insignificant with respect to reducing variability and adjusting the process average. This being the case, these factors become potential cost reduction factors. In other words, level settings for these factors could be determined by considering the lowest cost alternative.

Based on the optimum combination, the predicted S/N ratio value (η) is calculated as follows:

$$\hat{\eta} = (B_{+1} - \bar{T}_{SN}) + (A_+ - \bar{T}_{SN}) + \bar{T}_{SN}$$
$$= (52.96581 - 51.56434) + (52.41808 - 51.56434) + 51.56434$$
$$= 53.81955$$

Similarly, the predicted process average (μ) is as follows:

$$\hat{\mu} = (B_{-1} - \bar{T}_y) + (A_- - \bar{T}_y) + \bar{T}_y$$
$$= (0.003119 - 0.00272) + (0.00293 - 0.00272) + 0.00272$$
$$= 0.003329$$

We have also carried out a MINITAB analysis for the above engineering problem. The output table shows the linear analysis, ANOVA, and response tables for S/N ratio and mean.

MINITAB Taguchi Analysis: Roughness-I, Roughness-O versus A, B, C, D

```
Estimated Model Coefficients for SN ratios

Term           Coef    SE Coef         T       P
Constant    -51.5643   0.08762  -588.512   0.000
A 12          0.8537   0.08762     9.744   0.010
B 300         1.4015   0.08762    15.995   0.004
C 10          0.6520   0.08762     7.441   0.018
D No          0.1849   0.08762     2.110   0.169
A*B 12 300   -1.0219   0.08762   -11.664   0.007


S = 0.2478    R-Sq = 99.6%    R-Sq(adj) = 98.7%


Analysis of Variance for SN ratios

Source           DF    Seq SS    Adj SS    Adj MS        F       P
A                 1    5.8309    5.8309    5.8309    94.94   0.010
B                 1   15.7130   15.7130   15.7130   255.85   0.004
C                 1    3.4005    3.4005    3.4005    55.37   0.018
D                 1    0.2735    0.2735    0.2735     4.45   0.169
A*B               1    8.3548    8.3548    8.3548   136.04   0.007
Residual Error    2    0.1228    0.1228    0.0614
Total             7   33.6955
```

Linear Model Analysis: Means versus A, B, C, D

```
Estimated Model Coefficients for Means

Term            Coef    SE Coef         T       P
Constant    0.002720   0.000022   126.371   0.000
A 12        0.000210   0.000022     9.757   0.010
B 300       0.000399   0.000022    18.526   0.003
C 10        0.000196   0.000022     9.118   0.012
D No        0.000040   0.000022     1.858   0.204
A*B 12 300 -0.000269   0.000022   -12.486   0.006


S = 0.00006088   R-Sq = 99.7%    R-Sq(adj) = 99.0%


Analysis of Variance for Means

Source           DF     Seq SS     Adj SS     Adj MS        F       P
A                 1   0.000000   0.000000   0.000000    95.19   0.010
B                 1   0.000001   0.000001   0.000001   343.21   0.003
C                 1   0.000000   0.000000   0.000000    83.13   0.012
D                 1   0.000000   0.000000   0.000000     3.45   0.204
A*B               1   0.000001   0.000001   0.000001   155.90   0.006
Residual Error    2   0.000000   0.000000   0.000000
Total             7   0.000003


Response Table for Signal to Noise Ratios
Larger is better

Level       A        B        C        D
1       -50.71   -50.16   -50.91   -51.38
2       -52.42   -52.97   -52.22   -51.75
Delta     1.71     2.80     1.30     0.37
Rank         2        1        3        4


Response Table for Means

Level       A          B          C          D
1      0.002930   0.003119   0.002916   0.002760
2      0.002510   0.002321   0.002524   0.002680
Delta  0.000420   0.000798   0.000393   0.000080
Rank          2          1          3          4
```

Note The S/N ratio for nominal-the-best characteristic of the process is as follows:

$$S/N = 10x \log_{10}\left(\frac{\bar{y}^2}{s^2}\right) \tag{10.18}$$

and the S/N ratio for smaller-the-better of the process is as follows:

$$S/N = 10x \log_{10}\left[\text{sum}(y^2)/n\right] \tag{10.19}$$

As discussed earlier, the robust design philosophy, which is applied to basic technology behind product and process designs, enables the engineer to understand the technology to the point that it is flexible enough to be used on a variety of products or processes. In order to achieve robustness for the underlying technology of product/process designs, the first challenge for the engineers is to define the ideal function (a state of perfect action) associated with it. This is realized through the VOC. In order to guarantee that the customer will view the product/process function as ideal, the engineer must view the function from an engineering perspective. The engineer must translate the VOC into some meaningful and measurable engineering terms. To accomplish this, the engineer must translate the customer's intent into input or signal and perceived result into output response.

Once the ideal relationship between the signal factor and output response for the system is identified, the actual function of the system must be forced to approximate the ideal. This task of making the actual function of the design approach to the ideal state is the engineer's responsibility. He/she must identify the best level of product/process parameter which will enable actual performance to approach the ideal. To attain this goal, the designer needs to experiment on the engineered system. The engineered system is composed of four key elements, namely the signal factor, control factors, noise factors, and output response. These four goals can be achieved, if the engineer performs parameter design experiments.

Along with the importance of Taguchi designs and quality engineering, there are lots of criticisms as well. For complete details of the "triumphs and tragedies" of the methodology, see Pignatiello and Ramberg [24], Montgomery [18], and others.

10.8 Process Mapping for Improvement

Process maps are the most essential tools of Six Sigma, in which designing, measuring, managing, and improving processes are the primary focus. The process mapping at this stage helps one to identify the *bottlenecks* and *disconnects* that influence the improvement within the process activities. *Disconnects* are the points where handoffs from one group to another are poorly handled, or where a supplier and customer have not communicated clearly on one another's requirements. *Bottlenecks* are the points in the process where volume overwhelms capacity,

Fig. 10.15 Factors identified through an improved process

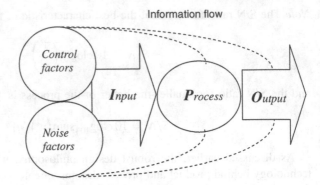

slowing the entire flow of work. Bottlenecks are the "weak link" in getting products and services to customers on time and in adequate quantities.

As you know the purpose of DOE is to identify the effect of controlled factors and noise factors, through different factorial designs, we have experimented some real-life situations identified factor effects and interaction effects. Figure 10.15 depicts an improved process model with the alignment of process output factors. Such models are important for model implementation and forecasting of future productions and inventories. We have also seen that the factors identified through designed experiments can be used for constructing regression models. Thus, an improved process model mapping enables one to look for few vital components among a large pool of critical components, which really contribute to the efficiency of the process.

The other problems associated with process mapping during the improvement stage are to identify the redundancies and coordinate the rework loops and suggesting appropriate time point of decisions/inspections. *Redundancies* are activities that are repeated at two points in the process, also can be parallel activities that duplicate the same result. *Rework loops* are the areas where a high volume of work is passed back up the process to be fixed, corrected, or repaired. The *decision* is necessary to identify potential delays in process where choices, evaluations, checks, or appraisal intervenes causes the process to perform poorly. Using appropriate set of statistical tools, one can overcome these problems.

10.8.1 *Improving a Process Data*

It has been observed that a scientifically processed data give rise to good statistical conclusions. During the preparation of samples, it is possible that the variability depends on the labeling procedure due to various physical characteristics of the samples. The process of data cleaning transforms the original data set by performing tasks such as removing errors, adjusting outliers, estimating missing values, encoding categorical variables, and standardizing variables. The transformation of data into a scale suitable for analysis is to remove as far as possible the effects of systematic sources of variation. The process of minimizing the effects of systematic sources of variation is referred to as *normalization*. The sole purpose of normalization of the data is to ensure that the variation in the expression values is indeed due to biological differences and not due to experimental artifacts (noises) (see [6, 20] for details).

We know that most of the statistical procedures assume an additive model for the data as follows:

$$Data = Model + Noise \text{ (or Error)} \qquad (10.20)$$

Transforming the data can sometimes help promote an additive structure by removing interaction effects between the model and error and stabilizing the error variance. In fact, the assumption of equal variances in analysis of variance model corresponds to an assessment of the spread versus level plot.

Noise Variables: Variables that may impact output, but which we have assumed, have minor impact. This assumption may be based on our C&E matrix or simply due to oversight. Noise variables are those which cannot be controlled completely. The noise variation may be classified into three main families of variation:

- *Positional*—Differences in variation due to similar processes across a production line

  - Reactor-to-reactor differences
  - Line-to-line differences
  - Press-to-press differences
  - Production location-to-location differences
  - Operator-to-operator differences

- *Temporal*—Differences in variation of a process over time

  - Shift-to-shift differences
  - Day-to-day differences
  - Week-to-week differences

- *Sequential*—Differences across a series of processes

  - If the output variable is affected by several different processes, we test process-to-process variability.

Essentially, the purpose of transformation (T), standardization (S), and normalization (N) is to make the data more accountable for computational aspects and its interpretations. Some common reasons for using either T or S or N are as follows:

- There is a spread-level effect across batches of samples,
- The distribution of a variable is strongly skewed,
- The residuals from a fitted model exhibit a systematic pattern, and
- The data do not satisfy the assumptions of a statistical procedure.

The main difficulty that arises in these situations is the presence of nonlinearity which can substantially increase the complexity of the statistical analysis. Another problem that can crop up here is the *multicollinearity* due to dependant structure of the observed random variables. By applying a nonlinear transformation to the data, we may be able to alleviate these problems to a great extent and produce meaningful analysis.

In a Six Sigma project context, it is very essential to identify the systematic variations present in the data set. In order to reveal the patterns existing in the data, the patterns in the data must be discerned, and associations and relationships must be studied and confirmed, which is sometimes called the exploratory data analysis (EDA). In fact, EDA is a visualization technique and is used for increasing the reliability of the information. The necessity of TSN can be judged based on a preliminary analysis of data like construction of histogram or through a box plot. They are the devices for communicating the information contained in a data set. Box plots are particularly effective for portraying comparisons among sets of observations. Side-by-side box plots can reveal differences and similarities in the data set. We now consider an example where we illustrate the utility of box plot and TSN.

Example 10.5 The data in Table 10.15 correspond to a measurement system analysis, where each observation is recorded based on a common location value. Identify the problem of unusual observations and process further for future analysis.

Solution In order to identify the unusual observations, we first carry out a box plot. The Fig. 10.16 shows that there are many unusual observations in three measurements. The observations lie below the bar are called the *inliers* and outside the plot are called *outliers*.

Table 10.15 A chemical measurement data

x1	x2	x3	x4	x5
1.847	−0.291	−0.245	1.681	−0.714
1.176	−1.482	−0.414	0.268	0.004
0.675	0.802	0.138	0.241	1.653
0.303	1.200	0.162	0.177	−0.244
0.574	0.705	−0.253	−0.655	3.002
−0.302	2.854	0.585	1.397	−1.109
−0.754	0.192	−1.312	0.164	0.043
−0.187	2.366	0.149	−0.281	−0.250
−0.205	−1.250	3.662	−1.018	−0.953
−0.051	−0.623	3.055	0.244	−0.436
−0.983	−0.297	−1.190	0.423	0.013
0.125	−0.796	−0.087	0.174	3.161
−1.030	−0.696	0.965	−0.072	−0.304
0.349	0.720	0.317	−1.914	2.016
2.910	0.460	1.155	0.809	−1.521

Fig. 10.16 Box plot of measurements

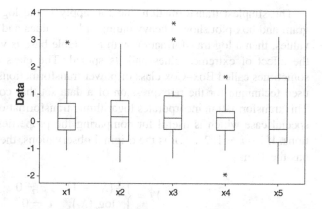

The normal probability plots of the data are shown in Fig. 10.17. It is interesting to see that the variables x_1 and x_4 have unusual observations, still they are normal (see p value is larger than 0.05), whereas the variable x_5 do not have any unusual observations, but normality is not accepted (p value is 0.010). In the case of x_3, the normality assumption does not hold because of many unusual observations. Hence, the process is not stable in general. So before subjecting the measurements for further analysis, it is better to process the data with some transformation or standardization.

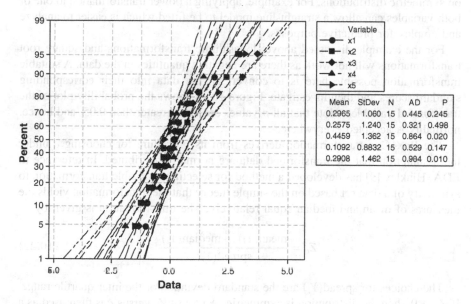

Fig. 10.17 Normal probability plot of measurements

The simplest transformation one can apply is the log transformation. If histogram and box plot show a heavy clump at low values and a very long tail at higher values, then a log transformation with a suitable base is very appropriate to reduce the effect of extreme values and its spread. The class of power transformation sometimes called Box–Cox class of power transformations is the other most widely used technique for the re-expression of a data set that consists of positive values. This transformation incorporates logarithmic transformation as discussed above as a special case which is useful for comparing the properties of different transformations. If X_i, $i = 1, 2, \ldots, n$ is the original observations, then a power transformation has the form

$$Y_i = \begin{cases} \frac{1}{c}(X_i^c - 1), & c \neq 0 \\ \log_e(X_i), & c = 0 \end{cases} \tag{10.21}$$

For various values of c, one gets the corresponding transformed data. For $c = 0$, the transformation will be logarithmic, and for $c = 0.5, 1, 2$, the transformation will be square root, identity, and square transformation, respectively. The purpose of the transformation is to try to induce symmetry in the distribution of the transformed data. For more discussion on transformation, we may refer to Cabrera and McDougall [6]. In fact, a transformation to the symmetry has the desirable property of providing a natural center to the data set which makes estimates of location conceptually easier to understand. Apart from its practical utility, it also enjoys some theoretical advantages associated with working with location estimates based on symmetric distributions. For example, applying a power transformation to one or both variables can allow a straight-line model to be fitted which is easier to analyze and employ for predictive purposes.

For the example discussed above, logarithmic transformations and square root transformations will not work as there are negative quantities in the data. A suitable transformation possible here is to convert all the data into their corresponding standard scores and then investigate. Figure 10.18 shows the probability plot of the transformed variables. Note that all p values are now greater than 0.05, and hence, normality is accepted everywhere.

The Box–Cox transformation derives from its procedure for choosing c under the assumption that the transformed data are normally distributed. In reference to EDA, Hinkley [9] has developed a method for selecting a suitable transformation to symmetry of a data set based on the simple theory that the transformation yields the measures of mean and median equal relatively. The transformation is given by

$$Z_c = \frac{\text{mean}(Y_i) - \text{median}(Y_i)}{\text{spread}(Y_i)} \tag{10.22}$$

The choices for $\text{spread}(Y_i)$ are the standard deviation or the inter-quartile range. If $Z_c = 0$, then the distribution is symmetric. A plot of Z_c versus c is then used as a diagnostic test.

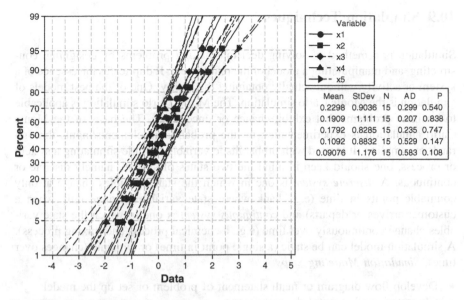

Fig. 10.18 Probability plot of transformed variables

In a Six Sigma project context, the above methods of data preprocessing have its own relevance. It can involve techniques such as *smoothing*, *aggregation*, *generalization*, *normalization*, and *attribute construction*. Smoothing is a form of data cleaning, whereas aggregation and generalization serve as forms of data reduction. Normalization is particularly useful for classification algorithms involving neural networks or distance measurements such as nearest-neighbor classification and clustering. The methods for data normalization employed here are min–max normalization, normalization by decimal scaling, and z-score normalization (or standardization as discussed above).

10.8.2 Improving a Stable Process

When improving a stable system, you do not single out one or two data points. You need to look at all data—not just high points or low points—not just the points you do not like—not just the latest point. All data are relevant. Improving a stable process is more complex than identifying a special cause. More time and resources are generally needed in the discovery process. Common causes of variation can hardly ever be reduced by attempts to explain the difference between individual points if the process is in statistical control. When dealing with special causes, you focus on a few data points. For common cause variation, you need to look at all the data points to fully understand the pattern. You should look at the entire system processes in statistical control that usually require fundamental changes for improvement.

10.9 Simulation Techniques

Simulation is a method of solving decision-making problems by designing, constructing, and manipulating a model of the real system. It duplicates the essence of a system or activity without actually obtaining the reality. One of the best models of simulation is the Monte Carlo simulation. The Monte Carlo simulation is applicable to business problems that exhibit chance or uncertainty. The basis of the Monte Carlo simulation is experimentation on the probabilistic elements/inputs through random sampling. It is used with probabilistic variables. While simulating a system or process, one should keep in mind the two states of system, namely discrete or continuous. A *discrete system* is one in which the state variables change at only countable points in time (e.g., bank where state variables change only when a customer arrives or departs) and *continuous system* is one in which the state variables change continuously over time (e.g., a chemical process and mixing process). A simulation model can be static (a single point in time) or dynamic (changes over time). *Simulation Modeling Steps*

- Develop flow diagram or math statement of problem or set up the model
- Characterize all steps of the process
- Determine input values and probability estimates and cumulative probability distribution
- Generate random number
- Determine appropriate measure scheme and simulate trials.

Advantages of Simulation

- Allows one to analyze large, complex problems for which analytical results are not available
- Allows the decision maker to experiment with many different policies and scenarios without actually changing or experimenting with the actual system.
- Allows one to compress time and enables study of interactions
- May require little or no complex mathematics and so may be intuitively more understandable.

Disadvantages of Simulation

- Numerical results are based on the specific set of random numbers used. These values represent only one of many possible outcomes
- To obtain more accurate results and minimize the likelihood of making a wrong decision, you should use a large number of trials in each simulation and/or repeat the entire simulation a large number of times
- A large number of repetitions can require significant amounts of computer time
- Each simulation requires its own special design to mimic the actual scenario under investigation and its own associated computer program. This may require significant development effort
- Does not generate optimal solutions as it is a trial-and-error approach

- Require managers to generate all conditions and constraints of real-world problem
- Each model is unique and not typically transferable to other problems.

We now present below a couple of examples to understand the practical importance of Monte Carlo simulation approach.

Example 10.6 Here, we assume two normal distribution with different mean and variance and then show that the sum of the variables is normal again. That is

If $X \sim N(50, 25)$ and $Y \sim N(100, 64)$, then to show that $X + Y \sim N(150, 89)$

Solution We use MINITAB to carry out this analysis. We first generated two normal random variables X and Y with respective mean and variance as given above for 25,000 times. The variable X + Y is then computed and stored in Z as the resulting variable. Figure 10.19 presents the histogram with normal curve of these random variables. The graph for Z1 corresponds to $N(150, 89)$. It is very clear that $Z = X + Y \sim N(150, 89)$.

10.9.1 Model Selection and Validation

In statistical research and management, the model selection problem is always crucial for any decision making. Among the choice of many competing models, how to decide the best is even more crucial for researchers. It is often believed that

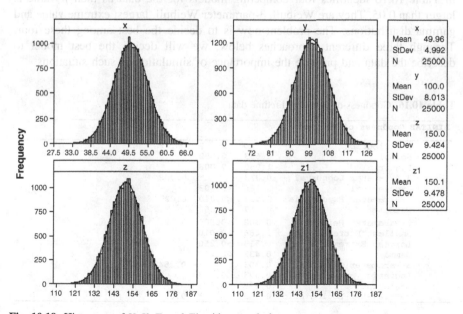

Fig. 10.19 Histogram of X, Y, Z, and Z1 with normal plot

only one reasonable model may be constructed for a given decision-making problem in market research. In empirical studies, one may be able to construct alternative models consistent with the hypothesis of the objectives [4, 7]. Many criteria may be used to compare quantitative marketing models. These criteria include such things as underlying assumptions, data requirements, and theoretical implications [14, 15, 22, 25].

Many methods for comparing structural forms of quantitative models have been proposed in the last three decades. These methods may be classified in terms of methodological emphasis as supermodel methods [2, 10], cross-validation methods [17, 28], likelihood methods [1, 8], or Bayesian methods [5, 25, 26]. The model validation is a procedure of model comparison, estimating the competing models and then compiling error statistics on the other [17, 28]. The interrelatedness of the above categories of methods is exemplified by the fact that using Bayesian arguments, Smith and Spiegelhalter [27] obtained criteria closely related to those of Akaike [1] and Schwarz information criterions.

Consider the following data set ($n = 40$) (see [3]):

85, 90, 92, 5, 10, 17, 54, 55, 58, 55, 58, 32, 33, 32, 82, 68, 34, 36, 92, 102, 103, 106, 146, 124, 142, 195, 65, 66, 68, 54, 55, 58, 143, 151, 158, 195, 114, 114, 116, and 57.

The problem is to decide the best probability model associated with the data. The sample statistics are mean = 83, median = 68, mode = 55, standard deviation = 47.58, and skewness = 0.614. The data are definitely not a normal one as it is skewed toward right. The goodness-of-fit test (MINITAB output) of the data shown in Table 10.16 identifies four competing models for the data as their p value is larger than 0.05. They are Weibull, 3-parameter Weibull, largest extreme value and gamma distributions. The problem now is to decide the best among these four. Through three different approaches below, we will decide the best model to describe the data and propose the importance of simulation in such situations.

Table 10.16 Goodness-of-fit test of lifetime data

MINITAB Goodness of Fit Test

Distribution	AD	P	LRT P
Normal	0.633	0.092	
Lognormal	1.050	0.008	
3-Parameter Lognormal	0.313	*	0.002
Exponential	3.186	<0.003	
2-Parameter Exponential	2.812	<0.010	0.027
Weibull	0.317	>0.250	
3-Parameter Weibull	0.308	>0.500	0.823
Smallest Extreme Value	1.486	<0.010	
Largest Extreme Value	0.313	>0.250	
Gamma	0.432	>0.250	
3-Parameter Gamma	0.303	*	0.254
Logistic	0.593	0.083	
Loglogistic	0.589	0.085	
3-Parameter Loglogistic	0.352	*	0.056

10.9.1.1 Model Selection: Information Theoretic Approach

A very important problem associated with model selection is the fitting of probable model to the data. In order to fit a model to a data, we need to estimate the parameters. This can be done using various methods such as percentile matching, method of moments, minimum distance method, minimum chi-square method, and maximum likelihood method. The first two are crude method and may be easy to implement but produce inferior estimates. The other three methods are more formal procedures with well-defined statistical properties. Although they produce reliable estimators, they can be complex sometimes. The maximum likelihood estimators of the models (MINITAB output) are shown in Table 10.17.

The Akaike information criterion (AIC) is a measure of the relative goodness of fit of a statistical model and was first published by Akaike [1]. It is grounded in the concept of information entropy, in effect offering a relative measure of the information lost when a given model is used to describe reality. It can be said to describe the trade-off between bias and variance in model construction, or loosely speaking between accuracy and complexity of the model. Two other information criterions similar to AIC are the Bayesian information criterion (*BIC*) and Schwarz information criterion (SIC), as they are used as criterion for model selection among a class of parametric models with different numbers of parameters. Choosing a model to optimize BIC is a form of regularization. When estimating model parameters using maximum likelihood estimation, it is possible to increase the likelihood by adding parameters, which may result in over fitting. The BIC resolves this problem by introducing a penalty term for the number of parameters in the model. This penalty is larger in the BIC than in the related AIC. The BIC was developed by Schwarz [29], who gave a Bayesian argument for adopting it. It is closely related to the AIC and hence is also referred to as the Akaike's Bayesian information criterion (ABIC). The BIC is an asymptotic result derived under the assumptions that the data distribution is in the exponential family. Another closely related information function is the Hannan–Quinn information criterion (HQIC).

Table 10.17 Maximum likelihood estimators of lifetime data

Distribution	Location	Shape	Scale	Threshold
Normal*	83.00000		47.58636	
Lognormal*	4.20502		0.76126	
3-Parameter Lognormal	4.98178		0.30774	-69.73051
Exponential			83.00000	
2-Parameter Exponential			78.05000	4.95000
Weibull		1.80328	93.12422	
3-Parameter Weibull		1.86288	95.42587	-1.82214
Smallest Extreme Value	107.70310		50.63953	
Largest Extreme Value	60.92960		38.63130	
Gamma		2.49243	33.30081	
3-Parameter Gamma		4.67124	22.26230	-20.99262
Logistic	79.53017		27.02753	
Loglogistic	4.28257		0.39303	
3-Parameter Loglogistic	4.81433		0.21261	-47.91801

* Scale: Adjusted ML estimate

The AIC, SIC, BIC, and HQIC are, respectively, given by

$$\text{AIC} = -2\log L(\hat{\Theta}) + 2p \tag{10.23}$$

$$\text{SIC} = -2\log L(\hat{\Theta}) + p\log(n) \tag{10.24}$$

$$\text{BIC} = -\log L(\hat{\Theta}) + \frac{0.5p\log(n)}{n} \tag{10.25}$$

and

$$\text{HQIC} = -\log L(\hat{\Theta}) + p\log(\ln(n)) \tag{10.26}$$

where $\log L(\hat{\Theta})$ is the maximum likelihood function and p is the number of free parameters that need to be estimated under the model. Schwarz [29] used the above criterion for estimating the dimension of a given model. This criterion is slightly different from Akaike [1], in which Schwarz's procedure leans more than Akaike's toward lower-dimensional models. For large number of observations, the procedures differ markedly from each other.

Given any two estimated models, the model with the lowest value of BIC is the one to be preferred if comparison is based on BIC. The BIC is an increasing function of σ_e^2 (variance) and an increasing function of p, where p is number of parameters of population under study. That is, unexplained variation in the dependent variable and the number of explanatory variables increase the value of BIC. Hence, lower BIC implies either fewer explanatory variables, better fit, or both. The BIC generally penalizes free parameters more strongly than does the AIC, though it depends on the size of n and relative magnitude of n and k. It is important to keep in mind that the BIC can be used to compare estimated models only when the numerical values of the dependent variable are identical for all estimates being compared. The models being compared need not be tested, unlike the case when models are being compared using an F test or likelihood ratio test.

In Table 10.18, we present the information criterion values for the four models under consideration. Note that the information criterion values corresponding to Weibull model are the lowest in comparison with the information criterion values of the other distributions, suggesting a strong support for Weibull distribution. The second competing model is the largest extreme value, as the information values are

Table 10.18 Information criterion values

Model	AIC	SIC	BIC	HQIC
Weibull	420.6829	424.0607	208.3876	210.9521
3-Weibull	422.6330	427.6997	208.3892	212.2325
Gamma	422.6819	426.0597	209.3871	211.9516
Largest extreme value	422.0297	425.4075	209.0610	211.6255

very close to Weibull. The last two models in preference are the 3-parameter Weibull and then gamma distribution. We now use a semi-parametric approach to decide the model preference.

10.9.1.2 Model Selection: Semi-parametric Approach

Suppose X be the decision variable under study and $f(x)$ the corresponding probability density function, then the limited expected value (LEV) function or the expected loss eliminated is defined as follows:

$$E[X; d] = \int_0^d x f(x) dx + d[1 - F(d)] \tag{10.27}$$

where $F(x)$ is the distribution function of the variable X. Another view of $E[X; d]$ is that it is the expected value of $Y = \min(X, d)$, that is, the mean of a random variable censored at d. The other quantity of interest is the loss elimination ratio (LER) which is the ratio of the expected loss eliminated to the expected value of X, that is, LER = $E[X; d]/E(X)$, provided $E(X)$ exists [12]. Note that $E[X; d]$ always exists. A quantity that needs to be compared with $E[X; d]$ is the empirical limited expected value (ELEV) function for a sample as

$$E_n(d) = \frac{1}{n} \sum_{i=1}^n \min(x_i, d) \tag{10.28}$$

For accepting any model as providing a reasonable description of the decision process, we should verify that $E[X; d]$ and $E_n(d)$ are essentially in agreement for all values of d. It is because as $d \to \infty$ $E[X; d] \to E(X)$, if it exists and $E_n(d) \to \overline{X}$. Thus, comparing $E[X; d]$ and $E_n(d)$ is like a method of moments approach in a restrictive way.

In life testing experiments, d may be sometimes called the truncation time or the censoring time. In this respect, another important characteristics of lifetime models are the mean residual life (MRL) at age $d > 0$ which is the conditional mean of $X - d$, given $X \geq d$, namely

$$e(d) = E[X - d | X \geq d] = \int_d^\infty (x - d) \frac{f(x)}{P(X \geq d)} dx \tag{10.29}$$

Table 10.19 Empirical estimates of expected values

X	$E_n(x)$	Limited expected value function			
		Weibull	3-Weibull	Gamma	Largest extreme value
5	5.00	4.78	4.80	4.88	4.78
10	7.50	8.45	8.45	9.78	8.33
20	11.67	10.98	10.12	12.43	10.76
40	18.75	20.21	22.14	25.98	21.04
80	31.00	33.28	43.56	44.56	35.02
120	45.83	49.32	50.18	54.64	49.98
150	60.71	62.12	63.43	66.19	63.05
200	71.88	70.17	73.25	75.02	71.25

Then, $E[X; d]$ and $e(d)$ are related through the equality

$$E(X) = E[X; d] + e(d)[1 - F(d)] \tag{10.30}$$

A plot of $e(d)$ can also give some indication as to the type of distribution that will model the data.

For the example discussed above, we compute the empirical and fitted LEV function as given in Eqs. (10.27) and (10.28) for the four competing distributions and see the agreement between (10.27) and (10.28). In Table 10.19, we present those empirical estimates considering the value of d as 150. From the table, it is seen that for Weibull distribution, the empirical expectation is closer to the LEV for all values of d. Hence, a Weibull distribution seems to be a good model.

As mentioned earlier, the use of $e(d)$ function can also give a similar conclusion. Since $e(d)$ is a part of LEV function, it is sufficient to work with LEV function only. The other methods such as minimum distance estimation and Bayes factor approach can also be used for identifying the best out of many alternatives (see [21] for details). Generally, these methods are not mathematically tractable, and hence, visualization becomes difficult. Below, we discuss the simulation approach to deal with these kinds of situations.

10.9.1.3 Model Selection: Simulation Approach

Here, we carry out a non-normal process capability for each model by simulating the distributions with their estimated value of the parameters given in Table 10.17. That is, by generating a large number of samples from Weibull (93.12422, 1.80328), 3-parameter Weibull (-1.82214, 95.42587, 1.86288), gamma (33.30081, 2.49243), and largest extreme value (38.6313, 60.9296), we compute the process capability for a given specification limits LSL = 30 and USL = 200. The model having large value of process capability is chosen as the best model. The model-wise process capability analysis is shown in Fig. 10.20a–d.

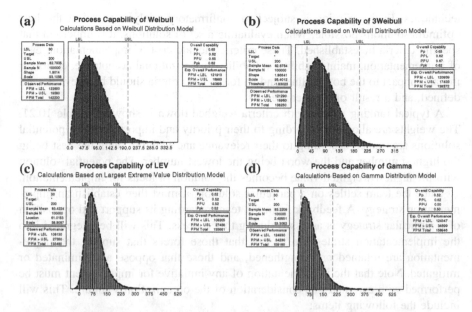

Fig. 10.20 a Process capability of Weibull distribution, **b** process capability of 3-parameter Weibull distribution, **c** process capability of largest extreme value distribution, **d** process capability of gamma distribution

Table 10.20 Model versus process capability	Model	Overall capability		Process average	Expected PPM total
		P_p	P_{pk}		
	Weibull	0.65	0.62	82.7935	140,895
	3-Weibull	0.65	0.62	82.8754	138,372
	LEV	0.52	0.52	83.4224	135,661
	Gamma	0.52	0.50	82.2209	158,946

The summarized process capability and expected PPM totals of each model are shown in Table 10.20. The capability analysis also gives a clear edge on Weibull models (both two and three parameter) as their capability is more than the other two models.

10.10 Implementation and Validation

It is very essential for any Six Sigma project to develop potential solutions and suggest improvement alternatives for a higher level of sigma. This is achieved through the implementation of pilot tests, prototype testing, DOE, and simulation techniques. Every solution is subject to confirmatory analysis to select the best optimum solutions possible. When evaluating a set of solutions, it is important that specific criteria be established. These criteria include cost of implementation, ease of implementation, maintainability, reliability, organizational acceptance, customer impact, impact to the bottom line, and so forth. The criteria should be relevant, well defined, and a result of team consensus.

A typical ranking matrix with criteria weighed down is shown in Table 10.21. The weights are allocated according to their priority and importance. The potential solutions are ranked according to their relevance and viability, with the best being the highest number and the worst being the lowest number. The potential solution with the highest weighted value becomes the solution of focus for implementation.

Once the team settles on a particular solution, it must then establish an implementation strategy. A feedback from the team regarding its support and opposition of a particular strategy is assessed during this process. This will be integrated into the implementation strategy to ensure that those forces that support the implementation are retained or strengthened, and those that oppose are eliminated or mitigated. Note that the implementation of any initiative for improvement must be performed with care and due consideration of the organization's culture. This will include the following items:

- *Infrastructure*—This includes everything necessary to ensure a successful implementation such as software, hardware, training, talent acquisition, organizational changes, and facilities
- *Communication plan*—A communication plan should be specific with regard to

  - Who, what, when, and where it communicates
  - Requirements of multiple communications, if necessary
  - Communications should reinforce and progress logically
  - Timely communications should occur before a phase is executed, not later

- *Competition for time, energy, and resources*—As with most organizations, the team's improvement initiative is not the only critical activity in the organization that demands resources. Therefore, it is essential to outline the requirements for implementation in advance and to secure approval by management.

Table 10.21 Ranking matrix with weighted criteria

Criteria	Weight	Potential solution				
		A	B	C	D	E
% process improvement	0.30	3	5	4	2	1
Cost of implementation	0.15	3	4.5	2	4.5	1
Reliability	0.05	2	1	4	5	3
Maintainability	0.10	5	4	3	2	1
Installation resources	0.20	2.5	2.5	4	1	5
Effects and impacts	0.20	4	1.5	3	5	1.5
Total		3.25	3.425	3.4	2.925	2

- *Management commitment*—This is pivotal to every successful implementation. Nothing will be accomplished without the management's support and commitment. The management must be visible, active, and engaged to ensure a successful implementation.

10.11 Improve Check Sheets

The improve check sheets should include the following:

- Prepare a list of innovative ideas for potential solutions
- Use the narrowing and screening techniques to further develop quality potential solutions
- Create a solution statement for at least two possible proposed improvements
- Make a final choice of solution based on success criteria
- Verify the present solution with the anticipated one
- Develop a plan for piloting and testing the solution, including an action plan, results assessment, and schedule
- Evaluate pilot results and confirm that the results are in tandem with the goal statement
- Identify and implement refinements to the solution based on lessons from the pilot study
- Consider potential problems and unintended consequences of the solution and develop preventive and contingent actions to address them.

10.12 Relevance for Managers

The methods discussed in the previous chapters are generally easy to apply and operationally viable. A minimum level of understanding about the process parameters and design specifications are enough to identify customer requirements and business objectives. This chapter addresses many important problems associated with engineering science, called design of experiments, simulation methods, model selection etc., which are considered to provide the most efficient and economical methods of reaching valid and relevant conclusions from the experiment. The importance of randomized designs, factorial designs, orthogonal arrays, and robust designs along with their practical significances are discussed in detail with number of real life examples. The importance of Taguchi design for quality engineering and designs for optimal level selections are also presented here. Taguchi calls for off-line quality and online quality control as two core experimental designs to improve quality and cost.

It is often found that, data comprises some model-based and noise (or error) based information. Hence improving such data by removing the presence of noise variation is a challenge for any improvement. Various methods for stabilizing the noise variation using normalization, standardization and transformations are discussed for better appreciation of a process. Through simulation, we have also studied many model based characteristics of a process. An exhaustive account of model selection and validation using parametric and nonparametric methods is also presented for immediate application. The implementation and validation of causes of variation is also addressed to complete the Improve phase activity. Hence improving quality is as good as improving a process, where an improved process will have less variation and all redundancies will be removed.

Exercises

10.1. Discuss various tasks and deliverables of Improve phase.
10.2. What is Balanced Score Card? Discuss various performance indicators associated with Balanced Score Card.
10.3. Discuss the importance of Kaizen events for quality improvements. How it is connected to Six Sigma project management?
10.4. What are the uses of 5S approach? Discuss each component in relevance to a Six Sigma project management.
10.5. Discuss various waste elimination methods. Show how Kanban philosophy is used to identify wastes in a process?
10.6. What is Design of Experiments? Discuss various principles associated with a design of experiments.
10.7. Distinguish between

  (i)   Experimental error and noise factor
  (ii) Randomization and Replication
  (iii) Partial confounding and total confounding
  (iv) Factorial designs and complete block designs

10.8. A 2^2 factorial experiment was conducted to determine whether the volume of two reagents had an effect on the ability of an assay method to measure levels of a specific drug in serum. The serum for each test was sampled from a serum pool spiked with a single dose of the drug. The data are shown in table with reagent volumes in ml.

Reagent		% drug recovered
A	B	
10	20	32
40	20	44
10	50	51
40	50	68
25	35	53

(i) Estimate the linear response equation. Are the linear effects of both reagents significant?

(ii) Also test for adequacy of fitted equation and give your conclusions about pure quadratic terms and interaction terms?

(iii) Carry out a residual analysis.

10.9. In a study of behavior of a fluidized bed reactor for the catalytic oxidation of benzene to maleic anhydride, a 2^4 factorial design was used to determine the effects of X_1 = temperature in degree centigrade (°C); X_2 = benzene/air ratio; X_3 = flow rate of reactants in m^3/h; and X_4 = height of the catalytic bed of particles at rest in cm on the conversion response Y, namely, the percentage of consumed benzene relative to the amount of moles of benzene introduced. The original levels of factors A_1, A_2, A_3, A_4 are given in below table.

Factors	Input variable	Low	High
A_1	Temperature, X_1 (°C)	420	460
A_2	Benzene/air ratio, X_2	1.5	2.5
A_3	Flow rate, X_3 (m^3/h)	4	6
A_4	Height, X_4	4	6

The experimental design and percentage conversion values are given below:

x_1	x_2	x_3	x_4	Y	
				Rep-1	Rep-2
420	1.5	4	4	63.03	65.83
460	1.5	4	4	62.19	63.77
420	2.5	4	4	64.01	64.57
460	2.5	4	4	61.60	60.23
420	1.5	6	4	58.95	59.78
460	1.5	6	4	78.34	79.45
420	2.5	6	4	45.75	48.79
460	2.5	6	4	72.66	71.12
420	1.5	4	6	46.36	49.56
460	1.5	4	6	68.62	68.78
420	2.5	4	6	35.16	36.47
460	2.5	4	6	59.24	60.32
420	1.5	6	6	71.62	70.45
460	1.5	6	6	84.01	86.45
420	2.5	6	6	61.18	60.18
460	2.5	6	6	77.78	76.56

(i)   Carry out the complete analysis of variance of this design.

(ii) Estimate the factor effects. Plot the effect estimates on a normal probability plot and select a tentative model.

(iii) Fit the model identified in part (ii) and analyze the residuals. Is there any indication of model inadequacy?

(iv) Repeat the analysis from parts (ii) and (iii) using $1/y$ as the response variable. Is there any indication that the transformation has been useful?

(v)   Carry out a half-replicate plan of this experiment and conclude.

10.10.   Discuss how a design can be made "Robust".

10.11.   What are the importance of parametric design and tolerance design? Bring out their basic differences.

10.12.   What are the characteristics of a Robust parametric design?

10.13.   Discuss how a process map can be used for improving a process?

10.14.   Discuss the necessity of transformation, standardization and normalization for improving a process data.

10.15.   Consider the following data:

43	34	43	32	87	35	71	65	12	52	19	48	17	24
52	65	40	54	62	45	2	13	18	49	57	21	64	71
45	81	52	40	35	78	43	45	44	55	79	37	19	14
31	71	51	35	27	74	22	8	22	15	20	23	35	37
17	21	52	48	9	22	12	12	85	40	39	30	42	8
66	74	19	4	30	6	19	31	25	33	32	51	68	43

Carry out the following:

(i)   Draw a Box-plot and check for any abnormalities

(ii) Construct a normal probability plot decide the quality of the data

(iii) Use a logarithmic transformation and then decide the quality of the data

(iv) Use MINITAB software and decide all the probable model for the data

(v)   Use Likelihood approach to decide the best model (assume Normal, Lognormal, Gamma, Weibull, and logistic probability distributions).

10.16   The following informations are obtained through a designed experiment for a chemical process:

Power (W)	Temperature (°C)	Moisture (% by weight)	Flow rate (L/min)	Particle size (mm)	Yield of oil (l)
415	25	5	40	1.28	63
550	25	5	40	4.05	21
415	95	5	40	4.05	36

(continued)

Power (W)	Temperature (°C)	Moisture (% by weight)	Flow rate (L/min)	Particle size (mm)	Yield of oil (l)
550	95	5	40	1.28	99
415	25	15	40	4.05	24
550	25	15	40	1.28	66
415	95	15	40	1.28	71
550	95	15	40	4.05	54
415	25	5	60	4.05	23
550	25	5	60	1.28	74
415	95	5	60	1.28	80
550	95	5	60	4.05	33
415	25	15	60	1.28	63
550	25	15	60	4.05	21
415	25	15	60	4.05	44
550	25	15	60	1.28	96

Discuss the whether the process yield is normal or not. If the process is not normal, discuss the methods to bring it to normal. Also propose other viable models for the variable yield of oil.

10.17.   What is simulation? State the advantages and disadvantages of simulation.

10.18.   For the data given in Q. 10.15, above simulate a suitable statistical model and study and compare the characteristics of the data.

10.19.   For the data on yield of oil given in Q. 10.16, above simulate a suitable statistical model and study and compare the characteristics of the data.

10.20.   Discuss the Information theoretic approach of model selection.

10.21.   For the data on yield of oil given in Q. 10.16, above discuss the Information theoretic approach of model selection and study the characteristics of the data.

10.22.   Discuss the semi-parametric approach of model selection.

10.23.   Discuss the simulation approach of model selection.

10.24.   If $X \sim \text{gamma}(10, 3)$ and $Y \sim \text{Weibull}(10, 3)$. Use simulation method to study the distribution of $X + Y$. Also try to study the distribution of $X + Y$ when $X \sim \text{gamma}(10, 1)$ and $Y \sim \text{Weibull}(10, 1)$.

References

1. Akaike, H.: A new look at the statistical identification model. IEEE Trans. Auto. Control **19**, 716–723 (1974)
2. Atkinson, A.C.: A method for discriminating between models. J. Roy. Stat. Soc. (B) **32**, 323–353 (1969)
3. Bain, L.J., Engelhardt, M.E.: Statistical Analysis of Reliability and Life-testing Analysis. Marcel Dekker, New York (1991)

4. Bass, F.: A simultaneous equation regression study of advertising and sales of cigarettes. J. Mark. Res. **6**, 291–300 (1969)
5. Blattberg, R.C., Sen, S.K.: An evaluation of the application of minimum Chi-square procedures to stochastic models of brand choice. J. Mark. Res. **10**, 421–427 (1975)
6. Cabrera, J., McDougall, A.: Statistical Consulting. Springer, Berlin (2002)
7. Carmone, F.H., Green, P.E.: Model misspecification in multi-attribute parameter estimation. J. Mark. Res. **18**, 87–93 (1981)
8. Cox, D.R.: Further results on tests of separate families of hypothesis. J. Roy. Stat. Soc. (B) **24**, 406–424 (1962)
9. Hinkley, D.: On quick choice of power transformations. Appl. Stat. **26**, 67–69 (1977)
10. Johnson, N.L., Kotz, S.: Urn Models and Their Applications. Wiley, New York (1977)
11. Kaplan, R.S., Norton, D.P.: The Balanced scorecard. Harvard Business School Press, Boston (1996)
12. Klugman, S.A., Panjer, H.H., Willmot, G.E.: Loss Models: From Data to Decisions, 3rd edn. Wiley Series in Probability and Statistics, Wiley, New York (2008)
13. Lee, S.S., Dugger, J.C., Chen, J.C.: Kaizen: an essential tools for inclusion in industrial technology curricula. J. Ind. Technol. **16**(1), 1–7 (2000)
14. Larreche, J.C., and Montgomery, D.B.: Framework for the comparison of marketing models: a Delphi study. J. Market. Res. **14**, 487–498 (1977)
15. Little, J.D.: Models and managers: the concept of a decision calculus. Manage. Sci. **16**, B466–B485 (1979)
16. Logothetis, N., Wynn, H.P.: Quality Through Design, Experimental Design, Off-Line Quality Control and Taguchi 's Contributions. Clarendon Press, Oxford (1989)
17. Mosteller, F., Tukey, J.W.: Data analysis, including statistics. In: Lindzey, F., Aronson, E., (eds.) Handbook of Social Psychology, vol. 2. Addison-Wesley, Reading (1968)
18. Montgomery, D.C.: Introduction to Statistical Quality Control, 3rd edn. Wiley, New York (1997)
19. Montgomery, D.C.: Design and Analysis of Experiments. Wiley India, New Delhi (2009)
20. Muralidharan, K.: A note on transformation, standardization and normalization. Int. J. Oper. Quant. Manage. IX(1 & 2), 116–122 (2010)
21. Muralidharan, K.: A teaching note for model selection and validation. Math. J. Interdiscip. Sci. **1**(2), 55–62 (2013)
22. Narasimhan, C., Sen, S.K.: New product models of test market data. J. Market. **47**, 11–24 (1983)
23. Park, R.: Value Engineering. St. Lucie Press, Boca Raton (1999)
24. Pignatiello, J.J., Ramberg, J.S.: Top ten triumphs and tragedies of Genichi Taguchi. Qual. Eng. **4**(2), 211–225 (1992)
25. Rust, R.T.: A probabilistic measure of model superiority. Working Paper 81–23. Graduate school of Business, University of Texas, Austin (1981)
26. Smith, A.F.M.: A general Bayesian linear model. J. Roy. Stat. Soc. (B) **2**, 213–220 (1973)
27. Smith, A.F.M., Spiegelhalter, D.J.: Bayes Factors and Choice Criteria for Linear Models. J. Roy. Stat. Soc. Ser. B (Methodol.) **42**(2), 213–220 (1980)
28. Stone, M.: Cross-valedictory choice and assessment of Statistical predictions. J. Roy. Stat. Soc. (B) **36**, 111–147 (1974)
29. Schwarz, G.: Estimating the dimensionality of a model. Ann. Stat **6**(2), 461–464 (1978)
30. Taguchi, G.: System of Experimental Design: Engineering methods to Optimize Quality and Minimize Cost. UNIPUB, White Plains, New York (1987)
31. Taguchi, G.: Introduction to Quality Engineering, Asian Productivity Organization. UNIPUB, White plains, New York (1991)
32. Taguchi, G., Wu, Y.: Introduction to Off-line Quality Control. Central Japan Quality Control Association, Nagoya (1980)
33. Womack, P., Jones, D.T.: Lean Thinking. Simon and Schuster, New York (1996)

Chapter 11
Control Phase

Control phase tasks and deliverables

- Document a new or improved process and measurement
- Use statistical process control (SPC) for validating collection systems
- Use SPC to ascertain the repeatability and reproducibility of metrics in an operational environment
- Define the control plan and its supporting plans such as communications plan of the improvements and operational changes to the customers and stakeholders, prepare implementation and risk management plan, consolidate cost-benefit, and change management plan
- Establish tracking procedures in an operational environment: monitor implementation, validate and stabilize performance gains, and jointly audit the results and confirm the financials
- Setup control plans for tolerances, controls, measures, and standard operating procedures
- Validate in-control process and benefits for process capability, MSA and Gage R&R, and documentation
- Document and the follow-up steps
- Exit review

11.1 Control Plans

So far, our attempt has been to identify the most important contributing variables of the process. Through the completed five phases, we have been trying to establish the best process model free of all kinds of variations. The method so far enabled us to identify a viable model and helped us to quantify the nature of the relationship between the important variables of the process output. The statistical process control technique can now be employed with considerable effectiveness for monitoring and surveillance of the process. Techniques such as control charts and capability assessment can be used to monitor the process output and detect when changes in the inputs are required to bring the process back to an in-control state.

© Springer India 2015
K. Muralidharan, *Six Sigma for Organizational Excellence*,
DOI 10.1007/978-81-322-2325-2_11

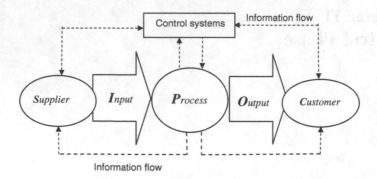

Fig. 11.1 A controlled system process model

The models that relate the influential inputs to process outputs facilitate the nature and magnitude of the adjustments required for such a controlled process. Figure 11.1 shows such a process model, where the information flow is monitored by a controlled system. This system enables the necessary adjustment in the process so that future values of the product characteristics will be approximately on target.

A control plan is like a living document that identifies critical input or output variables and associated activities that must be performed to maintain control of the process variation, products, and services in order to minimize deviation from the preferred values. The S-I-P-O-C model presented above helps maintain the spirit of the DMAIC philosophy under a responsible and accountable control plans.

The routine adjustment is often called engineering control, automatic control, or feedback control [14]. Statistical process control methods can be successfully integrated into a manufacturing system where engineering control can be fully utilized. At this stage, the Six Sigma project should address all dimensions of quality aspects, as an effective quality improvement can be instrumental in increasing productivity and reducing cost. The eight dimensions of quality are as follows:

- Performance
- Reliability
- Durability
- Serviceability
- Aesthetics
- Features
- Perceived quality
- Conformance to standards

Although quality improvement is a company-wide activity, and that every organizational unit must actively participate, it is the responsibility of Six Sigma project personnel to sustain the improvement and control for a long period of time. They are also responsible for evaluating and using quality cost information for identifying improvement opportunities in the system and for making these opportunities known to higher management. In the remaining part of this chapter, we study the control phase tasks and deliverables from quality improvement perspectives.

11.2 Statistical Process Control

We all know that variation is the cause of all defects and is inherent in all processes in varying amount. To make a product defect-free, it is essential to minimize the variation of a process. Hence, quantifying the amount of variation in a process is the first and the critical step toward improvement. This will further enhance the understanding of the type of causes and decide the course of action to reduce the variation. This action will have the lasting improvement on the quality of the product.

The ultimate goal of any Six Sigma project is improvement by reducing variation. The reduced variability ultimately leads to quality. According to Montgomery [30], quality is inversely proportional to variability. That is less the variability more the quality. Since variability can only be described by statistical terms, statistical methods play a central role in quality improvement efforts. The statistical tools and techniques offer a wide range of quality improvement tools for the detection of variations in the process.

There is a certain amount of variability in every product; consequently, no two things are exactly alike. For instance, a process is bound to vary from product to product, materials to materials, operation to operation, service to service, machine to machine, man to man, shift to shift, day to day, etc. However, if the variation is very large, the customer may perceive the unit to be undesirable and unacceptable. As mentioned earlier, quantifying the information is the first step toward the improvement of quality. The quantification is done on the basis of the characteristics under study. The characteristics can be either qualitative (or attributes) or quantitative (or variables) in nature. We will describe various tools and techniques for analyzing these characteristics in the chapters to come.

Significance of measuring variation

- Measuring variability determines the reliability of an average by pointing out as to how far an average is representative of the entire data.
- Helps to determine the nature and cause of variation in order to control the variation itself.
- Enables comparisons of two or more distributions with regard to their variability.

Understanding of the variation in values of a quality characteristic is of the primary importance in statistical process control (SPC). SPC, a sub-area of statistical quality control (SQC), consists of methods for understanding, monitoring, and improving process performance over time. In this chapter, we study various process control techniques in detail.

11.2.1 Describing Variations

As we know the amount of variation in a process tells us what that process is actually capable of achieving (tolerance), whereas specifications tell us what we want a process to be able to achieve. The first step in understanding variation should always be to plot a process data in time order. The time order can be hourly, daily, weekly, monthly, quarterly, yearly, or any suitable time period.

Generally, two types of variations are encountered in a process. They are *special cause* or *assignable cause variation* and the *random* or *irregular type of variations*. The goal of any process then will be to identify the causes and eliminate the causes to make an unstable process to a stable process. A process with only common cause variation is called stable. One purpose of control charting, the featured tool of SPC, is to distinguish between these two types of variation in order to prevent over-reaction and under-reaction to the process. The distinction between common causes and assignable causes is context dependent. A common cause today can be an assignable cause tomorrow. The designation could also change with a change in the sampling scheme. One wants to react, however, only when a cause has sufficient impact that it is practical and economic to remove it in order to improve quality.

Special Cause variation: The variations that are relatively large in magnitude and viable to identify are the special causes of variation or assignable causes of variation. They may come and go sporadically. Quite often, any specific evidence of the lack of statistical control gives a signal that a special cause is likely to have occurred. Hence, such variations are local to the process and are unstable. The best way of dealing with this kind of variation is to

- Get timely data so that special cause signal can identify easily.
- Take immediate action to remedy any damage.
- Immediately search for a cause. Find out what was different on that occasion. Isolate the deepest cause you can affect.
- Develop a longer term remedy that will prevent that special cause from recurring or, if results are good, retain that lesson.
- Use early warning indicators throughout your operation. Take data at the early process stages so you can tell as soon as possible when something has changed.

Common Cause variations: This type of variation is the sum of the multitude of effects of a complex interaction of random or common causes. It is common to all occasions and places and always present in some degree. Variation due to common causes will almost always give results that are in statistical control. These kinds of variations are normally stable and can be controlled with a proper treatment. This is ultimately achieved by the SQC and SPC techniques. Both these techniques involve the measurement and evaluation of variation in a process and the efforts made to limit or control such variation. In its most common applications, SPC helps an organization or process owner to identify possible problems or unusual incidents so that action can be taken promptly to resolve them.

The best way of analyzing the special cause variation and common cause variation is the control chart techniques. A control chart is a device intended to be used at the point of operation, where the process is carried out by the operators of that process. For every control chart, there are three zones, namely the stable zone (shows the presence of common causes of variation), warning zone, and action zone (both shows the presence of special causes of variation). A statistically controlled chart will be called a normal process whose process means (a measure of accuracy) and standard deviation (a measure of precision) will work according to the tolerance level suggested by the customer.

11.2.2 Control Charts

As mentioned at various places in the previous chapters, the main objective of Six Sigma is to identify the variation underlying the process and then control it to achieve a specific quality. We use control charts to separate common cause and special cause variation in a process. A control chart has three zones: a zone in which no action should be taken (presence of common causes); another zone that suggests more information should be obtained, and the third zone in which one requires some action to be taken (presence of special causes). Control limits are statistical limits set at ±3 standard deviations from the mean. As such, it has no relationship with specification limits, but may be the same or close to it. Any points falling outside statistical limits signal a special cause. The control chart technique was first introduced by Shewart in 1920 at the Bell Telephone Laboratories. It provides a common language for discussing process performance and can be used for almost any type of data collected over time.

One should use a control chart to (1) track performance over time, (2) evaluate progress after process changes/improvements, and (3) focus attention on detecting and monitoring process variation over time. Note that, the control limits have no simple relationship to upper and lower specification limits. They relate to the goals of charting to identify assignable causes and preventing over-control of systems. Also, charting methods generally include a *start-up phase* in which data are collected and the chart constants are calculated. Some people use control charts on 30 start-up or trial periods and some use 25 as the trail periods. Besides, all charting methods also include a steady-state phase in which the limits are fixed and the chart mainly contributes through (i) identifying the occasional assignable cause and (ii) discouraging people from changing the process input settings. When the charted quantities are outside the control limits, detective work begins to investigate whether something unusual and fixable is occurring. In some cases, production is shut down, awaiting detective work and resolution of any problems discovered.

Control charts are an essential tool of continuous improvement, and hence, it is necessary to record data on a regular basis. The type of control chart then depends

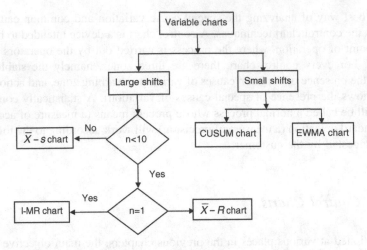

Fig. 11.2 Variable control chart road map

on the type of variables recorded. For continuous data, we have $\overline{X}$, range (R), and s-charts, whereas, if the data are attribute types, then we use attribute control charts. The road map for chart selection is depicted in Figs. 11.2 and 11.8, respectively, for variables and attributes. According to Montgomery [31], there are five reasons for using control charts. They are given as follows:

1. Control charts are a proven technique for improving productivity
2. Control charts are effective in defect prevention
3. Control charts prevent unnecessary process adjustment
4. Control charts provide diagnostic information
5. Control charts provide information about capability

Control charting techniques: Suppose that a quality characteristic is normally distributed with mean μ and standard deviation σ. If $x_1 x_2, \ldots, x_n$ is a sample of size n, then $\overline{x}$ is normally distributed with mean μ and standard deviation $\sigma/\sqrt{n}$. Suppose the standards mean and standard deviation are known for any process, then the control limits for mean are constructed using the 3σ-limit principle as follows:

$$\text{LCL} = \mu - 3\frac{\sigma}{\sqrt{n}} = \mu - A\sigma, \quad A = \frac{3}{\sqrt{n}}$$

$$\text{CL} = \mu \qquad\qquad\qquad\qquad\qquad\qquad (11.1)$$

$$\text{UCL} = \mu + 3\frac{\sigma}{\sqrt{n}} = \mu + A\sigma, \quad A = \frac{3}{\sqrt{n}}$$

For any given significance level α, one can construct the control limits for mean using the formulas

$$\text{LCL} = \mu - z_{\alpha/2} \frac{\sigma}{\sqrt{n}}$$

$$\text{CL} = \mu \qquad (11.2)$$

$$\text{UCL} = \mu + z_{\alpha/2} \frac{\sigma}{\sqrt{n}}$$

If a sample mean falls outside these limits, it is an indication that the process mean is no longer equal to μ. Further, the above results are still approximately correct, even if the underlying distribution is non-normal, because of central limit theorem. In practice, we usually will not know μ and σ. In that case, we have to estimate them from the sample of observations. These estimates should usually be based on at least 20–25 samples. This and other aspects of control charting techniques will be discussed in subsequent sections.

As pointed out by Woodall and Montgomery [58], process monitoring involves two phases. In Phase I, the practitioner collects a sample of time-ordered data from the process of interest. These Phase I data are used to gain process understanding. The practitioner must check for unusual or surprising results. In addition, the practitioner must assess the stability of the process, select an appropriate in-control model, estimate the parameter(s) of this model, and determine the design parameters to be used in Phase II. The monitoring method is then implemented with data collected successively over time in Phase II in order to detect changes in the process from the assumed in-control model.

In many applications, it is possible to adjust processes using feed-forward or feedback control. In order to do this, there must be an adjustment variable with some information available on its effect on the response of interest. For information and perspectives on this topic, one may refer to del Castillo [11], Box et al. [8], Box and Narasimhan [9] and Qin [42]. There have been some general overviews of statistical process monitoring. Stoumbos et al. [49] provided a review article on the more traditional methods for statistical process control (SPC) and change point detection. Recently, Woodall and Montgomery [58] reviewed some current directions on the theory and applications of statistical process monitoring. Nair et al. [35] and Woodall [56] identified two areas that have since received considerable attention, i.e., the monitoring of functions, now referred to as "profile monitoring," and surveillance with spatiotemporal data. Further, Frisén [20] provided an excellent overview of methods, evaluation criteria, and application areas.

11.2.3 Control Charts for Variables

Whenever one or two continuous variable or key output variable, summarize the quality characteristics of units or a system, it is advisable to use a variable charting approach. To control a process using variable data, it is necessary to keep a check on the current state of the accuracy (average) and precision (spread) of the distribution of the data. The $\overline{X}$-chart and R-charts are the most frequently used control chart for variables. Although they are used in both administrative and manufacturing applications, they are the tools of first choice in many manufacturing applications. The possible reasons are as follows:

- Most sensitive charts for tracking and identifying assignable cause of variation
- Based on control chart factors that assume a normal distribution within subgroups
- Subgroups allow for a precise estimate of "local" variability
- Changes in process variability can be distinguished from changes in process average
- Small shifts in process average can be detected very easily

The advantages of an $\overline{X}$-chart and R-chart disappear if systemic special causes occur in each subgroup. For example, suppose you are counting errors in orders received by phone and you have four operators taking orders. It would be natural to want to construct subgroups of 4, taking one order form from each operator. But if one operator is consistently better or worse than the others, you would be mixing special cause and common cause variation in the data. The chart will be useless, obscuring differences between operators and making it hard to detect changes in the process or variability. When data are collected in rational subgroups, it makes sense to use an $\overline{X}$-chart and R-chart. In the rational subgroup, we hope to have represented all the common causes of variation and none of the special causes of variation. $\overline{X}$-chart and R-charts allow us to detect smaller shifts than individual charts. Also, they allow us to clearly separate changes in process average from changes in process variability.

If $\overline{X}$-chart is a plot of the subgroup means, then R-chart is a plot of the subgroup ranges. Instead of plotting the subgroup ranges, if we plot the subgroup standard deviation along with the $\overline{X}$-chart, we get $\overline{X}$-s chart. Both $\overline{X}$-R charts and $\overline{X}$-s charts are two separate charts of the same subgroup data. They are considered to be the most sensitive charts for tracking and identifying assignable causes of variation based on control chart factors that assume a normal distribution within subgroups.

The control limits for $\overline{X}$ and R-chart are calculated as follows: If the sample number is k and the size of each sample is n, then

$$\overline{\overline{X}} = \frac{1}{nk} \sum_{i=1}^{n} \sum_{j=1}^{k} x_{ij}$$

and

$$\overline{R} = \frac{1}{k} \sum_{j=1}^{k} R_j$$

Then, the control limit for $\overline{X}$-chart is

$$\text{LCL} = \overline{\overline{X}} - A_2\overline{R}$$
$$\text{CL} = \overline{\overline{X}} \qquad (11.3)$$
$$\text{UCL} = \overline{\overline{X}} + A_2\overline{R}$$

and the control limit for R-chart is

$$\text{LCL} = D_3\overline{R}$$
$$\text{CL} = \overline{R} \qquad (11.4)$$
$$\text{UCL} = D_4\overline{R}$$

where the constants A_2, D_3 and D_4 are given in Appendix Table A.11 for various values of n. Note that, the centerline of $\overline{X}$-chart is $\overline{\overline{X}}$ and for R-chart, it is $\overline{R}$.

Further, if we use $\overline{\overline{X}}$ as an estimator of μ and $\frac{\overline{R}}{d_2}$ as an estimator of σ, then the 3σ-control limits according to (11.1) for $\overline{X}$-chart is

$$\text{LCL} = \overline{\overline{X}} - 3\frac{\overline{R}}{d_2\sqrt{n}} = \overline{\overline{X}} - A_2\overline{R}, \quad A_2 = \frac{3}{d_2\sqrt{n}}$$
$$\text{CL} = \overline{\overline{X}} \qquad (11.5)$$
$$\text{UCL} = \overline{\overline{X}} + 3\frac{\overline{R}}{d_2\sqrt{n}} = \overline{\overline{X}} + A_2\overline{R}, \quad A_2 = \frac{3}{d_2\sqrt{n}}$$

For R-chart, we estimate the estimates of standard deviation of R as $\hat{\sigma}_R = d_3\frac{\overline{R}}{d_2}$. This will ultimately lead to the control limits for R-chart as per given in (11.4). The constants A, A_2, D_3, D_4, d_2, and d_3 are tabulated for different values of n (see Appendix Table A.11 for details).

Note that, the control limits shown in (11.3) and (11.4) are *trial control limits*. They allow us to determine whether the process was in control when the n initial samples were taken. To test the hypothesis of past control, plot the values of $\overline{X}$ and R from each sample on the charts and analyze the resulting chart. If all points plot are inside the control limits and no systematic behavior is evident, then we can conclude that the process was in control in the past, and the trial control limits are suitable for controlling current or future production. This analysis of past data is sometimes referred to as a Phase I analysis (see Montgomery [31]).

As a practical significance, R-chart is drawn first. If R-chart is in control, only then it is meaningful to go ahead with the $\overline{X}$ chart. That is, target (mean chart) is controlled only when variation (R-chart) is under control. When R-chart is out of control, we often eliminate the out-of-control points and re-compute a revised value of $\overline{R}$. This value is then used to determine new limits for $\overline{X}$-chart and R-chart.

This will tighten the limits on both charts, making them consistent with a process standard deviation σ. This estimate of σ could be used as the basis of a preliminary analysis of process capability. The effective use of any control chart will require periodic revision of the control limits and centerlines. While revising control limits, it is highly recommended to use at least 25 samples or subgroups in computing control limits.

Interpreting $\overline{X}$-R-chart

- Use the signals of special causes on both charts
- Look at the R-chart first

  - If the range chart is unstable (has special causes), the limits on the $\overline{X}$-chart will be of no use
  - If the range chart is unstable, it is unsafe to draw conclusions about variation in the process average.
  - If the range chart is stable, proceed with the inspection of $\overline{X}$-chart

- Look for positive or negative correlations between the data points on the $\overline{X}$ and the R-chart (both move in the same direction or in opposite directions for every point). This happens when the data have a skewed distribution, and some conclusions may be affected.

Example 11.1 The thickness of a printed circuit board is an important quality parameter. Data on board thickness (in inches) are given in Table 11.1 along with their sample means, ranges, and standard deviations, for 25 samples of three boards each. Investigate whether the data are in control or not. Also, obtain the process capability of the process if the specification limits are 0.0631 ± 0.00047. Also, obtain the long-term sigma level of the process.

Solution According to the formula given above, the control limits for $\overline{X}$ and R-chart are calculated as follows:

$$\overline{\overline{X}} = \frac{1.5738}{25} = 0.063035$$

and

$$\overline{R} = \frac{0.023}{25} = 0.000975$$

and from Appendix Table A.11 for subgroup size $k = 3, A_2 = 1.023, D_3 = 0$ and $D_4 = 2.575$. Then, according to the formulas given in (11.3) and (11.4), we get

Control limits for $\overline{X}$-chart: LCL = 0.062038 and UCL = 0.063035
Control limits for R-chart: LCL = 0 and UCL = 0.0025106

Table 11.1 Data on thickness of printed circuit board

Sample number	X_1	X_2	X_3	$\overline{X}_i$	R_i	s_i
1	0.0629	0.0636	0.0640	0.0635	0.0011	0.00055678
2	0.0630	0.0631	0.0622	0.06277	0.0009	0.00049329
3	0.0628	0.0631	0.0633	0.06307	0.0005	0.00025166
4	0.0634	0.0630	0.0631	0.06317	0.0004	0.00020817
5	0.0619	0.0628	0.0630	0.06257	0.0011	0.00058595
6	0.0613	0.0629	0.0634	0.06253	0.0021	0.00109697
7	0.0630	0.0639	0.0625	0.06313	0.0014	0.00070946
8	0.0628	0.0627	0.0622	0.06257	0.0006	0.00032146
9	0.0623	0.0626	0.0633	0.06273	0.001	0.00051316
10	0.0631	0.0631	0.0633	0.06317	0.0002	0.00011547
11	0.0635	0.0630	0.0638	0.06343	0.0008	0.00040415
12	0.0623	0.0630	0.0630	0.06277	0.0007	0.00040415
13	0.0635	0.0631	0.0630	0.0632	0.0005	0.00026458
14	0.0645	0.064	0.0631	0.06387	0.0014	0.00070946
15	0.0619	0.0644	0.0632	0.06317	0.0025	0.00125033
16	0.0631	0.0627	0.0630	0.06293	0.0004	0.00020817
17	0.0616	0.0623	0.0631	0.06233	0.0015	0.00075056
18	0.0630	0.0630	0.0626	0.06287	0.0004	0.00023094
19	0.0636	0.0631	0.0629	0.0632	0.0007	0.00036056
20	0.0640	0.0635	0.0629	0.06347	0.0011	0.00055076
21	0.0628	0.0625	0.0616	0.0623	0.0012	0.0006245
22	0.0615	0.0625	0.0619	0.06197	0.001	0.00050332
23	0.0630	0.0632	0.0630	0.06307	0.0002	0.00011547
24	0.0635	0.0629	0.0635	0.0633	0.0006	0.00034641
25	0.0623	0.0629	0.0630	0.06273	0.0007	0.00037859

The control chart is presented in Fig. 11.3. Observing R-chart first, we could see that the variation is under control. There are no unusual patterns in the chart. The $\overline{X}$-chart also indicates that the process is in control as the individual means randomly vary about the centerline and do not fall outside the control limits.

Although $\overline{X}$ and R-charts are widely used, it is occasionally desirable to estimate the process standard deviation directly instead of indirectly through the use of the range R. Recall that the range method of estimating standard deviation loses statistical efficiency for moderate to large samples. Hence, the use of $\overline{X}$ and s-chart is recommended in those situations. Even for variable size samples, this chart is preferable.

If we define the sample standard deviation as (see also Chap. 5 for detailed estimation of standard deviation)

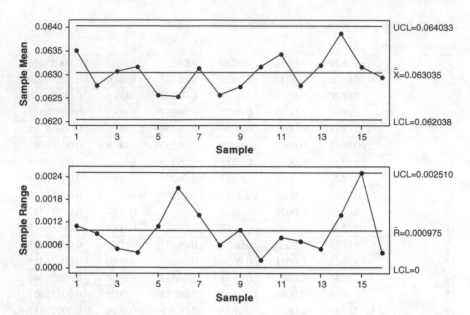

Fig. 11.3 $\overline{X}$ and R-chart for thickness

$$s = \sqrt{\frac{1}{n-1}\sum_{i=1}^{n}(X_i - \overline{X})^2}$$

Then,

$$E(s) = \sigma\left[\left(\frac{2}{n-1}\right)^{1/2}\Gamma\left(\frac{n}{2}\right)\right]\bigg/\Gamma\left(\frac{n}{2}-\frac{1}{2}\right)$$

$$= c_4\sigma$$

where

$$c_4 = \left[\left(\frac{2}{n-1}\right)^{1/2}\Gamma\left(\frac{n}{2}\right)\right]\bigg/\Gamma\left(\frac{n}{2}-\frac{1}{2}\right)$$

Hence, the control limits for s-chart are as follows:

$$\mathrm{LCL} = c_4\sigma - 3\sigma\sqrt{1 - c_4^2} = \left(c_4 - 3\sqrt{1 - c_4^2}\right)\sigma = B_5\sigma$$

$$\mathrm{CL} = c_4\sigma$$

$$\mathrm{UCL} = c_4\sigma + 3\sigma\sqrt{1 - c_4^2} = \left(c_4 + 3\sqrt{1 - c_4^2}\right)\sigma = B_6\sigma$$

(11.6)

The values of c_4, B_5, and B_6 are tabulated for various samples sizes in Appendix Table A.11. If no standards are given for σ, then it must be estimated by analyzing past data. Suppose that n preliminary samples are available, each of size k, and let s_i be the standard deviation of the ith sample. Then, the average of n standard deviation is

$$\bar{s} = \frac{1}{n}\sum_{i=1}^{n} s_i$$

Then, the statistic $\bar{s}/c_4$ is an unbiased estimator of σ. Therefore, the parameters of the s-chart are as follows:

$$\text{LCL} = \bar{s} - 3\frac{\bar{s}}{c_4}\sqrt{1 - c_4^2} = \left(1 - \frac{3}{c_4}\sqrt{1 - c_4^2}\right)\sigma = B_3\bar{s}$$

$$\text{CL} = \bar{s} \tag{11.7}$$

$$\text{UCL} = \bar{s} + 3\frac{\bar{s}}{c_4}\sqrt{1 - c_4^2} = \left(1 + \frac{3}{c_4}\sqrt{1 - c_4^2}\right)\sigma = B_4\bar{s}$$

Note that, $B_3 = B_5/c_4$ and $B_4 = B_6/c_4$. Since $\bar{s}/c_4$ is used to estimate σ, we may define the control limits on the corresponding $\bar{x}$-chart as

$$\text{LCL} = \bar{\bar{x}} - 3\frac{\bar{s}}{c_4\sqrt{n}} = \bar{\bar{x}} - A_3\bar{s}, \quad A_3 = \frac{3}{c_4\sqrt{n}}$$

$$\text{CL} = \bar{\bar{x}} \tag{11.8}$$

$$\text{UCL} = \bar{\bar{x}} + 3\frac{\bar{s}}{c_4\sqrt{n}} = \bar{\bar{x}} + A_3\bar{s}, \quad A_3 = \frac{3}{c_4\sqrt{n}}$$

The constants A_3, B_3, B_4 and c_4 are available in Appendix Table A.11.
   For the example discussed above,

$$\bar{s} = \frac{0.01195427}{25} = 0.000478$$

and hence, the control limits for $\bar{X}$-chart are as follows:

$$\text{LCL} = \bar{\bar{x}} - A_3\bar{s} = 0.063 - 1.954(0.000478) = 0.062066$$
$$\text{CL} = 0.063035$$
$$\text{UCL} = \bar{\bar{x}} + A_3\bar{s} = 0.063 + 1.954(0.000478) = 0.063934$$

For subgroup size 3, from Appendix Table A.11, the values of B_3 and B_4 are 0 and 2.568, respectively. Hence, the control limits for s-chart are as follows:

$$\text{LCL} = B_3\bar{s} = 0$$
$$\text{CL} = \bar{s} = 0.000478$$

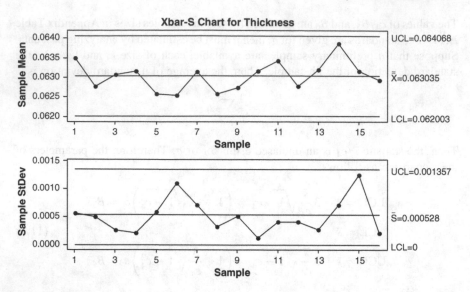

Fig. 11.4 $\overline{X}$ and s-chart for thickness

and

$$\text{UCL} = B_4\bar{s} = 0.00128$$

The plot of $\overline{X}$ and s-chart to assess the presence of assignable causes of variation is given in Fig. 11.4. The s-chart also clearly shows that the process variation is in control. Hence, the entire process is statistically in control.

Process capability of the process. In order to find the process capability of the process, we need to get the estimate of variation as follows:

$$\hat{\sigma} = \frac{\overline{R}}{d_2} = \frac{0.000975}{1.693} = 0.000576$$

Hence, the thickness follows a normal distribution with mean $\mu = 0.063035$ and $\sigma = 0.000576$. Now the specification limits are given to be 0.0631 ± 0.00047 (i.e., LSL = 0.06263 and USL = 0.06357). Hence, the probability of non-confirming thickness measurements are as follows:

$$
\begin{aligned}
P(X < \text{LSL}) &= P(X < 0.06263) \\
&= P\left(Z < \frac{0.06263 - 0.063035}{0.000576}\right) \\
&= P(Z < -0.70312) \\
&= 0.24099
\end{aligned}
$$

and

$$P(X > \text{USL}) = P(X > 0.06357)$$
$$= P\left(Z > \frac{0.06357 - 0.063035}{0.000576}\right)$$
$$= P(Z > 0.928819)$$
$$= 0.176491$$

Hence

$$P(\text{non-confirming thickness measurements}) = 0.24099 + 0.176491 = 0.41748.$$

That is, there are about 42 % non-confirming units outside the specification limits. One can also find the capability indices discussed in Chap. 8 (see Sect. 8.13 for details) as

$$C_p = \frac{\text{USL} - \text{LSL}}{6\sigma}$$
$$= \frac{0.06357 - 0.06263}{0.000576} = 1.63$$

Since C_p is larger than 1, the natural tolerances are well within the lower and upper specification limits. Consequently, relatively low number of non-confirming thickness measurements can be produced. The long-term sigma level corresponding to this C_p value is 3.4.

If measurements are not recorded according to the subgroup, then one may use *IMR*-charts for plotting individual data. *IMR*-charts are two separate charts of the same data. *IMR*-charts is a plot of the individual data, and *MR* chart is a plot of the moving range of the previous individuals. These charts are sensitive to trends, cycles, and patterns and most useful for destructive testing and batch processing.

Suppose x_i is the characteristics of interest, then the moving range is obtained as follows:

$$MR_i = |x_i - x_{i-1}|, \quad i = 2, 3, \ldots$$

Then, the control limits for *IMR*-charts is computed as follows:

$$\text{LCL} = \bar{x} - 3\frac{\overline{MR}}{d_2}$$
$$\text{CL} = \bar{x} \tag{11.9}$$
$$\text{UCL} = \bar{x} + 3\frac{\overline{MR}}{d_2}$$

Fig. 11.5 Box plot of
thickness measurements

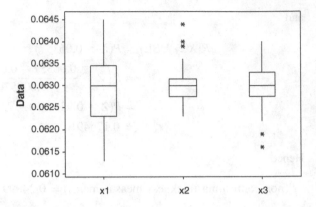

We now illustrate moving range chart concept through our ongoing example.
Figure 11.5 shows the box plot of the samples x_1, x_2, and x_3. Note that, the samples
x_2 and x_3 have unusual observations. Further, x_1 is normal (see Fig. 11.6) and x_2 and
x_3 not normal (not shown here). Hence, we have drawn a *IMR* control chart for x_1.

The control limits for mean are as follows:

$$LCL = 0.06284 - \frac{3 \times 0.000975}{1.128}$$

$$CL = \bar{x} = 0.06284$$

$$UCL = 0.06284 + \frac{3 \times 0.000975}{1.128} = 0.065433$$

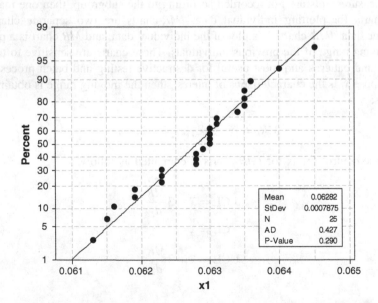

Fig. 11.6 Normal probability plot of x_1

and for moving range, the control limits are as follows:

$$LCL = 0$$
$$CL = \overline{MR} = 0.000975$$
$$UCL = D_4\overline{MR} = 3.267 \times 0.000975 = 0.003185$$

Figure 11.7 shows the *IMR*-chart for thickness based on the first sample (x_1) of observations. This chart shows that the process is in control, whereas the *IMR*-chart based on x_2 and x_3 is not statistically in control as there are few points (corresponding to the *outliers*) fall outside the control limits.

It is important to check the normality assumption when using the control chart for individuals. If a normal probability plot shows any kind of abnormality, then it may be rectified using a transformed variable to make it normal. Then, proceed with the moving range chart.

As a general check for all control charts, one must investigate whether all kinds of patterns exhibited in a control chart before concluding the process are in statistical control. It is possible that special causes of variation can be present, even if the process is in control. Some of the tests that need further investigations are as follows:

- Six points in a row, all increasing or all decreasing
- Nine points in a row on same side of centerline
- Fourteen points in a row, alternating up and down
- Two out of three points more than 2σ's from same side of the centerline
- Four out of five points more than 1σ from same side of the centerline

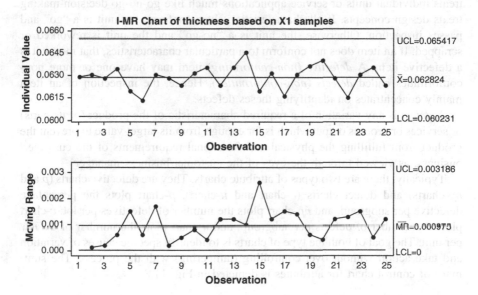

Fig. 11.7 *I-MR*-chart based on x_1 samples

- Eight points in a row more than 1σ from either side of the centerline
- Fifteen points in a row within 1σ from either side of the centerline

The Automotive industry Action Group (AIAG) identifies six rules in its SPC manual:

1. Points beyond the control limits
2. Seven points in a row on one side of the average
3. Seven points in a row that are consistently increasing (equal or greater than the preceding points) or consistently decreasing
4. Over 90 % of the plotted points are in the middle third of the control limit region (for 25 or more groups)
5. Less than 40 % of the plotted points are in the middle third of the control limit region (for 25 or more groups)
6. Obvious non-random patterns such as cycles

It is also true that most of the above charts generate two sets of control limits for single characteristics itself. If there are two continuous quality characteristics, there will be four charts. Hence, for monitoring more than two quality characteristics, one should go for a *multivariate control charting* method using *Hotelling's T^2*-statistics.

11.2.4 Control Charts for Attributes

Attributes are individual quality characteristics reflect acceptable or unacceptable outcomes. This assessment is done using go-no-go testing. The go-no-go testing treats individual units or service applications much like go-no-go decision-making treats design concepts. If all characteristic values conform, the unit is a "go" and passes inspection. Otherwise, the unit is a "no-go" and the unit is reworked or scrapped. If an item does not conform to a particular characteristics, that means it is a defective item. A *defective* (*non-conforming*) item may have one or more non-conformities called *defects* (*non-conformities*). Hence, the inspection of an item mainly concentrates on identifying theses defects.

A *defect* is any variation of a required characteristic of the product (or its parts) or services or process output which is far enough from its target value to prevent the product from fulfilling the physical and functional requirements of the customer/business, as viewed through the eyes of the customer/business manager.

Typically, there are two types of attribute charts. They are defective charts (*p* and *np*-charts) and defect charts (*c*-chart and *u*-chart). *p*-chart plots the percentage defective per subgroup, and *np*-chart plots the number of defectives per lot. *c*-chart plots the number of defects per subgroup, and *u*-chart plots the number of defects per unit. The goal of both the type of charts is to identify specific causes of variation and take action without over-controlling (tampering) with the process. The summary of control chart for attributes is depicted in Fig. 11.8.

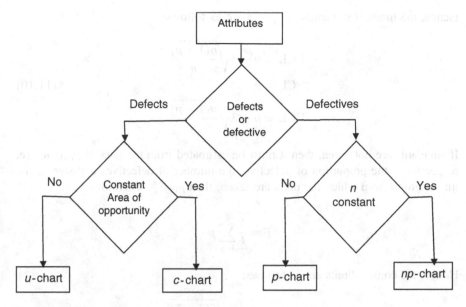

Fig. 11.8 Attribute control chart roadmap

Since *p*-chart and *np*-charts represent the proportion of non-conforming, it is expected that the sampling is done through a proper sample size and distribution to get required capability. The charting based on both these proportions generally requires the inspection of a much higher fraction of the total units to achieve system evaluation and monitoring goals effectively. For the purpose of analysis, it is suggested to have 20 or more subgroups. The probability law associated with these charting is done using binomial distribution (see Sect. 8.12.1 for details).

11.2.4.1 *p*-charts

The method of *p*-charting involves plotting results from multiple go-no-go tests. It generally requires the inspection of a much higher fraction of the total units to achieve required output. Intuitively, this follows because go-no-go output values generally provide less information on a per unit basis than counts of non-conformities or continuous quality characteristics values. The control limits for *p*-chart are calculated based on the binomial probability law. If $\hat{p} = \frac{x}{n}$ be the proportion of non-conforming units, then the number of defectives $X \sim BD(n, p)$, having mean and variance as *np* and $np(1 - p)$, respectively. Therefore,

$$\text{Var}(\hat{p}) = \text{Var}\left(\frac{X}{n}\right)$$

$$= \frac{np(1 - p)}{n^2} = \frac{p(1 - p)}{n}$$

Hence, the limits if standards are given are as follows:

$$LCL = p - 3\sqrt{\frac{p(1-p)}{n}}$$

$$CL = p \tag{11.10}$$

$$UCL = p + 3\sqrt{\frac{p(1-p)}{n}}$$

If standards are not given, then it must be estimated from the data. If p_i, x_i, n_i are, respectively, the proportion of defective, the number of defectives and size of the ith subgroup from which the units are taken, then $p_i = \frac{x_i}{n_i}$. Therefore,

$$\bar{p} = \frac{1}{k}\sum_{i=1}^{k} p_i$$

Hence, the control limits are as follows:

$$LCL = \bar{p} - 3\sqrt{\frac{\bar{p}(1-\bar{p})}{n}}$$

$$CL = \bar{p} \tag{11.11}$$

$$UCL = \bar{p} + 3\sqrt{\frac{\bar{p}(1-\bar{p})}{n}}$$

An ideal sample size for this charting can be estimated using the inequality $n * p_0 > 5$, where p_0 is the true fraction of non-conforming units when only common cause variation is present. Since the subgroup size does not have to be constant for p-chart, the control limits may vary from subgroup to subgroup based upon subgroup size. If the subgroup size is constant, then the control limits will be constant. Also, when the LCL is negative, the lower control limit may be set as zero.

The control limits shown above may be treated as trial control limits. If the process is under control, only then they may be used for future planning and decision making. Otherwise, the limits may be revised as per the requirements. As an illustration, we first consider an example for fraction non-conforming with constant sample size observed in a process.

Example 11.2 The data given in Table 11.2 give the number of non-conforming switches in a sample of fixed size 150 observed for 20 days. Construct a fraction non-conforming chart for these data. Does the process appear to be in statistical control? If not, assume that assignable causes can be found for all points outside the control limits and calculate the revised control limits.

Table 11.2 Data on the number of non-conforming switches

Day	x_i	p_i
1	8	0.05
2	1	0.01
3	3	0.02
4	0	0.00
5	2	0.01
6	4	0.03
7	0	0.00
8	1	0.01
9	10	0.07
10	6	0.04
11	6	0.053
12	0	0.007
13	4	0.020
14	0	0.000
15	3	0.013
16	1	0.027
17	15	0.000
18	2	0.007
19	3	0.067
20	0	0.040

Solution Here, $n = 150$ and $\bar{p} = 0.023$. Therefore,

$$LCL = \bar{p} - 3\sqrt{\frac{\bar{p}(1 - \bar{p})}{n}} = 0$$

and

$$UCL = \bar{p} + 3\sqrt{\frac{\bar{p}(1 - \bar{p})}{n}} = 0.059718$$

The p-chart is shown in Fig. 11.9. Note that two points correspond to days 9 and 17 go beyond the upper limit. Assuming that they are due to some assignable causes, and then, we remove those two points from the sample and revise the control limits. The revised control limits are LCL = 0, CL = 0.0163, and UCL = 0.04731. Accordingly, the chart is shown in Fig. 11.10.

The revised control chart still shows some assignable causes, which corresponds to day 1. We once again revise the control limit by excluding this point also from the samples. Figure 11.11 shows the revised control limit based on 17-day sample observations. Since all the points are now within the limit, we can conclude that the process is in control.

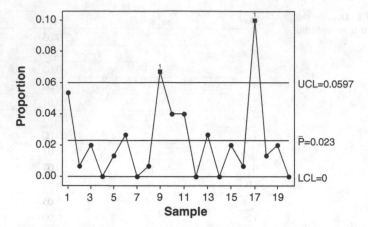

Fig. 11.9 *p*-chart for non-conforming switches

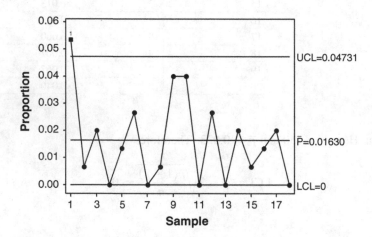

Fig. 11.10 Revised *p*-chart for non-conforming switches

Example 11.3 A paper mill uses a control chart to monitor the imperfection in finished rolls of paper. Production output is inspected for 20 days, and the resulting data are shown in Table 11.3. Use this data to set up a control chart for non-conformities per roll of paper. Does the process appear to be in statistical control?

Solution Here, standards are not given and the sample sizes are not constant for each subgroup. Also,

$$\bar{p} = \frac{0.667 + 0.778 + \cdots + 0.810}{20} = 0.699315$$

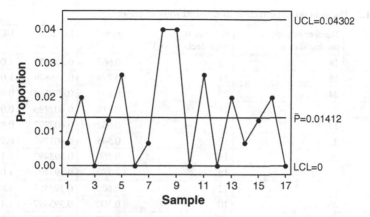

Fig. 11.11 Second revised *p*-chart for non-conforming switches

The control limits will be different for different subgroups and can be found using the formula given below

$$\text{LCL} = \bar{p} - 3\sqrt{\frac{\bar{p}(1-\bar{p})}{n_i}} \text{ and } \text{UCL} = \bar{p} + 3\sqrt{\frac{\bar{p}(1-\bar{p})}{n_i}}$$

Since the samples are varying, the control limits also will be different for different samples (see also Table 11.3 for control limit calculations). Note that for some cases, the upper control limit exceeds 1. Since the probability of non-conforming cannot exceed 1, the control limit in those cases may be set at 1. The chart is presented in Fig. 11.12.

11.2.4.2 *np*-charts

If *p* denotes the proportion of defectives found in a group of items, then *np*-charts measure the number of non-conforming units in the sample. Like in *p*-chart, here also each proportion is a subgroup of samples. Preferably large constant subgroup is required to have a good decision. In this case, the control limits will be constant. The control limits for *np*-chart are also calculated based on the binomial probability law, and the limits are obtained as follows:

$$\begin{aligned} \text{LCL} &= n\bar{p} - 3\sqrt{n\bar{p}(1-\bar{p})} \\ \text{CL} &= n\bar{p} \\ \text{UCL} &= n\bar{p} + 3\sqrt{n\bar{p}(1-\bar{p})} \end{aligned} \tag{11.12}$$

Table 11.3 The data on imperfection in finished rolls of paper

Sample number	Number of rolls produced (n_i)	Number of imperfections (x_i)	$p_i = \frac{x_i}{n_i}$	LCL	UCL
1	18	12	0.667	0.376879	1.097298
2	18	14	0.778	0.376879	1.067878
3	24	20	0.833	0.420263	0.937903
4	22	18	0.818	0.407793	0.994521
5	22	15	0.682	0.407793	0.955428
6	22	12	0.545	0.407793	0.997298
7	20	11	0.550	0.393497	1.114933
8	20	15	0.750	0.393497	1.055428
9	20	12	0.600	0.393497	1.097298
10	20	10	0.500	0.393497	1.135151
11	18	18	1.000	0.376879	1.024521
12	18	14	0.778	0.376879	1.067878
13	18	9	0.500	0.376879	1.158651
14	20	10	0.500	0.393497	1.135151
15	20	14	0.700	0.393497	1.067878
16	20	13	0.650	0.393497	1.081739
17	24	16	0.667	0.420263	0.944164
18	24	18	0.750	0.420263	0.924521
19	22	20	0.909	0.407793	0.997903
20	21	17	0.810	0.4009	1.033909

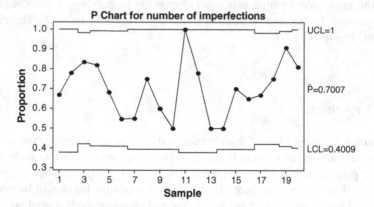

Fig. 11.12 p-chart for number of imperfections

where

$$\bar{p} = \frac{\text{total number of defective items}}{n} = \frac{\sum_{i=1}^{k} x_i}{n}$$

The *np*-chart monitors the number of defectives rather than the proportion of defectives. It is considered by many to be preferable over the *p*-chart because the number of defective is easier for quality technicians and operators to understand than is the proportion of defective. The centerline of *np*-chart is fixed at $n\bar{p}$.

Example 11.4 Consider the data given in Table 11.2, above on the number of non-conforming switches. We have $n = 150$ and $\bar{p} = 0.023$. Therefore, the control limits are as follows:

$$\text{LCL} = n\bar{p} - 3\sqrt{n\bar{p}(1-\bar{p})} = 150(0.023) - 3\sqrt{150(0.023)(0.977)} = -2.0578$$

and

$$\text{UCL} = n\bar{p} + 3\sqrt{n\bar{p}(1-\bar{p})} = 150(0.023) + 3\sqrt{150(0.023)(0.977)} = 8.957799$$

The *np*-chart is shown in Fig. 11.13. Since LCL is negative, it is set at zero. Here, also two points correspond to days 9 and 17 go beyond the upper limit, and hence, the process is not in control. One may need to revise the control limits by removing those out-of-control points from the study. Note that, both *p*-chart and *np*-chart give identical charts, and hence, the method of controlling the variation remains the same and hence is not discussed here.

Example 11.5 A process that produces titanium forgings for automobile turbo-charger wheels is to be controlled through the use of a fraction non-conforming chart. One sample of size 150 is taken each day for 20 days, and the results shown below are observed. Construct a fraction non-conforming control chart for these data (Table 11.4). Does the process appear to be in control?

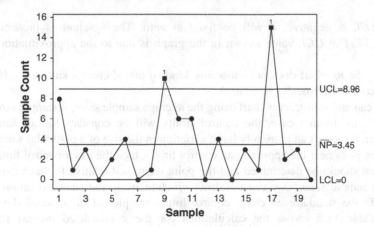

Fig. 11.13 *np*-chart of non-conforming switches

Table 11.4 Data on non-conforming forgings

Day	x_i
1	3
2	2
3	4
4	2
5	5
6	2
7	1
8	2
9	0
10	5
11	2
12	4
13	1
14	3
15	6
16	0
17	1
18	2
19	3
20	2

Solution Here, $n = 150$.

$$\overline{p} = \frac{50}{150*20} = 0.0167 \text{ and}$$

$CL = n\overline{p} = 2.5$. The LCL and UCL are -2.2036 and 7.2036 respectively.

Since *LCL* is negative, it will be fixed at zero. The *np*-chart is presented in Fig. 11.14. The *UCL* value shown in the graph is due to the approximation in $\overline{p}$ value.

Here, the *np*-chart does not show any kind of out-of-control situations. Hence, the process is statistically in control.

One can also construct *p*-chart using the average sample sizes, if sample sizes are not uniform. In that case, the control limits will be constant for all samples. However, if there is an unusually large variation in the size of a particular sample or if a point plots near the approximate control limits, then the exact control limits for that point should be determined and the point examined relative to that value. For varying sample sizes, one can construct *np*-chart using the standardization technique. In the standardized chart, control limits are plotted in standard deviation units. Table 11.5 shows the calculations for the standardized *p*-chart for the example discussed in Example 11.3 above.

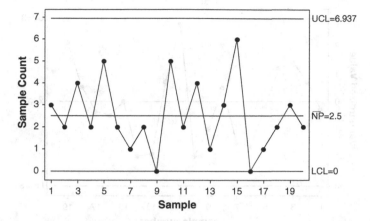

Fig. 11.14 np-chart for non-conforming units

Table 11.5 Calculation for standardized control chart for p, $\bar{p} = 0.699315$

Sample number	n_i	x_i	$p_i = \frac{x_i}{n_i}$	$\sigma_p = \sqrt{\frac{\bar{p}(1-\bar{p})}{n_i}}$	$\sigma_p = \frac{p_i - \bar{p}}{\sigma_{pi}}$
1	18	12	0.667	0.108	−0.299
2	18	14	0.778	0.108	0.728
3	24	20	0.833	0.094	1.428
4	22	18	0.818	0.098	1.214
5	22	15	0.682	0.098	−0.177
6	22	12	0.545	0.098	−1.579
7	20	11	0.550	0.103	−1.457
8	20	15	0.750	0.103	0.494
9	20	12	0.600	0.103	−0.969
10	20	10	0.500	0.103	−1.944
11	18	18	1.000	0.108	2.782
12	18	14	0.778	0.108	0.728
13	18	9	0.500	0.108	−1.844
14	20	10	0.500	0.103	−1.944
15	20	14	0.700	0.103	0.006
16	20	13	0.650	0.103	−0.481
17	24	16	0.667	0.094	−0.346
18	24	18	0.750	0.094	0.541
19	22	20	0.909	0.098	2.145
20	21	17	0.810	0.100	1.106

For a standardized control chart, the centerline will be at zero, and the lower and upper control limits will be at −3 and 3, respectively. The standardized p-chart in Fig. 11.15 does not show any unusual pattern and hence is statistically in control.

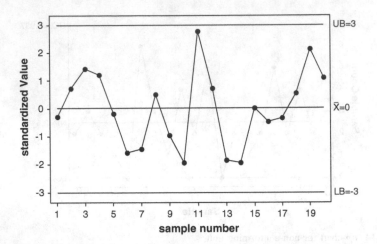

Fig. 11.15 Standardized control chart for p

Note that, using the original control limits discussed above, we could bring the process in control after revision of limits for a couple of times. Hence, there is some gain in using a standardized control chart for p, if the data are collected based on varying samples.

11.2.4.3 *c*-charts

Consider the situations where the number of events, defects, errors, or non-conformities can be counted, but there is no information about their presence, then binomial distribution may not be a good model to fit them. If we measure the count of non-conforming defects such as holes in a fabric, number of spots detected in a film coat, number of imperfections on a painted door, errors in a typed document, number of sales calls made, etc., the suitable applied to these cases will be Poisson distribution (for details see Sect. 8.12.2). *c*-chart plots the number of defects per subgroup. Each count is a subgroup of samples, and area of opportunity must be constant (lot, unit, invoice, etc.). Therefore, the control limits will be constant.

The design of *c*-chart is similar to *np*-chart. The centerline of the chart is fixed at $\bar{c} = \frac{1}{k}\sum_{i=1}^{k} c_i$, where c_i is the number of defects on the ith unit and k is the number of units examined. The control limits are, respectively, given by

$$LCL = \bar{c} - 3\sqrt{\bar{c}}$$
$$CL = \bar{c} \tag{11.13}$$
$$UCL = \bar{c} + 3\sqrt{\bar{c}}$$

Like in *np*-chart, the LCL will be fixed at zero, if $\bar{c} - 3\sqrt{\bar{c}} < 0$. If standards are given, then the centerline may be replaced by the mean of Poisson distribution.

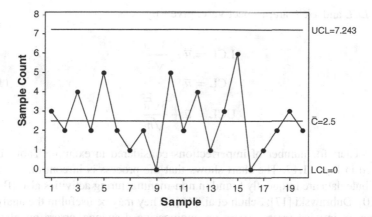

Fig. 11.16 *c*-chart for non-conformities

In Fig. 11.16, we present the *c*-chart for non-conformities considered in Example 11.5 above. Here, the centerline is set at $\bar{c} = 2.5$. The chart does not show any abnormalities although two points are falling on the LCL. Hence, the variation is statistically in control.

11.2.4.4 *u*-charts

u-chart plots the number of defects per unit. For example, length of pieces of material and the area examined may be allowed to vary, and the *u*-chart will show the number of defects per unit area, either in per square meter or any other suitable measure of interest. Thus, *u*-chart is suitable for controlling the number of defects or non-conformities per unit or time period, and the sample size can be allowed to vary. The goal of both *c*-chart and *u*-chart is to identify specific causes of variation and take action without over-controlling (tampering) with the process.

Since the defects per unit will follow a Poisson distribution, *u*-chart works with the principle of Poisson law. The design of *u*-chart is similar to *p*-chart for proportion of defective. The control limit will vary according to each sample size, and the centerline is kept at

$$\bar{u} = \frac{\text{Total number of defects}}{\text{total sample inspected}} = \frac{\sum_{i=1}^{k} x_i}{\sum_{i=1}^{k} n_i}, \qquad (11.14)$$

where x_i is the number of defects in the *i*th sample. If the sample sizes vary for each unit, then the LCL and UCL will differ for each unit. Therefore, for practical purpose, we may keep it constant, if sample sizes remain with 25 % either side of the average sample size $\bar{n}$.

The *LCL* and *UCL* are, respectively, given by

$$LCL = \bar{u} - \frac{3\sqrt{\bar{u}}}{\sqrt{n}}$$

$$CL = \bar{u} \tag{11.15}$$

$$UCL = \bar{u} + \frac{3\sqrt{\bar{u}}}{\sqrt{n}}$$

A plot *u*-chart for number of imperfections considered in example Table 11.3 is presented in Fig. 11.17. The chart shows that the process is in control.

Attribute data are frequently found in non-manufacturing activities also (Burkom et al. [10]; Dubrawski [17]; Schuh et al. [46]). They may be useful in the analysis of absenteeism, invoice errors, errors on engineering drawings, errors on plans and documents, etc. Activity sampling is a technique based on the binomial theory and is used to obtain a realistic picture of time spent on particular activities. For example, in activities such as discriminating the item as good or bad, rejected or accepted, and defective or non-defective, we are interested in finding the proportion of non-conforming items and their analysis, for which *p*-chart and *np*-chart are found to be suitable. When we are concerned with the number of non-conformities or defects in a product rather than simply determining whether the product is defective or non-defective, we use a *c*-chart or a *u*-chart. Both *c*-chart and *u*-chart use Poisson distribution to model the non-conformities.

Example 11.6 An automobile manufacturer wishes to control the number of non-conformities in a subassembly area producing manual transmissions. The inspection unit is defined as four transmissions, and data from 20 samples are shown in Table 11.6.

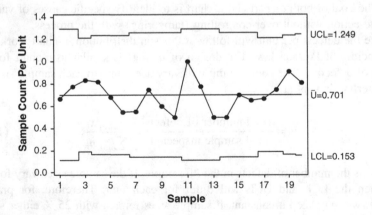

Fig. 11.17 *u*-chart for number of imperfections

Table 11.6 Data on non-conformities in manual transmissions

Sample number	Number of non-conformities (x_i)	$u_i = \frac{x_i}{4}$
1	1	0.25
2	3	0.75
3	2	0.50
4	1	0.25
5	0	0.00
6	2	0.50
7	1	0.25
8	3	0.75
9	5	1.25
10	2	0.50
11	1	0.25
12	0	0.00
13	1	0.25
14	1	0.25
15	3	0.75
16	2	0.50
17	4	1.00
18	2	0.50
19	1	0.25
20	2	0.50

Perform the following analysis

(a) Set up a control chart for non-conformities per unit
(b) Do these data come from a controlled process? If not, revise the control limits
(c) Suppose the inspection unit is redefined as eight transmissions. Design an appropriate control chart for monitoring future productions

Solution

(a)

$$\bar{u} = \frac{\sum_{i=1}^{20} u_i}{20} = \frac{9.25}{20} = 0.4625$$

Therefore

$$LCL = \bar{u} - \frac{3\sqrt{\bar{u}}}{\sqrt{n}} = -0.558$$

and

$$UCL = \bar{u} + \frac{3\sqrt{\bar{u}}}{\sqrt{n}} = 1.483$$

(b) The control chart is shown in Fig. 11.18. Since all the points are within the control limits, the process is in control.
(c) The control chart is shown in Fig. 11.19 and is non-conformities based on unit of eight transmissions. Here, also all the points are within the control limits, and the process is in control.

Attribute control charts have many advantages, as several quality characteristics can be considered jointly, and the unit may be classified as conforming or non-conforming. On the other hand, if several quality characteristics are treated as variables, then each one must be measured, and variable control chart must be used

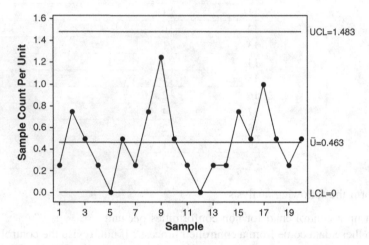

Fig. 11.18 *u*-chart of non-conformities

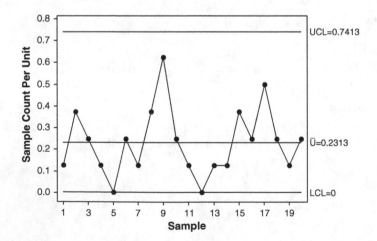

Fig. 11.19 *u*-chart of non-conformities (based on 8 transmissions)

to control each one of them. Still, the variable control chart provides more useful information about process performance than does an attribute chart.

Control charts based on the geometric and exponential distributions have not been included as part of the basic core of tools for quality practitioners, but trends are changing. There are more and more applications where one deals with a "rare event," and use of the more traditional p-chart and c-chart is ineffective ([58]). MINITAB 16, for example, now includes the g-chart and a control chart method for time-between-event data based on the Weibull distribution, a generalization of the exponential distribution, primarily to address needs in healthcare monitoring.

11.2.5 Cumulative Sum Chart

The charts discussed so far are highly effective to detect sharp, intermittent changes to a process. However, if one is interested in a small, sustained shift in a process, other types of control charts may be preferred, for example, the cumulative sum (CUSUM) chart. It is a chronological plot of the CUSUM of deviations of a sample statistic from a reference value (or nominal or target value). The CUSUM value focuses on target value rather than on the actual average of process data. Another chart is the *moving average* chart. This chart is a chronological plot of the moving average, which is calculated as the average value updated by dropping the oldest individual measurement and adding the newest individual measurement. Thus, a new average is calculated with each individual measurement. With exponential increase, an exponentially weighted moving average (EWMA) chart may be appropriately used (see Champ and Woodall [12] and Montgomery [31]). These charts are also called time-weighted control charts.

Let x_i be the value of interest for the ith subgroup (sample). Let T be the target value, $T = \mu_0$ (the mean level of the process) for the mean chart, and $T = d_2\sigma$ for the range chart, when the process is in control. The cumulative sum of the first k deviations of the variable of interest from the target value is as follows:

$$C_k = \sum_{i=1}^{k} (x_i - T) \tag{11.16}$$

The CUSUM chart is a plot of C_k values against the sample number k. Now, for $r < s$,

$$C_s - C_r = \sum_{i=1}^{s} (x_i - T) - \sum_{i=1}^{r} (x_i - T)$$
$$= \sum_{i=r+1}^{s} x_i - (s - r)T \tag{11.17}$$

Hence,

$$\frac{\sum_{i=r+1}^{s} x_i}{s-r} = \frac{C_s - C_r}{s-r} + T$$

$$= \beta(s,r) + T \tag{11.18}$$

That is the average value of interest from sample number $(r+1)$ to s is equal to the average slope of the CUSUM graph between these two sample numbers plus the target (T). If the production process remains in control, then the average value of the variable of interest will be close to T, and hence, the slope $\beta(s,r)$ will be close to zero for all (s, r). If the process goes out of control owing to the increase or decrease in the variable of interest, the CUSUM graph will show a positive or negative slope. The average slope $\beta(s,r)$ of the CUSUM graph can be used to estimate the extent of departure from the target value.

Example 11.7 A machine produces spokes for the wheels of bicycles. When the process is in control, the machine produces spokes whose lengths are normally distributed about mean 25 cm and standard deviation 0.02 cm. The sample values along with the cumulative sums for the fifteen machines are shown in Table 11.8. Use CUSUM chart to investigate whether the machine is correctly set.

Solution As per the information provided, $T = 25$ cm. The cumulative sums are then calculated and shown in Table 11.7. The CUSUM plot of C_k against k is shown in Fig. 11.20.

From the chart, it is seen that from about the eighth set of data, there appears to be some linearity. Thereafter, the trend develops in an upward way, indicating

Table 11.7 Data on machine spokes

Sample (i)	x_i	$x_i - 25$	$C_k = \sum_{i=1}^{k} (x_i - 25)$
1	24.998	−0.002	−0.002
2	25.016	0.016	0.014
3	25.024	0.024	0.038
4	25.022	0.022	0.06
5	24.984	−0.016	0.044
6	25.012	0.012	0.056
7	25.024	0.024	0.08
8	24.978	−0.022	0.058
9	25.012	0.012	0.07
10	25.017	0.017	0.087
11	25.024	0.024	0.111
12	25.026	0.026	0.137
13	25.027	0.027	0.164
14	25.028	0.028	0.192
15	25.028	0.028	0.21

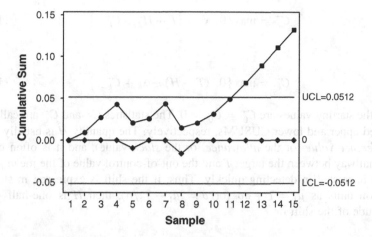

Fig. 11.20 CUSUM chart of machine spokes

evidence of a shift in the process average. The change in the expected value of length of spokes from its target value can be estimated through Eq. (11.18). That is, the mean length of spokes after eighth set of data is estimated as follows:

$$\frac{C_s - C_r}{s - r} + T = \frac{0.210 - 0.058}{15 - 8} + 25$$
$$= 25.022$$

Since there is no computation of any control limits in this chart, the CUSUM chart is not considered to be a control chart in its complete sense. Therefore, the decision rules based on cumulative sums can be achieved through some subjective considerations. Generally, the tabular and graphical or *V-mask* procedures are used for decision making in this chart. We explain the tabular procedure below:

Writing

$$C_k = \sum_{i=1}^{k} (x_i - T)$$
$$= \sum_{i=1}^{k-1} (x_i - T) + (x_k - T) \qquad (11.19)$$
$$= C_{k-1} + (x_i - T)$$

Then, the tabular procedure works by accumulating derivations from T that are above target with one statistic C^+ and accumulating derivations from T that are below target with one statistic C^-, where

$$C_k^+ = \max\{0,\ x_k - (T+H) + C_{k-1}^+\} \qquad (11.20)$$

and

$$C_k^- = \max\{0,\ (T-H) - x_k + C_{k-1}^-\} \qquad (11.21)$$

where the starting values are $C_0^+ = C_0^- = 0$. The statistics C^+ and C^- are called the on-sided upper and lower CUSUMs, respectively. The quantity H is usually called the *reference value* (or the *allowance*, or the *slack* value), and it is often chosen about halfway between the target T and the out-of-control value of the mean μ_1 that we are interested in detecting quickly. Thus, if the shift is expressed in standard deviation units as $\mu_1 = T + \delta\sigma$ (or $\delta = |\mu_1 - T|/\sigma$), then H is one-half of the magnitude of the shift or

$$H = \frac{\delta}{2}\sigma = \frac{|\mu_1 - \mu_0|}{2} \qquad (11.22)$$

If either C_k^+ or C_k^- exceeds the decision interval L, the process is considered to be out of control, where L is a function of the process standard deviation, usually taken as five times the process standard deviation (σ). We now illustrate the tabular CUSUM for the example discussed in Table 11.8.

For the computation of the statistics C_k^+ and C_k^- given in Eqs. (11.20) and (11.21), respectively, we use the following specifications:

Table 11.8 Tabular CUSUM analysis table

Sample (i)	$\bar{x}_i - 25.01$	C_k^+	N^+	$24.99 - \bar{x}_i$	C_k^-	N^-
1	−0.012	0	0	−0.008	0	0
2	0.006	0.006	1	−0.026	0	0
3	0.014	0.02	2	−0.034	0	0
4	0.012	0.032	3	−0.032	0	0
5	−0.026	0.006	4	0.006	0.006	1
6	0.002	0.008	5	−0.022	0	0
7	0.014	0.022	6	−0.034	0	0
8	−0.032	0	0	0.012	0	1
9	0.002	0.002	1	−0.022	0	0
10	0.007	0.009	2	−0.027	0	0
11	0.014	0.023	3	−0.034	0	0
12	0.016	0.039	4	−0.036	0	0
13	0.017	0.056	5	−0.037	0	0
14	0.018	0.074	6	−0.038	0	0
15	0.018	0.092	7	−0.038	0	0

$$T = 25$$
$$\sigma = 0.02$$
$$\mu_1 = T + 1\sigma = 25 + 0.02 = 25.02$$
$$H = \frac{\sigma}{2} = 0.01$$
$$L = 5\sigma = 0.1$$

The complete calculations are shown in Table 11.8.

Since none of the C_k^+ or C_k^- value exceeds the decision level $L(=0.1)$, the process seems to be in control. Note that at sample number, the value of C_k^+ becomes zero first time indicating the change in process average.

11.2.6 EWMA Chart

In the mean and range charts, the decision signal obtained largely depends on the last point plotted. The exponentially weighted moving average (EWMA) chart is a type of moving mean chart in which an "exponentially weighted mean" is calculated each time a new result becomes available. That is, if $x_1, x_2, \ldots$ are the independent sample observations, then

$$z_t = \lambda x_t + (1 - \lambda) z_{t-1} \tag{11.23}$$

where $0 < \lambda \le 1$ and $z_0 = \mu_0$. Sometimes, the average of the preliminary data is used as the starting value of the EWMA chart, so that $z_0 = \bar{x}$. This chart was introduced by Roberts [43].

Now from (11.23), we get

$$
\begin{aligned}
z_t &= \lambda x_t + (1 - \lambda) z_{t-1} \\
&= \lambda x_t + (1 - \lambda)[\lambda x_{t-1} + (1 - \lambda) z_{t-2}] \\
&= \lambda x_t + \lambda(1 - \lambda) x_{t-1} + (1 - \lambda)^2 [\lambda x_{t-2} + (1 - \lambda) z_{t-3}] \\
&= \ldots \\
&= \lambda \sum_{k=0}^{t-1} (1 - \lambda)^k x_{t-k} + (1 - \lambda)^t z_0 \\
&= \lambda \sum_{k=0}^{t-1} (1 - \lambda)^k x_{t-k} + (1 - \lambda)^t \mu_0
\end{aligned}
\tag{11.24}
$$

In EWMA chart, while giving weights, the most recent data are given the highest weight. If $\lambda = 0.4$, then the weight assigned to the current sample mean is 0.4 and the weights given to the preceding means are 0.24, 0.144, 0.0864, and so forth.

Thus, the weight $\lambda(1 - \lambda)^t$ decreases geometrically with the age of the sample means.

If x_1's are independent random variable with mean μ_0 and variance σ^2, then

$$
\begin{aligned}
E(Z_t) &= \lambda \sum_{k=0}^{t-1} (1 - \lambda)^k E(x_{t-k}) + (1 - \lambda)^t \mu_0 \\
&= \lambda \sum_{k=0}^{t-1} (1 - \lambda)^k \mu_0 + (1 - \lambda)^t \mu_0 \\
&= \lambda \mu_0 \sum_{k=0}^{t-1} (1 - \lambda)^k + (1 - \lambda)^t \mu_0 \\
&= \lambda \mu_0 [1 + (1 - \lambda) + (1 - \lambda)^2 + \cdots + (1 - \lambda)^{t-1}] + (1 - \lambda)^t \mu_0 \\
&= \lambda \mu_0 \frac{[1 - (1 - \lambda)^t]}{[1 - (1 - \lambda)]} + (1 - \lambda)^t \mu_0 \\
&= \mu_0
\end{aligned}
$$

(11.25)

and

$$
\begin{aligned}
\mathrm{Var}(Z_t) &= \mathrm{Var} \left[\lambda \sum_{k=0}^{t-1} (1 - \lambda)^k E(x_{t-k}) + (1 - \lambda)^t \mu_0 \right] \\
&= \mathrm{Var} \left[\lambda \sum_{k=0}^{t-1} (1 - \lambda)^k E(x_{t-k}) \right] \\
&= \sum_{k=0}^{t-1} \left[\lambda (1 - \lambda)^k \right]^2 \mathrm{Var}(x_{t-k}) + 2 \sum_{k}^{t-1} \sum_{l}^{t-1} \lambda (1 - \lambda)^k \mathrm{Cov}(x_{t-k}, x_{t-l}) \\
&= \sum_{k=0}^{t-1} \left[\lambda (1 - \lambda)^k \right]^2 \sigma^2 \\
&= \lambda^2 \sigma^2 \sum_{k=0}^{t-1} (1 - \lambda)^{2k} \\
&= \lambda^2 \sigma^2 [1 + (1 - \lambda)^2 + \cdots + (1 - \lambda)^{2(t-1)}] \\
&= \sigma^2 \left(\frac{\lambda}{2 - \lambda} \right) [1 - (1 - \lambda)^{2t}]
\end{aligned}
$$

(11.26)

Hence, the control limits for EWMA chart are as follows:

$$LCL = \mu_0 - L\sigma\sqrt{\left(\frac{\lambda}{2-\lambda}\right)\left[1-(1-\lambda)^{2t}\right]}$$

$$CL = \mu_0 \tag{11.27}$$

$$UCL = \mu_0 + L\sigma\sqrt{\left(\frac{\lambda}{2-\lambda}\right)\left[1-(1-\lambda)^{2t}\right]}$$

where the factor L is the width of the control limits. In Example 11.8, we illustrate the EWMA chart.

Example 11.8 The following data represent individual observations on molecular weight taken hourly from a chemical process.

1045	1055	1037	1064	1095	1008	1050	1087	1125	1146
1139	1169	1151	1128	1238	1125	1163	1188	1146	1167

The target value of molecular weight is 1050, and the process standard deviation is thought to be about 25. Use an EWMA chart with $\lambda = 0.2$ and $L = 2.7$ and analyze the data.

Table 11.9 EWMA analysis for molecular weight

t	x_t	$z_t (=0.2x_t + 0.8z_{t-1})$	LCL	UCL
1	1045	1049.000	1036.500	1063.500
2	1055	1050.200	1032.712	1067.288
3	1037	1047.560	1030.673	1069.327
4	1064	1050.848	1029.474	1070.526
5	1095	1059.678	1028.742	1071.258
6	1008	1049.343	1028.287	1071.713
7	1050	1049.474	1028.000	1072.000
8	1087	1056.979	1027.819	1072.181
9	1125	1070.583	1027.704	1072.296
10	1146	1085.667	1027.630	1072.370
11	1139	1096.333	1027.583	1072.417
12	1169	1110.867	1027.553	1072.447
13	1151	1118.893	1027.534	1072.466
14	1128	1120.715	1027.522	1072.478
15	1238	1144.172	1027.514	1072.486
16	1125	1140.337	1027.509	1072.491
17	1163	1144.870	1027.506	1072.494
18	1188	1153.496	1027.504	1072.496
19	1146	1151.997	1027.502	1072.498
20	1167	1154.997	1027.501	1072.499

Solution Here, we have $\mu_0 = 1050$, $\sigma = 25$, $\lambda = 0.2$, and $L = 2.7$. The computation of z_t, LCL, and UCL is shown in Table 11.9.

Note that, the control limits increase in width as t increases from $t = 1, 2, \ldots$ until they stabilize at $t = 7$ onward (see Table 11.10). This is because the term $[1 - (1 - \lambda)^{2t}]$ approaches to unity, as t gets larger. This means that the EWMA control chart has been running for several time periods, and the control limits will approach steady-state values given by

$$\text{LCL} = \mu_0 - L\sigma\sqrt{\left(\frac{\lambda}{2 - \lambda}\right)}$$

and

$$\text{UCL} = \mu_0 + L\sigma\sqrt{\left(\frac{\lambda}{2 - \lambda}\right)}$$

respectively. For this problem, the steady-state values are 1027.5 and 1072.5, respectively. The EWMA chart and the moving average chart are presented in Figs. 11.21 and 11.22, respectively. From the figure, it is seen that the chart is not in control.

The performance of EWMA chart is approximately equivalent to that of the CUSUM chart, and in some ways, it is easier to set up and operate [14]. The design parameters of the chart are the multiple of sigma used in the control limits and the value of λ. The EWMA control chart is often used with individual measurements. However, if rational subgroups of size $n > 1$ are taken, then simply replace x_t by $\bar{x}_t$ and σ with $\sigma/\sqrt{n}$. For further discussion on EWMA chart and the computation of the optimum levels of these parameters, one may refer to Lucas and Saccucci [27] and Montgomery [31] and references contained therein.

11.2.7 Economic Design of Control Charts

Control charts are generally used to establish and maintain statistical control of a process. They are also effective devices for estimating process parameters,

Table 11.10 FAR and ARL$_0$ of a sign control chart for various values of n and t

n	5	6	7	8	9	10
t	5	6	7	8	9	10
$P(T \geq t \vert IC)$	0.0312	0.0156	0.0078	0.0039	0.0020	0.0010
FAR(α)	0.0624	0.0312	0.0156	0.0078	0.0040	0.0020
ARL$_0$	16	32	64	128	256	512

Source Chakraborti and Graham [15]

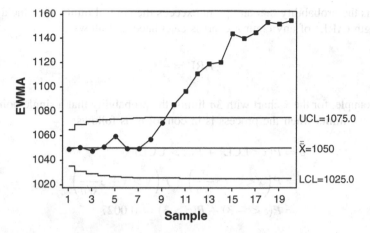

Fig. 11.21 EWMA chart of the molecular weight

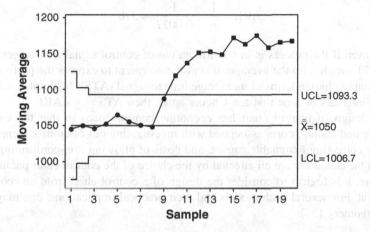

Fig. 11.22 Moving average chart of the molecular weight

particularly in process capability studies. The use of a control chart requires that the engineer or analyst selects a sample size, a sampling frequency or interval between samples, and the control limits for the chart. Selection of these three parameters is usually called the design of the control chart [14]. Traditionally, the control chart is designed with respect to statistical criteria only. This usually involves selecting the sample size and control limits such that the average run length of the chart to detect a particular shift in the quality characteristic and the average run length of the procedure when the process is in control are equal to specified values. The frequency of sampling is rarely treated analytically, and usually, the practitioner is advised to consider such factors as the production rate, the expected frequency of shifts to and out-of-control state, and the possible consequences of such process shifts in detecting the sampling interval.

If p is the probability that any point exceeds the control limits, then the average run length (ARL) of any control chart is calculated as follows:

$$\text{ARL} = \frac{1}{p} \qquad (11.28)$$

If for example, for the $\bar{x}$-chart with 3σ limits, the probability that a single point falls outside the limits when the process is in control is as follows:

$$p = P(x < \text{LCL}) + P(x > \text{UCL})$$
$$= P\left(x < \bar{x} - 3\frac{\sigma}{\sqrt{n}}\right) + P\left(x > \bar{x} + 3\frac{\sigma}{\sqrt{n}}\right)$$
$$= P(z < -3) + P(z > 3) = 0.0027$$

Therefore, the ARL of $\bar{x}$-chart when the process is in control is as follows:

$$\text{ARL} = \frac{1}{p} = \frac{1}{0.0027} = 370$$

That is even if the process is in control, an out-of-control signal will be generated every 370 samples, on the average. It is also convenient to express the performance of the control chart in terms of its average time to signal (ATS). If samples are taken at fixed intervals of time that are k hours apart, then $\text{ATS} = k * \text{ARL}$.

The design of control chart has economic consequences in that the costs of sampling and testing, costs associated with investigating out-of-control signals and possibly correcting assignable causes, and costs of allowing non-conforming units to reach the consumer are all affected by the choice of the control chart parameters. Therefore, it is logical to consider the design of a control chart from an economic viewpoint. For several such models and their practical implications, one may refer to Montgomery [31].

11.2.8 Role of Process Monitoring

Within the DMAIC process, process monitoring plays a very important role. It is useful in the measure phase to assess current performance and to monitor the performance of measurement systems. It is also useful in the control phase to monitor input variables so that gains from the improve phase can be maintained over time. From a practical perspective, it is important to design monitoring methods so that there is reasonably low number of false alarms. If a scheme signals too often that a process change has occurred, but there has either been no process change or an unimportant change, then there will be a tendency for the user to ignore signals altogether. The action after a signal from a control chart can vary, depending on the field and the application. In some cases, one might simply pay more attention to the process, and in others, one might reset and recalibrate equipment. Many companies have out-of-control action plans that specify the actions to be taken after a signal is given.

In health care, the focus is usually on monitoring the health of individual patients or the performance of physicians or hospitals. Some of the literature on monitoring of health characteristics of individual patients with chronic health problems, such as asthma, was reviewed by Tennant et al. [50]. Thor et al. [51], on the other hand, reviewed some of the literature on the use of control charts to improve healthcare delivery. A general overview of health-related surveillance, including risk-adjusted monitoring, was presented in Woodall et al. [57]. For information on the use of control charts in healthcare monitoring, one may refer to the books by Winkel and Zhang [54] and Faltin et al. [18]. The book by Faltin et al. [18] contains a chapter on the use of Six Sigma in healthcare improvement.

Woodall and Montgomery [58] strongly support the use of Six Sigma in health care, but presently find that there is much less focus on the use of metrics in health care than with Six Sigma in industry. In public health surveillance, one is quite often interested in monitoring disease or mortality rates. Sonesson and Bock [47] gave an excellent review of prospective public health surveillance, but much work has been done since. An updated review was given by Sparks [48]. In addition, Tsui et al. [52] gave a review of temporal and spatiotemporal surveillance methods for disease surveillance in public health.

In syndromic surveillance, data from disparate sources including non-traditional sources are combined to detect bioterrorism or a disease outbreak (Lombardo et al. [28]; Fricker [19]). For example, one could consider data on over-the-counter drug sales, absenteeism rate, and emergency room visits in order to obtain an early warning of an attack or outbreak. Spatiotemporal data consist of events over time, but each event is associated with a location in a region of interest. Sometimes, the exact location is known, but most often, one only knows in which of several subregions the events are located. Most of the work on prospective spatiotemporal monitoring has been done in public health cluster detection applications, where the goal is to detect emerging clusters where disease rates are higher than expected [see, for example, the books by Lawson and Kleinman [25] and Rogerson and Yamada [45], and the review paper by Robertson et al. [44].

Data are often collected on products using laser scanners and coordinate measurement machines. Thus, numerous measurements are available on each item at various locations. Some of the issues in monitoring with these types of data were discussed by Wells et al. [53]. It can be very helpful to use engineering knowledge to better monitor and find root causes of changes in variation patterns. Apley and Lee [5] discussed some of the issues involved with this type of approach and provided some useful references. A related area is the use of images for process monitoring. Most often, images have been used in industry for product inspection, where non-conforming items are separated from the conforming items, but we see opportunities for detecting more subtle quality changes before non-conforming items are produced. Megahed et al. [29] reviewed work on the use of control charts with image data.

In an increasing number of industrial applications, the quality of a process or product is best described by a function called a "profile." In these applications, a response variable is related to one or more explanatory variables. In these cases, changes in the profile over time are of interest ([55, 56]). Various types of models have been used to represent profiles, including simple linear regression, nonlinear regression, multiple regression, nonparametric regression, mixed models, and wavelet models. For an up-to-date overview of profile monitoring studies, one may refer to the book by Noorossana et al. [37] and the references contained therein.

For industrial applications, involving multiple applications and processes, monitoring is done on the basis of time-to-event data. In such cases, the observations are more likely to be auto-correlated. Knoth and Schmid [23], Psarakis and Papaleonida [41], and Prajapati and Singh [39] reviewed the extensive literature on process monitoring with auto-correlated data. Much of the research in monitoring with auto-correlated data is based on the assumption of a known time series model (Ledolter and Bisgaard [26]). For other research and issues on profile monitoring, we recommend the article by Woodall and Montgomery [58]. One of the developing areas on profile monitoring is the nonparametric control charting methods, which is described in detail below.

11.2.9 Nonparametric Control Charts

Statistical quality (process) control (SQC/SPC) methods help improve quality of products by identifying and possibly reducing any assignable sources of variation that might be present. In this context, the distribution of chance (common) causes is often assumed to follow some parametric distribution such as normal. Hence, they are called parametric SQC/SPC. In parametric SPC, the choice of the plotting statistics and the calculation of the control limits are calculated from the data which can be individual or subgroups (sample) of observations. When there is uncertainty about the underlying distributions, the suitability and performance of these methods become questionable. Thus, there is a need for some flexible and robust control charts that do not require normality or any other specific distributional assumptions. Nonparametric (or distribution-free) control charts can serve this broader purpose and can thus offer useful alternatives to parametric control charts.

Bakir and Reynolds [7] probably is the first paper to discuss nonparametric procedures for process control for within-group samples. From early 2000s onward, there is a surge of nonparametric control charting techniques for various process characteristics. Chakraborti et al. [1] are credited to have given an overview of such studies and have reviewed nonparametric control charts up to the 2000, for the first time. Chakraborti and Graham [15] and Chakraborti et al. [12] have further reviewed the literatures through 2010. There is now a vast collection of nonparametric control charts in the literature, including nonparametric control charts equivalent to Shewhart-type chart, CUSUM chart, EWMA chart, and other variants of these charts. These methods have also been shown to perform well compared to their parametric counterparts.

A nonparametric control chart is defined in terms of its in-control (IC) run-length distribution. If the IC run-length distribution is the same for every continuous distribution, the chart is called nonparametric or distribution-free. This means that the IC properties of nonparametric charts are the same of all continuous distributions. Two of the most important problems in parametric SPC are monitoring the *process mean* and/or *process variation* (range/standard deviation). In the nonparametric setting, a broader view is taken to consider monitoring the center or location and/or the scale parameter of a *process distribution*. Hence, median is generally used for controlling the location parameter in nonparametric setup. Also, in nonparametric SPC (NSPC), the underlying distribution is assumed to have either (i) a location model, with a cumulative distribution function as $F(x - \theta)$, where θ is the location parameter or (ii) a scale model, with a cumulative distribution function as $F(x|\tau)$, where $\tau > 0$ is the scale parameter or (iii) a location-scale model, with a cumulative distribution function as $F[(x - \theta)/\tau]$, where θ and τ are the location and scale parameter, respectively. The goal is to monitor the parameter θ or τ (or both), based on individual measurements or random samples (rational subgroups) usually taken are equally spaced time points.

A nonparametric control chart works under two phases:

- Phase I or the *retrospective phase*: Here, the preliminary analysis is done to ensure that the process is in IC. This means that the process is operating at or near some acceptable target value along with some natural variation and no special causes of concern are present. Once this state is achieved, reference data are collected, any unknown parameters are estimated, and SPC moves to the next phase.
- Phase II or the *prospective phase* or *monitoring phase*: Here, the process monitoring continues based on new incoming data, irrespective of whether the process characteristic is known or unknown.

The nonparametric charts have been typically based on the sign (SN) and the Wilcoxon signed-rank (SR) test statistics [1]. These well-known distribution-free tests have their own advantages. For example, the simple SN test can be used under the most minimal of assumptions in order to make inference on any percentile of a continuous distribution, but the SR test can only be used for the median of a symmetric continuous distribution (see Sect. 9.6.2 of Chap. 9 for further details of these tests). The SR test is almost as efficient as the t test even under the normal assumption, but the SN test is more efficient than the SR test when the distribution is heavier tailed and symmetric, such as the double exponential. The test statistics of these and other distribution-free tests are used as building blocks to create suitable plotting statistics for the nonparametric control charts. Below, we illustrate two nonparametric control charts for location parameter, namely Shewhart-type sign charts and Shewhart-type signed-rank charts.

11.2.9.1 Shewhart-Type Sign Charts

Sign control charts are based on the well-known *sign test* (see Sect. 9.6.2.2 for details). The sign test is one of the simplest and most widely applicable nonparametric tests that can be used for statistical inference on any percentile of a continuous distribution. Suppose that the median θ of a continuous process needs to be maintained at an IC value specified to be θ_0. Amin et al. [4] considered Shewhart-type charts for this problem using what are called "within-group sign" statistics: This is called the SN chart, and it works as follows:

Let $X_{i1}, X_{i2}, \ldots, X_{in}$ for $i = 1, 2, \ldots$ be the ith sample of independent observations of size $n(n > 1)$. The plotting statistic for the SN chart is as follows:

$$SN_i = \sum_{j=1}^{n} \text{sign}(X_{ij} - \theta_0) \tag{11.29}$$

where

$$\text{sign}(x) = \begin{cases} -1, & \text{if } x < 0 \\ 0, & \text{if } x = 0 \\ 1, & \text{if } x > 0 \end{cases}$$

If T_i = the sum of positive counts (number of sample observations greater than θ_0), then $SN_i = 2T_i - n$. Since T_i follows a binomial distribution with parameters n and $p = 0.5$, it follows that the IC distribution of SN_i is symmetric about 0, and hence, the control limits of the two-sided nonparametric Shewhart-type sign chart are given by

$$\begin{aligned} \text{LCL} &= -c \\ \text{CL} &= 0 \\ \text{UCL} &= c \end{aligned} \tag{11.30}$$

where c is the smallest integer such that

$$P(SN_i \geq c|IC) \leq \alpha/2 \tag{11.31}$$

If the plotting statistic SN_i falls on or outside one of the control limits, that is, if $SN_i \leq -c$ or $SN_i \geq c$, the process is declared to be out of control (OOC). The charting constant c is obtained for a specified ARL_0 which is equal to the reciprocal of the nominal false alarm rate (FAR), denoted by α. In Table 11.10, we reproduce some values of $P(T \geq t|IC)$ along with the values of $\text{FAR}(\alpha)$ and ARL_0. The charting constant c is obtained using the relation $c = 2t - n$. The procedure is illustrated through an example from Montgomery [31] below.

Example 11.9 Here, we consider data from Montgomery [31] (Tables 5.1 and 5.2) on the inside diameters of piston rings manufactured by a forging process. The

Table 11.11 Inside diameter measurements (mm) on forged piston rings

Sample	Observations					SN_i values	SR_i values
26	74.012	74.015	74.030	73.986	74.000	2	8
27	73.995	74.010	73.990	74.015	74.001	1	4
28	73.987	73.999	73.985	74.000	73.990	−4	−14
29	74.008	74.010	74.003	73.991	74.006	3	7
30	74.003	74.000	74.001	73.986	73.997	0	−3
31	73.994	74.003	74.015	74.020	74.004	3	9
32	74.008	74.002	74.018	73.995	74.005	3	10
33	74.001	74.004	73.990	73.996	73.998	−1	−6
34	74.015	74.000	74.016	74.025	74.000	3	12
35	74.030	74.005	74.000	74.016	74.012	4	14
36	74.001	73.990	73.995	74.010	74.024	1	4
37	74.015	74.020	74.024	74.005	74.019	5	15
38	74.035	74.010	74.012	74.015	74.026	5	15
39	74.017	74.013	74.036	74.025	74.026	5	15
40	74.010	74.005	74.029	74.000	74.020	4	14

samples 1–25 of Table 5.1 are shown to be statistically in control. The remaining samples, from 26–40, are considered to be the prospective samples (Table 5.2) each of five observations, in our analysis. Assume that the underlying distribution is symmetric with a known median $\theta_0 = 74$ mm. The samples along with the sign test statistics are shown in Table 11.11.

Solution Here, $n = 5$, $t = 5$, and hence $c = 5$. The sign chart is shown in Fig. 11.23 with control limits at ±5. Note that, the observations 12, 13, and 14 lie on the UCL

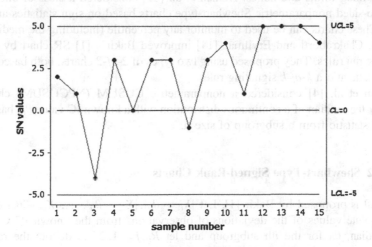

Fig. 11.23 Shewhart-type sign control chart for Montgomery [31] piston data

which indicates that the process is OOC starting at observation 12 (i.e., sample 37), indicating that the process median has shifted upward from the in-control value of 74 mm.

To study the sensitivity and performance of the nonparametric control charts, many authors have studied signaling rules for OOC by constructing warning limits, run rules, and signaling rules. A runs-rule is a signaling rule based on a run of the plotting statistic that describes when a control chart should signal an OOC situation [15, 31]. In nonparametric SPC, Amin et al. [4] considered Shewhart-type sign charts with warning limits and run rules. The two-sided control chart has the CL at 0 and symmetrically placed at $\pm a$. The warning limits are drawn at $\pm w$ with $0 \leq w < a$. A signal is indicated if r consecutive plotting statistics fall in the region or zone $[w, a)$ or in the zone $(-a, -w]$ or if any plotting statistic falls on or outside $\pm a$. The ARL of the two-sided chart can be calculated from the ARLs for the two one-sided (upper and lower) charts, respectively. One may refer to Amin et al. [4], and Chakraborti and Graham [15] for more details on these aspects.

In addition to the warning limits or zones, run rules have also been used to improve the performance of control charts. The rules considered include the following:

(a) A single plotting statistic plots on or outside the control limits (the *1-of-1* rule)
(b) k consecutive plotting statistics plot on or outside the control limits (the *k-of-k* rule)
(c) k of the last w $(k \leq w)$ plotting statistics plot on or outside the control limits (the *k-of-w* rule)

The *1-of-1* rule in (a) is the simplest, which corresponds to the basic form of a control chart. Rules (a) and (b) are special cases of rule (c). Note that, rules (b) and (c) have also been used in the context of supplementing the Shewhart-type charts with warning limits and zones. Human et al. [22] considered a class of one-sided and two-sided nonparametric Shewhart-type charts based on sign statistics and run rules. These charts can be used to monitor any percentile (including the median) of interest. Chakraborti and Eryilmaz [14] improved Bakir's [1] SR chart by incorporating run rules. They proposed using two types of *2-of-2* charts, both based on th SR statistic and a *k-of-k* signaling rule.

Amin et al. [4] considered a nonparametric CUSUM (NPCUSUM) chart to monitor the median of a continuous distribution with a known IC value θ_0, based on the SN statistic from a subgroup of size n.

11.2.9.2 Shewhart-Type Signed-Rank Charts

This test is proposed by Bakir [1]. Let $|X_{i1} - \theta_0|, |X_{i2} - \theta_0|, \ldots, |X_{in} - \theta_0|$ denote the absolute values of the deviations of observations from the known IC value of the median, θ_0 for the ith subgroup and let $R_{ij}, j = 1, 2, \ldots$ denote the rank of

$|X_{ij} - \theta_0|$ among the n absolute deviations. Then, the plotting statistic for the Shewhart-type SR chart is given by

$$SR_i = \sum_{j=1}^n \text{sign}(X_{ij} - \theta_0)R_{ij}, \quad i = 1, 2, \ldots \quad (11.32)$$

where

$$\text{sign}(x) = \begin{cases} -1, & \text{if } x < 0 \\ 0, & \text{if } x = 0 \\ 1, & \text{if } x > 0 \end{cases}$$

Note that, the statistic SR_i is the difference between the sum of positive and negative deviations, respectively, and is linearly related to the well-known Wilcoxon signed-rank statistic W^+ through the relationship $SR_i = 2W^+ - n(n+1)/2$, where W^+ is the sum of the ranks of the absolute values corresponding to the positive deviations (see Sect. 9.6.2.3 for details). Since the IC distribution of SR_i is symmetric about 0, the control limits of the two-sided nonparametric Shewhart-type sign chart are given by

$$\begin{aligned} \text{LCL} &= -d \\ \text{CL} &= 0 \\ \text{UCL} &= d \end{aligned} \quad (11.33)$$

where d is a positive integer between (and including) 1 and $n(n+1)/2$. If the plotting statistic SR_i falls on or outside one of the control limits, that is, if $SR_i \leq -d$ or $SR_i \geq d$, the process is declared to be OOC. The charting constant d is obtained for a specified ARL_0 which is equal to the reciprocal of the nominal FAR. Bakir [1] calculated the FAR (p_0^+) and the in-control average run length (ARL_0^+) of the upper one-sided Shewhart-type signed-rank (SR^+) control chart by using the null distribution of the Wilcoxon SR statistic. For two-sided SR chart, the FAR and the ARL_0 can be obtained from $p_0 = 2p_0^+$ and $ARL_0 = ARL_0^+/2$, respectively.

In Table 11.11, we present the values of SR_i for each sample. For $n = 5$, the control limits are set at ± 5. Accordingly, the Schewart-type signed-rank control chart is shown in Fig. 11.24. Note that, this chart also signals the change in median at observation number 12 (i.e., sample 27). For further discussion of other nonparametric charts based on signed-rank and false alarm policies, one may refer to Chakraborti et al. [12]. The authors also discuss number of benefits of nonparametric control chart over their parametric counterparts as well.

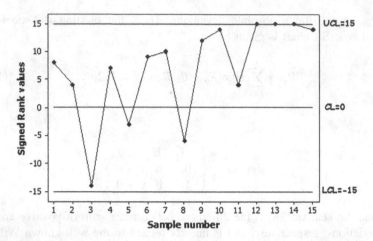

Fig. 11.24 Shewhart-type signed-rank control chart for Montgomery [30] piston data

11.2.9.3 Nonparametric Control Charts for Scale

As both Chakraborti et al. [1] and Chakraborti and Graham [15] have noted, not much work is currently available on nonparametric charts for scale. In their 1995 paper, Amin et al. [4] considered a nonparametric Shewhart-type sign chart for variability based on the number of observations that fall between or outside the first and the third quartiles, which are assumed known. They compared their chart to the parametric S^2 chart; the overall conclusion is that the nonparametric chart can provide a useful alternative when normality is in doubt.

Thus, in general, nonparametric control charts are simpler to devise in case the standards are known because the IC distribution of the plotting statistics is often known or simple to obtain. In practice, however, the process parameters are often unknown or unspecified. In such cases, a Phase I analysis is typically used to bring the process IC and Phase II control limits are calculated using a set of data taken from the IC process (see examples above). Such IC data are referred to as *reference data* or *calibration data*. Usually, the criteria to design Phase I and II control charts are different. In Phase I, the false alarm probability (FAP), which is the probability of at least one false alarm, is controlled to be at a small nominal value, whereas in Phase II, a large nominal value of the ARL_0 is usually used for the metric.

There has been other work on nonparametric control charts. Among these, Albers and Kallerberg [1] studied conditions and remedies under which the non-parametric charts become viable alternatives to their parametric counterparts. They consider Phase II charts for individual observations in case the process parameters are unknown based on empirical quartiles or order statistics. The basic problem is that for the very small FAR, typically used in the industry, a very large reference sample size is usually necessary to set up the chart.

Another area that has received some attention is control charts for variable sampling intervals (VSI). In a typical control charting environment, the time interval between two successive samples is fixed, and this is called a fixed sampling interval (FSI) scheme. VSI schemes allow the user to vary the sampling interval between taking samples. This idea has intuitive appeal since when one or more charting statistics fall close to one of the control limits but not quite outside, it seems reasonable to sample more frequently, whereas when charting statistics plot closer to the centerline, no action is necessary and only few samples might be sufficient. Amin and Widmaier [1] compared the Shewhart $\overline{X}$-charts with sign control charts, under the FSI and VSI schemes, on the basis of ARL for various shift sizes and several underlying distributions like the normal and distributions that are heavy-tailed and/or asymmetric like the double exponential and the gamma. It is seen that the nonparametric VSI sign charts are more efficient than the corresponding FSI sign charts.

Apart from detecting shifts or changes in certain parameters of a distribution such as the location or the scale, some authors have considered detecting change in the entire distribution. A nonparametric goodness-of-fit test, capable of detecting changes of a general nature in a distribution, has been adapted to develop a control chart for such problems. For a general introduction to such tests, one may refer to Gibbons and Chakraborti [1] and Zou and Tsung [59].

According to Chakraborti et al. [1], the advantages of nonparametric control charts are as follows:

- Simple and easy to chart
- No need to assume a particular distribution for the underlying process
- The IC run-length distribution is the same for all continuous distributions
- More robust and outlier resistant
- More efficiency in detecting changes when the true distribution is markedly non-normal
- No need to estimate the variance while constructing charts for the location parameter

11.3 Process Capability Studies

During the analyze phase, we have carried out a detailed process capability analysis (see Sect. 8.13 for details) and studied the significance of common causes affecting the strength of the process. A process capability shows whether the remaining variations are acceptable and whether the process will generate products or services which match the specified requirements. At the control stage, the capability studies focus on the controlled processes and their relevance of maintaining common causes of variations for future course of actions.

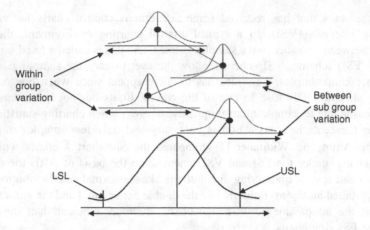

Fig. 11.25 Cumulative long-term capability. *Source* Muralidharan and Syamsundar [34]

Note that, the common cause variations are inherent in the process and special cause variations are due to real changes in the process. Most often, these variations happen because of shift in the process average and spread in the data (see Fig. 11.25 for long-term variability and Fig. 11.26 for short-term variability). Long-term variation is the combination of short-term variation, process shifts, and long-term process drifts. In order to analyze these shifts, we use control chart to detect change in the performance of a process. Control charts with limits may be used to assist in the interpretation of data (see Oakland [38] for further explanations of long-term and short-term variability encountered in a process).

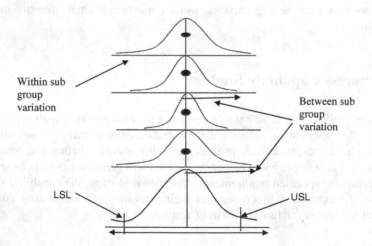

Fig. 11.26 Cumulative short-term capability. *Source* Muralidharan and Syamsundar [34]

According to Alwan [1], there are cases where autocorrelation within data sets force the charted quantities go outside the control limits, often indicating the standard operating procedures (SOP) are either not in place or not being followed. Like SOPs, charts encourage consistency which only indirectly relates to producing outputs that conform to specifications. When a process is associated with charted quantities within the control limits, it is said to be "*in control*" even if it generates a continuous stream of non-conformities.

For process capability studies, variable control charts are always preferred over attribute control chart. One of the most important advantage of $\overline{X}$ and R-charts is that they often provide an indication of impending trouble and allow operating personnel to take corrective action before any defectives are actually produced. Thus, $\overline{X}$ and R-charts are leading indicators of trouble, whereas p-charts (or c and u-charts) will not react unless the process has already changed so that more non-conforming units are being produced (see Montgomery [31] for more details).

A C_p index is defined as the ratio of maximum allowable characteristics to the normal variation of the process (6σ) and is given by

$$C_p = \frac{\text{USL} - \text{LSL}}{6\sigma} = \frac{2T}{6\sigma} = \frac{T}{3\sigma} \tag{11.34}$$

where T is the natural tolerance of the process. The capability of the process is decided by the value of C_p. The value of $C_p = 1$ corresponds to a short-term sigma level of 3, where the natural tolerances coincide with the specification limits. And $C_p = 2$ corresponds to a short-term sigma level of 6 (see Appendix Table A.12). Figure 11.27 depicts the process fallout and C_p values of a normal process. Note that, C_p offers a simple comparison of total variation with tolerances.

As discussed in previous section, $\overline{X}$ and R-chart provide information about the performance or capability of the process. For the computation of C_p statistics, we need to estimate the standard deviation which can be done through $\overline{R}$. An estimate of standard deviation is $\hat{\sigma} = \frac{\overline{R}}{d_2}$. As an example, consider the following information provided by a controlled process: The specification limits are 74 ± 0.05 mm, and the average range is $\overline{R} = 0.023$. Then, $\hat{\sigma} = \frac{0.023}{2.326} = 0.0099$, and the estimated capability index is $\hat{C}_p = \frac{74.05 - 73.95}{6(0.0099)} = 1.68$. This value indicates that the natural tolerance limits in the process are well inside the lower and upper specification limits. Consequently, a relatively low number of non-conforming piston rings will be produced.

Example 11.10 The following data were collected from a process manufacturing power supplies. The variable of interest is output voltage and $n = 5$.

Sample no.	$\bar{x}$	R
1	103	4
2	102	5
3	104	2
4	105	8
5	104	4
6	106	3
7	102	7
8	105	2
9	106	4
10	104	3
11	105	4
12	103	2
13	102	3
14	105	4
15	104	5
16	105	3
17	106	5
18	102	2
19	105	4
20	103	2

1. Compute centerline and control limits for controlling future production.
2. Assume that the quality characteristic is normally distributed and estimate the process standard deviation.
3. What are the natural tolerance limits of the process?
4. Estimate the process fraction non-conforming if the specifications on the characteristic were 103 ± 4?
5. Obtain the capability of the process

Solution

1. The control limits for $\bar{x}$-chart are CL $= \bar{\bar{x}} = 104.05, \text{LCL} = 99.01$, and UCL $= 109.09$, and the control limits for R-chart are CL $= \bar{R} = 3.8$, LCL $= -2.36$, and UCL $= 9.96$.
2. $\hat{\sigma} = \frac{3.8}{2.326} = 1.6337$
3. The natural tolerance limits of the process are $\bar{\bar{x}} \pm 3\hat{\sigma} = 104.05 \pm 4.9011$

Fig. 11.27 Process fallout
and the process capability.
a $C_p < 1$, **b** $C_p = 1$, **c** $C_p > 1$

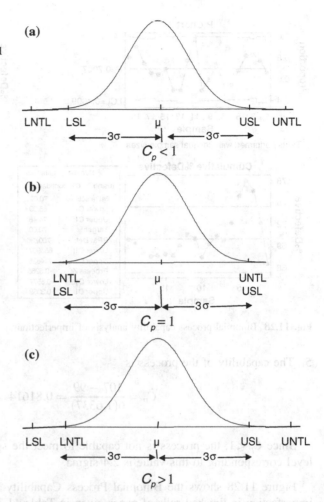

4. Let p be the fraction non-conforming. Then, to find

$$p = P(X < \text{LSL}) + P(X > \text{USL})$$
$$= P(X < 99) + P(X > 107)$$
$$= P\left(Z < \frac{99 - 104.05}{1.6337}\right) + P\left(Z > \frac{107 - 104.05}{1.6337}\right)$$
$$= P(Z < -3.09114) + P(X > 1.805717)$$
$$= 0.0365 = 0.0365$$

That is, about 3.5 % of the time, the voltage unit will be out of specifications.

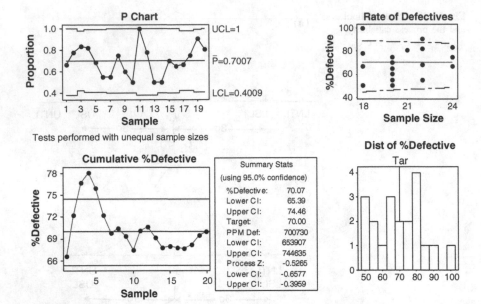

Fig. 11.28 Binomial process capability analysis of imperfections

5. The capability of the process is

$$\hat{C}_p = \frac{107 - 99}{6(1.6337)} = 0.81614$$

Since $\hat{C}_p < 1$, the process is not capable to meet the specifications. The sigma level corresponding to this value is 2.4 sigma.

Figure 11.28 shows the Binomial Process Capability Analysis of the data on imperfection in finished rolls of paper given in Table 11.3.

The C_p value discussed above is the ability to meet customer specification limits in respect of the voice of process. Hence, in a simple term, the capability of the process can be expressed as

$$C_p = \frac{\text{USL} - \text{LSL}}{6\sigma} = \frac{\text{VOC}}{\text{VOP}} \qquad (11.35)$$

For a controlled process, it is better to use the quantity C_{pk} (see Sect. 8.13 for details) instead of C_p, as C_{pk} takes care of the centering of the process relative to the specification limits. Also, the sigma level of the process can be easily calculated as

$$\sigma_{level} = 3C_{pk} \qquad (11.36)$$

The quantities C_p and C_{pk} represent long-term capability (see Fig. 11.20). For short-term capability, we compute the performance indices such as P_p and P_{pk}. The performance indices represent the process ability to meet customer specification limits on characteristics of interest. The difference between the use of C_p (or C_{pk}) and P_p (or P_{pk}) is that the process characteristic is demonstrated to be in control in the former case, whereas this may not be true with the latter.

Once the process has been improved, it needs to be monitored closely to maintain the gain. Two tools for holding the gain are the total productive maintenance (TPM) and *visual controls*. TPM is a methodology pioneered at Toyota group that works to ensure that every machine in a production process is always able to perform its required tasks so that production is never interrupted. TPM maximizes equipment effectiveness by using a preventive maintenance program throughout the life of the equipment. A TPM system uses historical data, manufacturer's recommendations, diagnostic tests, and other techniques to schedule maintenance activity so that machine downtime can be minimized. An effective TPM system includes continuous improvement initiatives as it seeks more effective and efficient ways to predict and diagnose maintenance-related problems.

Visual controls are approaches and techniques that permit one to visually determine the status of a system, factory, or process at a glance and prevent or minimize process variations. Visual controls work to provide a constant focus and attention on the process. The level of attention can help stabilize variation at the improved level of a process. This control can be integrated with Poke yoke system, which is discussed below.

11.4 Poke Yoke

Poke Yoke is also called "mistake proofing" which is another powerful method of eliminating waste, controlling the inventories needed for a process. This philosophy allows building quality into the process and simplifies the identification of root causes of a particular problem.

> Defects arise because errors are made; the two have a cause and effect relationship....
> Yet errors will not turn into defects if feedback and action take place at the error stage
>   Shigeo Shingo

Poke Yoke recognizes that as long as humans are involved in process, there will be errors and that if proper analysis is done, you can implement Poke-Yoke to prevent the error. Otherwise, you believe either humans are infallible or that there is nothing you can do about mistakes.

Poke yoke systems create a process in which a worker cannot create an error. Parts and/or processes are designed so that desired results are inevitable. True Poke yoke systems are ones that recognize where an error is most likely to occur in a process and introduce some solution that prevents the error from happening, regardless of what the human does.

Some of the Poke yoke systems type of errors and their *safeguards* are as follows:

- *Forgetfulness*: Forgetting a step or a part (Safeguards: use checklist, visual standard operating procedure, etc., to recall)
- *Errors due to Misunderstanding*: Not very familiar with the required operation (Safeguards: continuous training, visual SOP, etc.)
- *Errors in Identification*: Problems in identification or clearness of required steps or parts (Safeguards: training, visual training, standardization, etc.)
- *Errors due to Lack of Experience*: New employees (Safeguards: skill building and training, work standardization, etc.)
- *Errors due to Lack of Standards*: No clear way to perform the task or job (Safeguards: standard operations, visual instructions, etc.)
- *Errors due to Machine Readability*: Machine out of specifications (Safeguards: TPM, critical parts list, maintain equipment, history list, etc.)

The main principles of Poke Yoke are as follows:

- Build quality into the process
- All inadvertent errors and defects can be eliminated
- Stop doing it wrong and start doing it right now
- Do not think of excuses, think about doing it right
- An 80 % chance of success is good enough—Focus on implementation
- Attack defects and errors as a team
- Ten heads are better than one—Brainstorm ideas, discuss, and think
- Seek out the root cause, use the 5 "Whys"

Thus, Poke Yoke activities enable process control by eliminating source of variation and also are helpful in reducing the occurrence of rare events. Apart from the above tools, there are many other simple techniques which can help to reduce waste in a process. Even the *Pull* and *Push* system also helps the management to produce the inventories in an equilibrium manner and a zero waste condition. The greater importance of waste reduction methods has now developed into Lean thinking of manufacturing.

11.5 Designed for Six Sigma

Designed for Six Sigma (DFSS) describes the application of Six Sigma tools to product development and process design efforts with the goal of "designing in" Six Sigma performance capability. DFSS is a methodology by which new products and services can be designed and implemented. In many of our discussions done previously, we have seen Six Sigma as a process improvement philosophy and methodology, whereas DFSS is centered on designing new products and services. Further, Six Sigma focuses on one or two CTQ metrics, looks at processes, and aims to improve the CTQ performance by about one process-sigma at a time. In contrast, DFSS focuses on every single CTQ that matters to every customer, looks at products and services as well as the processes by which they are delivered, and aims to bring forth a new product/service with a performance of about 4.5 sigma or better.

Other differences are that DFSS projects are often much larger and take longer duration and are often based on a long-term business need for new products, rather than a short-term need to fix a customer problem. If Six Sigma works with the philosophy of DMAIC, then DFSS works with the philosophy of DMADV (define, measure, analyze, design, and verify).

- *Define* the project goals and customer's (internal and external) requirements.
- *Measure* and determine customer needs and specifications; benchmark competitors and industry.
- *Analyze* the process options to meet the customer's needs.
- *Design* (detailed) the process to meet the customer's needs.
- *Verify* the design performance and ability to meet customer's needs.

There are many other versions of DFSS, namely DCCDI (define, customer concept, design, and implement). Here, customer requirements are used to generate ideas. These ideas will be put in practice to meet the customer and business specifications and commercialization of the product/service. Identify, design, optimize, and validate (IDOV) is another well-known design methodology, especially used in the manufacturing world. Here, after identifying the customers and their specifications, design phase translates the customer CTQs into functional requirements and into solution alternatives. A best solution is selected to optimize the design and performance through rigorous and advanced statistical tools. The validate phase will make sure that the design you have developed will meet the customer CTQs. Define, measure, explore, develop, and implement (DMEDI) is another type of DFSS methodology. If DFSS is to work successfully, it is important that it covers the full life cycle of any new product or service. This begins when the organization formally agrees to the requirement for something new and ends when the new product/service is in full commercial delivery.

Throughout the DFSS process, it is important that the following points be kept in mind:

- Is the product concept well identified?
- Are your customers real?
- Will customers buy your product?
- Can the company make this product at a competitive cost?
- Are the financial returns acceptable?
- Does this product fit in with the overall business strategy?
- Is the risk assessment acceptable?
- Can the company make this product better than the competitor?
- Can product reliability and maintainability goals be met?
- Has a plan for transfer to manufacturing been developed and verified?

For additional information on DFSS, one may refer to Perry and Bacon [1], and Montgomery and Woodall [15].

11.6 Quality Function Deployment

Quality function deployment (QFD) is a "method to transform user demands into design quality, to deploy the functions forming quality, and to deploy methods for achieving the design quality into subsystems and component parts, and ultimately to specific elements of the manufacturing process." QFD is designed to help planners focus on characteristics of a new or existing product or service from the viewpoints of market segments, company, or technology development needs. The technique yields charts and matrices. QFD helps transform VOC into engineering characteristics (and appropriate test methods) for a product or service, prioritizing each product or service characteristic while simultaneously setting development targets for product or service.

The concept was originally developed by Dr. Yoji Akao in Japan in 1966, who combined his work in quality assurance and quality control points with function deployment used in value engineering. QFD is applied in a wide variety of services, consumer products, military engineering, and emerging technology products. The technique is also included in the new ISO 9000:2000 standard which focuses on customer satisfaction. This concept also draws some parallel with the concept of QFD-associated *Hoshin Kanri process*. It resembles more of management by objectives (MBO), but adds a significant element in the goal setting process, called "catchball." Use of these Hoshin techniques by US companies such as Hewlett Packard has been successful in focusing and aligning company resources to follow stated strategic goals throughout an organizational hierarchy (Wikipedia information). Since the early introduction of QFD, the technique has been developed to shorten the time span and reduce the required group efforts.

The QFD makes use of the following techniques and tools:

- *House of Quality.* House of Quality is a tool for depicting the relationship between quality metrics. The matrix includes all aspects of quality parameters extending beyond the technological solutions to management requirements housed on a particular performance barometer. For instance, quality to cost, technology, reliability, function, parts, technology, manufacturing, and service deployments are associated with their importance of quality. In addition, the same technique can extend the method into the constituent product subsystems, configuration items, assemblies, and parts. From these detailed level components, fabrication and assembly process, QFD charts can be developed to support statistical process control techniques.
- *Pugh concept selection.* Pugh concept selection can be used in coordination with QFD to select a promising product or service configuration from among listed alternatives.
- *Modular Function Deployment.* Modular function deployment uses QFD to establish customer requirements and to identify important design requirements with a special emphasis on modularity.

There are also other minor differences between the applications of QFD in modular function deployment as compared to House of Quality, for example, the term "customer attribute" is replaced by "customer value," and the term "engineering characteristics" is replaced by "product properties." But the terms have similar meanings in the two applications.

11.7 Standardization

Every successful Six Sigma project ultimately leads to some standards in the process. They can be both tangible and non-tangible. The standard can be a statement, specification, or quantity of material against which measured outputs from a process may be judged as acceptable or unacceptable. Standardization is therefore a method of creating best practices followed in the organization. A standard work (also called standardized work) includes the specification of cycle time, task time, task sequence, and the minimum inventory of parts on hand required to conduct the activity.

It is expected that the improvements in cycle time reduction, reduction in cost of poor quality (CoPQ), and constancy in lead time and increase in return of investment are achieved as per the VOC requirements. By a good data and performance standard, one can use these levels for future standards. A Six Sigma project's success depends on the ability to deliver flexible rather than rigid standards of specifications as expected by the stakeholders.

11.8 Standard Operating Procedures and Work Instructions

According to EPA QA/G-6 (April 2000), a standard operating procedure (SOP) is a set of written instructions that document a routine or repetitive activity followed by an organization. The development and use of SOPs are an integral part of a successful quality system as it provides individuals with the information to perform a job properly and facilitates consistency in the quality and integrity of a product or end result. The term "SOP" may not always be appropriate, and terms such as protocols, instructions, worksheets, and laboratory operating procedures may also be used. For this document, "SOP" will be used.

SOPs can also provide employees with a reference to common business practices, activities, or tasks. New employees use a SOP to answer questions without having to interrupt supervisors to ask how an operation is performed. The international quality standard ISO 9001 essentially requires the determination of processes (documented as standard operating procedures) used in any manufacturing process that could affect the quality of the product.

The development and use of SOPs minimizes variation and promotes quality through consistent implementation of a process or procedure within the organization, even if there are temporary or permanent personnel changes. SOPs can indicate compliance with organizational and governmental requirements and can be used as a part of a personnel training program, since they should provide detailed work instructions. It minimizes opportunities for miscommunication and can address safety concerns. When historical data are being evaluated for current use, SOPs can also be valuable for reconstructing project activities when no other references are available. In addition, SOPs are frequently used as checklists by inspectors when auditing procedures. Ultimately, the benefits of a valid SOP are reduced work effort, along with improved comparability, credibility, and legal defensibility.

The people associated with a Six Sigma project should now sit, discuss, and formulate a thorough working plan for operating procedures of their project. All quality control (QC) activities should be made a part of it. It should also allow self-verification of the quality and consistency of the work. Specific criteria for each should be included that are required to demonstrate successful performance of the method. Describe the frequency of required calibration and quality control checks and discuss the rationale for decisions. Finally, it should also describe the limits/criteria for QC data/results and actions required and describe the procedures for reporting QC data and results.

11.9   Process Dashboards

In management information systems, a *dashboard* (or *control panel*) is "an easy to read, often single page, real-time user interface, showing a graphical presentation of the current status (snapshot) and historical trends of an organization's key performance indicators to enable instantaneous and informed decisions to be made at a glance."

For example, a manufacturing dashboard may show key performance indicators related to productivity such as number of parts manufactured or number of failed quality inspections per hour. Similarly, a human resources dashboard may show KPIs related to staff recruitment, retention, and composition, for example, number of open positions, or average days or cost per recruitment. Dashboards often provide at-a-glance views of KPIs relevant to a particular objective or business process (e.g., sales, marketing, human resources, or production). The term dashboard originates from the automobile dashboard where drivers monitor the major functions at a glance via the instrument cluster. Dashboards give signs about a business letting you know something is wrong or something is right. The corporate world has tried for years to come up with a solution that would tell them if their business needed maintenance or if the temperature of their business was running above normal. Dashboards typically are limited to show summaries, key trends, comparisons, and exceptions. There are four key elements to a good dashboard.

1. Simple and communicates easily
2. Minimum distractions, otherwise, it could cause confusion
3. Supports organized business with meaning and useful data
4. Applies human visual perception to visual presentation of information

There are many types of dashboards. Among all, the digital dashboards dominate the business exchange. Digital dashboards may be laid out to track the flows inherent in the business processes that they monitor. Graphically, users may see the high-level processes and then drill down into low-level data. This level of detail is often buried deep within the corporate enterprise and otherwise unavailable to the senior executives.

Three main types of digital dashboard dominate the market today: stand-alone software applications, Web browser-based applications, and desktop applications also known as desktop widgets. The last are driven by a widget engine. Specialized dashboards may track all corporate functions. Examples include human resources, recruiting, sales, operations, security, information technology, project management, customer relationship management, and many more departmental dashboards.

Digital dashboard projects involve business units as the driver and the information technology department as the enabler. The success of digital dashboard projects often depends on the metrics that were chosen for monitoring. Key performance indicators, balanced scorecards, and sales performance figures are some of the content appropriate in business dashboards.

Like a car's dashboard (or control panel), a software dashboard provides decision makers with the input necessary to "drive" the business. Thus, a graphical user interface may be designed to display summaries, graphics (e.g., bar charts, pie charts, and bullet graphs), and gauges (with colors similar to traffic lights) in a portal-like framework to highlight important information.

Digital dashboards allow managers to monitor the contribution of the various departments in their organization. To gauge exactly how well an organization is performing overall, digital dashboards allow you to capture and report specific data points from each department within the organization, thus providing a "snapshot" of performance.

Benefits of using digital dashboards include

- Visual presentation of performance measures
- Ability to identify and correct negative trends
- Measure efficiencies/inefficiencies
- Ability to generate detailed reports showing new trends
- Ability to make more informed decisions based on collected business intelligence
- Align strategies and organizational goals
- Saves time compared to running multiple reports
- Gain total visibility of all systems instantly
- Quick identification of data outliers and correlations

11.10   Change Management and Resistance

The stakeholder analysis is a good tool for managing resistance from outside your team. According to Muir [12], the GRPI (goals, roles, and responsibilities; processes and procedures; and interpersonal relationships) tool helps manage a successful team over the length of the project. The GRPI model may also be useful as a diagnostic tool when the team is not working well and you are not sure what is wrong. Initially, the goals, roles, and processes are the most important to define, but as the project progresses, the interpersonal state of the team can quickly become the greatest concern for continued success.

Note that, resistance is a human behavior and should not be suppressed for any reason. For a Six Sigma project implementation, they help stakeholders to learn their own mistakes and drawbacks. When you are steering a project team, you will encounter resistance at all stages. You will probably not get the same type of resistance from the same people as you progress through your project, so you must continue to monitor and manage it. This will pave way for information clarity and establish a good communication. This will ultimately lead to success.

Muir [12] gives the following resistance types and strategies:

- *Cognitive*. People who have self-proclaimed reasons for believing the success/failure of the project
- *Ideological*. People who believe that the proposed change violates fundamental values that have made the organizations what it is
- *Power driven*. People who perceive that for them, the proposed change will lead to a loss of power, autonomy, status, and self-control
- *Psychological*. People who have difficulties to learn, adopt, or assimilate new concepts and new behaviors
- *Fear*. New situations and unfamiliar processes can lead to a general, unverbalized fear of changing the ways things are done. It will eventually lead to one of the above types if not addressed properly

It is true that the resistance can crop up any time during the DMAIC phases. More often, it leads to the betterment of the process, if the resistance is taken in a positive way. The chances of knowing the problem, its causes, and the effects will reduce the burden of improving and controlling the process. Hence, to sustain the overall success, one should face some resistance from the stakeholders.

11.11   Documentation

The SOPs discussed above are documents and procedures necessary for carrying out a particular process with minimum variation. The purpose of these documents is to make certain that the activity is performed the same way over a long time. This is especially important when multiskilled, cross-trained personnel move into a variety of positions. The development and updating of these documents involve the people who perform the work. The documentation process is begun by listing the major steps, then examining for its adaptability and utility, and then breaking down into sub-processes and then continued until maximum utility is derived.

Once the process has been improved, it must be monitored to ensure the gains are maintained and to determine when additional improvements are required. Several tools and methods are available to assist in this regard. They include

- *Control charts*. Used to monitor the stability of the process and determine when a special cause is present and when to take appropriate action. The choice of a particular chart depends on the nature of the process. Control chart is the resultant of VOP.
- *Process capability studies*. Provide us with the opportunity to understand how the VOP (control limits) compares with the VOC (specification limits) and help us determine whether the process average must be shifted or recentered or the variation reduced.
- *Process metrics*. This includes a wide variety of in-process and end-of-process metrics that measure the overall efficiency and effectiveness of the process. Examples include cycle times, takt time, response rate, work in process, backlog, defect rates, rework rates, and scrap rates.

Collectively, these tools and methods help us gauge the overall health of a process and provide triggers for reevaluating a process for further improvement (see also Kubiac and Benbow [24]).

11.12 Control Check Sheets

- Ensure that the data analysis supports the goal statement as per the project charter
- Select ongoing measures to monitor performance of the process and confirmed effectiveness of our solutions
- Prepare all essential documents of the revised process, including key procedures and process maps
- Identify an "owner" of the process who will take over responsibility for our solution and for managing continuing operations
- Develop process management charts detailing requirements, measures, and responses to problems in the process
- Prepare storyboard documenting the team's work and data collected during the project
- Forwarded other issues and opportunities which were not able to address to senior management
- Celebrate the hard work and successful efforts of the team

11.13   Relevance for Managers

The final phase activities of Six Sigma project are discussed in this chapter. Statistical process quality control is used for validating collection systems, to ascertain the repeatability and reproducibility of metrics in an operational environment, monitor implementation, and validate and stabilize performance gains. We use control chart to track performance over time, evaluate progress after process changes/improvements, and focus attention on detecting and monitoring process variation over time. Both variable charts ($\overline{X}$, R, and σ) and attribute control charts (p, np, c, and u) are studied in detail in this chapter. These charts have much statistical significance in managerial decision making. There are charts which can detect small and sustained shift in a process. Two such types of control charts are the Cumulative Sum (CUSUM) chart and Exponentially Weighted Moving Average (EWMA) charts.

The underlying principle of every control chart is that data come from a normal process. This ensures that any out-of-control signal is attributed to assignable causes of variation. Also, most of the charts discussed above either use some process parameters or some process characteristics. In the absence of these two assumptions, one cannot use the above control charts for monitoring variation. Hence, the use of nonparametric control is encouraged, which is generally defined in terms of its in-control (IC) run-length distribution. This is a relatively new area of research and has a lot of potential in industrial application. This chapter takes care of a detailed discussion on nonparametric control charts and many new directions in this area. Literature is now expanding in this area.

The process capability study, which is considered to be an essential managerial requirement, is also presented in detail. The philosophy of Poke Yoke, for building quality into the process and Quality Function Deployment for achieving design quality into sub-systems and component parts are also integrated in this chapter. The use of standardization, work place management, process dashboards, and many other management tools for documenting the quality are discussed with reference to a Six Sigma project.

Exercises

11.1. What are the eight dimensions of quality?
11.2. What is the significance of measuring variation? Discuss various methods of describing variation.
11.3. Explain various components of a control chart. Discuss how out of control points are detected in a control chart?
11.4. Discuss the importance and applications of variable control chart. Explain how the control limits are constructed for a variable chart?
11.5. What is trial control chart? How are they implemented?
11.6. Explain the procedure of constructing control limits for $\overline{X}$-chart and R-chart.

11.7. Samples of size 5 are taken from a process every hour. The $\overline{X}$ and R values for a particular quality characteristic are determined. After 25 samples have been collected, it was found $\overline{X} = 20$ and $\overline{R} = 4.56$. Assume that the quality characteristic is normally distributed.

   (i)   Compute 3-sigma control limits for $\overline{X}$ and R-charts.
   (ii) Both charts are assumed to exhibit control. Estimate the process standard deviation.
   (iii) When the process standard deviation is constant, if the process mean shifts to 24 units. What is the probability of (a) not detecting a shift on the first subsequent sample (b) detecting a shift by the third subsequent sample?
   (iv) What would be the appropriate control limits for the $\overline{X}$-chart if the type I error probability was to be 0.05.
   (v)   Obtain 2-sigma control limits for $\overline{X}$ and R-charts.
   (vi) Suppose an s-chart is to be used instead of R-chart, what would be the appropriate control limits of s-chart?

11.8. The net weight (in oz) of a dry bleach product is to be monitored by $\overline{X}$ and R control charts using a sample size of $n = 5$. Data for 20 samples are shown as follows.

Sample number	x_1	x_2	x_3	x_4	x_5
1	15.8	16.3	16.2	16.1	16.6
2	16.3	15.9	15.9	16.2	16.4
3	16.1	16.2	16.5	16.4	16.3
4	16.3	16.2	15.9	16.4	16.2
5	16.1	16.1	16.4	16.5	16.0
6	16.1	15.8	16.7	16.6	16.4
7	16.1	16.3	16.5	16.1	16.5
8	16.2	16.1	16.2	16.1	16.3
9	16.3	16.2	16.4	16.3	16.5
10	16.6	16.3	16.4	16.1	16.5
11	16.2	16.4	15.9	16.3	16.4
12	15.9	16.6	16.7	16.2	16.5
13	16.4	16.1	16.6	16.4	16.1
14	16.5	16.3	16.2	16.3	16.4
15	16.4	16.1	16.3	16.2	16.2
16	16.0	16.2	16.3	16.3	16.2
17	16.4	16.2	16.4	16.3	16.2
18	16.0	16.2	16.4	16.5	16.1
19	16.4	16.0	16.3	16.4	16.4
20	16.4	16.4	16.5	16.0	15.8

(a) Set up $\overline{X}$ and R control charts using these data. Does the process exhibit statistical control? Revise the control limits if necessary.

(b) Estimate the process mean and process standard deviation.

(c) If the specifications are at 16.2 ± 0.5 oz what conclusions would you draw about process capability?

(d) What fraction of containers produced by this process is likely to be below the lower specification limits of 15.7 oz?

11.9. The data below are $\overline{X}$ and R values for 24 samples of size $n = 5$ taken from a process producing bearings. The measurements are made on the inside diameter (in mm) of the bearing, with only deviations from 0.50 in multiples of 10,000 are recorded.

Sample number	$\overline{X}$	R
1	34.5	3
2	34.2	4
3	31.6	4
4	31.5	4
5	35.0	5
6	34.1	6
7	32.6	4
8	33.8	3
9	34.8	7
10	33.6	8
11	31.9	3
12	38.6	9
13	35.4	8
14	34.0	6
15	37.1	5
16	34.9	7
17	33.5	4
18	31.7	3
19	34.0	8
20	35.1	4
21	33.7	2
22	32.8	1
23	33.5	3
24	34.2	2

(i) Set up $\overline{X}$ and R-charts on this process. Does the process seem to be in statistical control? If necessary revise the trail control limits. Estimate process mean and process standard deviation.

     (ii) If specifications on this diameter are 0.5030 ± 0.0010, find the percentage of nonconforming bearings produced by this process. Assume that the diameter is normally distributed.

11.10. Discuss the relevance of Automotive Industry Action Group (AIAG) tests of rules in identifying out of control points in a control chart.

11.11. Distinguish between a defect and a defective. Give examples for each.

11.12. Discuss the application areas of attribute control chart. Explain how the control limits are constructed for an attribute control chart?

11.13. Discuss how binomial and Poisson distributions are used in constructing attribute control charts?

11.14. Discuss the methods of constructing p- and np-charts.

11.15. Data on the number of nonconformities in samples of 100 printed circuit boards are given below:

Sample number	Number of nonconformities
1	21
2	24
3	16
4	12
5	15
6	5
7	28
8	20
9	31
10	25
11	20
12	24
13	16
14	19
15	10
16	17
17	13
18	22
19	18
20	39
21	30
22	24
23	16
24	19
25	17
26	15

Construct a p-chart np-chart and interpret.

11.16. The data for a control chart for fraction nonconforming with variable sample size are given below:

Sample number (i)	Sample size (n_i)	Number of nonconforming units (Di)
1	100	12
2	80	8
3	80	6
4	100	9
5	110	10
6	110	12
7	100	11
8	100	16
9	90	10
10	90	6
11	110	20
12	120	15
13	120	9
14	120	8
15	110	6
16	80	8
17	80	10
18	80	7
19	90	5
20	100	8
21	100	5
22	100	8
23	100	10
24	90	6
25	90	9

Construct a p-chart and interpret the results.

11.17. Discuss the methods of constructing c- and u-charts.

11.18. Data on number of nonconformities in personal computers are given below:

Sample number, i	Sample size n	Total Number of nonconformities, (c_i)
1	5	10
2	5	12
3	5	8
4	5	14
5	5	10
6	5	16
7	5	11
8	5	7
9	5	10
10	5	15
11	5	9
12	5	5
13	5	7
14	5	11
15	5	12
16	5	6
17	5	8
18	5	10
19	5	7
20	5	5

Construct a c-chart, u-chart, and interpret.

11.19. What is a CUSUM chart? How is it implemented?

11.20. In the process of producing shaft seals used in automobiles engines, grease is applied to the area of the seal surface, which will be in contact with the shaft. The amount of grease is critical since it provides the initial lubrication in the start-up of a new engine before the oil has a chance to fully circulate. For correct operation, the amount of grease applied to the seal should be 0.10 gram in weight. Too much grease can cause the seal to vulcanize, resulting in malfunction, while too little grease will cause the seal to run dry, resulting in seal damage. Departures in the mean level by more than 0.01 gm are considered important enough for immediate corrective action. To monitor the process of grease deposition, current practice is to take two pre-weighted clean (greaseless) seals and run them through the machine and then have them reweighed. The change in weight reflects the amount of grease on the seal. Grease deposition is monitored hourly on a routine basis. The following table shows the results of 35 consecutive hourly samples of size 2.

Sample number	X_1	X_2
1	0.091	0.096
2	0.108	0.105
3	0.117	0.079
4	0.105	0.12
5	0.114	0.083
6	0.11	0.108
7	0.087	0.097
8	0.109	0.099
9	0.105	0.094
10	0.1	0.117
11	0.089	0.091
12	0.076	0.091
13	0.119	0.101
14	0.093	0.111
15	0.115	0.101
16	0.125	0.107
17	0.101	0.107
18	0.093	0.093
19	0.108	0.104
20	0.109	0.112
21	0.092	0.105
22	0.092	0.098
23	0.096	0.075
24	0.082	0.081
25	0.106	0.083
26	0.088	0.104
27	0.085	0.076
28	0.069	0.098
29	0.094	0.099
30	0.073	0.092
31	0.097	0.095
32	0.077	0.087
33	0.084	0.093
34	0.108	0.088
35	0.091	0.098

(i) Implement a tabular CUSUM scheme for the process mean using h = 4.77. Interpret the chart results.

(ii) If the CUSUM chart signals out of control, is there a suggestion to increase or decrease in the application of amount of grease.

(iii) If the CUSUM chart signals out of control, estimate the time of the onset of the shift and also estimate new mean grease level.

(iv) Design an equivalent V-mask scheme to the tabular chart. Plot the cumulative sums and place the V-mask on the sampling period that first signaled out of control with the tabular chart. Demonstrate that the V-mask also signals out of control at this same sampling period.

11.21. Discuss the importance of exponentially weighted moving average (EWMA) chart. Explain the method of constructing EWMA chart.

11.22. The data below are temperature readings from a chemical process in degrees centigrade, taken every two minutes. (Read the observation down from left)

953 952 972 945 975 970 969 973 940 936 985 973 955 950 948 957 940 933 965 973 949 941 966 966 934 937 946 952 935 941 937 946 954 935 941 933 960 968 969 956

(i) Estimate the mean and standard deviation of the process.

(ii) Set up and apply EWMA control chart to these data using $\lambda = 0.1$ and $L = 3$.

(iii) Apply moving average control chart with $\omega = 4$.

11.23. Viscosity measurements of a polymer are made every 10 min by an online viscometer. Thirty-six observations are shown below (read down from left). The target viscosity for this process is $\mu_0 = 3200$.

3169	3205	3185	3188
3173	3203	3187	3183
3162	3209	3192	3175
3154	3208	3199	3174
3139	3211	3197	3171
3145	3214	3193	3180
3160	3215	3190	3179
3172	3206	3183	3175
3175	3203	3197	3174

(i) Estimate the process standard deviation.

(ii) Set up and apply EWMA control chart to these data using $\lambda = 0.1$ and $L = 2.7$.

11.24. Distinguish between average run length (ARL) and average time to signal (ATS).

11.25. Discuss how process monitoring is done through nonparametric control charts?

11.26. State the advantages of nonparametric control charts.

11.27. Describe the implementation of Shewhart-type sign charts.

11.28. Describe the implementation of Shewhart-type signed-rank charts.

11.29. For the data given in 11.22, construct a variable control charts based on the first 30 observations. Also implement a Shewhart-type sign chart for the last 10 observations.

11.30. For the data given in 11.9, carry out the following:

   (i) Compute center line and control limits for controlling future production.

   (ii) Assume that the quality characteristic is normally distributed, estimate the process standard deviation.

   (iii) What are the natural tolerance limits of the process?

   (iv) Estimate the process fraction nonconforming if the specifications on the characteristic were 103 ± 4?

   (v) Obtain the capability of the process

11.31. What is Poke-Yoke? Discuss the main principles of Poke-Yoke.

11.32. What is Designed for Six Sigma (DFSS)? How does its implementation impact the quality of a Six Sigma project?

11.33. Distinguish between quality function deployment (QFD) and modular function deployment (MFD).

11.34. Distinguish between standard operating procedure (SOP) and management by objectives (MBO).

11.35. Discuss how a process dashboards helps in identifying key process inputs of a Six Sigma process?

11.36. What are the key elements of process dashboards? What the benefits of dashboards?

References

1. Albers, W., Kallenberg, W.C.M.: Empirical nonparametric control charts: estimation effects and corrections. J. Appl. Stat. **31**, 345–360 (2004)
2. Alwan, L.C.: Effects of autocorrelation on control charts. Commun. Stat. Theory Methods **21** (1992)
3. Amin, R.W., Widmaier, O.: Sign control charts with variable sampling intervals. Commun. Stat. Theory Methods **28**, 1961–1985 (1999)
4. Amin, R.W., Reynolds Jr, M.R., Bakir, S.T.: Nonparametric quality control charts based on the sign statistic. Commun. Stat. Theory Methods **24**, 1597–1623 (1995)
5. Apley, D.W., Lee, H.Y.: Simultaneous identification of premodeled and unmodeled variation patterns. J. Qual. Technol. **42**(1), 36–51 (2010)
6. Bakir, S.T.: A distribution-free Shewhart quality control chart based on signed-ranks. Qual. Eng. **16**, 613–623 (2004)
7. Bakir, S.T., Reynolds Jr, M.R.: A nonparametric procedure for process control based on within-group ranking. Technometrics **21**, 175–183 (1979)

8. Box, G., Luceño, A., Paniagua-Quiñones, C.: Statistical Control by Monitoring and Feedback Adjustment, 2nd edn. Wiley, Hoboken, NJ (2009)
9. Box, G.E.P., Narasimhan, S.: Rethinking Statistics for Quality Control. Qual. Eng. **22**, 60–72 (2010)
10. Burkom, H.S., Elbert, Y., Feldman, A., Lin, J.: Role of Data Aggregation in Biosurveillance Detection Strategies with Applications from ESSENCE. Morbidity and Mortality Weekly Report **53** (Suppl), 67–73 (2004)
11. Castillo, D.E.: Statistical Process Adjustment for Quality Control. Wiley, Hoboken, NJ (2002)
12. Champ, C.W., Woodall, W.H.: Exact results for Shewhart control charts with supplementary runs rules. Technometrics **29** (1987)
13. Chakraborti, S., Van der Laan, P., Bakir, S.T.: Nonparametric control charts: an overview and some results. J. Qual. Technol. **33**, 304–315 (2001)
14. Chakraborti, S., Eryilmaz, S.: A nonparametric Shewhart-type signed-rank control chart based on runs. Commun. Stat. Simul. Comput. **36**, 335–356 (2007)
15. Chakraborti, S., Graham, M.A.: Nonparametric control charts. In: Encyclopedia of Statistics in Quality and Reliability, vol. 1, pp. 415–429. Wiley, New York (2007)
16. Chakraborti, S., Human, S.W., Graham, M.A.: Nonparametric (distribution-free) quality control charts. In: Balakrishnan, N. (ed.) Handbook of Methods and Applications of Statistics: Engineering, Quality control, and physical sciences, pp. 298–329. Wiley, New York (2011)
17. Dubrawski, A.: The Role of Data Aggregation in Public Health and Food Safety Surveillance. In: Biosurveillance—Methods and Case Studies, Kass-Hout, T., Zhang, X. (eds.), Boca Raton (chapter 9), FL: CRC/Taylor & Francis Group (2011)
18. Faltin, F.W., Kenett, R.S., Ruggeri, F. (eds.): Statistical Methods in Healthcare. West Sussex, UK, Wiley (2012)
19. Fricker Jr, R.D.: Introduction to Statistical Methods for Biosurveillance with an Emphasis on Syndromic Surveillance. UK, Cambridge University Press, Cambridge (2013)
20. Frisén, M.: Optimal Sequential Surveillance for Finance, Public Health, and Other Areas (with Discussion). Sequential Analysis **28**, 310–337 (2009)
21. Gibbons, J.D., Chakraborti, S.: Nonparametric Statistical Inference, 4th edn (Revised and expanded). Marcel Dekker, New York (2003)
22. Human, S.W., Chakraborti, S., Smit, C.F.: Nonparametric Shewhart-type control charts based on runs. Commun Stat. Theory Methods **39**, 2046–2062 (2010)
23. Knoth, S., Schmid, W.: Control charts for time series: a review. In: Lenz, H.-J., Wilrich, P-Th (eds.) Frontiers in Statistical Quality Control, vol. 7, pp. 210–236. Physica-Verlag, Heidelberg, Germany (2004)
24. Kubiak, T.M., Benbow, D.W.: The Certified Six Sigma Black Belt Handbook, 2nd edn. Dorling Kindersley Pvt Ltd., India (2010)
25. Lawson, A.B., Kleinman, K. (eds.): Spatial & Syndromic Surveillance for Public Health. Wiley, Hoboken, NJ (2005)
26. Ledolter, J., Bisgaard, S.: Challenges in constructing time series models from process data. Qual. Reliab. Eng. Int. **27**(2), 165–178 (2007)
27. Lucas, J.M., Saccucci, M.S.: Exponentially weighted moving average Control Schemes: Properties and Enhancements. Technometrics, **32** (1990)
28. Lombardo, J. et al.: A Systems Overview of the Electronic Surveillance System for the Early Notification of Community-Based Epidemics (ESSENCE II). Journal of Urban Health: Bull. N. Y. Acad. Med. 80(Supplement 1), pp. i32–i42 (2003)
29. Megahed, F.M., Wells, L.J., Camelio, J.A., Woodall, W.H.: A spatiotemporal method for the monitoring of image data. Qual. Reliab. Eng. Int. **28**, 967–980 (2012)
30. Montgomery, D.C.: Introduction to Statistical Quality Control, Wiley, India (2003)
31. Montgomery, D.C.: Introduction to Statistical Quality Control. Wiley, India (2004)
32. Montgomery, D.C., Woodall, W.H.: An overview of Six Sigma. Int. Stat. Rev. **76**(3), 329–346 (2008)
33. Muir, A.: Lean Six Sigma Way. McGraw Hill, New York (2006)

34. Muralidharan, K., Syamsundar, A.: Statistical Methods for Quality. Reliability and Maintainability. PHI, New Delhi (2012)
35. Nair, V., Hansen, M., Shi, J.: Statistics in Advanced Manufacturing. J. Am. Stat. Assoc. **95**, 1002–1005 (2000)
36. Nelson, L.S.: The Shewhart control chart: tests for special causes. J. Qual. Technol. **16**, 237–239 (1984)
37. Noorossana, R., Saghaei, A., Amiri, A. (eds.): Statistical Analysis of Profile Monitoring. Wiley, Hoboken, NJ (2011)
38. Oakland, J.S.: Statistical Process Control. Elsevier, New Delhi (2005)
39. Prajapati, D.R., Singh, S.: Control charts for monitoring the autocorrelated process parameters: a literature review. Int. J. Prod. Qual. Manag. **10**, 207–249 (2012)
40. Perry, R.C., Bacon, D.W.: Commercializing Great Products with Design for Six Sigma. Prentice-Hall, Upper Saddle River (2007)
41. Psarakis, S., Papaleonida, G.E.A.: SPC procedures for monitoring autocorrelated processes. Qual. Technol. Quant. Manag. **4**, 501–540 (2007)
42. Qin, S.J.: Statistical Process Monitoring: Basics and Beyond. J. Chemom. **17**, 480–502 (2003)
43. Roberts, S.W.: Control chart tests based on Geometric moving average. Technometrics **1**(23), 134–145 (1959)
44. Robertson, C., Nelson, T.A., MacNab, Y.C., Lawson, A.B.: Review of Methods for Space-Time Disease Surveillance. Spatial and Spatiotemporal Epidemiology **1**, 105–116 (2010)
45. Rogerson, P., Yamada, I.: Statistical Detection and Surveillance of Geographic Clusters. FL, CRC Press, Boca Raton (2009)
46. Schuh, A., Woodall, W.H., Camelio, J.A.: The Effect of Aggregating Data When Monitoring a Poisson Process J. Qual. Technol. **45**(3), 260–271 (2013)
47. Sonesson, C. Bock, D.: A Review and Discussion of Prospective Statistical Surveillance in Public Health. J. Roy. Stat. Soc. A **166**, 5–21 (2003)
48. Sparks, R.: Challenges in designing a disease surveillance plan: What we have and what we need. IIE Trans. Healthc. Eng. **49**, 794–806 (2013).
49. Stoumbos, Z.G., Reynolds, M.R., Jr., Ryan, T. P., Woodall, W.H.: The State of Statistical Process Control as We Proceed into the 21st Century. J Am. Stat. Assoc. **95**, 992–998 (2000)
50. Tennant, R., Mohammed, M.A., Coleman, J.J., Martin, U.: Monitoring Patients Using Control Charts: A Systematic Review. International Journal for Quality in Health Care **19**, 187–194 (2007)
51. Thor, J., Lundberg, J., Ask, J., Olsson, J., Carli, C., Härenstam, K.P., Brommels, M.: Application of Statistical Process Control in Healthcare Improvement: Systematic Review. Quality and Safety in Health Care **16**, 387–399 (2007)
52. Tsui, K.-L., Wong, S.Y., Jiang, W., Lin, C.-J.: Recent Research and Developments in Temporal and Spatiotemporal Surveillance for Public Health. IEEE Trans. Reliab. **60**(1), 49–58 (2011)
53. Wells, L.J., Megahed, F.M., Camelio, J.A., Woodall, W.H.: A Framework for Variation Visualization and understanding in Complex Manufacturing Systems. J. Intell. Manuf. **23**, 2025–2036 (2012)
54. Winkel, P., Zhang, N.F.: Statistical Development of Quality in Medicine. Wiley, Hoboken, NJ (2007)
55. Woodall, W.H.: Use of Control Charts in Health-Care and Public-Health Surveillance (with Discussion). J. Qual. Technol. **38**(2), 89–104 (2006)
56. Woodall, W.H.: Current Research in Profile Monitoring. Producão **17**, 420–425 (2007)
57. Woodall, W.H., Marshall, J.B., Joner Jr., M.D., Fraker, S.E., Abdel-Salam, A.G.: On the use and evaluation of prospective scan methods in health-related surveillance. J. Roy. Stat. Soc.: Ser. A **171**(1), 223–237 (2009)
58. Woodall, W.H., Montgomery, D.C.: Some Current Directions in the Theory and Application of Statistical Process Monitoring. J. Qual. Technol. **46**(1), 78–94 (2014)
59. Zou, C., Tsung, F.: Likelihood ratio-based distribution-free EWMA control charts. J. Qual. Technol. **42**, 174–196 (2010)

Chapter 12
Sigma Level Estimation

According to Six Sigma philosophy, what you do every day in the business directly determines what the customers see. If Y is the output of the process and X's are the business processes, then the focus of the Six Sigma project is to establish the mathematical relationship between the defects produced by a business process and the causes of those defects. This is expressed as

$$Y = f(x_1, x_2, \ldots, x_n)$$

Note that the number of defects generated by the current process will be measured by counting (for discrete data) them or by making a physical measurement (for continuous data) and then examining the distribution of the data and calculating the process capability. The process capability depends on customer's upper specification limit (USL) and lower specification limit (LSL). A process capability index [7] is a measure relating the actual performance of a process to its specified performance, where processes are considered to be a combination of the equipment, materials, people, plant, methods, and the environment.

12.1 Sigma Level for Normal Process

The calculation of process capability depends on the type of data under investigation. For attribute data system, the capability is defined in terms of pass/fail or good/bad as the case may be. They are calculated in terms of defects per million opportunities (DPMO) often called parts per million (PPM). Whereas for continuous data system, the capability is defined in terms of defects under the curve and outside of the specification limits. The process capability of the defect level expressed in DPMO is the sigma level (or Z units or sigma rating). Corresponding to each DPMO, we have sigma level table available and can be consulted. There are two kinds of sigma ratings —the *short-term sigma* and the *long-term sigma*. It is a measure which compares process variability vis-a-vis the requirements [1, 5]. Thus knowing the sigma rating we can establish the rejections that can be expected from the process. The average,

© Springer India 2015
K. Muralidharan, *Six Sigma for Organizational Excellence*,
DOI 10.1007/978-81-322-2325-2_12

the standard deviation, and the specification limits are required to get the sigma rating.

The sigma value of a process describes the quality level of that process. A quality level of K sigma exists in a process when the half tolerance of the measured product characteristic is equal to K times the standard deviation of the process:

$$K * \text{process standard deviation} = \text{half tolerance of specification} \qquad (12.1)$$

However, this definition alone does not account for the centering of a process. A process is centered when $X = T$, where X is the process average or mean and T is the target value, which is typically the midpoint between the customer's USL and the LSL. A process is off-centered when the process average, X, is not equal the target value T. The off-centering of a process is measured in standard deviations or sigma.

Now the calculation of DPMO, which indicates how many defects, would arise if there were one million opportunities involved in the calculation of defects per opportunity (DPO). DPO expresses the proportion of defects over the total number of opportunities in a group. That is,

$$DPO = \frac{\text{Number of defects}}{\# \text{ of units} * \# \text{ of opportunities}}$$

If defects per unit (DPU) is the ratio of number of defects to the number of units, then

$$DPO = \frac{DPU}{\# \text{ of opportunities}}$$

In Table 12.1, we present the sigma conversion table with a 1.5 sigma shift correspond to different DPMO.

Example 12.1 On inspection of 500 purchase orders, it is found that 10 of them are defectives. Suppose there are 5 opportunities for the defects to occur, then

$$DPU = \frac{\text{Number of defects}}{\# \text{ of units}}$$

$$= \frac{10}{500} = 0.020$$

$$DPO = \frac{DPU}{\# \text{ of opportunities}}$$

$$= \frac{0.02}{5} = 0.004$$

Table 12.1 Six Sigma conversion table (includes a 1.5 sigma shift)

Sigma level	DPMO	Sigma level	DPMO	Sigma level	DPMO
0.00	933,193	4.05	5386	5.05	193
0.50	841,345	4.10	4661	5.10	159
0.75	773,373	4.15	4024	5.15	131
1.00	691,462	4.20	3467	5.20	108
1.25	401,294	4.25	2980	5.25	89
1.50	500,000	4.30	2555	5.30	72
1.75	401,294	4.35	2186	5.35	59
2.00	308,537	4.40	1866	5.40	48
2.25	226,627	4.45	1589	5.45	39
2.50	158,655	4.50	1350	5.50	32
2.75	105,650	4.55	1144	5.55	26
3.00	66,807	4.60	968	5.60	21
3.25	40,059	4.65	816	5.65	17
3.50	22,750	4.70	687	5.70	13
3.60	17,865	4.75	577	5.75	11
3.70	13,904	4.80	483	5.80	9
3.75	12,225	4.85	404	5.85	7
3.80	10,724	4.90	337	5.90	5
3.90	8198	4.95	280	5.95	4
4.00	6210	5.00	233	6.00	3.4

Then, DPMO is

$$DPMO = DPO \times 10^6 = 4000$$

This 4000 DPMO is corresponding to 4.15 sigma level.

Example 12.2 Suppose out of 350 loan applications processed 45 of them are found to have some errors. If 8 opportunities per application can be possible, then

$$DPU = \frac{45}{350} = 0.12857$$
$$DPO = \frac{0.12857}{8} = 0.016$$

and

$$DPMO = DPO \times 10^6 = 16,000$$

This 16,000 DPMO is corresponding to 3.65 sigma level.

Now, consider an example based on a continuous variable.

Example 12.3 Piston rings used in automobiles are coated with hard chrome plating. For a particular stage of production, the requirement of plating thickness is 195 ± 20 mm. A random sample of 50 piston rings is obtained as

189	181	183	191	180	182	187	188	189	189
177	191	192	180	184	190	179	198	197	179
171	188	165	191	176	172	199	185	183	183
187	189	186	191	185	172	182	184	185	201
189	187	186	183	178	173	172	193	184	183

(a) Perform normality checks on the data.
(b) Calculate the sigma level of the process, if LSL = 168 and USL = 198.

Solution

(a) The probability plot clearly supports a normal distribution as $p > 0.05$ (Fig. 12.1).
(b) The mean ($\bar{x}$) and standard deviation (σ) of the variable here are 184.58 and 7.52, respectively. Now

$$Z_L = \frac{\text{LSL} - \bar{x}}{\sigma}$$

$$= \frac{168 - 184.58}{7.52} = -2.2$$

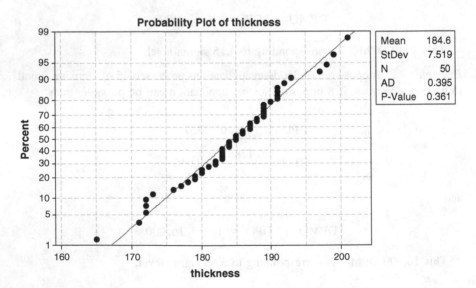

Fig. 12.1 Probability plot of thickness

$$Z_U = \frac{USL - \bar{x}}{\sigma}$$

$$= \frac{198 - 184.58}{7.52} = 1.78$$

Therefore,

$$P(\text{non conformance}) = P(Z < Z_L) + P(Z > Z_U)$$
$$= 0.0139 + 0.0375 \text{ (according to standard Normal table)}$$
$$= 0.0514$$

The P(non-conformance) $= 0.0514$ is corresponding to 1.63 according to the inverse standard normal distribution, which is the long-term sigma level (Z_{LT}) of the process. The short-term sigma level (Z_{ST}) is obtained as $Z_{LT} + 1.5 = 3.13$. Thus, the sigma level of the process is 3.13.

Figure 12.2 presents the process capability of thickness as per MINITAB software. The PPM total corresponds to the overall performance is 52009.27, which according to DPMO table (Appendix Table A.13) has 3.13 level of sigma.

Process Capability of thickness

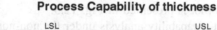

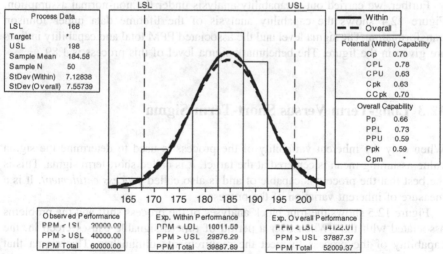

Process Data	
LSL	168
Target	*
USL	198
Sample Mean	184.58
Sample N	50
StDev (Within)	7.12838
StDev (Overall)	7.55739

Potential (Within) Capability	
Cp	0.70
CPL	0.78
CPU	0.63
Cpk	0.63
CCpk	0.70

Overall Capability	
Pp	0.66
PPL	0.73
PPU	0.59
Ppk	0.59
Cpm	*

Observed Performance		Exp. Within Performance		Exp. Overall Performance	
PPM < LSL	20000.00	PPM < LOL	10011.59	PPM < LSL	14122.01
PPM > USL	40000.00	PPM > USL	29876.29	PPM > USL	37887.37
PPM Total	60000.00	PPM Total	39887.89	PPM Total	52009.37

Fig. 12.2 Process capability of the process

12.2 Sigma Level for Non-normal Process

In the above example, the plating thickness followed a normal process and hence we could manually obtain the DPMO using the normal distribution tables. If the project is focused on cycle time reduction, then we will have USL for the cycle time for the individual process. These limits may have come from competitive bench-marking studies, VOC surveys, or the company's strategic plans. In these cases, we calculate the DPMO by assuming any other positive skewed distributions such as exponential, Weibull, Gamma, and lognormal distributions. In the next example, we consider one such situation.

Example 12.4 Consider the (page 118 of Muir) following data on measurement of lifetimes of an electronic item. The value zero indicates the item failed upon installation.

34.35	2.28	16.15	0.88	0.22	146.39
12.91	398.79	66.83	102.72	0.0	0.75
0.46	1.64	21.11	58.44	3923.66	0.0
0.05	26.31	1638.8	11.58	21.01	4.27
72.2	247.41	1895	0.1	71.64	18.32

Carry out a Capability study.

Solution Since the data do not follow normal distribution, we first carry out a distribution identification test. It is found that Weibull distribution fits well to the data and hence a Weibull plot is presented in Fig. 12.3.

Further, we carried out a capability analysis under the non-normal assumption. Figure 12.4 shows the capability analysis of the lifetime data under Johnson transformation. The sigma level and the associated PPM total and capability indices are given in the figure. The benchmark sigma level of this process is 1.89.

12.3 Long-Term Versus Short-Term Sigma

When only the inherent variability of the process is used to determine the sigma value assuming mean is centered at the target, it is called short-term sigma. This is the best that the process is capable of and is also called *process entitlement*. It is a measure of inherent variation of the process.

Figure 12.5 shows the four block analyses of the process shift and the problems associated with that. When over a period of time assignable causes creep in, the capability of the process to meet the requirements diminishes. This sigma that represents the capability of the process to meet the requirements over a period of time, considering that extraneous conditions cause process shifts from that at which it was set, is called the long-term sigma, denoted as Z_{LT}. To compensate for the

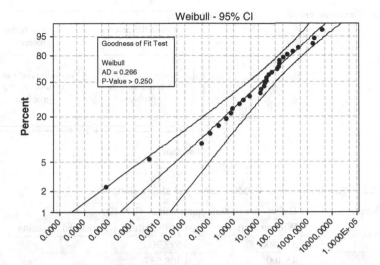

Fig. 12.3 Weibull plot of lifetime data

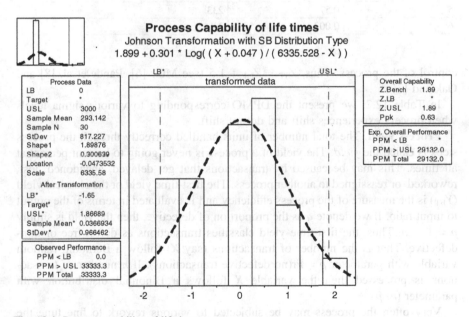

Fig. 12.4 Process capability of lifetimes

inevitable consequences associated with process centering errors, the distribution mean is offset by 1.5 standard deviations. This adjustment provides a more realistic idea of what the process capability will be over repeated cycles, and hence short-term sigma, denoted as Z_{ST}, is obtained. It is a measure of inherent variation plus

Fig. 12.5 Variation versus
control

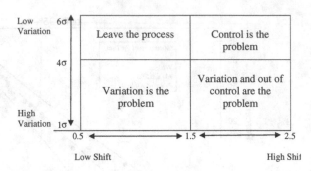

Table 12.2 Sigma levels versus DPMO (before and after a shift)

Sigma levels	Defects per million opportunities (DPMO)	
	Without shift	With shift of 1.5 standard deviation
1	317,400	697,700
2	45,400	308,537
3	2700	66,807
4	63	6210
5	0.57	233
6	0.002	3.4

control of the process. Thus, $Z_{ST} = Z_{LT} + 1.5$ (see Muir [6], Pande et al. [8] and Oakland [7]).

In Table 12.2, we present the DPMO corresponding to various sigma levels when process experiences shift and do not shift.

Process yield. The total number of units handled correctly through the process steps is called the *yield*. The yield of a process is never going to be cent percent at all times. This may be caused by transactions that get delayed, abandoned, lost, reworked, or reassigned to another process. The first-time yield or the classical yield (Y_{FT}) is the measure of the process efficiency and is evaluated in terms of the output to input ratio. If we denote q as the proportion of defective, then the yield is simply $p = 1 - q$. Thus, the first-time yield classifies transactions as defect-free or non-defective. That is the number of transactions (say X) follow a Bernoulli random variable with parameter $p = $ Pr(no defective transactions). If n number of transactions is processed then the variable X follows a Binomial distribution with parameter (n, p).

Very often the process may be subjected to various rework to fine tune the process which would involve additional cost of manpower, material, handling, creating excess capacity, etc. and many a times it is preferred to use rolled yield rather than the classical yield. In this case, the extent of rework that the process generates can also be a measure of the efficiency of the process and is known as *rolled throughput yield* (Y_{RT}).

If errors are calculated in terms of discrete counts, then the number of counts follows a Poisson distribution with DPU as the mean. Then, the efficiency (or yield) of the process is simply the probability of no errors in the rework process and is obtained as

$$Y_{RT} = \Pr(\text{there is no defective})$$
$$= P(X = 0)$$
$$= e^{-DPU}$$

Table 12.3 shows the sigma level corresponds to different yields and the equivalent DPMO.

Example 12.3 There are 6 boxes each having 10 cups. Three cups are found to have a scratch mark. What is the probability that a box does not have any scratch mark? Also obtain the corresponding yield.

Table 12.3 Yield versus sigma level

Yield (%)	DPMO	Sigma level	Yield (%)	DPMO	Sigma level
6.68	933,200	0	94.79	52,100	3.125
8.455	915,450	0.125	95.99	40,100	3.25
10.56	894,400	0.25	96.96	30,400	3.375
13.03	869,700	0.375	97.73	22,700	3.5
15.87	841,300	0.5	98.32	16,800	3.625
19.08	809,200	0.625	98.78	12,200	3.75
22.66	773,400	0.75	99.12	8800	3.875
26.595	734,050	0.875	99.38	6200	4
30.85	691,500	1	99.565	4350	4.125
35.435	645,650	1.125	99.7	3000	5.25
40.13	598,700	1.25	99.795	2050	4.375
40.025	549,750	1.375	99.87	1300	4.5
50	500,000	1.5	99.91	900	4.625
54.975	450,250	1.625	99.94	600	4.75
59.87	401,300	1.75	99.96	400	4.875
64.565	354,350	1.875	99.977	230	5
69.15	308,500	2	99.982	180	5.125
73.405	265,950	2.125	99.987	130	5.25
77.34	226,600	2.25	99.992	80	5.375
80.92	190,800	2.375	99.997	30	5.5
84.13	158,700	2.5	99.99767	23.35	5.625
86.97	130,300	2.625	99.99833	16.7	5.75
89.44	105,600	2.75	99.999	10.05	5.875
91.545	84,550	2.875	99.99966	3.4	6
93.32	66,800	3			

Solution

n Number of cups per box = 10
q P(scratches per cup) = 3/60 = 0.05
p P(no scratch per cup) = 1 − 0.05 = 0.95

Thus, the number of defective cups (say X) follow a Binomial distribution with parameter $n = 10$ and $p = 0.95$.

Therefore,

$$\text{Yield} = P(\text{none of the cups having a scratch})$$
$$= \Pr(X = 0)$$
$$= \binom{10}{10}(0.95)^{10}(0.05)^{10-10}$$
$$= 0.95^{10} = 0.5987$$

Alternatively, if we assume the number of scratches on a cup as Poisson with mean as

DPU = the average number of defects in a given box = 10 * 0.05 = 0.5
Hence,

$$\text{Yield} = P(\text{none of the cups having a scratch})$$
$$= e^{-\text{DPU}} = e^{-0.5}$$
$$= 0.606$$

which is similar to the one obtained through binomial assumption. The sigma level corresponds to this yield is 1.75. If we obtain the sigma level through DPMO conversion corresponds to a single opportunity (i.e., DPU = DPO = 0.5), the DPMO is 500,000, which corresponds to 1.5 sigma level. The difference in sigma level is due to the approximation of binomial distribution to Poisson distribution. If the sample size is large (that is, $n \to \infty$) and the value of p is small, then same sigma level can be achieved in both the cases.

Further, if we assume there are about 6 opportunities to select the boxes, then the sigma level can be obtained as follows:

$$\text{DPU} = 0.5$$
$$\text{DPO} = \frac{0.5}{6} = 0.0833$$

and

$$\text{DPMO} = \text{DPO} \times 10^6 = 83333.33$$

This 83333.33 DPMO is corresponding to 2.9 sigma level.

Example 12.4 A transaction process involves 1000 units, which was completed with 994 units by the end of the process. A rework was initiated to fine tune the process yielding the following results:

Defects:	0	1	2	3
No. of units Processed:	818	164	16	2

Compute the classical yield, first-time yield, and rolled throughput yield. Also obtain the sigma level if the number of opportunities for handling the transaction process is (i) 1 and (ii) 4.

Solution

$$\text{Classical yield} = \frac{\text{output}}{\text{input}}$$

$$= \frac{994}{1000} = 99.4\%$$

$$\text{First time yield} = \frac{\text{number with zero defects}}{\text{total number of defects}}$$

$$= \frac{818}{1000} = 81.8\%$$

$$\text{DPU} = \text{mean} = \frac{0(818) + 1(164) + 2(16) + 3(2)}{1000}$$

$$= \frac{202}{1000} = 0.202$$

and

$$\text{Rolled throughput yield} = e^{-\text{DPU}} = e^{-0.202} = 81.7\%$$

which is corresponding to 2.3 sigma level.

Now

(i) Sigma level, if the number of opportunities for handling the transaction process is 1:

$$\text{DPO} = \text{DPU} = 0.202$$

and

$$\text{DPMO} = \text{DPO} \times 10^6 = 202,000,$$

which is corresponding to 2.3 sigma level.

(ii) Sigma level, if the number of opportunities for handling the transaction process is 4:

$$\text{DPO} = \frac{\text{DPU}}{4} = \frac{0.202}{4} = 0.0505$$

and

$$\text{DPMO} = \text{DPO} \times 10^6 = 50,500,$$

which is corresponding to 3.14 sigma level.

In the absence of sigma conversion table, the sigma quality level for a given DPMO (or PPM) as can also be approximately determined using the equation (Schmidt and Launsby [9]):

$$\text{Sigma level} = 0.8408 + \sqrt{29.37 - 2.221 \times \ln(\text{DPMO})}$$

12.4 Cost of Poor Quality

An important performance dimension not captured by defects or sigma measures is the cost of impact of defects, often called cost of poor quality (CoPQ). The CoPQ quantifies the money lost as the result of defects and problems. It is essentially the cost of the defect or problem that has been identified in the process. It also refers to the overall cost of whatever defects are present in the process. Unlike sigma or DPMO, CoPQ speaks a language which almost everyone understands: money. CoPQ measures help one to select problems with clear bottom line benefits. For example: Suppose an existing process has 400 units of opportunities to carry out and each costs $10 per units. Suppose the initial process yield is calculated at 0.35. Then the current cost of manufacturing one unit is $1400 (=400 × 10 × 0.35). If after an improvement process, the yield improved to 0.67, then the cost of manufacturing one unit is $2680 (=400 × 10 × 0.35). The CoPQ associated with this process is $1280.

The CoPQ provides the justification for making investments to improve the process by quantifying the unseen cost of operating the process as it currently performs. This data is crucial, because many people tend to think things are optimal the way they are and resist investments to change things. With good data and process performance measures—such as DPMO, Sigma, Yield, and CoPQ—the organization can look for areas of greatest "gap" or concern. Also, these measures are a great starting point to track improvement down the road, allowing you to document gains and performance enhancements based on hard data versus anecdotes.

Thus, the main components of the CoPQ are as follows:

- Cost of non-conformities (or quality costs)

  - Internal failure costs—the cost of deficiencies discovered before delivery that is associated with the failure to meet explicit requirements or implicit needs of customers.
  - External failure costs—the cost associated with deficiencies that are found after the customer receives the product. Also included are lost opportunities for sales revenue.
  - Appraisal cost—costs that are incurred to determine the degree of conformance to quality requirements.
  - Prevention costs—costs that are incurred to keep failure and appraisal costs to a minimum.

- Cost of inefficient processes
- Cost of lost opportunities for sales revenue

In Defeo and Barnard [3], the CoPQ is appropriately renamed as the cost of poorly performing processes (CoP^3). This is to emphasize the fact that the CoPQ is not limited to quality, but is essentially the cost of waste associated with poor performance of processes. One may refer to Gryna et al. [4] and Campanella [2] for various other aspects of these failure components and categories.

12.5   Relevance for Managers

The success of every Six Sigma project is measured in terms of its sigma level. Like sigma estimation for measuring the variation, the sigma level estimate shows the potential capability of a process. The sigma value (or rating) of a process describes the quality level of that process, which compares process variability vis-a-vis the requirements. Generally, these specifications differ for normal and non-normal processes. This chapter emphasizes these aspects and offers sigma level computation for normal and non-normal processes. It also makes an assessment about the log-term and short-term capability of a process. Thus, knowing the sigma rating, we can establish the rejections that can be expected from the process. To get the sigma rating, process average, standard deviation, and specification limits are required.

An important performance dimension not captured by defects or sigma measures is the cost of impact of defects, often called CoPQ. The CoPQ quantifies the money lost as the result of defects and problems. It provides justification for making investments to improve the process by quantifying the unseen cost of operating the process as it currently performs. With this knowledge, the management is now able to decide whether their project has really contributed to the quality and improvement of their organization.

Exercises

12.1. Distinguish between

      (i)   Tolerance limits and specification limits.
     (ii)   Long-term sigma and short-term sigma.
    (iii)   Defects per unit and defects per opportunity.

12.2. The efficiency measurement (measured in percentage) of 50 workers during a
       workout is given below:

94	83	78	76	88	86	93	80	91	82
89	97	92	84	92	80	85	83	98	103
87	88	88	81	95	86	99	81	87	90
84	97	80	75	93	95	82	82	89	72
85	83	75	72	83	98	77	87	71	80

      (i)   Perform a normality check on the data.
     (ii)   Calculate the sigma level of the process, if lower and upper efficiency
            levels are given as 76 and 93, respectively.
    (iii)   Compute various process capability measures.

12.3. Discuss how yield is related to DPMO and sigma level of a process?

12.4. A manufacturing process is found to produce 4 % defective parts of the total
       products. In a sample of 30 items produced, obtain the DPU, DPMO, and
       yield. Assess the sigma level of the process.

12.5. The number of defective items produced by a manufacturing process is found
       to be according to Poisson distribution with mean 6. What is the probability
       that a randomly selected item is (i) non-defective and (ii) two defectives?
       Also obtain the yield of the process.

12.6. In question above, if the defective item produced by a manufacturing process
       is according to binomial distribution, then in a sample of 20 items, what is
       the probability that a randomly selected item is (i) non-defective and (ii) two
       defectives? Also obtain the corresponding yield. Also obtain the sigma level
       of the manufacturing process.

12.7. What is CoPQ? Discuss various components of CoPQ.

References

1. Breyfogle, III F.W.: Implementing Six Sigma-Smarter Solutions Using Statistical Methods.
   Wiley, Inc, London
2. Campanella, J.: Principles of Quality Costs, 3rd edn. ASQ, Milwaukee (1999)
3. Defeo, J.A., Barnard, W.: Juran's Six Sigma Breakthrough and Beyond, McGraw Hill, New
   York (2004)

4. Gryna, F.M., Chua, R.C.H., and Defeo, J. (2007). Juran's Quality Planning & Analysis for Enterprise Quality, Tata McGraw-Hill, New Delhi
5. Hahn, G.J., Doganaksoy, N., Hoerl, R.: The Evolution of six sigma. Qual. Eng. **12**(3), 317–326 (2000)
6. Muir, A.: Lean Six Sigma way. McGraw-Hill, New York (2006)
7. Oakland, J.S.: Statistical Process Control. Elsevier, New Delhi (2005)
8. Pande, P.S., Newuman, R.P., Cavanagh, R.R.: The Six Sigma Way. Tata McGraw Hill, New Delhi (2003)
9. Schmidt, S.R., Launsby, R.G.: Understanding Industrial Designed Experiments. Air Academy Press, Colorado Springs, CO (1997)

Chapter 13
Continuous Improvement

Among all the frameworks of quality improvement, the oldest management system is the total quality management (TQM), which is a strategy for implementing and managing quality improvement activities across organizations. TQM began in the early 1980s, influenced by the philosophies of W. Edward Deming, Joseph Juran, and others. It evolved into a broader spectrum of concepts and ideas, involving participative organizations and work culture, customer focus, supplier quality improvement, and many other activities to focus all elements of the organization around the quality improvement goal. Typically, organizations that have implemented TQM employ quality councils or high-level teams that deal with strategic quality initiatives, workforce-level teams that focus on routing production or business activities, and cross-functional teams that address specific quality improvement issues.

Many organizations saw the mission of TQM as one of the training for reducing variability of a process. Therefore, TQM achieved only moderate success for a variety of reasons like the following [1]:

- Lack of top-down, high-level management commitment and involvement
- Inadequate use of statistical methods and insufficient recognition of variability reduction as a prime objective
- General, as opposed to specific, business-results-oriented objectives
- Too much emphasis on widespread training as opposed to focused technical education.

Another reason for the erratic success of TQM is that many managers and executives regarded it as just another "gimmick" to improve quality. During 1950s and 1960s, programs such as "zero defects" and "value engineering" were widely deployed, but they had little impact on quality and productivity improvement.

Process improvements often focus on incremental changes or solutions to eliminate or reduce defects, costs, or cycle time. They leave basic design and assumptions of a process intact. This is achieved through specific opportunities, called problems, which are being identified or recognized. A focus on improvement opportunities should lead to the creation of teams whose membership is determined

© Springer India 2015
K. Muralidharan, *Six Sigma for Organizational Excellence*,
DOI 10.1007/978-81-322-2325-2_13

by their work on and detailed knowledge of the process, and their ability to take improvement action. The team must then be provided with good leadership and the right tools to tackle the job.

By using reliable method and creating a favorable environment for team-based problem solving, this can be achieved. Some of the graphical tools like histogram, Pareto analysis, scatter diagram, and box plots discussed in previous chapters are very useful in problem identification and solving. Even the use of SPC and control charts constitutes the ideal way of monitoring current process performance, predicting future performance and suggesting the need for corrective action. Below, we discuss some specific quality improvement methods in detail, from their philosophical and technical contribution point of view.

13.1 Deming's Quality Philosophy

The most important and widely used theory of quality improvement is of W. Edwards Deming's (October 14, 1900–December 20, 1993) quality philosophy introduced in early 1920s. Most of his theories are based on the logic that most product defects were a result of management shortcomings rather than careless workers and that inspection after the fact was inferior to designing processes that would produce better quality. He argued that enlisting the efforts of willing workers to do things properly the first time and giving them the right tools were the real secrets of improving quality and not the team of inspectors.

Although the core of his method to improve quality was the use of statistics to detect flaws in production processes, he developed a broader management philosophy that emphasized problem solving based on cooperation. He exhorted managers to "drive out fear" so that workers would feel free to make improvements in the workplace. This philosophy is carried out in four different stages, namely plan–do–check–act, which describes the basic logic of data-based process management [3].

- *Plan* Review current performance for issues and gaps. Gather data on key problems. Identify and target root causes of problems. Devise possible solutions, and plan a test implementation of the highest potential solution
- *Do* Pilot the planned solution
- *Check* (or study) Measure the results of the test to see whether the intended results are being achieved. If problems arise, look into the barriers that are obstructing your improvement efforts
- *Act* Based on the test solution and evaluation, refine and expand the solution to make it permanent, and incorporate the new approach wherever applicable

Companies that sought to improve their quality by adopting Deming's methods often found they had to change their entire culture. To convince workers that managers really did want to enlist them as partners, many companies eliminated cherished management perquisites like special parking spaces and executive dining rooms because shop-floor workers found them offensive. Deming is most famous for articulating the ideas captured in his fourteen points and seven deadly diseases. The points were first presented in his book *Out of the Crisis*. Although Deming does not use the term in his book, it is credited with launching the TQM movement. He revised his fourteen points in 1990 from the original version published in 1986. The revised fourteen versions of his quality philosophy are detailed below:

1. Create and publish to all employees a statement of the aims and purposes of the company or other organization. The management must demonstrate constantly their commitment to this statement.
2. Learn the new philosophy, top management, and everybody.
3. Understand the purpose of inspection, for improvement of processes and reduction of cost.
4. End the practice of awarding business on the basis of price tag alone.
5. Improve constantly and forever the system of production and service.
6. Institute training.
7. Teach and institute leadership.
8. Drive out fear. Create trust. Create a climate for innovation.
9. Optimize toward the aims and purposes of the company the efforts of teams, groups, and staff areas.
10. Eliminate exhortations for the work force.
11. a. Eliminate numerical quotas for production. Instead, learn and institute methods for improvement.
    b. Eliminate management by objectives (MBO). Instead, learn the capabilities of the system and how to improve them.
12. Remove barriers that rob the hourly worker of his right to be proud of workmanship.
13. Encourage education and self-improvement for everyone.
14. Take action in the company to work, and to accomplish the transformation. The transformation is everybody's job.

According to Deming, the "Seven Deadly Diseases" include the following:

1. Lack of constancy of purpose
2. Emphasis on short-term profits
3. Evaluation by performance, merit rating, or annual review of performance
4. Mobility of management
5. Running a company on visible figures alone
6. Excessive medical costs
7. Excessive costs of warranty, fueled by lawyers who work for contingency fees.

"A Lesser Category of Obstacles" includes the following:

1. Neglecting long-range planning
2. Relying on technology to solve problems
3. Seeking examples to follow rather than developing solutions
4. Excuses, such as "our problems are different"
5. Obsolescence in school that management skill can be taught in classes

6. Reliance on quality control departments rather than management, supervisors, managers of purchasing, and production workers
7. Placing blame on workforces who are only responsible for 15 % of mistakes where the system designed by management is responsible for 85 % of the unintended consequences
8. Relying on quality inspection rather than improving product quality.

Deming observes that companies employing annual ratings of their workers from their superiors are tempted to work for themselves, not for the company. Rewards and punishments do not motivate people to do their best; in fact, they do exactly the opposite, and pride in workmanship is lost. Managers, in effect, are managers of defects, and the organization becomes the loser. The job of management is inseparable from the welfare of the company. If individuals are to work jointly toward a common end, they require time to build trust and synergy. Teamwork— along with team effectiveness—develops over time. According to Deming, for companies that provide health insurance, medical costs represent the largest single expenditure. For example, medical costs have a multiplier effect for any company because they are embedded in the costs of the products and services of every other firm with which it does business.

13.2 Crosby's Quality Philosophy

In the quality management (QM) circles, Philip Crosby (June 18, 1926–August 18, 2001) is known as a successful quality deployment champion. Crosby promoted the concept of "zero defects" and "cost of quality (COQ)." Crosby emphasizes on the delivery of defect-free products and services that conform to specifications and defined quality as conformance to requirements. Crosby's response to the quality crisis was the principle of "*doing it right the first time.*" He was recognized by corporations around the globe (especially in the USA and Europe) as a QM guru, a business philosopher, and an innovator. He adds that quality is measured by the COQ given by a simple formula: COQ = (Prevention + Appraisal + Failure) cost, which is the expense of non-conformation cost of doing things wrong. Crosby's quality philosophy can be summarized as follows [2]:

1. Make it clear that management is committed to quality.
2. Form quality improvement teams with senior representatives from each department.
3. Measure processes to determine where current and potential quality problems lie.
4. Evaluate the COQ and explain its use as a management tool.
5. Raise the quality awareness and personal concern of all employees.
6. Take actions to correct problems identified through previous steps.
7. Establish progress monitoring for the improvement process.

8. Train supervisors to actively carry out their part of the quality improvement program.
9. Hold a Zero Defects Day to reaffirm management commitment.
10. Encourage individuals to establish improvement goals for themselves and for their group.
11. Encourage employees to tell management about obstacles toward improving quality.
12. Recognize and appreciate those who participate.
13. Establish Quality Councils to communicate on a regular basis.
14. Do it all over again to emphasize that the quality improvement process never ends.

Crosby believes that QM is all about the systematic way of guaranteeing that organized activities happen the way they are planned. Hence, his medicine for management to prevent poor quality is the following:

- *Integrity*: Treat quality seriously throughout the whole business organization from top to bottom so that the company's future will be judged on its quality performance.
- *Systems*: Appropriate measures and systems should be put in place for quality costs, education, quality, performance, review, improvement, and customer satisfaction.
- *Communication*: The communication systems are of paramount importance to communicate requirements and specifications and improvement opportunities around the organization. Customers and operators know what needs to be put in place to improve and listening to them will give you the edge.
- *Operations*: Work with and develop suppliers. Processes should be capable, and improvement culture should be the norm.
- *Policies*: Must be clear and consistent throughout the business.

As an author, Crosby is famous in the management field. Some of his books are best sellers even today. Three of them are "*Quality without tears: the art of Hassle free management*," "*Quality is free: The Art of Making Quality Certain*," and "*Managing the Quality Revolution*." The other details of his contributions (accessed through Wikipedia online library) are

1. *Cutting the COQ*. Boston, Industrial Education Institute OCLC 616899 (1967).
2. *The strategy of situation management*. Boston, Industrial Education Institute OCLC 13761 (1969).
3. *Quality is Free*. New York: McGraw-Hill. ISBN 0-07-014512-1 (1979).
4. *The Art of Getting Your Own Sweet Way*. McGraw-Hill. ISBN 0-07-014527-X (1981).
5. *Quality Without Tears*. New York: McGraw-Hill. ISBN 0-07-014511-3 (1984).
6. *Running things*. New York: McGraw-Hill. ISBN 0-07-014513-X (1986).
7. *The Eternally Successful Organization*. New York: McGraw-Hill. ISBN 0-07-014533-4 (1988).
8. *Let's talk quality*. New York: McGraw-Hill. ISBN 0-07-014565-2 (1989).

9. *Leading, the art of becoming an executive.* New York: McGraw-Hill. ISBN 0-07-014567-9 (1990).

10. *Completeness: Quality for the 21st Century.* Plume. ISBN 0-452-27024-3. (1994).

11. *Philip Crosby's Reflections on Quality.* McGraw-Hill. ISBN 0-07-014525-3 (1995).

12. *Quality is still free: Making Quality Certain in Uncertain Times.* McGraw-Hill. ISBN 0-07-014532-6 (1996).

13. *The Absolutes of Leadership (Warren Bennis Executive Briefing).* ISBN 0-7879-0942-4 (1997).

14. *Quality and Me: Lessons from an Evolving Life.* ISBN 0-7879-4702-4 (1999).

13.3   Juran's Quality Philosophy

In the field of quality, Dr. Joseph M. Juran (December 24, 1904–February 28, 2008) is known as "the architect of quality." He propounded the universal approach to "managing for quality" through his "quality trilogy" philosophy. He is one of the key persons in Japan's quality transformation after World War II. The underlying concept of the quality trilogy is that managing for quality consists of three basic quality-oriented processes: quality planning, quality control, and quality improvement [14].

Quality planning

- Identify the customers, both external and internal
- Determine the needs of those customers
- Establish Quality Goals that meet the needs of customer and suppliers alike
- Develop a product that can respond to those needs
- Optimize the product features so as to meet our needs and customer needs
- Prove process capability.

Quality control

- Choose control subjects
- Choose units of measurement
- Establish measurement
- Monitor actual performance
- Measure actual performance
- Interpret the differences—actual versus standard
- Take corrective action for differences.

Quality improvement

- Identify specific projects for improvement
- Improve the process through projects and sustain gains
- Organize for diagnosis for discovery of causes

- Diagnose to find causes
- Prove that remedies are effective under operating conditions
- Provide control to hold the gains.

Apart from this, Juran also drew parallels for quality to financial processes. According to him, the financial process involves budgeting, cost control (or expense control), and cost reduction (or profit improvement). The best way of financial improvement is setting proper business goals and the cost reduction. This can be done through mergers and acquisitions and is the latest trend happening in business these days. Dr. Joseph Juran, known as Father of quality, passed away on February 28, 2008, at the age of 103.

13.4 Feigenbaum's Quality Philosophy

Dr. Armad V. Feigenbaum originated the concept of total quality control (TQC) in his book TQC (1991), first published in 1951. The book has been translated into many languages, including Japanese, French, and Spanish. Feigenbaum is an ASQ honorary member and served as ASQ president for two consecutive terms. According to him, the three steps to quality are:

1. Quality leadership
2. Modern quality technology
3. Organizational commitment.

Dr. Feigenbaum emphasizes the concept of TQC throughout all functions of an organization. TQC means both planning and control. He urges creating a quality system to provide technical and managerial procedures that ensure customer satisfaction and an economical COQ [5].

Dr. Feigenbaum is more concerned with organizational structure and a system's approach to improve quality than he is with statistical methods. He encouraged management commitment over compartmentalization of departments and processes for quality improvement.

13.5 Ishikawa Quality Philosophy

Ishikawa [13] developed the cause-and-effect diagram. He worked with Deming through the Union of Japanese Scientists and Engineers (JUSE). The quality philosophy proposed by him includes the following:

- Quality first, not short-term profit
- Consumer orientation, not producer orientation
- The next process is your customer, breaking down the barrier of sectionalism

- Using facts and data to make presentations, utilization of statistical methods
- Respect for humanity as a management philosophy, full participatory management
- Cross-function management.

13.6 Taguchi Quality Philosophy

The man who brought quality engineering (QE) and robust design in the modern era is the famous statistician Dr. Genichi Taguchi (January 1, 1924–June 2, 2012) in the late 1940s. Essentially, QE deals with activities performed for the purpose of reducing variability in product/process function. The methodologies and philosophies developed by Taguchi for optimizing product and process function are based on engineering concerns, not purely scientific or statistical ones. They facilitate the engineer in his/her quest to find the answer to the question of "how" to maintain function not necessarily "why" function varies. The objective is to assist the engineer in the synthesis of phenomenon rather than its analysis in order that robust product/process functions is achieved. These methodologies enable the engineer to efficiently and rapidly obtain the technological capability required to keep an organization profitable and competitive in today's market.

According to Taguchi, there are two forms of QE: off-line and online QE. Both are used for reducing the variability in product/process function.

Off-line QE: Here, activities are performed upstream in the life of a product or process design. The need is to optimize the functionality of an engineered system through design strategies. This concept forces the engineer to focus on "how" to divert energy into ideal functioning, rather than concentration on specific customer-related problems. This helps the engineer in conducting more efficient industrial research to enhance technological capability.

Online QE: Here, activities are performed downstream in the manufacturing and field service environments. Their purpose is to control the behavior of product or process through control mechanisms, inspection, or preventive maintenance. The objective is to balance the relationship between monitoring quality and the cost associated with this monitoring.

Dr. Taguchi relates the situation where the design of a product/process produces consistent, high-level performance, despite being subjected to a wide range of changing customer and manufacturing conditions, to a state called *robustness*. The lack of robustness is due to the presence of noise which also means variability. Factors that cause variation are referred to as noise factors. They are factors which are uncontrollable either from a practical or cost point of view. For example, in an injection molding process, ambient temperature wreaks havoc on the process output. In this case, ambient temperature would be considered a noise factor causing variability in output.

Noise and noise factors can be classified as outer-, inner-, and between-type product noises. Outer noises are conditions that affect performance from an external standpoint. Inner noises are conditions causing the variation in performance that are internal to the product or process design. And, finally, between product noises relate to the variability in performance seen in products made exactly the same in the manufacturing environment. A robust parameter design is the best way of tracking the control factors in the presence of noise factors. It also facilitates the identification of different factors and factor levels to achieve robustness in a product/ process. See also Sect. 10.7 for the detailed analysis of parameter designs.

The following information about Taguchi is accessed through Wikipedia online library.

Career: In 1950, he joined Electrical Communications Laboratory (ECL) of the Nippon Telegraph and Telephone Corporation at a time when statistical quality control was beginning to become popular in Japan, under the influence of W. Edwards Deming and the Japanese Union of Scientists and Engineers (JUSE). ECL was engaged in rivalry with Bell Labs to develop cross-bar and telephone switching systems, and Taguchi spent twelve years there in developing methods for enhancing quality and reliability. Even at this point, he was being consulted widely in Japanese industry, with Toyota being an early adopter of his ideas.

During the 1950s, he collaborated widely with Indian scientists and statisticians, and in 1954–1955 was a visiting professor at the Indian Statistical Institute, where he worked with C.R. Rao, Ronald Fisher, and Walter A. Shewhart. While working at the SQC Unit of ISI, he was introduced to the orthogonal arrays invented by C.R. Rao—a topic which was to be instrumental in enabling him to develop the foundation blocks of what is now known as Taguchi methods.

Contributions: Taguchi has made very influential contributions to industrial statistics. Key elements of his quality philosophy include the following:

1. Taguchi loss function, which measures financial loss to society resulting from poor quality;
2. The philosophy of *off-line quality control*, designing products and processes so that they are insensitive ("robust") to parameters outside the design engineer's control; and
3. Innovations in the statistical design of experiments, notably the use of an outer array for factors that are uncontrollable in real life, but are systematically varied in the experiment.

Honors: Taguchi has received number of the honors from various organizations and nations. Some of them are

- Indigo Ribbon from the Emperor of Japan
- Willard F. Rockwell Medal of the International Technology Institute
- Honorary member of the Japanese Society of Quality Control and of the American Society for Quality
- Honored as a Quality Guru by the British Department of Trade and Industry (1990)
- Shewhart Medal of the American Society for Quality (1995).

13.7 Management Systems Standards

As described earlier, the Deming's PDCA practice is applied in several management system standards in order to achieve continual enhancement. These elements have their foundation in the goal-setting theory [15] and MBO [4]. For modern business, this principle has become an effective tool for execution and implementation of corporate social responsibility (CSR). CSR [7] is used as a framework for measuring an organization's performance against economic, social, and environmental parameters. The standards most relevant for CSR are ISO 14001, ISO 50001, OHSAS 18001, ISO 9001, SA 8000, and ISO 26000.

ISO is the International Organization for Standardization based at Geneva, Switzerland, and was formed in 1947. The objectives of ISO covers the promotion of standards in the world with a view to facilitate international exchange of goods and services and to develop cooperation in the sphere of intellectual, scientific, technological, and economic activity including transfer of technology to developing countries. It also aims at harmonization of standards at the international level with a view to minimize trade and technical barriers, that is, to eliminate country-to-country differences, to eliminate terminology confusion, and to increase quality awareness between organizations.

ISO 9000 encourages a documented quality system. Hence, a quality system manual is important for getting an ISO certification. Management must have a written policy statement of their commitment to quality, and it must be communicated to and understood by all employees. Management must clearly define quality-related organizational responsibilities and interrelationships. A management representative must be assigned to oversee the implementation and continuous improvement of the quality system. At the same time, senior management must continually review the system for any kind of omissions and commissions.

An ISO 9000 company must identify all processes that directly affect the quality of the product or service and ensure that these processes are carried out under controlled conditions, including formal approval of process design and equipment, documented work instructions, documented quality criteria, development of quality plans describing how the process is to be monitored, and creating a suitable working environment (see Muralidharan and Syamsundar [18] for details).

The ISO 14001 standard is part of the ISO 14000 series, which has been developed to provide organizations with a structure for handling environmental impact. ISO 14000:2004 sets out the criteria for an environment management system (EMS) and can be certified to. It does not state requirements for environmental recital, but maps out a framework that a company or organization can follow to set up an effective environmental management system. It can be used by any organization irrespective of its activity or sector. Using ISO 14000:2004 can provide assurance to company management and employees as well as external stakeholders that environmental influence is being measured and improved.

ISO 50001 supports organizations in all sectors to use energy resource fully, through the development of an energy management system (EnMS) and can be certified. An EnMS is a set of consistent or interrelating elements to establish an energy policy and energy objectives, and processes and procedures to achieve those objectives. ISO 50001 is based on the PDCA cycle that makes it easy to assimilate the EnMS into other management systems. The EnMS requirements are divided under the headings: energy policy, planning, implementation and operation, checking, and management appraisal. The issues under each heading are the same as for ISO 14001 with the exclusion that the energy policy is a statement of the overall purposes and direction related to the energy performance and that the system is based on the energy aspects of environmental ones.

OHSAS 18001 (officially "BS OHSAS 18001:2007") is an internationally applied British Standard for occupational health and safety management systems. OHSAS endorses a safe and healthy working environment by proving a framework that allows an organization to reliably identify and control its health and safety risks and improve overall show (BSI 2013). BS OHSAS 18001:2007 is the most common standard for OHSAS in the world, and it has a structure that corresponds well with other ISO standards. The requirements are divided under the headings: policy and planning; hazard identification, risk evaluation and risk control; legal and other requirements; objectives, targets and management programs; implementation and operation; checking and corrective action; and management review.

The intent of SA 8000 is to provide an average based on international human rights norms and national labor laws that will protect and empower the personnel and can be certified to. The necessities regarding the issues are as follows: child labor; forced and compulsory labor; health and safety at the workplace; freedom of association and right to collective bargaining; discrimination; disciplinary practices; working hours; compliance with laws and industry standards; and remuneration (SAI 2012). There are many similarities to that already described for management systems and standards. The management system is divided into the following: policy; management representation; SA 8000 worker representation; management review; planning and implementation control of suppliers/subcontractors and sub-suppliers; addressing concerns and taking corrective action; outside communication and stakeholder management; access for verification; and records (SA 8000:2008).

ISO 26000:2010 provides guidance on how businesses and organizations can operate in a socially responsible way. This means acting in an ethical and transparent way that contributes to the health and welfare of society. ISO 26000 provides guidance rather than requirements, so it cannot be certified, unlike the other standards mentioned above. Instead, it helps clarify what social responsibility is, helps businesses and organizations translate principles into effective actions, and shares best practices relating to social responsibility globally. It is aimed at all types of organizations regardless of their activity, size, or location (ISO 2013). The standard is based on seven fundamental principles: accountability, transparency, ethical behavior, respect for stakeholder interests, respect for the rule of law, respect for international norms and behavior, and respect for human rights (ISO 2010).

Further, ISO 26000 is a comprehensive standard covering a large number of subjects and issues for social responsibility. It also embraces the target areas in the previous described system standards. Therefore, ISO 26000 would fit well as a framework for exploring CSR practice. As globalization rushes and large corporations serve as global providers, these corporations have increasingly recognized the welfares of providing CSR programs in their various locations [20].

13.8 Six Sigma Quality Philosophy

In Chap. 1, we have a detailed discussion on Six Sigma philosophy and its relevance to modern-day quality improvement. The implementation of Six Sigma in business organizations has boosted the role of statistics and statisticians in the process. Some of these topics have been discussed by Hahn et al. [9], Snee [19], and Hahn and Doganaksoy [8]. People advocate the use of Six Sigma because of the following reasons:

- *Philosophy*: The philosophical perspective views all work as processes that can be defined, measured, analyzed, improved, and controlled (DMAIC).
- *Set of tools*: Six Sigma as a set of tools includes all the qualitative and quantitative techniques used by the Six Sigma expert to drive process improvement.
- *Methodology*: The methodological view of Six Sigma recognizes the underlying and rigorous approach known as DMAIC. It defines the steps a Six Sigma practitioner is expected to follow, starting with identifying the problem and ending with implementing long-lasting solutions.
- *Metrics*: In simple terms, Six Sigma quality performance means 3.4 defects per million opportunities with 1.5 sigma shift in the process mean.

Since its origins, there have been three generations of Six Sigma implementations. Generation I Six Sigma focused on defect reduction and basic variability reduction, primarily in manufacturing. Motorola is a classic example of Generation I Six Sigma. In Generation II Six Sigma, the emphasis was on variability reduction and defect elimination with focused efforts to projects and activities that improved business performance through improved product design and cost reduction. General Electric is often classified as the leader of Generation II Six Sigma. In Generation III, Six Sigma has the additional focus of creating value throughout the organization and for its stakeholders. Creating value can take many forms, such as increasing stock prices and dividends, job retention or expansion, expanding markets for company products/services, developing new products/services that reach new and broader markets, and increasing the levels of customer satisfaction throughout the range of products and services offered. Caterpillar and Bank of America are good examples of Generation III Six Sigma companies, because their implementations are focused on value creation for all stakeholders in the broad sense. See Hahn et al. [10] and Montgomery and Woodall [16] for a good discussion of evolutions of Six Sigma.

The key elements in a DMAIC project are team discipline, structured use of metrics and tools, and execution of a well-designed project plan that has clear goals and objectives. If a process cannot be improved as it is currently designed, then DMADV process is used to fundamentally redesign a process (see Chap. 1 for details).

Generally, Six Sigma has been far more successful than TQM was. There are several reasons for this, principal among them being the strong focus on projects that positively impact business financial performance. When quality improvement projects result in real savings, expanded sales opportunities, or documented improvements in customer satisfaction, upper management pays attention. Then, business leaders are more likely to be fully involved, to commit the resources needed to train personnel, and to make Six Sigma positions full time, using these positions as stepping stones to higher positions of responsibility in the organization. The level of technical training in Six Sigma is generally deeper and more extensive than in the typical TQM programs [11, 12].

13.9 Lean Six Sigma

Lean Six Sigma (LSS) modifies the DMAIC approach by emphasizing speed. Lean focuses on streamlining a process by identifying and removing non-value-added steps. A leaned production process eliminates waste. Target metrics include zero wait time, zero inventory, scheduling using customer pull, cutting batch sizes to improve flow, line balancing, and reducing overall process time. LSS's goal is to produce quality products that meet customer requirements as efficiently and effectively as possible.

The operating philosophy of Lean is to eliminate waste through continuous improvement (see Womack and Jones [21]). By waste, we mean unnecessarily long cycle times, or waiting times between value-added work activities. Waste can also include rework, scrap, and excess inventory. Rework and scrap are often the result of excess variability, so there is an obvious connection between Six Sigma and Lean. As we know, a business process is a series of interconnected sub-processes. The tools and focus of Six Sigma is to fix processes, whereas Lean concentrates on the interconnection between the processes. The central theme of Six Sigma is defect reduction. The root causes of the defects are examined, and improvement efforts are focused on those causes. In LSS, the emphasis is indirect in that defects can cause delays to occur. George [6] provided a good introduction to how Lean and Six Sigma work together. According to Muir [17], the two philosophies can be summed up as follows: "Reduce the time it takes to deliver a defect-free product or service to the customer." It is not the question of whether the customer wants it right or quickly, but both.

The value-added activities in a Lean manufacturing can be achieved through

- Defining value from the client's perspective
- Identify the value stream
- Only make what the client pulls
- Keep the flow moving continuously and
- Always improve the process.

The important metrics associated with LSS are

- Process cycle efficiency—the capacity of the process to complete a particular step in a given time.
- Process cycle time—the average time for a particular step to complete one item.
- Work in progress—the material that has been input to the process but that has not reached the output stage. This includes the material being processed by the various steps and the material waiting to be processes by one or more steps.
- Throughput rate—the number of items or amount of material output from the process in a given period of time.

As discussed earlier, the Six Sigma process improvement goes with the philosophy of DMAIC techniques. In LSS, it develops with the identification of the problems and ends with retaining the benefits of the program. Thus, the steps followed in LSS are R-DMAIC-S, where R stands for recognize and S stands for sustain. It is important to identify the significant gaps in a business problem by articulating the business strategy. This will decide what the management wants to deliver. Once a Six Sigma project completes through the DMAIC procedure, the management will be able to identify the tangible benefits coming out of the project and its immediate impact can be assessed for retaining similar such projects.

In order to achieve the required target, a process flow is necessary to establish. A perfect process has continuous flow as products, services, and knowledge are transformed continuously without delay from step to step. A flow is created by eliminating queues and stops and improving process flexibility and reliability. The other aspects of establishing the continuous flow is by identifying the value-added and non-valued activities in a process. Any activities that add no value to the client are by definition "waste." The mapping of all activities and steps according to its time of occurrence required to bring a product, service or capability to the client is called *Value Stream Mapping* (VSM).

Thus, implementing Lean will address waste and its root causes. Waste points us to problems within the system. Lean also focuses on efficient use of equipment and people and minimizes issues by standardizing work. A Lean cost model says that decreased cost always leads to increased profit. Hence, the tools for both philosophies differ depending on the business issue to be resolved. The characteristics specific to LSS are

- Speed and flexibility
- Involves all employees
- Positive results in short time frame
- Focused on smaller-scale projects
- Less scientific: Often trial and error.

When all the three methods like Six Sigma, LSS, and DFSS are integrated into one, the Deming's system of profound knowledge can be achieved. Thus, the quality philosophies discussed so far have their own merits and relevance in quality improvement and process/product/service developments. Even today, the management is finding these philosophies relevant and essential for all types of quality improvements and services to the customers.

13.10 Relevance for Managers

Quality improvement is not instant, but continuous. One should experiment all types of improvement programs to reach a particular level. This chapter summarizes some of the widely used quality philosophies practiced in industries and organizations. They include Deming's fourteen good principles of management, Crosby's zero defect concepts, Juran's quality trilogy, Feigenbaum's TQC, Taguchi's quality engineering, Ishikawa's cause-and-effect diagram, and management system standards. They are tested and tried for years, some of them being relevant even today.

Most of the above quality philosophies work in isolation and do not promote technicalities in the organizational activities. The only tool, which advocates extensive use of statistical methods, is the Six Sigma philosophy. It integrates all the above quality philosophies and promotes a structured problem-solving approach to reduce variation in a process, in the process of improving quality across organizations. We also have another improvement program called LSS, which modifies the DMAIC approach by emphasizing speed. This is more popular than Six Sigma these days, because of its less dependency on technical expertise. Since the operating philosophy of Lean is to eliminate waste through continuous improvement, one can implement LSS with all ease. However, a dedicated implementation of SS project always yields an unmatchable ROI as compared to a LSS project.

Exercises

13.1 What is TQM? Discuss the relevance of TQM in the new age management?

13.2 Discuss the quality philosophy of Edward Deming, citing the best principles of management.

13.3 Discuss the Crosby's quality philosophy in reference to quality management.

13.4 Discuss the importance of "Quality trilogy" of Juran, in reference to improving a manufacturing process.

13.5 What is the relevance of total quality control (TQC) as proposed by Feigenbaum. How does it compare with TQM?

13.6 Differentiate the off-line quality engineering and on-line quality engineering as proposed by G. Taguchi. How are they connected with the robust design concept?

13.7	Discuss various ISO standards and their quality versions.
13.8	Discuss how Six Sigma is connected with TQM philosophy.
13.9	Carry out a comparative study on Lean and Six Sigma philosophies.
13.10	What are the characteristics of LSS?

References

1. Ahmad, S., William, J.D., Charles, H.A.: Process performance in product development: measures and impacts. Eur. J. Innov. Manage. **7**(3), 205–217 (2004)
2. Crosby, P.: Quality is Free. McGraw-hill, New York (1979)
3. Deming, W.E.: Out of the Crisis. MIT Center for Advanced Engineering Study, Cambridge (1986)
4. Drucker, P.: The man who changed the world. *Bus. Rev. Wkly.* p. 49 (1997)
5. Feigenbaum, A.: Total quality control, 3rd ed. revised, McGraw-Hill, New York (1991)
6. George, M.L.: Lean Six Sigma. McGraw-Hill, New York (2002)
7. Greening, D.W., Turban, T.B.: Corporate social performance as a competitive advantage in attracting a quality workforce. Bus. Soc. **39**, 254–280 (2000)
8. Hahn, G.J., Doganaksoy, N.: The Role of Statistics Business and Industry. Hoboken, NJ, Wiley (2008)
9. Hahn, G.J., Hill, W.J., Hoerl, R.W., Zinkgraf, S.A.: The impact of Six Sigma improvement: a glimpse into the future of statistics. Am. Stat. **53**(3), 208–215 (1999)
10. Hahn, G.J., Doganaksoy, N., Hoerl, R.W.: The evolution of Six Sigma. Qual. Eng. **12**(3), 28–34 (2000)
11. Harry, M., Schroeder, R.: Six Sigma: The Breakthrough Management Strategy Revolutionizing the World's Top Corporations. Doubleday, New York (2000)
12. Harter, D.E., Krishnan, M.S., Slaughter, S.A.: Effects of process maturity on quality, cycle time, and effort in software product development. Manage. Sci. **46**(4), 451–466 (2000)
13. Ishikawa, K.: What is Total Quality Control? Prentice-Hall, Upper Saddle River NJ (1985)
14. Juran, J.M., Gryna F.M., Jr.: Quality Planning and Analysis, 2nd edn. McGraw-Hill, New York (1980)
15. Locke, E., Gary, L.: New directions in goal-setting theory. Assoc. Psychol. Sci. **15**(5), 265–268 (2006)
16. Montgomery, D.C., Woodall, W.H.: An overview of Six Sigma. Int. Stat. Rev. **76**(3), 329–346 (2008)
17. Muir, A.: Lean Six Sigma Statistics. McGraw-Hill, New York (2006)
18. Muralidharan, K., Syamsunder, A.: Statistical Methods for Quality, Reliability and Maintainability. PHI India Ltd., New Delhi (2012)
19. Snee, R.: Why should statisticians pay attention to Six Sigma? An examination for their role in the Six Sigma methodology. Qual. Prog. **32**(9), 100–103 (1999)
20. Williams, H. E.: CSR activities through the PDCA cycle approach for emerging companies. Proceedings of Corporate Governance: Contemporary issues and Challenges in Indian economic environment, pp. 216–229. Gujrat Technological University (2014)
21. Womack, J.P., Jones, D.T.: Lean Thinking: Banish Waste and Create Wealth in Your Organization. Simon and Schuster, New York (1996)

Chapter 14
Marketing Six Sigma

According to Pande et al. [16], a Six Sigma organization is *"An organization that is actively working to build the themes and practices of Six Sigma into its daily management activities and is showing significant improvements in process performance and customer satisfaction."* The objective of Six Sigma is to discover the non-value-added activities hiding in working system. Even though there are a lot of effective tools used to improve and to reduce the loss in process, but Six Sigma solutions is the most popular tool and it has been judged as the world-class strategy for quality improvement [10]. The objectives of quality improvement project using Six Sigma could be the financial benefits increment, the operational performance increment, and to better the company's image [1, 5, 22, 23].

Companies achieving Six Sigma should see the results in their bottom lines and with their customers—and should not oversell their efforts. Performance against plan is how a business typically defines success. Business gauge success by a multitude of metrics: revenue, income, profit, customer satisfaction, market share, return on investment, return on assets, and so on. Bottom line, planned success means reaching and sustaining goals over time, usually growth goals (see also Muralidharan and Neha [15] for more details). According to Creveling et al. [3], companies that effectively implement Six Sigma tools, methods, and best practices find the following benefits:

- *Systematic innovation*: Generate and define ideas linked with market opportunities in a structured way
- *Manage risks better*: Identify critical issues early in the commercialization process such that plans can be developed to mitigate or eliminate the risk of going forward
- *Higher return yield from a project portfolio*: Avoid over boarding resources with too many low-risk, small gain projects through a discriminating selection process

The objective of any Six Sigma project leads to how to evaluate the accomplishment of its improvement [4, 6, 11, 17–20]. The economic index would be determined if the project aims to increase the financial benefit. To tailor Six Sigma to marketing, one should start with an overview of how it works. It is found that

© Springer India 2015
K. Muralidharan, *Six Sigma for Organizational Excellence*,
DOI 10.1007/978-81-322-2325-2_14

marketing professionals rarely view their own work as process oriented. According to American Marketing Association (AMA), marketing is "a set of processes for creating, communicating and delivering value to customers and … managing customer relationships in ways that benefit the organization and stakeholders." The American Heritage Dictionary describes the process as "a series of actions, changes, or functions bringing about a result," and according to Moorman and Rust [13], the marketing function should play a key role in managing several important connections between the customer and critical firm elements, including connecting the customer to (i) the product, (ii) service delivery, and (iii) financial accountability.

14.1 What is Six Sigma Marketing?

According to Reidenbach [21], a Six Sigma Marketing (SSM) is a fact-based data-driven disciplined approach to growing market share by providing targeted product/ markets with superior value. It is organized around the following elements:

- Customer value is the driving strategic metric. It replaces the emphasis on customer satisfaction embraced by both SS and marketing and provides a much stronger link to market share gains and revenue increases
- SSM has a unique set of powerful tools designed around the idea of customer value to concentrate the organization's efforts on both acquisition of new customers and retention of current customers
- It strives to make marketing a more effective and efficient factor within the organization
- It uses a flexible procedure of DMAIC philosophy understood by the customer and marketing peoples
- Its goal is defect reduction
- SSM expands the traditional view of marketing to include emphasis not only on pricing, product, promotion, and distribution, but also on processes

SSM takes elements from both SS and marketing and forges them into a powerful and focused discipline designed to increase the enterprise's market share and topline revenues. It is a structured approach that addresses the need for more effective and efficient marketing activities in order to achieve value proposition goals at lower costs.

According to Webb and Gorman [24], traditional approaches to managing, marketing, and sales make it difficult for managers to know with any certainty what changes they should make. As a result, managers often rely on gut instinct and experience instead of systematic thinking and hard data. Companies are beginning to recognize that process improvement can be used to increase revenue. This focus on sales process improvement has five underlying premises or principles:

1. Creating value toward customers
2. Managing on the basis of data and facts
3. Analyzing cause and effect
4. Minimizing waste, errors, and defects
5. Setting the context for collaboration

The end result of any marketing and sales activities is a series of customer actions, ultimately resulting in orders. These actions can and should be measured in terms of raw material, work-in-progress, and finished goods inventories. While designing a sales process, managers should aim to

- create legitimate, verifiable value for customers;
- organize the company's resources, the workflow between people, and the work steps themselves; and
- measure at least some of the actions customers take.

A process approach enables one to identify bottlenecks and their effects; so that prioritization of improvement efforts can be made more effectively. The key to measuring the right data in a sales process is to understand the value perceived by the customer in each step. This facilitates one to understand the customer's behavior, which is a prerequisite for solving sales problems. A *customer value mapping* (CVM) is the best tool for doing this. It can also estimate how the customers value the products and services. Similar to this is the *product value mapping* (PVM) developed by the client describing the "features," "benefits," and "So what?" The last question prompts one to articulate why the benefits would be important to the prospect and is similar in intent to those derived from customer value maps.

A process approach helps managers employ their initiatives by doing the following:

- Providing baseline measures of the process that point to the root causes of problems
- Considering the expected ROI of decisions, based on expected costs and benefits as calculated from reliable data about sales operations
- Enabling managers to measure—or at least consider—the productive capacity of the marketing and sales process, or of sub-processes, before and after a decision
- Showing how a decision about one area of the sales process might affect other areas

Given today's pressure, most managers will not even consider an initiative that does not support their business goals. In marketing and sales, the main goal is almost always to increase sales, usually while increasing profits. According to Webb and Gorman [24], the four basic strategies for increasing sales are to

1. Sell more existing products to existing customers
2. Sell more new products to existing customers
3. Sell more existing products to new customers
4. Sell more new products to new customers

and the six major strategies for increasing profitability are to

1. Reduce the cost of bringing on new customers
2. Reduce the cost of servicing existing customers
3. Increase customer loyalty
4. Drop unprofitable customers or increase their profitability
5. Increase prices while maintaining or reducing product costs
6. Sell more of the most profitable products, fewer of the less profitable ones

Thus, SS process improvement can apply in sales and marketing organizations in three basic ways:

1. Improving your ability to assist customers on their buyer's journey
2. Applying process improvement to customers in the way you sell products to them
3. Providing in-site SS teams to customers

Creveling et al. [3] contemplate the SS marketing as three process arenas of marketing for enabling a business to attain a state of sustainable growth as follows:

- *Strategic marketing process*—product or service portfolio renewal
- *Tactical marketing process*—product or service commercialization
- *Operational marketing process*—post-launch product or service line management

The strategic and tactical areas are internally focused and hence they are called *inbound* marketing areas. External data are critical to successful portfolio definition and development, and product commercialization. However, the output of those processes is intended for internal use. These process outputs are not yet ready for external consumption. The outputs that are ready for prime-time market exposure are part of *outbound* marketing. The operational processes involving post-launch product marketing, sales, and support are customer-facing activities. Given the different customers of inbound and outbound marketing, the requirements for each differ. These requirements ultimately define the success (or failure) of the deliverables.

According to the authors, the natural flow of marketing work starts with strategic renewal of the offering portfolios, to the tactical work of commercializing new offerings, and finally to the operational work of managing the product and services lines in the post-launch sales, support, and service environment. The method to guide marketing strategic work is called IDEA (*Identify, Define, Evaluate, and Activate*). The approach for tactical work is called UAPL (*Understand, Analyze, Plan and Launch*), and the method to direct marketing's operational work is called LMAD (*Launch, Manage, Adapt and Discontinue*).

The methods described above are product/service specific and are useful for commercialization of the projects. Each of these processes feature distinct phases in which sets of tasks are completed. Each task can be enabled by one or more tools, methods, or best practices that give high confidence that the marketing team will develop the right data to meet the task requirements for each phase of work.

Marketing processes and their deliverables must be designed for efficiency, stability, and most importantly, measureable results and hence the importance of Six Sigma. Therefore, the phases discussed in IDEA, UAPL, and LMAD processes ensure measurable results that fulfill any company's requirements.

The benefit of integrating SS into marketing process includes better information (management by fact) to make better decisions. Using the more robust approach reduces the uncertainty inherent in marketing—a creative, dynamic discipline. One way to maintain growth over time is to focus on "leading indicators" of your desired goal. Leading indicators are factors that precede the occurrence of a desired result. The other indicators which monitor the leading indicators are the "lagging indicators." We describe these indicators a little more in detail below.

14.2   The Leading and Lagging Indicators

The *leading indicators* are the indicator that precedes the occurrence of something. It is used to signal the upcoming occurrence of an event. By tracking leading indicators, one can prepare for or anticipate the subsequent event and be proactive. For example, barometric pressure and radar positions for the surrounding region indicate the upcoming weather. One way to maintain growth over time is to focus on leading indicators. A powerful leading indicator is customer satisfaction before a sales transaction. Another leading indicator is the distribution channel's satisfaction with a product (or samples). The leading indicators help to anticipate whether your process will be in target or not.

An indicator that follows the occurrence of something is called the *lagging indicators*. It is used to determine the performance of an occurrence or an event. By tracking lagging indicators, one reacts to the results. Examples are the high and low temperatures, precipitation, and humidity for a given day. Lagging indicators can include functional performance measures such as unit manufacturing cost (UMC), quality measures such as defects per million opportunities (DPMO), and time-based measures of reliability such as mean time between failures (MTBF).

The leading and lagging indicators can also be described in terms of cause and failure of a physical system. For example, obesity and heart attacks are *lagging indicators*, as you can react to them only after they are detected. Food and exercise are *leading indicators* that can tell you when to make changes to help avoid obesity and heart attacks. So do not wait for lagging indicators to scare you into action once they have developed to the point where you can detect them. Measuring fundamentals of human behavior and decision making will set you on a course of action that lets you prevent process failures. Thus, SS project must evolve methods for identifying these indicators and prevent the indicators to dominate each other in any process activities.

14.3 Measurement-Based Key Marketing Indicators

With respect to a Six Sigma methodology, the *key marketing indicators* (KMI) play a very important role in judging the overall performance of the organization. It provides a series of measures against which internal managers and external investors can judge the business and how it is likely to perform over the short term and long term. Obviously, KMIs cannot operate in a vacuum. One cannot establish a KMI without a clear understanding of what is possible—so we have to be able to set upper and lower limits of the KMI in reference to the market and how the competition is performing. This means that an understanding of benchmarks is essential to make KMIs useful, as they put the level of current performance in context—both for start ups and established enterprises—though they are more important for the latter. Benchmarks also help in checking what other successful organizations see as crucial in building and maintaining competitive advantage, as they are central to any type of competitive analysis. The best possible way to achieve a competitive edge is through the use of measurement-based KMIs, as they can establish current performance, benchmark, and target levels. For each monitoring module, the organization can then establish what the current level of performance is in a measurable and understandable way. This current performance can then be marketed with guaranteed results [2].

As different individuals and organizations will put a different emphasis on each item of information, a definitive list of what is and what is not a KMI will depend on individual decisions and will vary considerably according to the stage of company development. Start-up enterprises need to place their emphasis on structural factors and established companies on operational performance. Some of the crucial measurement-based KMI are as follows:

- *Gross profit* is one key measure to the success of the organization. Research shows that survival rates are linked to levels of gross profit; gross profit margins above that of the competition provide clear evidence of competitive advantage.
- *Return on capital* (ROC) employed is another key measure of the success of the organization. The ability to use investment effectively is central to effective long-term development.
- *Z-score* is a measure of the liquidity of the enterprise and clearly defines positive or negative trends.

Both gross profit and return on capital employed are part of the model-based scorecard for overall objectives. Other components within the financial reporting module that might be considered as KMIs are factors such as the debt/equity ratio (DER), project success rates, bad debt rates, and free cash flow (FCF). Including time, budget, and specification to project reporting would also be a natural addition in the measurement-based KMIs.

In addition to the creation of the enterprise balanced scorecard, in which gross profit, return on capital, and Z-scores are standard elements, the identification of KMIs in each of the operational areas or knowledge centers also assists the enterprise in plan development. These KMIs will change over time, but their creation as part of the initial creation of each knowledge center will focus and direct their operational activities. One of the most valuable contributions that KMI analysis can deliver is an understanding of the true nature of the competitive environment.

The KMIs are the essential part of any effective *management information system* (MIS). In a decentralized planning system focused around knowledge centers, the choice of key performance indicators is the first stage in the re-evaluation of the information system to make it more valuable and relevant to the operating unit rather than one that is centrally provided. Thus, the choices of KPI determine what will drive that part of the enterprise and what information must be collected to analyze and manage it. Such information gathering or software choices create information networks that are relevant and provide data which is used specifically for operational purposes and reducing information overload. The information system also helps the management to focus on

- Action planning and implementation with an emphasis on *management by objectives* (MOB) which will include a standardized rate of return and detailed project control
- Training as part of a companywide approach to focusing staff and management on essential operational requirements
- Business planning as a core part of the business plan outline
- Identification of necessary actions in changes in management, exit planning, survival planning, and recovery planning
- Set priorities for investment appraisal, and the choice of emphasis that should be given to the main strategies within the golden circle, consolidation (including cost cutting), market penetration, market development, and product development

We now discuss the importance of supply chain metrics in marketing placing emphasis on process capability and sigma level during and after the launch of a Six Sigma project.

14.4 Relevance of Supply Chain Metrics in Marketing

Supply chain measurements or metrics such as inventory turns, cycle time, DPMO, and fill rate are used to track supply chain performance. Commonly used supply chain management metrics can help you to understand how your company is operating over a given period of time. Supply chain measurements can cover many areas including procurement, production, distribution, warehousing, inventory, transportation, and customer service like the area of logistics [12, 25]. However, a good performance in one part of the supply chain is not sufficient. The solution is for you to focus on the key metrics in each area of your supply chain. Tracking your metrics allows you to view your performance over time and guides you on how to optimize your supply chain. It allows management to identify problem areas. It also allows for comparison with other companies through the characteristics of industry benchmarking. The supply chain metrics improves the logistics operations in the following ways:

- The first step is to identify the metrics that you want to use. Do not use every metric available. Rather, focus on the vital measurements necessary for your business. These can be considered your KPIs
- Next, you need to understand the meaning of these metrics. It is not enough for the management to simply view these measurements; they must also understand the meaning behind them
- The next step is to learn the mechanics behind measurements. What drives them toward positive and negative? Try to understand the various factors that influence your results
- Using this information, identify the weakness or areas of improvement in your current processes
- Set goals based on these improvement areas. The goals should be aggressive, but yet obtainable. Goals can be based on benchmarking against like companies or goals can be set to reflect a specific percentage improvement over past performance
- Put corrective action in place to improve your processes. Make sure that these corrective actions do not negatively affect other areas. Also, check that all affected areas have a clear understanding of the changes
- Monitor your results. Did your corrective actions yield your desired results? If so, what is your next area for improvement? If you did not get the desired results, what went wrong? Try to identify the root cause of your undesired results

All Six Sigma projects are generally marketed as per the sigma level of the process or project. It is not mandatory to use the Sigma scale for this. Besides Sigma, DPMO, and Yield, there are other valid ways to express and measure the performance of a process or product/service. One can also use methods such as control charts and process capability indicators, for this purposes.

Some of the "logistical" issues that surround Six Sigma measures are as follows:

- Establish guidelines for Six Sigma measures to be applied effectively across an organization—to ensure consistency
- Six Sigma measures are not "static." As customer requirements change, Sigma performance will also change. It will be a good idea to continue calculating with the old requirement, simultaneously with the new requirement, at least for some time. This will make transition smoother, and the project team will be judged fairly for their work

- Set priorities on what can and should be measured. No one should expect accurate Sigma performance data for every part of a company in a short time
- Sigma measures (or any other method of measures) by itself will not improve the performance. They are just report cards or milestones to show where the company is on its journey toward excellence. To bring about improvement, methods of analysis and tools are required to be used

14.5 Importance of Data in Marketing

Effective control of the LMAD phases depends on what you choose to measure and how often. Many business processes have been designed and fitted with a measurement that is loaded with things that are easy and convenient to measure. Often easy to obtain data is least effective in telling you anything fundamental about what is really going on in your marketing and sales environments. Easy-to-measure data is almost always a lagging indicator of what has just happened. Lagging a measure is a function of how often you choose to take data and when you get around to analyze it.

To make a statistically sound decision, ensure that you collect right data and collect enough of it. If you measure attribute data, you have to gather large amounts of data to make a sound decision. If you measure continuous data, typically associated with a fundamental dynamic, relatively small sample of data is needed to make a decision. A good example of a poor measure of sales-force performance influence on revenue is the number of sales professionals sent out to interact with customers. Sheer numbers of people is a gross measure that is easy to quantify, but it cannot provide behavioral relationship data in the context of human interaction. Qualitative relationship dynamic data are just plain hard to gather, document, and communicate, but it is one parameter that is fundamental to sales. If you can get the data, it will show a fundamental cause-and-effect relationship.

Data are the result of taking measures while doing the work of marketing and sales. Data are produced and gathered by conducting tasks using specific tools, methods, and best practices. Hence, the important features of data from marketing and sales point of view are as follows:

- Data characteristics
- Data quality
- Data reliability
- Data relevancy

As Lean and Six Sigma mature, the impact of business will expand and deepen [5, 7–10, 14]. Both Lean and SS can help create a competitive advantage with better information and more proactive management. In this respect, Creveling et al. [3] predict the following business trends:

- The formation of marketing *centers of excellence* to promote continuous improvement and the standardization and "right-sizing" of marketing tools, tasks, and deliverables
- Enriched and expanded understanding, on the part of marketing, as to the best practices and methods of executing tools, tasks, and deliverable workflows. As executives who want sustained growth become aligned with the proper structuring (right-sizing) of marketing work, the efficiency and performance of marketing teams will become far more predictable. Critical marketing functions executed with rigor and discipline produce deliverables with greater certainty
- Differentiated marketing workflows will be categorized as strategic, tactical, and operational processes similar to those in R&D, product design, and production/service support engineering organizations. The jack-of-all trade approach will migrate toward more marketing specialization to improve execution excellence that assimilates analytical marketing tools into its best practices
- Six Sigma concepts will serve as the foundation for a universal marketing language. A greater investment will be made in applying its concepts and implementing specific tool-enabled marketing activities to produce more predictable, successful growth
- Increased focus on the fundamental marketing variables critical to customer behavioral dynamic $(y = f(x))$. Better definition of what underwrites a real cause-and-effect relationship, and how to measure and control critical marketing parameters that prevent problems. Measuring variables that signal impending failure rather than measuring failures and reacting to them. Marketing teams will stop measuring what is easy and convenient if it is not fundamental to true cause-and-effect relationship within and across marketing variables
- A shift from DMAIC Six Sigma for problem solving and cost control to a phase-gate approach aimed at problem prevention and investment in properly designed marketing workflows to enable growth projects
- Growing accountability of marketing professionals to drive growth and, with demonstrated success, a rebalancing of marketing and technical/engineering professional in an enterprise so that marketing can better and more completely perform the expected tasks. This will reflect the designed balance between an enterprise's marketing and technical innovation strategy
- Improved collaboration between technical and marketing professionals across strategic, tactical, and operational environments
- Transformed thinking to a platform and modular design will help marketing design and monitor the flow of product and service offerings in a balanced portfolio deployment context. This modular approach reduces the intensity of risk for a single launch and spreads risk across multiple launches
- Elevated use of an integrated set of scorecards to measure marketing risk to improve decision making

The benefits of applying Lean and Six Sigma to marketing make it worth the investment. Using the Lean and Six Sigma approaches gives decision makers better information and helps drive uncertainty out of marketing. Companies can better align product ideas with solid market opportunities and better balance and manage their offerings portfolio by using Lean and Six Sigma to significantly increase the probability of marketplace success. These approaches offer a common structure and language that will facilitate communication throughout the tactical product development and commercialization process between marketing and engineering.

14.6 Six Sigma Marketing Value Tools

SSM is driven by the premise of providing customers and markets with superior value—the best quality at the most competitive price. Apart from the usual DMAIC tools, SSM is dependent upon several new and unique value tools. Reidenbach [21] details the following value addition tools for marketing SS:

- *The competitive value model.* This model captures the VOM for every targeted product/market in which an organization chooses to compete. A competitive value model provides two critical pieces of information. First, it clearly identifies the trade-off between quality and price based on how the market views these two components of value. Second, the value model identifies the CTQs that make up the market's definition of value and prioritizes them in terms of their importance to the market. This eliminates a focus on a relatively unimportant CTQ that has little opportunity for enhancing an organization's competitive value proposition
- *The competitive value matrix.* SSM is all about gaps—value gaps. The competitive value matrix depicts the gaps between competitive products, brands, and services on the basis of two key drivers of value, quality, and price, and how the market evaluates each competitive offering. The competitive value matrix provides the first step to competitive strategy based on these gaps. The organization can choose to lead, to challenge, to follow, or to niche
- *The competitive vulnerability matrix.* It is a powerful customer acquisition tool that identifies the low-hanging fruit—those customers who are not getting the value necessary to cement their loyalty to a competitor—and what is the basis of this poor value? This tool provides detailed and actionable information for prying loose those customers at a low cost of acquisition
- *The customer loyalty matrix.* Market share is a function of three factors: customer satisfaction, customer retention, and the increased buying of current customers. How loyal is your customer base? How vulnerable are they to defection? What are the quality or price factors on which your performance is based that are driving them away? Are these factors systemic (based on a sales problem, a distribution problem, a product problem, a process problem, a pricing problem, or a quality problem)? This and other types of information are available from the customer loyalty matrix

Webb and Gorman [24] consider marketing and sales as a production process. According to them, quality and process improvement methods have roots in manufacturing where they add value in very obvious and tangible ways. In contrast, sales add value by providing information and resolving issues, both of which are intangible. Further, there are few analogies between sales and manufacturing:

- *Leads are analogous to raw materials in manufacturing.* Advertising and promotion produce leads, people who may need the company's products and services or who actively seek information about them
- *Sales opportunities are analogous to work-in-process in manufacturing.* Sales people qualify sales opportunities, people who are considering a purchase and who fit the company's qualification criteria
- *Customers are analogous to finished goods.* Salespeople close deals with people who buy their products and services and thus become customers

Hence, a customer centric data-based approach is essential for an effective marketing. In order to survive the competition, company should continuously promote activities which add value to their customers through their sales process. An effective analysis calls for learning what prospects and customers are doing at each stage of the marketing and sales process and what results are occurring at these stages and why. This entails gathering data on what marketing and sales people are doing and on the results they are producing. We must discern cause and effect ($y = f(x)$) and then decide what needs to be changed and how to change it to improve the result. Developing a right sales process can overcome this problem to a great extent.

The right sales process is the one that adds the most value for both the customer and the company. The sales process that will achieve this depends on the industry, the company, and the customer base, and it will change as markets, products, and business conditions develop. Most companies do not currently have the right sales process in place, because few of them have taken a process-oriented approach. A mind-set to change should be developed first for this. To change the sales process, one must make the new process the path of least resistance, particularly for your salespeople. Let the salespeople make their own hard data and place the DMAIC tool play between your process and your problem. As you learn more and clarify your process and your problem, you may find yourself working back-and-forth between the measure and analyze steps. As you try to develop and test a theory how to create an improvement, you may have to go back-and-forth between the analyze and improve phases as well.

The process improvement, with its definitions and measures of value, can help management put a process or parts of a process in place that will keep everyone honest when it comes to adding value for customers. Promoting more sophisticated Lean manufacturing is another value-adding phase of marketing activity. The fundamental idea behind Lean manufacturing, as pioneered by Toyota and documented by Womack and Jones in *The Machine that changed the world: The story of Lean production*, comes down to doing more with less while giving customers exactly what they want. The six basic steps of Lean, which have become widely accepted in supply chain management, are to

1. Implement basic 5S housekeeping principles
2. Identify customer value and map the value stream
3. Remove waste, which is anything the customer would not pay for
4. Reduce batches to the smallest possible size and ensure smooth handoffs and production flows
5. Have customers "pull" the product rather than pushing it on them and
6. Pursue perfection by continually reducing errors, mistakes, and waste

To make a marketing process meaningful, one must make common sense a common practice. There is nothing strange in it. Much of process improvement is just codified common sense. Define something before you measure it. Measure it before you analyze it. Analyze it before you decide how to improve it. Improve it and then control it to the new level of performance. Six Sigma aims to make common sense common management practice. Managing with Six Sigma may not be as exciting for some people as traditional management practices. Managing data, facts, measurement, analysis, and experiments minimizes the politics, power struggles, personality clashes, and drama of organizational life. It works better, but mere improved performance does not appeal to some people. Others fear the transparency that data and facts bring to the table. Submitting to the rule of data demands maturity of mind, respect for reality, and dedication to standards is nothing but common sense.

14.7 Relevance for Managers

This is relatively a new and emerging concept of Six Sigma. Having completed Six Sigma project, the focus now shifts to show case quality improvement and market the sigma level of the project. This is the essence of SSM. Generally, management does not go beyond setting up a goal or successful implementation of a SS project. SSM is a fact-based data-driven disciplined approach to growing market share by providing targeted product/markets with superior value. For a business to sustain, one should focus on three aspects of SSM, namely strategic, tactical, and operational marketing process. They facilitates efficiency, stability, and most importantly measureable results of Six Sigma. The benefit of integrating SS into marketing process includes better information (management by fact) to make better decisions.

Once again data form the essential part of fact-based decisions, whether it is marketing or sales. The relevance of data characteristics, data quality, data reliability, and data relevancy is an important feature of marketing and sales. We have also discussed in this chapter many value addition tools for marketing SS such as competitive value model, competitive value matrix, competitive vulnerability matrix, and customer loyalty matrix. The relevance of supply chain metrics in marketing, measurement-based KMI, leading and lagging indicators, etc., is also highlighted with reference to SSM.

Exercises

14.1. What are the objectives of a Six Sigma project?
14.2. What is SSM? Describe various elements associated with a SSM.
14.3. State various focus areas of SSM and its benefits.
14.4. Distinguish between CVM and PVM.
14.5. Describe how sustainable business practices are achieved through SSM.
14.6. Describe various aspects of IDEA, UAPL, and LMAD, as the marketing strategies of Six Sigma.
14.7. Describe how the leading and lagging indicators facilitate the efficiency of SSM?
14.8. State some of the KMI used for enhancing the potential of SSM. How are they connected with a good MIS?
14.9. Discuss the relevance of supply chain metrics in marketing of Six Sigma.
14.10. What are the "logistical" issues that surround Six Sigma measures?
14.11. What are the main marketing value tools used in SSM?
14.12. State the Lean principles widely accepted in supply chain management.

References

1. Chakravorty, S.S.: Six Sigma programs: an implementation model. Int. J. Prod. Econ. **119**, 1–16 (2009)
2. Cox, R.F., Issa, R.R., Ahrens, D.: Management's perception of key performance indicators for construction. J. Constr. Eng. Manage. **129**(2), 142–151 (2003)
3. Creveling, C.M., Hambleton, L., McCarthy, B.: Six Sigma for Marketing Processes: An Overview for Marketing Executives, Leaders and Managers. Pearson Education, Boston (2006)
4. General Electric: General Electric Company 1997 Annual Report (1997).
5. Hahn, G.J., Doganaksoy, N., Hoerl, R.W.: The evolution of Six Sigma. Qual. Eng. **12**(3), 317–326 (2000)
6. Hendricks, C.A., Kelbaugh, R.L.: Implementing Six Sigma at GE. Assoc. Qual. Participation **21**(4), 48–53 (1998)
7. Juran, J.M.: Juran on Planning for Quality. Free Press, New York (1988)
8. Kaplan, R.S., Norton, D.P.: The balanced scorecard—measures that drive performance. Harvard Bus. Rev. **70**(1), 71–79 (1992)
9. Kaplan, R.S., Norton, D.P.: Putting the balanced scorecard to work. Harvard Bus. Rev. **71**(5), 138–140 (1993)
10. Kumar, D.U., Nowicki, D.: On the optimal selection of process alternatives in Six Sigma implementation. J. Prod. Econ. **111**, 456–467 (2008)
11. LG Electronics: Six Sigma case studies for quality improvement, prepared for the national quality prize of Six Sigma for 2000 by LG Electronics/Digital Appliance Company (2000).
12. Mentzer, J.T.: Defining supply chain management. J. Bus. Logistics **22**(2), 1–25 (2001)
13. Moorman, C., Rust R.T.: The Role of Marketing. J. Mark. **63** (Special Issue 1999), 180–197 (1999)
14. Muir, A.: Lean Six Sigma Statistics. McGraw-Hill, New York (2006)

15. Muralidharan, K., Neha, R.: Six Sigma: some marketing essentials. Int. J. Mark. Hum. Resour. Manage. **4**(2), 1–12 (2013)
16. Pande, P., Neuman, R., Cavanagh, R.: The Six Sigma Way: How GE, Motorola, and Other Top Companies are Honing Their Performance. McGraw Hill Professional, New York (2000)
17. Park, S.H.: Six Sigma for quality and productivity promotion. Productivity series 32, The Asian productivity Organization (2003)
18. Pyzdek, T.: The six sigma handbook: a complete guide for green belts, black belts and managers at all levels. Tata McGraw-Hill, Noida (2010)
19. Pyzdek, T.: The Complete Guide to Six Sigma, p. 431. Quality Publishing, Tucson (1999)
20. Pyzdek, T.: Six Sigma and lean production. Qual. Digest, p. 14 (2000)
21. Reidenbach, R.E.: Six Sigma Marketing: From cutting costs to growing market share. ASQ Quality press, Milwaukee, Wisconsin (2009)
22. Su, C.T., Chou, C.-J.: A systematic methodology for creation of Six Sigma project: a case study of semiconductor foundry. Expert Syst. Appl. **34**, 2693–2703 (2008)
23. Wang, C.H., Hsu, Y.: Enhancing rubber component reliability by response model. J. Comput. Ind. Eng. **57**, 806–812 (2009)
24. Webb, M.J., Gorman, T.: Sales and Marketing the Six Sigma Way. Kaplan Publishing, Fort Lauderdale (2006)
25. Supply chain metric (n.d.). Retrived from http://www.supplychainmetric.com/

Chapter 15
Green Six Sigma

15.1 Introduction

Every organization needs to improve their processes for better management, quality, and return on investment (ROI). The Six Sigma philosophy is a fact-based management approach of reducing variation in a process and thereby improving the quality [1, 3, 4, 7]. At its best, Six Sigma improves the way managers manage by establishing the context in which, and the means by which, they can decide what is important and what is not, and what requires attention and resources and what does not. Process improvement does not do away with management principles such as responsibility, authority, chain of command, and span of control. Nor does it ignore management practices such as goal setting, planning, delegating, and controlling to plan. Instead, it enables at all levels to adhere to those principles and employs those practices. It does this by keeping them focused on the customer, on the organization's purpose, and on the facts of a business situation.

The cooperation among organizations to minimize the logistical impact of manpower utilization, product movement, and material flows ultimately leads to maximization of the resources in the organization. It is nothing but the greening of alignment and integration of environmental management within the organizational management. It borders the firm's environmental impact and corporate boundaries within which the organization functions. Irrespective of the functional area, every business organization must function within the purview of governmental regulations and sustainable business practices. In particular, a Six Sigma organization must focus on the greening of business not only from the end user perspective, but also from the environmental hazards which affect the general nature of the business.

The Green Six Sigma (GSS) can be defined as the qualitative and quantitative assessment of the direct and eventual environmental effects of all processes and products of an organization. The activities involve the systematic usage of infrastructure and manpower, and optimum use of technology and accountability of sustainable business practices (see Refs. [5, 6]). The activities can be further

© Springer India 2015
K. Muralidharan, *Six Sigma for Organizational Excellence*,
DOI 10.1007/978-81-322-2325-2_15

classified into two broad categories as qualitative and quantitative assessment of environmental effects.

Some of the specific qualitative assessments that need special attention are as follows:

- Source of information regarding raw materials and machinery
- Strength of the company in terms of its size, manpower, machinery, and methods
- Major breakthrough achievements in the past and present
- Availability of the complete information on innovations carried out by the company
- Availability of environment friendly and sustainable developments carried out by the company
- Waste disposal schemes employed
- Sources of energy conservation and fuel consumption
- Methods of reducing noise pollution
- Environmental performance measurement
- Process control and quality control measures practiced in the company
- Tax paid and company liabilities specific to environmental issues
- Reactive rather and proactive maintenance activities

Apart from the above qualitative information, there are many expected and unexpected environmental effects taking place in an organization, which can be measured in some quantitative scale. Some of them are given as follows:

- number of collaborations between companies and governments
- number of major accidents and catastrophes that have taken place
- number of projects conducted on healthy business practices and sustainable business practices
- number of bugs reported in computer programs and software
- number of training programs held for educating the employees
- amount and of level of green house gas (GHG) emissions
- amount of carbon released by the company
- amount of waste produced and disposed in a day

Companies involved with Six Sigma projects should maintain the above information to its fullest as a part of green activities in the organization. The governing body should consider it to be a compliance issue and ensure that it is practiced and promoted by every company. This issue should be the initial barometer for evaluating any Six Sigma project, and hence, the responsibility of such organization is very high in delivering the services. The government should promote and reward such organization looking at their overall contribution to nature and society. This will promote the green and clean technology concept, which is the need of the hour.

The driving force of GSS can be anything such as improving public relations, improving customer relations, improving investor relations, improving corporate relations, reduce logistic costs, satisfy customer requirements, understand competitors and government compliance. An organization, therefore, must encourage manufacturing of green products and involve green methods while designing their product. The best business practice will then become an integral part of such an organization.

15.2 Green Six Sigma Tools and Techniques

As discussed in the previous section, SS philosophy works with a structured problem-solving approach spanned through Define, Measure, Analyze, Improve, and Control, in short DMAIC tools of problem solving. A GSS also works with the same principle, but with the inclusion of environmental concern in each phase. The green and clean use of technology, knowledge, and management of infrastructure will also be included in the process while suggesting improvements. While identifying the GSS project and its impact on sustainable issues, two fundamental questions that must impact the business must be asked. They are given as follows:

- Where is the problem? In the process or in the product?
- What exactly is causing the problem? The general features of the problem? Or the suitability and adaptability of the problem?
- How do the environmental issues impact on the output of the process?
- How does the service mechanism address the green issues?
- How frequent innovations are required in the process?
- Is the organization responding for the first time to environmental issues?
- What is the frequency of customer demand for green improvements in product and services?

The environmental issues and sustainable business practices are the results of continuous improvement. This also happens when innovations are incorporated in the process. To make the process or product green and clean, the company should always focus on those relevant activities, which can deliver and include the voice of customer (VOC) in the process. Hence, the objective of the GSS project will start with the definition of the hypothesis such as:

- Whether there are any influences of environmental regulations on economic performance of the company
- Whether there are any influences of environmental regulations on innovation performance of the company
- Whether there are any influences of innovation performance on economic performance of the company

The hypothesis can be process specific, product specific, service specific, and area specific to the project of the organization. A good data collection plan depends on these aspects. A data collection plan should be prepared specifying the information needed and the required format of the data and which holds or has access to the relevant records. Before requesting data from supply chain partners, it is considered a good practice to introduce and explain the objectives of the project to them and, hopefully, gets their buy-in and active support. When data are outside the organization, it is useful to have designated people in all companies involved, so that they can coordinate and manage the data collection process internally. Instead of focusing more on VOC, it is better to focus on voice of process (VOP) to get a good data supported by green issues of business.

The DMAIC principle of GSS works as follows: At the measure phase, the data suitable to the VOP may be assessed in coordination with the VOC. Through a process mapping or timeline, one can establish the flow of inputs (green measures) from the supplier side to the flow of outputs to the customer. All the physical flow accounts for the general variations (due to materials, methods, machines, people, environment, measures, etc.) encountered in a process. These variations are generally classified as the assignable causes of variations and random (or chance) causes of variations, whereas the information flow helps to identify the chance causes which are affecting the environmental variations in the model. Hence, one of the main objectives of GSS will be to tackle the influence of chance variations in the process.

Specifically, a measurement system analysis of GSS will include the following aspects:

- Identifying and defining the various components of measurement error
- Listing and discussing the sources of error in measurement systems used in processes
- Listing the parameters influencing the consistency of the measurement
- Identifying random effects which contributes the excess and redundant use of materials
- Identifying factors affecting green performance measures
- Identifying the priorities for engineering tolerance and process capability of the process
- Assessing of environmental conditions

A GSS project personnel should always consider environmental friendly materials, working condition, and green technologies to minimize overall execution time, cycle time, and wastage of materials. They should identify those opportunities where rework and rerun of the activities are low to avoid excess energy consumptions. The aim should not be confined to minimization of the infrastructure only, but also to minimize all types of pollution (noise, sound, gas, electricity, etc.) arising from an activity.

During the analysis and control phases, care should be taken to induct the most important input variables (X's), which will contribute positively to the green environment. The variables impacting green aspects should always be considered for any improvement activities. Such an output will sustain long and will have lasting effect on quality of the business.

15.3 Sustainability Issues of Green Six Sigma

For any business to sustain a strong business leadership, manpower knowledge and technology must be present. That too, people working on Six Sigma projects should bring complete transparency on the positive and negative aspects of the environmental issues they are addressing through the project. The management should take note of environmental management standards (EMS) as a response to new environmental regulations being imposed on companies. Environmental management has both short-term and long-term consequences, affecting the current performance and long-term sustainability of businesses. Carbon management is relatively a new part of this process, gaining significance in light of climate change threat [8]. It is important to create a comprehensive environmental strategy and understand the potential trade-offs between its constituent parts. Thus, EMS is a structured framework for managing an organization's significant impact on the environment. These impacts can include waste, emissions, energy use, transport and consumption of materials, and, increasingly, climate change factors.

It is observed that all projects are certified for quality and standards by some agencies. Generally, the SS projects are evaluated through SS organizations or certifying bodies. By and large, most of the companies are certified through international agencies such as International Standards Organization (ISO), American Society for Quality (ASQ), Quality Councils (QC), or *Bureaus of Standards*. The two organizations that are accountable for quality standards in India are the Bureau of Indian Standards (BIS) and Quality Council of India (QCI). The objectives of ISO cover the promotion of standards in the world with a view to facilitate international exchange of goods and services and to develop cooperation in the sphere of intellectual, scientific, technological, and economic activities including transfer of technology to developing countries. It also aims at harmonization of standards at the international level with a view to minimize trade and technical barriers. That is, to eliminate country to country differences, to eliminate terminology confusion, and to increase quality awareness are the objectives of these standards.

ISO 9000 encourages a documented quality system. Hence, a quality system manual is important for getting ISO certification. The management must have a written policy statement of their commitment to quality, and it must be communicated to and understood by all employees. The management must clearly define quality-related organizational responsibilities and interrelationships. A management representative must be assigned to oversee the implementation and continuous improvement of the quality system. At the same time, senior management must continually review the system for any kind of omissions and commissions.

An ISO 9000 company must identify all processes that directly affect the quality of the product or service and ensure that these processes are carried out under controlled conditions, including formal approval of process, design and equipment, documented work instructions, documented quality criteria, and development of quality plans describing how the process is to be monitored and create a suitable working environment.

Besides ISO 9000, for companies that want certification of their environmental credentials, there exists a series of international standards such as ISO 14000–14001 series. These are a set of voluntary standards and guidelines for companies aiming to minimize their environmental impact. This series was published in the year 1996 and is the only standard in the ISO 14000 series for which certification by an external authority is available and concerns the specification of requirements for a company's environmental management system.

Eco-management and Audit Scheme (EMAS) is another voluntary European-wide standard introduced by the European Union and applied to all European countries. It was formally introduced into the UK in April 1995. According to the Institute of Environmental Management Assessment (IEMA, 2008, accessed online: www.iemanet/ems/emas), the aim of EMAS is "to recognize and reward those organizations that go beyond minimum legal compliance and continuously improve their environmental performance." Participating organizations must regularly produce a public environmental statement, checked by an independent environmental verifier that reports on their environmental performance.

15.4 Benefits of Green Six Sigma

The benefits of GSS can be summarized as follows:

- Improves sustainable business practices
- Reduces overall business costs and thereby improves profits
- Reduces GHG emissions
- Encourages green energy conservations in companies
- Reduces large amount of wastages
- Promotes cost-effective technology and raw material usages
- Improves visibility of green drivers
- Improves brand image
- Improves customer satisfaction
- Improves employee satisfaction

15.5 Green Six Sigma: Some Quality Guidelines

As mentioned earlier, SS project execution is a highly responsible job. Greening a SS project is further a coordinated effort comes through the management and stakeholders including the customers. In order to develop a roadmap for policy makers to improve the efficiency and quality, we propose the following guidelines for green quality certifications. These guidelines will help improve the efficiency of the environmental regulations and also help to assist individual firms in adopting best innovative practices to comply with environmental regulations and improve performance. These guidelines are stated on the basis of various categories such as environmental regulations, sustainable business practices, voluntary actions, pressures from stakeholders, business innovations, and performance (financial and environmental), and they are briefly stated below:

- Prepare environmental mission and vision statements
- Set a defined standard of environmental regulations through economic parameters
- Involve pollution control and waste reduction in every production processes
- Check for harmful emissions and unnecessary energy consumption frequently
- Specify emission target, greenhouse gases, and carbon budgets for each SS projects
- Implement economically viable waste resource for collection and disposal
- Implement environmentally friendly practices such as recycling and remanufacturing
- Record all public environmental statement
- Record and act upon all environmental compliance violations
- Develop innovative processes and IT savvy technical know-how's
- Promote own sources of electricity generations
- Practice sustainable business practices to save energy consumption
- Establish a clear environmental management system to collect data and responsibilities
- Promote eco-friendly vehicles and transportations for goods movement
- Assess achievements of the company in terms of energy conservation, cost saving through sustainable business practices
- Assess performance of the company in terms of the growth and market share on sustainable practices
- Assess product reach and geographical area covered due to sustainable practices
- Any other specific methods adopted by the company for energy recovery and saving

The above-stated guidelines may not be exhaustive in nature. Depending on the company environment, one may revise or modify the guidelines to suit one's purpose. Studies have shown that environmental performance, financial performance, and business innovations are strongly correlated; rather, business innovations always contribute to environmental performance and financial performances. Thus, the regulations stated above are very important for the business to sustain and improve the quality of the organizations. Some of the above guidelines have a strong base of quantitative and qualitative characteristics and, therefore, can be evaluated for assessment anytime in the organization. And it should be made mandatory for any quality certification.

15.6   Green Six Sigma: Moving Toward Excellence

An excellent organization, by definition, is one that has successfully worked out its integrated portfolio of requirements, processes, outcomes, and competence. Moving an organization toward such an excellence requires discipline and commitment. It involves the management of innovation and change and a systematic approach to assess and design excellence into an enterprise. This presupposes a clear understanding of an organization's concept of excellence and its acceptance as a basis for integrating and synergizing essential processes into an overall scheme of things that balances on the one hand, and the nexus among the key variables of requirements, processes, outcomes, and competence on the other hand. It helps to create an organization that through enterprise planning, position planning, and competency development, and dynamically balances conflicting social, organizational, and personnel demands in such a way that facilitates sustainable adaptation, growth, and success.

For excellence, Six Sigma cannot be treated as yet another stand-alone activity. It requires adherence to a whole philosophy rather than just the usage of a few tools and techniques of quality improvement [2]. Six Sigma projects must be targeted for process and product improvements that have a direct impact on both financial and operation goals. Integrating environmental initiatives and sustainable business practices into corporate management can lead to increased business, improved business performance, and further enhancement of the company's credibility with stakeholders. The enhanced environmental concerns necessitate performance measurement and reporting systems catering to green initiatives. An effective, balanced, and dynamic performance measurement system is, therefore, critical for monitoring, controlling, and improving a GSS. This is achieved through a systematic implementation of phase-wise tools and execution of DMAIC philosophies discussed in previous chapters.

15.7   Relevance for Managers

This chapter introduces the best business practices that can be achieved through GSS implementation. GSS also works with/on the same principle as a SS project, but with the inclusion of environmental concern in each phase. Hence, every project is bound to maintain discipline and commitment toward their business output adhering to environmental concerns. The chapter, therefore, is meant for business organizations promoting green activities, professionals working in SS projects, consultants, and academicians. We have proposed many tools and techniques of implementing GSS projects and suggested various quality guidelines to promote sustainable business practices in the organization. We recommend and hope that these guidelines should be made an integral part of every project.

Exercises

15.1. What is GSS? State the qualitative and quantitative characteristics of GSS.
15.2. What are the implementation tools and techniques of SSM?
15.3. What are the management aspects associated with the measurement system analysis of GSS?
15.4. Describe some of the vital sustainability issues of GSS.
15.5. Describe how GSS is different from a LSS?
15.6. What are the benefits of GSS?
15.7. State some of the important quality guidelines of GSS.
15.8. Describe how GSS can be used for organizational excellence.

References

1. Chaudhary, S.: The Power of Six Sigma. Dearborn Trade. A Kaplan Professional Company, Chicago (2001)
2. Dale, B.: Marginalization of quality: is there a case to answer. TQM Mag. 12(4), 266–274 (2000)
3. Eckes, G.: Six Sigma for Everyone. Willey, New Jersey (2003)
4. Harry, M., Schroeder, R.: Six Sigma: The Breakthrough Management Strategy. Doubleday (A Division of Random House, Inc.), New York (1999)
5. Muralidharan, K.: Green statistics: some quality guidelines. Int. J. Ind. Eng. Res. 6(9), 15–20 (2013)
6. Muralidharan, K., Ramanathan, R.: Green statistics: a management perspective. Indian Inst. Ind. Eng. J (2013) (to appear)
7. Pandey, P.S., Neuman, R.P., Cavanagh, R.P.: The Six Sigma Way. McGraw-Hill, New York (2000)
8. Walley, N., Whitehead, B.: It is not easy being green. Harvard Bus. Rev. 72(3), 54–64 (1994)

Chapter 16
Six Sigma: Some Pros and Cons

16.1 Introduction

Six Sigma is a business strategy that seeks to identify and eliminate causes of errors or defects or failures in business processes by focusing on outputs that are critical to customers [17, 18]. It is also a measure of quality that strives for near elimination of defects using the application of statistical methods. It propagates an evidence-based decision-making approach by specifying tools and techniques to measure and monitor processes for changes in variation, i.e., quality. Through quantifying financial return on quality improvement projects, Six Sigma helps to realize the impact of such projects on an organization's bottom line. At first, this methodology was met with some skepticism from managers and organizations because of its statistical complexity. However, after working with and tweaking the system, business world has now realized the potential beneficial of this concept [7, 8, 10, 11, 14, 16].

Six Sigma as a problem-solving philosophy has been well recognized as an alternative for many existing quality improvement programs for achieving and sustaining operational and service excellence. While the original focus of Six Sigma was on manufacturing, today it has been widely accepted in areas like service, healthcare and transactional processes etc. It provides a rigorous analytical framework and a clear and consistent data management and analysis capability for the implementation of quality projects [3, 5].

It is noted that Six Sigma fundamentally depends on linear systems for its function that is discussed in one of the earlier chapters. If variation can enter a complex system randomly and episodically, Six Sigma is a sub-optimal analytical tool. The DMAIC process begins with "define," and complex systems are generally difficult to define. Defining measurement on variables as well as attributes is easy in transactional and service processes, while it is altogether different in service and healthcare delivery systems. This points to a requirement for a clearly stated measurement system for the better implementation of Six Sigma in healthcare,

© Springer India 2015
K. Muralidharan, *Six Sigma for Organizational Excellence*,
DOI 10.1007/978-81-322-2325-2_16

scientific, and academic institutions, and many other places. One can see that the following aspects of the Six Sigma strategy are not emphasized in the earlier quality improvement initiatives:

- The Six Sigma strategy places a clear focus on achieving measurable and quantifiable financial returns to the bottom line of an organization.
- The Six Sigma methodology of problem solving integrates human elements and process elements of improvement.
- The Six Sigma strategy emphasizes importance of strong and passionate leadership, which is required for its successful deployment.
- Six Sigma creates an infrastructure for accountable and responsible officials that include champions, master black, black belts, green belts, and yellow belts.
- Six Sigma methodologies use tools and techniques for fixing problems in business processes in a sequential and disciplined fashion. Each tool and technique within the Six Sigma methodology has a role to play, and when, where, why, and how these tools or techniques are applied will decide the success and failure of a Six Sigma project.
- Six Sigma emphasizes the importance of data and decision making, based on facts and data rather than assumptions. Six Sigma forces people to put measurements in place. Measurement must be considered as a part of the cultural change.
- Six Sigma uses the concept of statistical thinking and encourages the application of proven statistical tools and techniques for defect reduction through process variability reduction methods.

In spite of having a formal methodology, it is loaded with number of advantages and disadvantages in equal perspectives. We discuss this issue below.

16.2 Six Sigma: Advantages and Disadvantages

The main advantage of Six Sigma compared to other approaches to quality control is that Six Sigma is customer driven. Six Sigma addresses the entire process behind the production of an item or completion of a service, rather than just the final outcome. It is proactive rather than reactive, as it sets out to determine how improvements can be made even before defects or shortcomings are found. It is particularly valuable to a manufacturing concern that produces precision goods, such as medical technology, where quality is the utmost customer priority and the customer expects to bear the cost of the Six Sigma process. Even businesses that are unable to implement Six Sigma due to cost or practicality may benefit from the philosophy of customer satisfaction that underlies Six Sigma. In addition, improvements are measured using statistical ratios that can be empirically modified to reflect financial results. Financial results mean an increase in shareholders, which further benefits an organization and its employees. A recent trend reveals that many companies are being requested by shareholders to use the Six Sigma system prior to purchasing stock in their company.

Some other practical advantages of Six Sigma implementation are the following:

- Six Sigma targets variation in the processes and focuses on the process improvement rather than final outcome.
- Implementation of Six Sigma methodology leads to rise of profitability and reduction in costs. Thus, improvements achieved are directly related to financial results.
- Six Sigma embarks on effective use of scientific techniques and precise tools. It is a prospective methodology as compared to other quality programs as it focuses on prevention of defects rather than fixing it.
- Six Sigma quality refers to having 3.4 defects per million opportunities or product samples. It is a challenge to achieve this quality standard, and meeting Six Sigma quality standards is a strong selling point for companies that achieve it.
- Six Sigma is successfully implemented in almost/nearly every business category including return on sales, return on investment, employment growth, and stock value growth.
- Six Sigma increases information transparency across the organization. Adopting Six Sigma process improvement places a continual process improvement methodology at all levels of an organization. Once Six Sigma is embedded in corporate culture, the business processes will continue to improve. Furthermore, new problems will be quickly identified and corrected due to close monitoring of Six Sigma improvements.
- Six Sigma infuses upper management with passion and dedication. Training a few Six Sigma black belts creates a core of process improvement experts who will train green belts and yellow belts as well as new hires at all levels.
- Six Sigma is customer driven and thus aims to achieve maximum satisfaction by minimizing the defects. It targets customer delight and new innovative methods to exceed customer expectations.
- By focusing on defect prevention over fixing defects, companies can realize major and continuing savings over prior rework, scrap, and return costs.
- Six Sigma is data driven. No changes are made until the current process is thoroughly understood, documented, and measured. The revised process is similarly measured and verified. If the Six Sigma project does not deliver what was intended, the Six Sigma team is available to correct new-found problems or to study what went wrong.
- When Six Sigma projects are thoroughly documented, the lessons learned from those projects are readily shared among experts and can be searched by other employees looking for ideas to improve on their own sites or assembly lines.
- Six Sigma is attentive to the entire business processes, and training is integral to the management system where the top down approach ensures that every good thing is capitalized and every bad thing is quickly removed.

Application of Six Sigma to all aspects of production and planning process may create rigidity and bureaucracy that can create delays and stifle creativity. In addition, its customer focus may be taken to extremes, where internal quality control measures that make sense for a company are not taken because of the overlying goal of achieving the Six Sigma-stipulated level of consumer satisfaction. For example, an inexpensive measure that carries a risk of a slightly higher defect rate may be rejected in favor of a more expensive measure that helps to achieve Six Sigma, but adversely affects profitability. The other disadvantages which can go against the philosophy are stated as follows:

- Six Sigma is a quality initiative, not a targeted process to reduce costs. Dramatically improving quality may save money by eliminating defective products, expensive rework, and returned products. Yet the cost savings are a by-product of the process improvements to any quality improvements. Six Sigma projects may not yield any cost savings at all. Improving product quality can generate capital costs and long-term overhead costs in terms of more quality personnel.
- Applicability of Six Sigma is being argued among the Six Sigma critics. According to them, quality standards should depend on specific task, and measuring 3.4 defects per million as a standard leads to more time spent in areas which are less profitable.
- Six Sigma qualities are not synonymous with lean manufacturing. Adding new quality checks, improved test equipment, or better manufacturing equipment with tighter tolerances may actually increase manufacturing footprints, use more material, or consume more resources. Lean Six Sigma attempts to hybridize these two concepts, but it requires far more data and analysis to compare potential projects and their outcomes.
- Six Sigma does not work well with intangible results. Six Sigma projects are best for physical products that are out of specification, either too large or too small. They can be applied to business processes that generate measurable outputs such as calls handled per hour or customer wait time. Six Sigma projects do not work well with goals like improving customer satisfaction or lifting employee morale.
- Six Sigma gives emphasis on the rigidity of the process that basically contradicts the innovation and kills the creativity. The innovative approach implies deviations in production, the redundancy, the unusual solutions, and insufficient study which are opposite to Six Sigma principles.
- Six Sigma may be misquoted as any another continuous improvement program. It thus promotes outsourcing of improvement projects with lack of accountability.
- While converting the theoretical concepts into practical applications, there are a lot of real-time barriers which need to be resolved.
- Not all products or operations need to meet Six Sigma quality standards. Medical manufacturers and aerospace companies should meet Six Sigma quality standards because lives literally depend upon the reliability of their products. Cell phones and computers that meet Six Sigma quality standards are more reliable than competition, but may be more expensive.
- Employee dedication and skilled manpower is a must for Six Sigma implementation. The absence of these can kill the success of a Six Sigma project.
- Employee confidence can take a beating if any Six Sigma project fails to give expected returns.

A successful Six Sigma organization does not fall under any management hype or customer grading. If the management is keen in improving their process, then the organization size and financial status hardly matter in proceeding with a Six Sigma project. In fact, it is true that Six Sigma activity is extremely expensive for many small business organizations to implement a project. Employees must obtain training from certified Six Sigma institutes in order that an enterprise receives Six Sigma certification. Even if a firm wishes to implement Six Sigma without formal certification, some essential training is necessary to understand the system and its application to particular business processes. However, many small businesses cannot afford such training even for a single employee. In addition, small businesses that need to remain nimble and creative often find the Six Sigma system of process analysis stifling, bureaucratic, and overly time-consuming. This is a serious concern for Six Sigma.

16.3 Six Sigma: Limitations

Projects which are directed for implementation are selected by organizations subjectively rather than objectively, which means that goals may be mistakenly thought of as attainable and favorable when in fact they may eventually be a waste of resources and time. Also, researchers investigating the trend have noticed that some individuals calling themselves "experts" in Six Sigma methodology actually do not comprehend the techniques and complex tools necessary to effectively implement the quality control process in an organization. Thus, these companies hiring "experts" are being treated to a substandard version of the principles which will do nothing to help their company, instead only lend a warped perspective of what it is supposed to do. In order to keep this from happening further, the Six Sigma community must come together and stand up for strict training and certification standards to be issued for any organization. Such limitation really comes as a hindrance for the success of Six Sigma. Few other important limitations are stated as follows [1, 11]:

- The challenge with Six Sigma is the availability of good data. A poor data can lead to the wrong direction and wasted effort.
- The assumption of normality for applying DMAIC tool no more exists in complex transactional processes. This limits the area of applications to simple and linear systems.
- The right selection and prioritization of projects is one of the critical success factors of a Six Sigma program. The prioritization of projects in many organizations is still based on pure subjective judgment. Very few powerful tools are available for prioritizing projects, and this should be major thrust for concern in the future.

- The identification of critical-to-quality characteristic also poses challenge to service and transactional processes because of poor visibility of customer requirement. Due to dynamic market demands, the CTQs of today will not necessarily be meaningful tomorrow. All CTQs should be critically examined at all times and refined as necessary. Very little research has been done on the optimization of multiple CTQs in Six Sigma projects.
- The relationship between cost of poor quality (COPQ) and Six Sigma process quality level requires more justification.
- The start-up cost for institutionalizing Six Sigma into a corporate culture can be a significant investment. This particular feature may discourage many small and medium-sized enterprises from the introduction, development, and implementation of the Six Sigma strategy.
- The calculation of defect rates or error rates is based on the assumption of normality. The calculation of defect rates for non-normal situations is not yet properly addressed in the current Six Sigma literature.
- Assumption of 1.5 sigma shift for all service processes does not make much sense. This particular issue should be the major thrust for future research, as a small shift in sigma could lead to erroneous defect calculations.
- The start-up cost for institutionalizing Six Sigma into a corporate culture can be a significant investment. This particular feature would discourage many small and medium-sized enterprises from the introduction, development, and implementation of Six Sigma strategy.
- Many Six Sigma projects are limited to engineering expertise only. Unless this expertise is integrated with statistician's expertise, a Six Sigma project can derail any time. This is happening in many organizations.
- There is overselling of Six Sigma by too many consulting firms. Many of them claim expertise in Six Sigma when they barely understand the tools, techniques, and the Six Sigma roadmap.
- Six Sigma can easily digress into a bureaucratic exercise if the focus is on issues such as the number of trained black belts and green belts and number of projects completed instead of bottom-line savings.
- Research has shown that the skills and expertise developed by black belts are inconsistent across companies and are dependent to a great extent on the certifying body. Black belts believe they know all the practical aspects of advanced quality improvement methods such as design of experiments, robust design, response surface methodology, statistical process control, and reliability, when in fact they have barely scratched the surface.

16.4 Six Sigma: Dos and Don'ts

Two of the important influencing factors of success of any Six Sigma project are the type of project one involved and the person who is responsible for the implementation of the project. This will entail a flexible environment for the journey toward excellence. These will address the issues on project selection, target setting, and infrastructure usage and manpower utilization. One should focus positively on activities that directly add value to customers, stay at a high esteem, and involve a mix of people. A good-quality culture is what you are going to establish through these simple methods. There is nothing wrong, if one uses commonsense and creativity for anticipating probable causes and effects and then suggests suitable methods of eliminating variation [14, 19].

Further, it is expected that the following positive aspects of project management may be taken seriously while readying for a Six Sigma project:

- Ensure the necessity of conducting Six Sigma type of projects
- Let the project selection be on a valid criteria
- Assess the rationale behind a project on a consensus basis
- Align projects with key goals as per the potential of the organization
- Pilot the project, if necessary
- Challenge the status quo and set up a proper hypothesis
- Listen to VOC and VOP and target the solutions
- State clearly tangible and non-tangible results
- Build confidence within the project team
- Think big in terms of results, benefits, and scale of improvement
- Place faith in data and data analysis
- Set measurement priorities that match your resources
- Practice continuous improvement of your measurements
- Avoid spontaneous use of computational methods
- Involve the owner and stakeholders
- Unleash everyone's potential
- Create measurement reports that convey information quickly and simply
- Build openness and establish proper communication
- Be ready to revise your plan based on the current learning
- Monitor the improvement continuously
- Be bold in reporting and correcting the demerits of the project
- Celebrate success

The following points may be treated as negative aspects of a Six Sigma project that can harm during the execution of the project:

- Too many projects in hand at one time
- Failure to explain the reasons for the projects chosen
- Not allowing enough time for an ongoing project to yield
- Indifference in deciding on leadership
- Turning a blind eye to customer information
- Focusing on isolated areas
- Overloading the process with inputs and outputs
- Looking upon the core processes as unchangeable
- Overconfidence in delivering the output
- Overtraining and practice

- Blindly believing your measurement systems
- Expecting data to confirm your assumptions
- Starting to analyze every problem in detail
- Confusion in control and specification limits
- Leaving documents to gather dust
- Exaggerating opportunity counts
- Not leveraging technology

The above-mentioned dos and don'ts are stated in a general perspective. For phasewise dos and don'ts of a Six Sigma project, the readers may refer to Pande et al. [14]. Besides these, one should also see the checklist provided at the end of each chapter.

16.5 Six Sigma: The Future

Six Sigma is now recognized as a powerful approach to achieve process improvements and sustained quality in a business environment. While the original focus of Six Sigma was on manufacturing, today it has been widely accepted in areas like service, healthcare and transactional processes etc [1]. Recently, a number of articles have focused on the importance of Six Sigma for services and the challenges of applying this quality improvement methodology to service operations [2, 4, 6, 9, 12, 13, 15]. Kumar et al. [12] debunks seven most common myths about Six Sigma as:

- it is a fad
- it is all about statistics
- it is good only for manufacturing processes
- it is effective in large organizations
- it is the same as TQM
- it requires strong infrastructure and massive training, and
- it is not cost-effective

When manufacturers realize that they need to improve a product implementing Six Sigma or go out of business, they may standardize manufacturing processes. Process monitoring and data collections are put in place. As product quality improves, tighter quality standards may be adopted or the process is left as it is, but closely supervised. Yet data collection and analysis takes up time, resources, and the cost of achieving high-quality needs to be balanced against other business objectives. This should be made a continuous and an ongoing activity to sustain Six Sigma benefits. However, the enthusiasm through which it starts does not remain for a long period either because of over confidence of success, hard work experienced through the project, or some other reasons. The success of a Six Sigma also includes retaining the success for a long time, initiating new projects, and retaining employees and customers as well. If collective wisdom does not prevail, the future of Six Sigma will be in danger.

According to Snee [17], statistical thinking can be defined as thought processes, which recognize that variation is all around us and present in everything we do. All work is a series of interconnected processes, and identifying, characterizing, quantifying, controlling, and reducing variation provide opportunities for improvement. The above principles of statistical thinking within Six Sigma are robust, and therefore, it is fair to say that Six Sigma will continue to grow in forthcoming years.

Another important danger of Six Sigma is that it is by and large promoted by engineers and managers with limited knowledge of statistical methods. It is true that Six Sigma uses most of the statistical tools forgetting the necessity of statisticians who can deliver better. Too much dependency on black belts and master black belts can only make Six Sigma a publicity stunt rather than a sensible activity. As long as the philosophy goes in tandem with proper technical assistance, Six Sigma has bright future.

As a reminder, one should also understand that data explosion problem is threatening the business organizations enormously. This has paved way for a new analytics area called the "Big data analytics," which is now posing a challenge simultaneously for data collection and data analysis. The issue of quality is completely alien in this area. People are only concerned with the automated data collection tools and converting such data into knowledge and information. Probably, Six Sigma can offer many aspects of quality improvement and speed up the process for an "information-jargon-free" implementation.

Although the total package may change as part of the evolutionary process, the core principles of Six Sigma will continue to grow in the future. Six Sigma has made a huge impact on industry, and yet the academic community lags behind in its understanding of this powerful strategy. It will therefore be incumbent on academic fraternity to provide well-grounded theories to explain the phenomena of Six Sigma. In other words, Six Sigma lacks theoretical underpinnings, and hence, it is our responsibility as academicians to bridge the gap between the theory and practice of Six Sigma [1]. Readers are also recommended to see online resources available for many interesting articles and facts about Six Sigma. Some of the sources are mentioned in the reference section.

16.6 Relevance for Managers

This chapter addresses the pros and cons of the Six Sigma philosophy. An exhaustive account of advantages and disadvantages, limitations, misconceptions, and dos and don'ts of Six Sigma are presented here. The issues surrounding future of Six Sigma are also elicited. These issues will form the basis for the acceptability and adaptability of the Six Sigma concept.

References

1. Antony, J.: Pros and cons of Six Sigma: an academic perspective. TQM Mag. **16**(4), 303–306 (2004)
2. Antony, J.: Six Sigma for service processes. Bus. Process Manage. J. **12**(2), 234–248 (2006)
3. Antony, J., Bañuelas, R.: Key ingredients for the effective implementation of Six Sigma program. Meas. Bus. Excell. **6**(4), 20–27 (2002)
4. Antony, J., Antony, F.J., Kumar, M., Cho, B.R.: Six sigma in service organizations: benefits, challenges and difficulties, common myths, empirical observations and success factors. Int. J. Qual. Reliab. Manage. **24**(3), 294–311 (2007)
5. Breyfogle III, F.W.: Implementing Six Sigma: smarter solutions using statistical methods. Wiley, New York (1999)
6. Chakrabarty, T., Tan, K.C.: The current state of six sigma application in services. Manag. Serv. Qual. **17**(2), 194–208 (2007)
7. Goh, T.N.: A strategic assessment of Six Sigma. Qual. Reliab. Eng. Int. **18**(2), 403–410 (2002)
8. Harry, M.J., Schroeder, R.: Six Sigma: the breakthrough management strategy revolutionizing the world's top corporations. Doubleday, New York (1999)
9. Hensley, R.L., Dobie, K.: Assessing readiness for six sigma in a service setting. Manag. Serv. Qual. **15**(1), 82–101 (2005)
10. Hoerl, R.W.: Six Sigma and the future of the quality profession. Qual. Prog. **31**(6), 35–42 (1998)
11. Hoerl, R.W.: Six Sigma black belts: what do they need to know? J. Qual. Technol. **33**(4), 391–406 (2001)
12. Kumar, M., Antony, J., Madu, C.N., Montgomery, D.C., Park, S.H.: Common myths of six sigma demystified. Int. J. Qual. Reliab. Manage. **25**(8), 878–895 (2008)
13. Nakhai, B., Neves, J.S.: The challenges of Six Sigma in improving service quality. Int. J. Qual. Reliab. Manage. **26**(7), 663–684 (2009)
14. Pande, P., Neuman, R., Cavanagh, R.: The Six Sigma way. McGraw-Hill, New York (2001)
15. Patton, F.: Does six sigma work in service industries? Qual. Prog. **38**(9), 55–60 (2005)
16. Pyzdek, T.: Six Sigma and lean production. Qual. Digest, p. 14 (2001)
17. Snee, R.D.: Statistical thinking and its contribution to total quality. Am. Stat. **44**(2), 116–121 (1990)
18. Snee, R.D.: Impact of Six Sigma on quality engineering. Qual. Eng. **12**(3), 9–14 (2000)
19. Snee, R.D., Hoerl, R.W.: Leading Six Sigma companies. FT Prentice-Hall, Upper Saddle River (2003)
20. http://www.improvementandinnovation.com/
21. http://www.smallbusiness.chron.com/
22. http://www.tamarawilhite.hubpages.com/
23. http://www.sixsigmaonline.org/
24. http://www.qualitygurus.com/
25. http://www.processexcellencenetwork.com/
26. http://www.dummies.com/

Chapter 17
Six Sigma: Some Case Studies

17.1 Introduction

The objective of this chapter is to make students and academics to familiar with business problems and Six Sigma projects. Cases studies are valuable for several reasons, as it provide one, with experience of organizational problems that one probably have not had the opportunity to experience at firsthand. In a relatively short period of time, one will have the chance to appreciate and analyze the problems faced by many different companies and to understand how managers tried to deal with them. There are three case studies presented in this chapter. All the cases are presented from an academic perspective. Information for the case studies was obtained from company consultations and personal contacts. At the time of preparing this chapter, these organizations have successfully completed several Six Sigma projects. The confidentiality about the company's name and the people associated with the projects is kept confidential on request.

The first case study focuses on an improvement project of a leading tyre manufacturing company in India. According to the case presented, the company has a record of implementing Six Sigma projects in almost all areas of work. The second organization is an electrical appliances manufacturing company, and the project aims to reduce defects in a manufacturing process. The last case is on a service quality-related project implemented by a nationalized leading banking and insurance company headquartered in Bengaluru. The theory and concepts help reveal what is going on in the companies studied and allow one to evaluate the solutions that specify companies adopted to deal with their problems. All the three projects strictly follow the DMAIC procedure of Six Sigma project philosophy and are presented in sequential form. Some of the technical terms and abbreviations are company-specific and is not explored here.

17.2 Case Study-1: Reduction in Extruder-Specific Power Consumption in Duplex

The proposed tyre manufacturing company experiences high power consumption in one of its Extruder. The specific power consumption (SPC) in Duplex is causing lot of power sharing problem for other plants and locations for the company. Through a brainstorming session, the company's think-tank has finalized a Six Sigma project for reducing the power consumption. The team members include four functional experts, two engineers from operation and one engineer from power and planning department besides four worker-level officials. One of the engineers is a MBB, one is BB, and the other two are GB's. After scoping the project, following metrics are taken for power consumption reduction in Duplex extruder.

Project Baseline

"Power consumption is 0.247 kWh/kg in Duplex Extruder."

Project Statement

"Reduction in SPC kWh/kg in duplex extruder from 0.247 to 0.199 kWh/kg by February, 2014."

Goal/Objective Statement

"The project objective was reduction in power consumption in Duplex extruder from 0.247 to 0.199 kWh/kg."

For other details of the project, see Table 17.1.

The operational definition for metrics is shown in Table 17.2. The measurement of SPC is monitored through energymeter and is measured in kWh/kg.

Through a detailed C and E analysis and FMEA analysis, the project team was able to identify a number of CTQ variables, which include both continuous and discrete variables. The first-level C and E analysis has identified the following causes:

- Man–Machine idle running condition, mill running during set up time, mill running during size change time, stock loader running after finish up compound, excess usage of machines
- Machine—Low utilization of machines, idle condition line running, cushion extruder TCU running in idle condition, cushion calendar TCU running in idle condition, empty screw running on idle condition, main extruder TCU running on idle condition, mill conveyor running in idle condition, hydraulic motor "ON" during idle condition machine, blander Bar run when feed mill start, idle condition marking conveyor blower motor running, table heater running in Idle condition, and cushion extruder screw running in idle condition takes high current
- Process—Line start in advanced, loose connection at power connections panel, excess usage of machines
- Material—Milling over load, warming up times more
- Measure—Temperature set point kept high.

Table 17.1 Six Sigma project definition worksheet

Step	S. no.	Element	Actual
Describe the business problem	1	What is the specific problem affecting the success of your business?	High power consumption of extruder area
	2	Identify the critical to customer (CTX) category (quality, delivery or cost) associated with this problem	Cost
	3	Name the business metric associated with this problem (existing management performance indicator)?	Power consumption and specific power consumption
	4	When was the problem first observed (specify month/year)?	April 2013–Sept 2013
	5	How much? What is the extent or magnitude of the problem as measured by your business metric?	High extruder power consumption 13,406 kWh/day at 66,073 kg production
	6	How do you know this is a problem? What target is not being met?	Management expectation: 12,406 kWh/day
Scope business problem	7	What is the output product or service delivered to the customer related to the business problem?	Reduction in power consumption
	8	Name the business process delivering the product or service. Think in terms of a process that can be mapped	Extruder power
	9	Identify the scope of the project in the above process map	As per Pareto
Process metrics and objective statement	10	What is wrong with the features named above (i.e., what is the defect on this measurable characteristic that does not meet some requirement)?	High power consumption of extruder
	11	Name the primary metric which measures the identified defect. The primary metric will be used to measure the success of the six sigma project	As per Pareto
	12	UOM (unit of measure) of the primary metric	kWh/kg (SPC)
	13	Estimate the baseline performance level of this primary metric	0.247 Wh/kg (5027 kWh/day at 20,366 kg)
	14	Can you estimate the entitlement (best short term observed historical performance) of this primary metric? Yes or no? If so, what is the entitlement performance level?	0.178 kWh/kg
	15	Prescribed target of the project (?0 % improvement from the baseline toward the entitlement)	0.199 kWh/kg (4050 kWh/day at 20,366 kg)
	16	S.M.A.R.T. objective statement	Reduction in specific power consumption kWh/kg in duplex extruder from 0.247 to 0.199 kWh/kg by 12 February 2014

Table 17.2 Operational definition of metrics

Sampling of data
Power reading data in kWh—daily 6.30 am power consumption data are captured from energy meter
Production data in kg—daily take production from SAP, tread in nos and sidewall in meter
Measurement method
Power reading data in kWh—power consumption data are to be captured from energy meter and dump into main PC through energy management system software. Accuracy need to be verified
Production data in kg—daily take production from SAP, tread in numbers and sidewall in meter. Accuracy need to be verified
Calculation of metric
Total power in kWh/production in kg, kWh/kg
Reporting of metric
Cumulative specific power shall be tracked on daily basis. Weekly report shall be circulated to process owner with copy to MBB
Consequential metric
Extruder utilization

The summary of treatment of identified X's, classified according to its importance, is shown in Table 17.3. The middle bucket X's are subject to immediate action on X to reduce RPN to less than 125, and for the top bucket, the variables were subject to graphical and statistical analysis.

Number of exploratory data analysis such as Box plot, Main Effects plot, Interactions plot, ANOVA are carried out to identify the significant X's. Among many of those X's identified, the variable X (=line speed) measured in 10–15 mpm is found to be causing a serious quality issue of the power consumption. This was later ascertained through a two-sample t-test for 5 % significance level. On further analysis and monitoring, this problem was rectified during the control phase of the project.

Figure 17.1 shows the Box plot of the SPC before and after the improvement. The figure clearly shows the improvement. The average SPC is now 0.19276 kWh/kg, which as per the goal statement is almost equal. The six-pack process capability study of SPC is shown in Fig. 17.2.

This project has yielded both non-financial benefits and organizational benefits. They include the 100 % achievement of reduction in power consumption SPC

Table 17.3 FMEA summary of variables

S. no.	Bucket allocation	RPN and severity	Number of X's
1	Top bucket	RPN—301 to 1000 or severity >7	57
2	Middle bucket	RPN—126 to 300 and severity <7	82
3	Lower bucket	RPN—1 to 125 and severity <7	144

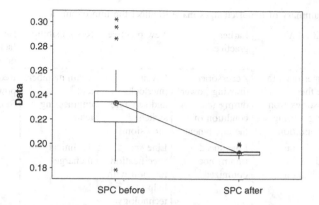

Fig. 17.1 SPC before and after the improvement

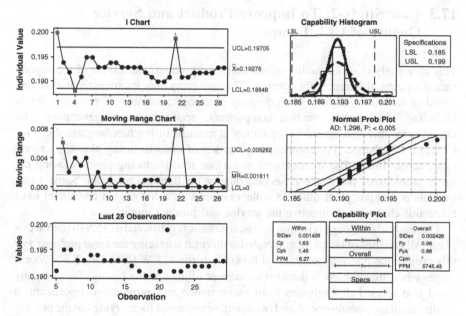

Fig. 17.2 Process capability six-pack of SPC after improvement

(kWh/kg) as targeted. Further, the project has given exposure to all the members associated with it, a systematic approach to problem solving and also on the usage of various quality tools. It has also helped in sharing of learning among the members and inspired many to become green belts and black belts. The summary of major changes made to affect the improvement is given in Table 17.4 to sustain the improvement.

Table 17.4 Summary of major changes made to affect the improvement

S. no.	Identified X	Earlier practice	New practice	Responsibility	Benefit (achieved or expected)
1	No interlocks to cease the accessories from running during idle condition	Accessories drawing power during idle condition of the equipment	Given interlocks and separate switch provision	Team member and engineering team	Reduction in power SPC (kWh/kg)
2	Lack of line speed optimization	Line speed (mpm) not optimized	Line speed specification revision with help of technology	Technical incharge	

17.3 Case Study-2: To Improve Product and Service Quality of CFL Lamps

This case study is about quality improvement of product and service of a company manufacturing different light sources. The company has its own reputation and brand in India and abroad. The company produces a number of electronic items including transformers, traction components, appliances, and generators. The organization has also earned its reputation in manufacturing tubular glass shells for FTL and CFL lamps. The company has number of manufacturing plants in various cities across India. The management at various manufacturing plant is facing the same problem of product and service quality on lighting products. Some smart officials at Nagpur plant office in collaboration with a Six Sigma consultant have taken this challenge to improve the service and product quality.

Currently, the average rejection for last 6 months is estimated as 15,980 ppm at this plant. All varieties of lamps manufactured at this unit are facing the same problem. For this project, the part number selected for study is the "15 W CFL–DF–2U". Process stages where the problem is detected are cement filling and hot basing. The company hired two black belt engineers from maintenance and production department, a quality assurance professional, and two middle executives for carrying out the project. The overall supervision is under a MBB who is serving with the Six Sigma consultancy. The problem statement is as follows:

Problem Statement

"Reduction in glass crack for CFL (Compact Fluorescent Lamp) during manufacturing."

The main concern for the management is to fix the rejection on the above-mentioned unit of lamps. Many ad hoc-level customer satisfaction survey and appraisals were carried out to understand the situation, which really turned out to be a matter of quality concern for the company. Hence, the objective of the project is to achieve less than 1000 ppm in market return and less than 2000 ppm in life test. This target can address the current quality problem for the rejection of CFL tubes. The situation can be ascertained from the Pareto diagram (see Fig. 17.3), which shows roughly 74 % variation are due to electronic failure and 14 % variation are due to glass crack.

The manufacturing process of lamps in general include cutting, bending, washing, coating and drying, wiping and brushing, baking, sealing, fusion, exhaust, aging, cap (cement) filling, and hot basing. As mentioned earlier, the proposed project is identified for cement filling and hot basing, as they are the main reason for glass crack. The detailed process map is shown in Fig. 17.4, identifying properly the value added and non-value added activities.

The causes affecting the crack were identified through a C and E diagram. A detailed process of FMEA is also carried out to identify the failure modes and potential causes. Table 17.5 gives the concise information about the causes and failure modes for crack.

Besides, the usual FMEA and C and E, the project team has validated the causes and their effects based on other statistical studies. Finally, it has been found that the cement weight and basing are critical parameters for glass crack. To find out the critical parameters in basing machine, which are responsible for glass crack, a half-replicate DOE was carried with the probable parameters: Y_1—glass crack, X_1—blower speed, and X_2—temperature profile at three different zones. Table 17.6 shows the experiment of the design.

The analysis also tries to control effects of loose base (basing) along with the crack. After running the analysis for main effects and interaction effects, it has been found that the cement weight and zone temperatures are critical for both glass crack and loose base. Further, the optimization study on cement weight on glass crack and loose base resulted in the following inferences:

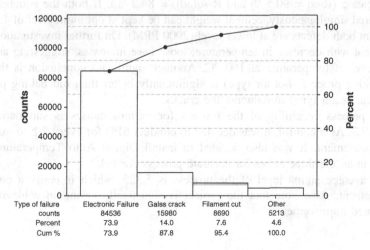

Type of failure	Electronic Failure	Galss crack	Filament cut	Other
counts	84536	15980	8690	5213
Percent	73.9	14.0	7.6	4.6
Cum %	73.9	87.8	95.4	100.0

Fig. 17.3 Pareto chart for CFL failures

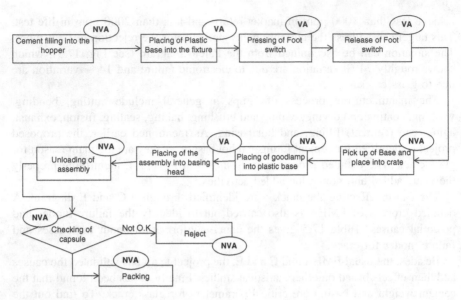

Fig. 17.4 Process map for crack identification

$$\text{Crack} = 11969 - 7224 * \text{cement weight} + 1187 * \text{cement weight}^2$$
$$\text{Loose base} = 71223 - 21279 * \text{cement weight} + 1539 * \text{cement weight}^2$$

From the first polynomial equation, it is observed that as cement weight increases, glass crack increases (R-sq = 92.4 % and R-sq(adj) = 87.3 %). It is therefore, recommended to keep lesser cement weight; however, the second equation shows that if we reduce cement weight below 4 g, then loose cap defect will increase (R-sq = 90.5 % and R-sq(adj) = 84.2 %). If both the equations are considered simultaneously, cement weight can be kept at optimum level of 4.5 g, at this point both defects are at lower level (2000 PPM). On further investigation, it is found that with decrease in temperature, loose base increases, suggesting an optimum level of temperature at 160 °C. Another significant conclusion is that the new baking process (hot air type) is significantly better than old baking process (electrical heater type) in reducing the cracks.

The process capability of the process for various causes is summarized in Table 17.7. As a control, it was decided to conduct SPC for every 2 h to maintain the improvement. It was also decided to install Digital Auto Temperature controllers in all the three zones for accurate process control.

The average sigma level of the process is 5.325, which is really a potential improvement. Also note that the project is statistically strong in achieving the anticipated improvement.

Table 17.5 FMEA of glass crack

S. no.	Process function (step)	Potential failure modes (process defects) (Y's)	Potential failure effects (Y's)	SEV	Potential causes of failure (X's)	OCC	Current process controls	DET	RPN
1	Cap filling	Excess or lower cement in cap	Glass crack	8	cement weight	7	Weighing scale	6	336
2	Basing	Uneven baking		8	Temp profile	6	With indicator	6	288
3	Basing	Direct or radiated temp		8	Type of baking	6	Electrical heaters	6	288
4	Basing	Less or excess expansion in based capsule		8	Cement expansion	4	With scale	6	192
5	Basing	Variation in blowing speed		8	Blower speed	3	With controller	2	48

Table 17.6 DOE table for glass crack

Std order	Run order	Center Pt	Blocks	Zone-1	Zone-2	Zone-3	Cement wt	B3	Crack PPM	Loose base in Nos/50	Expansion
4	1	1	1	190	190	130	3	4	1000	32	169.3
5	2	1	1	130	130	190	3	3	1340	35	167.9
1	3	1	1	130	130	130	3	4	100	45	172.3
11	4	1	1	130	190	130	5.5	4	568	36	168.7
7	5	1	1	130	190	190	3	4	2500	28	184
10	6	1	1	190	130	130	5.5	4	357	33	179.4
14	7	1	1	190	130	190	5.5	3	5789	3	160.7
3	8	1	1	130	190	130	3	3	609	36	167.6
15	9	1	1	130	190	190	5.5	3	7860	2	165.8
12	10	1	1	190	190	130	5.5	3	8750	4	173.7
9	11	1	1	130	130	130	5.5	3	675	42	158.1
2	12	1	1	190	130	130	3	3	307	37	165.3
17	13	0	1	160	160	160	4.25	3.5	1089	4	173
8	14	1	1	190	190	190	3	3	3450	26	153.7
16	15	1	1	190	190	190	5.5	4	9887	2	152.8
6	16	1	1	190	130	190	3	4	507	30	161
13	17	1	1	130	130	190	5.5	4	348	36	162.5

Table 17.7 Process capability of glass crack

Causes	C_{pk}	Short-term sigma level
Cement weight	1.57	4.70
Temperature: Zone-1	1.81	5.45
Temperature: Zone-2	1.85	5.55
Temperature: Zone-3	1.87	5.60

17.4 Case Study-3: Customer Complaint Resolution Through Re-engineering Debit Card and PIN Issuance Process

Here, we discuss a case of Six Sigma improvement project on complaint resolution through service efficiency. The case refers to a nationalized bank committed to the welfare of rural people. The bank offers both retail and wholesale banking services and has a strong customer base spread across the nation. All branches are linked online and work on real-time mode, and therefore encourage information technology-based service to meet diverse customer needs. The reason behind the project is to resolve the customer complaints on their debit card issuance service, which is impacting the overall quality of their service. Table 17.8 shows the data on the type of complaints and their percentage contribution received for a period of six months in the past year.

The bank believes in holistic approach to tackle the issue and started the VOC technique for capturing the data for necessary improvement. This exercises boiled down to the conclusions on high turnaround time (TAT) and high first-time-not-right (FTNR) complaints. The immediate attention is now to reduce the complaints to the card ratio, which is calculated as

$$\text{Complaints to the card ratio} = \frac{\text{Total complaints received}}{\text{Number of cards issued}}$$

Table 17.8 Data on complaints types

No.	Top complaints for January–June, 2014	Percentage of total complaints received (%)
1	Delay/non-delivery of debit card/PIN	13
2	Account activation related	9
3	Cheque deposited at branch not credited/delayed	7
4	Address change request given at branch and not done	6
5	Delay in closure of account	5
6	AQB charges/reversal related	4
7	Delay/non-receipt of welcome kit	3
8	Alerts sent to non-account holder	3
9	Others	50

Fig. 17.5 The complaint
metrics

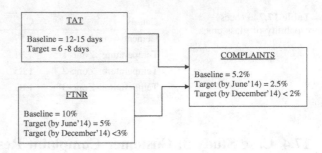

Figure 17.5 gives the metrics associated with the complaints, which will address
a major segment of the problem on the complaint resolution.

The VOC information was very effective in classifying the customer complaints
into two categories: (1) the complaints resolution necessary at the office level and
(2) the complaints resolution at the customer relationship level. Some of the issues
identified through the descriptive analysis are

- Customer unable to place request for new card without a telephone identification
  number (TIN)
- Customer had to place re-issuance request on same day or else routed to branch
- Staff not aware of TAT to be committed to customer
- Rework at branch and operations due to FTNR
- Delay in processing due to FTNR
- Customer has no active card in case new card is not delivered within 10 days,
  which leads to card swallow by ATM
- Cards not delivered due to various reasons including courier vendor
  performance
- SMS reaches customer before returned card is received at branch, which leads to
  customer harassment.

However, the root cause analysis led to the following reasons for the customer
dissatisfaction:

- *Staff*—incorrect filling of forms, eligibility not verified, address not validated at the
  time of filling the form, lack of awareness of TAT, lack of awareness of backend
  processes, discrepancy in communicating the address to courier boys, etc.
- *Process and policy*—delivery of card takes too long, unable to establish request
  through phone, old card deactivated before new card issuance, etc.
- *Measurement*—incorrect TAT communicated to customer, end to end TAT not
  monitored, incorrect quoting of complaints, etc.
- *System*—card status not monitored properly, customer address not captured
  completely, customer alternate address not captured, card re-direct request not
  captured, system updating not done frequently, etc.
- *Situation*—address not available, customer door locked, address outside delivery area, incomplete address, customer shifted to new address, etc.

Table 17.9 Prioritization of the causes

Process step	Potential failure modes	Potential effects	SEV (S)	Potential causes	OCC (O)	Current controls	DET (D)	RPN
Hot listing and re-issuance	TIN not available	Customer re-directed to branch	8	Customer forgot TIN/low TIN usage	7	None	10	560
	Re-issuance not placed on the same day	Customer re-directed to branch	7	Customer unable to locate TIN on same day	7	None	10	490
				Customer unsure about wanting to re-issue card	4	None	10	280
Card upgrade at branch	Request not processed	Delay in card processing, customer complaint	8	FTNR: incorrect or incomplete details furnished by customer	8	Cure FTNR by calling customer	5	320
Request processing by back office	Request not processed	Delay in card processing, customer complaint	8	FTNR: incorrect or incomplete form filled by branch	5	Cure FTNR by calling branch staff	4	160
Card printing and embossing	Request not processed	Delay in card processing, customer complaint	8	Shortage of row materials	3	Inventory maintenance	5	120
Card dispatch	Card not received	Customer without card/unable to transact	8	Return to origin due to door lock, incorrect address, absence of customer	7	Card returned to origin	9	504
			9	Old card deactivated	5	Card returned to origin	8	360
	Complaints logged		9	Card not received within committed TAT	8	Card returned to origin	7	504
			9	Customer communicated incorrect TAT	8	Card returned to origin	8	576

On further investigation, it is found that 75 % of FTNR reasons were attributed to staff discipline and demanded simpler process. Interestingly, 80 % of reasons are customer-dependent and difficult to control. The summary of prioritization of the causes and the risk assessment is shown in Table 17.9.

Based on the risk assessment report, the management has finally taken the challenge to contact every individual personally, and a cleanup campaign was started to place the things in order. At the office level, the following challenges have been mooted:

- System enhancement with easy browsing
- System enhanced to capture redirection
- Deferred SMS on returned cards
- Courier SLA revision
- Established in-house training for executives and below-level officers.

Finally, the management has implemented the following control factors to monitor the improvement:

- Re-engineer delivery mechanism
- Form simplification/upfront form scrutiny
- Redefine E2E TAT and develop TAT utility
- Reissue through phone channel
- Old card deactivation post delivery of new card
- Introduced green PIN.

As a global service improvement, the bank has simultaneously initiated the implementation of KAIZEN concepts and 5S (housekeeping) concept for reducing the TAT at all the transactional-level premises. Intensive training program for sales representative and executive level officials is made mandatory for permanent employees working at both urban and at rural offices.

17.5 Relevance for Managers

This chapter is very essential for all managers, Six Sigma professionals, academicians, and students to understand the real awakening of problems and solutions in business organizations. All the three projects are unique in their product and service, and hence, the quality concept also is unique. Since the delivery mechanism and the customer appraisal are different in all the three project areas, the approach to problem solving also will be seen differently. The cases presented are all practically relevant and open up many improvement areas for sustaining quality and growth.

Appendix

See Tables A.1, A.2, A.3, A.4, A.5, A.6, A.7, A.8, A.9, A.10, A.11, A.12, A.13, A.14, A.15 and A.16.

© Springer India 2015
K. Muralidharan, *Six Sigma for Organizational Excellence*,
DOI 10.1007/978-81-322-2325-2

Table A.1 Tables of binomial probability sums $\Pr(x \leq K) = \sum_{x=0}^{K} \binom{n}{x} p^x (1-p)^{n-x}$

```
n = 2
K \ p =.1  .2   .3   .4   .5   .6   .7   .8   .9
-------------------------------------------------
0 |  0.81 0.64 0.49 0.36 0.25 0.16 0.09 0.04 0.01
1 |  0.99 0.96 0.91 0.84 0.75 0.64 0.51 0.36 0.19
2 |  1.00 1.00 1.00 1.00 1.00 1.00 1.00 1.00 1.00


n = 3
K \ p =.1    .2    .3    .4    .5    .6    .7    .8    .9
-------------------------------------------------------
0 |  0.729 0.512 0.343 0.216 0.125 0.064 0.027 0.008 0.001
1 |  0.972 0.896 0.784 0.648 0.500 0.352 0.216 0.104 0.028
2 |  0.999 0.992 0.973 0.936 0.875 0.784 0.657 0.488 0.271
3 |  1.000 1.000 1.000 1.000 1.000 1.000 1.000 1.000 1.000


n = 4
K \ p =.1     .2     .3     .4     .5     .6     .7     .8     .9
----------------------------------------------------------------
0 |  0.6561 0.4096 0.2401 0.1296 0.0625 0.0256 0.0081 0.0016 0.0001
1 |  0.9477 0.8192 0.6517 0.4752 0.3125 0.1792 0.0837 0.0272 0.0037
2 |  0.9963 0.9728 0.9163 0.8208 0.6875 0.5248 0.3483 0.1808 0.0523
3 |  0.9999 0.9984 0.9919 0.9744 0.9375 0.8704 0.7599 0.5904 0.3439
4 |  1.0000 1.0000 1.0000 1.0000 1.0000 1.0000 1.0000 1.0000 1.0000


n = 5
K \ p =.1      .2      .3      .4      .5      .6      .7      .8      .9
------------------------------------------------------------------------
0 |  0.59049 0.32768 0.16807 0.07776 0.03125 0.01024 0.00243 0.00032 0.00001
1 |  0.91854 0.73728 0.52822 0.33696 0.18750 0.08704 0.03078 0.00672 0.00046
2 |  0.99144 0.94208 0.83692 0.68256 0.50000 0.31744 0.16308 0.05792 0.00856
3 |  0.99954 0.99328 0.96922 0.91296 0.81250 0.66304 0.47178 0.26272 0.08146
4 |  0.99999 0.99968 0.99757 0.98976 0.96875 0.92224 0.83193 0.67232 0.40951
5 |  1.00000 1.00000 1.00000 1.00000 1.00000 1.00000 1.00000 1.00000 1.00000


n = 6
K \ p =.1      .2      .3      .4      .5      .6      .7      .8      .9
------------------------------------------------------------------------
0 |  0.53144 0.26214 0.11765 0.04666 0.01562 0.00410 0.00073 0.00006 0.00000
1 |  0.88574 0.65536 0.42018 0.23328 0.10938 0.04096 0.01094 0.00160 0.00006
2 |  0.98415 0.90112 0.74431 0.54432 0.34375 0.17920 0.07047 0.01696 0.00127
3 |  0.99873 0.98304 0.92953 0.82080 0.65625 0.45568 0.25569 0.09888 0.01585
4 |  0.99994 0.99840 0.98906 0.95904 0.89062 0.76672 0.57982 0.34464 0.11426
5 |  1.00000 0.99994 0.99927 0.99590 0.98438 0.95334 0.88235 0.73786 0.46856
6 |  1.00000 1.00000 1.00000 1.00000 1.00000 1.00000 1.00000 1.00000 1.00000


n = 7
K \ p =.1      .2      .3      .4      .5      .6      .7      .8      .9
------------------------------------------------------------------------
0 |  0.47830 0.20972 0.08235 0.02799 0.00781 0.00164 0.00022 0.00001 0.00000
1 |  0.85031 0.57672 0.32942 0.15863 0.06250 0.01884 0.00379 0.00037 0.00001
2 |  0.97431 0.85197 0.64707 0.41990 0.22656 0.09626 0.02880 0.00467 0.00018
3 |  0.99727 0.96666 0.87396 0.71021 0.50000 0.28979 0.12604 0.03334 0.00273
4 |  0.99982 0.99533 0.97120 0.90374 0.77344 0.58010 0.35293 0.14803 0.02569
5 |  0.99999 0.99963 0.99621 0.98116 0.93750 0.84137 0.67058 0.42328 0.14969
6 |  1.00000 0.99999 0.99978 0.99836 0.99219 0.97201 0.91765 0.79028 0.52170
7 |  1.00000 1.00000 1.00000 1.00000 1.00000 1.00000 1.00000 1.00000 1.00000
```

(continued)

Table A.1 (continued)

```
n = 8
 K \ p =.1      .2       .3       .4       .5       .6       .7       .8       .9
-----------------------------------------------------------------------------------
 0 | 0.43047 0.16777 0.05765 0.01680 0.00391 0.00066 0.00007 0.00000 0.00000
 1 | 0.81310 0.50332 0.25530 0.10638 0.03516 0.00852 0.00129 0.00008 0.00000
 2 | 0.96191 0.79692 0.55177 0.31539 0.14453 0.04981 0.01129 0.00123 0.00002
 3 | 0.99498 0.94372 0.80590 0.59409 0.36328 0.17367 0.05797 0.01041 0.00043
 4 | 0.99957 0.98959 0.94203 0.82633 0.63672 0.40591 0.19410 0.05628 0.00502
 5 | 0.99998 0.99877 0.98871 0.95019 0.85547 0.68461 0.44823 0.20308 0.03809
 6 | 1.00000 0.99992 0.99871 0.99148 0.96484 0.89362 0.74470 0.49668 0.18690
 7 | 1.00000 1.00000 0.99993 0.99934 0.99609 0.98320 0.94235 0.83223 0.56953
 8 | 1.00000 1.00000 1.00000 1.00000 1.00000 1.00000 1.00000 1.00000 1.00000


n = 9
 K \ p =.1      .2       .3       .4       .5       .6       .7       .8       .9
-----------------------------------------------------------------------------------
 0 | 0.38742 0.13422 0.04035 0.01008 0.00195 0.00026 0.00002 0.00000 0.00000
 1 | 0.77484 0.43621 0.19600 0.07054 0.01953 0.00380 0.00043 0.00002 0.00000
 2 | 0.94703 0.73820 0.46283 0.23179 0.08984 0.02503 0.00429 0.00031 0.00000
 3 | 0.99167 0.91436 0.72966 0.48261 0.25391 0.09935 0.02529 0.00307 0.00006
 4 | 0.99911 0.98042 0.90119 0.73343 0.50000 0.26657 0.09881 0.01958 0.00089
 5 | 0.99994 0.99693 0.97471 0.90065 0.74609 0.51739 0.27034 0.08564 0.00833
 6 | 1.00000 0.99969 0.99571 0.97497 0.91016 0.76821 0.53717 0.26180 0.05297
 7 | 1.00000 0.99998 0.99957 0.99620 0.98047 0.92946 0.80400 0.56379 0.22516
 8 | 1.00000 1.00000 0.99998 0.99974 0.99805 0.98992 0.95965 0.86578 0.61258
 9 | 1.00000 1.00000 1.00000 1.00000 1.00000 1.00000 1.00000 1.00000 1.00000


n = 10
 K \ p = .1     .2       .3       .4       .5       .6       .7       .8       .9
-----------------------------------------------------------------------------------
 0 | 0.34868 0.10737 0.02825 0.00605 0.00098 0.00010 0.00001 0.00000 0.00000
 1 | 0.73610 0.37581 0.14931 0.04636 0.01074 0.00168 0.00014 0.00000 0.00000
 2 | 0.92981 0.67780 0.38278 0.16729 0.05469 0.01229 0.00159 0.00008 0.00000
 3 | 0.98720 0.87913 0.64961 0.38228 0.17188 0.05476 0.01059 0.00086 0.00001
 4 | 0.99837 0.96721 0.84973 0.63310 0.37695 0.16624 0.04735 0.00637 0.00015
 5 | 0.99985 0.99363 0.95265 0.83376 0.62305 0.36690 0.15027 0.03279 0.00163
 6 | 0.99999 0.99914 0.98941 0.94524 0.82812 0.61772 0.35039 0.12087 0.01280
 7 | 1.00000 0.99992 0.99841 0.98771 0.94531 0.83271 0.61722 0.32220 0.07019
 8 | 1.00000 1.00000 0.99986 0.99832 0.98926 0.95364 0.85069 0.62419 0.26390
 9 | 1.00000 1.00000 0.99999 0.99990 0.99902 0.99395 0.97175 0.89263 0.65132
10 | 1.00000 1.00000 1.00000 1.00000 1.00000 1.00000 1.00000 1.00000 1.00000


n = 11
 K \ p - .1     .2       .3       .4       .5       .6       .7       .8       .9
-----------------------------------------------------------------------------------
 0 | 0.31381 0.08590 0.01977 0.00363 0.00049 0.00004 0.00000 0.00000 0.00000
 1 | 0.69736 0.32212 0.11299 0.03023 0.00586 0.00073 0.00005 0.00000 0.00000
 2 | 0.91044 0.61740 0.31274 0.11892 0.03271 0.00592 0.00058 0.00002 0.00000
 3 | 0.98147 0.83886 0.56956 0.29628 0.11328 0.02928 0.00429 0.00024 0.00000
 4 | 0.99725 0.94959 0.78970 0.53277 0.27441 0.09935 0.02162 0.00197 0.00002
 5 | 0.99970 0.98835 0.92178 0.75350 0.50000 0.24650 0.07822 0.01165 0.00030
 6 | 0.99998 0.99803 0.97838 0.90065 0.72559 0.46723 0.21030 0.05041 0.00275
 7 | 1.00000 0.99976 0.99571 0.97072 0.88672 0.70372 0.43044 0.16114 0.01853
 8 | 1.00000 0.99998 0.99942 0.99408 0.96729 0.88108 0.68726 0.38260 0.08956
 9 | 1.00000 1.00000 0.99995 0.99927 0.99414 0.96977 0.88701 0.67788 0.30264
10 | 1.00000 1.00000 1.00000 0.99996 0.99951 0.99637 0.98023 0.91410 0.68619
11 | 1.00000 1.00000 1.00000 1.00000 1.00000 1.00000 1.00000 1.00000 1.00000


n = 12
 K \ p = .1     .2       .3       .4       .5       .6       .7       .8       .9
-----------------------------------------------------------------------------------
 0 | 0.28243 0.06872 0.01384 0.00218 0.00024 0.00002 0.00000 0.00000 0.00000
 1 | 0.65900 0.27488 0.08503 0.01959 0.00317 0.00032 0.00002 0.00000 0.00000
 2 | 0.88913 0.55835 0.25282 0.08344 0.01929 0.00281 0.00021 0.00000 0.00000
 3 | 0.97436 0.79457 0.49252 0.22534 0.07300 0.01527 0.00169 0.00006 0.00000
 4 | 0.99567 0.92744 0.72366 0.43818 0.19385 0.05731 0.00949 0.00058 0.00000
 5 | 0.99946 0.98059 0.88215 0.66521 0.38721 0.15821 0.03860 0.00390 0.00005
 6 | 0.99995 0.99610 0.96140 0.84179 0.61298 0.33479 0.11785 0.01941 0.00054
 7 | 1.00000 0.99942 0.99051 0.94269 0.80615 0.56182 0.27634 0.07256 0.00433
 8 | 1.00000 0.99994 0.99831 0.98473 0.92700 0.77466 0.50748 0.20543 0.02564
```

(continued)

Table A.1 (continued)

```
 9 | 1.00000 1.00000 0.99979 0.99719 0.98071 0.91656 0.74718 0.44165 0.11087
10 | 1.00000 1.00000 0.99998 0.99968 0.99683 0.98041 0.91497 0.72512 0.34100
11 | 1.00000 1.00000 1.00000 0.99998 0.99976 0.99782 0.98616 0.93128 0.71757
12 | 1.00000 1.00000 1.00000 1.00000 1.00000 1.00000 1.00000 1.00000 1.00000
```

n = 13

K \ p =	.1	.2	.3	.4	.5	.6	.7	.8	.9
0	0.25419	0.05498	0.00969	0.00131	0.00012	0.00001	0.00000	0.00000	0.00000
1	0.62134	0.23365	0.06367	0.01263	0.00171	0.00014	0.00000	0.00000	0.00000
2	0.86612	0.50165	0.20248	0.05790	0.01123	0.00132	0.00007	0.00000	0.00000
3	0.96584	0.74732	0.42061	0.16858	0.04614	0.00779	0.00065	0.00002	0.00000
4	0.99354	0.90087	0.65431	0.35304	0.13342	0.03208	0.00403	0.00017	0.00000
5	0.99908	0.96996	0.83460	0.57440	0.29053	0.09767	0.01822	0.00125	0.00001
6	0.99990	0.99300	0.93762	0.77116	0.50000	0.22884	0.06238	0.00700	0.00010
7	0.99999	0.99875	0.98178	0.90233	0.70947	0.42560	0.16540	0.03004	0.00092
8	1.00000	0.99983	0.99597	0.96792	0.86658	0.64696	0.34569	0.09913	0.00646
9	1.00000	0.99998	0.99935	0.99221	0.95386	0.83142	0.57939	0.25268	0.03416
10	1.00000	1.00000	0.99993	0.99868	0.98877	0.94210	0.79752	0.49835	0.13388
11	1.00000	1.00000	1.00000	0.99986	0.99829	0.98737	0.93633	0.76635	0.37866
12	1.00000	1.00000	1.00000	0.99999	0.99988	0.99869	0.99031	0.94502	0.74581
13	1.00000	1.00000	1.00000	1.00000	1.00000	1.00000	1.00000	1.00000	1.00000

n = 14

K \ p =	.1	.2	.3	.4	.5	.6	.7	.8	.9
0	0.22877	0.04398	0.00678	0.00078	0.00006	0.00000	0.00000	0.00000	0.00000
1	0.58463	0.19791	0.04748	0.00810	0.00092	0.00006	0.00000	0.00000	0.00000
2	0.84164	0.44805	0.16084	0.03979	0.00647	0.00061	0.00003	0.00000	0.00000
3	0.95587	0.69819	0.35517	0.12431	0.02869	0.00391	0.00025	0.00000	0.00000
4	0.99077	0.87016	0.58420	0.27926	0.08978	0.01751	0.00167	0.00005	0.00000
5	0.99853	0.95615	0.78052	0.48585	0.21198	0.05832	0.00829	0.00038	0.00000
6	0.99982	0.98839	0.90672	0.69245	0.39526	0.15014	0.03147	0.00240	0.00002
7	0.99998	0.99760	0.96853	0.84986	0.60474	0.30755	0.09328	0.01161	0.00018
8	1.00000	0.99962	0.99171	0.94168	0.78802	0.51415	0.21948	0.04385	0.00147
9	1.00000	0.99995	0.99833	0.98249	0.91022	0.72074	0.41580	0.12984	0.00923
10	1.00000	1.00000	0.99975	0.99609	0.97131	0.87569	0.64483	0.30181	0.04413
11	1.00000	1.00000	0.99997	0.99939	0.99353	0.96021	0.83916	0.55195	0.15836
12	1.00000	1.00000	1.00000	0.99994	0.99908	0.99190	0.95252	0.80209	0.41537
13	1.00000	1.00000	1.00000	1.00000	0.99994	0.99922	0.99322	0.95602	0.77123
14	1.00000	1.00000	1.00000	1.00000	1.00000	1.00000	1.00000	1.00000	1.00000

n = 15

K \ p =	.1	.2	.3	.4	.5	.6	.7	.8	.9
0	0.20589	0.03518	0.00475	0.00047	0.00003	0.00000	0.00000	0.00000	0.00000
1	0.54904	0.16713	0.03527	0.00517	0.00049	0.00003	0.00000	0.00000	0.00000
2	0.81594	0.39802	0.12683	0.02711	0.00369	0.00028	0.00001	0.00000	0.00000
3	0.94444	0.64816	0.29687	0.09050	0.01758	0.00193	0.00009	0.00000	0.00000
4	0.98728	0.83577	0.51549	0.21728	0.05923	0.00935	0.00067	0.00001	0.00000
5	0.99775	0.93895	0.72162	0.40322	0.15088	0.03383	0.00365	0.00011	0.00000
6	0.99969	0.98194	0.86886	0.60981	0.30362	0.09505	0.01524	0.00078	0.00000
7	0.99997	0.99576	0.94999	0.78690	0.50000	0.21310	0.05001	0.00424	0.00003
8	1.00000	0.99922	0.98476	0.90495	0.69638	0.39019	0.13114	0.01806	0.00031
9	1.00000	0.99989	0.99635	0.96617	0.84912	0.59678	0.27838	0.06105	0.00225
10	1.00000	0.99999	0.99933	0.99065	0.94077	0.78272	0.48451	0.16423	0.01272
11	1.00000	1.00000	0.99991	0.99807	0.98242	0.90950	0.70313	0.35184	0.05556
12	1.00000	1.00000	0.99999	0.99972	0.99631	0.97289	0.87317	0.60198	0.18406
13	1.00000	1.00000	1.00000	0.99997	0.99951	0.99483	0.96473	0.83287	0.45096
14	1.00000	1.00000	1.00000	1.00000	0.99997	0.99953	0.99525	0.96482	0.79411
15	1.00000	1.00000	1.00000	1.00000	1.00000	1.00000	1.00000	1.00000	1.00000

n = 16

K \ p =	.1	.2	.3	.4	.5	.6	.7	.8	.9
0	0.18530	0.02815	0.00332	0.00028	0.00002	0.00000	0.00000	0.00000	0.00000
1	0.51473	0.14074	0.02611	0.00329	0.00026	0.00001	0.00000	0.00000	0.00000
2	0.78925	0.35184	0.09936	0.01834	0.00209	0.00013	0.00000	0.00000	0.00000
3	0.93159	0.59813	0.24586	0.06515	0.01064	0.00094	0.00003	0.00000	0.00000
4	0.98300	0.79825	0.44990	0.16657	0.03841	0.00490	0.00027	0.00000	0.00000
5	0.99670	0.91831	0.65978	0.32884	0.10506	0.01914	0.00157	0.00003	0.00000
6	0.99950	0.97334	0.82469	0.52717	0.22725	0.05832	0.00713	0.00025	0.00000
7	0.99994	0.99300	0.92565	0.71606	0.40181	0.14227	0.02567	0.00148	0.00001

(continued)

Table A.1 (continued)

```
 8 | 0.99999 0.99852 0.97433 0.85773 0.59819 0.28394 0.07435 0.00700 0.00006
 9 | 1.00000 0.99975 0.99287 0.94168 0.77275 0.47283 0.17531 0.02666 0.00050
10 | 1.00000 0.99997 0.99843 0.98086 0.89494 0.67116 0.34022 0.08169 0.00330
11 | 1.00000 1.00000 0.99973 0.99510 0.96159 0.83343 0.55010 0.20175 0.01700
12 | 1.00000 1.00000 0.99997 0.99906 0.98936 0.93485 0.75414 0.40187 0.06841
13 | 1.00000 1.00000 1.00000 0.99987 0.99791 0.98166 0.90064 0.64816 0.21075
14 | 1.00000 1.00000 1.00000 0.99999 0.99974 0.99671 0.97389 0.85926 0.48527
15 | 1.00000 1.00000 1.00000 1.00000 0.99998 0.99972 0.99668 0.97185 0.81470
16 | 1.00000 1.00000 1.00000 1.00000 1.00000 1.00000 1.00000 1.00000 1.00000
```

```
n = 17
K \p  P= .1      .2      .3      .4      .5      .6      .7      .8      .9
----------------------------------------------------------------------------
 0 | 0.16677 0.02252 0.00233 0.00017 0.00001 0.00000 0.00000 0.00000 0.00000
 1 | 0.48179 0.11822 0.01928 0.00209 0.00014 0.00000 0.00000 0.00000 0.00000
 2 | 0.76180 0.30962 0.07739 0.01232 0.00117 0.00006 0.00000 0.00000 0.00000
 3 | 0.91736 0.54888 0.20191 0.04642 0.00636 0.00045 0.00001 0.00000 0.00000
 4 | 0.97786 0.75822 0.38869 0.12600 0.02452 0.00252 0.00010 0.00000 0.00000
 5 | 0.99533 0.89430 0.59682 0.26393 0.07173 0.01059 0.00066 0.00001 0.00000
 6 | 0.99922 0.96234 0.77522 0.44784 0.16615 0.03481 0.00324 0.00008 0.00000
 7 | 0.99989 0.98907 0.89536 0.64051 0.31453 0.09190 0.01269 0.00049 0.00000
 8 | 0.99999 0.99742 0.95972 0.80106 0.50000 0.19894 0.04028 0.00258 0.00001
 9 | 1.00000 0.99951 0.98731 0.90810 0.68547 0.35949 0.10464 0.01093 0.00011
10 | 1.00000 0.99992 0.99676 0.96519 0.83385 0.55216 0.22478 0.03766 0.00078
11 | 1.00000 0.99999 0.99934 0.98941 0.92827 0.73607 0.40318 0.10570 0.00467
12 | 1.00000 1.00000 0.99990 0.99748 0.97548 0.87400 0.61131 0.24178 0.02214
13 | 1.00000 1.00000 0.99999 0.99955 0.99364 0.95358 0.79809 0.45112 0.08264
14 | 1.00000 1.00000 1.00000 0.99994 0.99883 0.98768 0.92261 0.69038 0.23820
15 | 1.00000 1.00000 1.00000 1.00000 0.99986 0.99791 0.98072 0.88178 0.51821
16 | 1.00000 1.00000 1.00000 1.00000 0.99999 0.99983 0.99767 0.97748 0.83323
17 | 1.00000 1.00000 1.00000 1.00000 1.00000 1.00000 1.00000 1.00000 1.00000
```

```
n = 18
K \ p = .1      .2      .3      .4      .5      .6      .7      .8      .9
----------------------------------------------------------------------------
 0 | 0.15009 0.01801 0.00163 0.00010 0.00000 0.00000 0.00000 0.00000 0.00000
 1 | 0.45028 0.09908 0.01419 0.00132 0.00007 0.00000 0.00000 0.00000 0.00000
 2 | 0.73380 0.27134 0.05995 0.00823 0.00066 0.00003 0.00000 0.00000 0.00000
 3 | 0.90180 0.50103 0.16455 0.03278 0.00377 0.00021 0.00000 0.00000 0.00000
 4 | 0.97181 0.71635 0.33265 0.09417 0.01544 0.00128 0.00004 0.00000 0.00000
 5 | 0.99358 0.86708 0.53438 0.20876 0.04813 0.00575 0.00027 0.00000 0.00000
 6 | 0.99883 0.94873 0.72170 0.37428 0.11894 0.02028 0.00143 0.00002 0.00000
 7 | 0.99983 0.98372 0.85932 0.56344 0.24034 0.05765 0.00607 0.00016 0.00000
 8 | 0.99998 0.99575 0.94041 0.73684 0.40726 0.13471 0.02097 0.00091 0.00000
 9 | 1.00000 0.99909 0.97903 0.86529 0.59274 0.26316 0.05959 0.00425 0.00002
10 | 1.00000 0.99984 0.99393 0.94235 0.75966 0.43656 0.14068 0.01628 0.00017
11 | 1.00000 0.99998 0.99857 0.97972 0.88106 0.62572 0.27830 0.05127 0.00117
12 | 1.00000 1.00000 0.99973 0.99425 0.95187 0.79124 0.46562 0.13292 0.00642
13 | 1.00000 1.00000 0.99996 0.99872 0.98456 0.90583 0.66735 0.28365 0.02819
14 | 1.00000 1.00000 1.00000 0.99979 0.99623 0.96722 0.83545 0.49897 0.09820
15 | 1.00000 1.00000 1.00000 0.99997 0.99934 0.99177 0.94005 0.72866 0.26620
16 | 1.00000 1.00000 1.00000 1.00000 0.99993 0.99868 0.98581 0.90092 0.54972
17 | 1.00000 1.00000 1.00000 1.00000 1.00000 0.99990 0.99837 0.98199 0.84991
18 | 1.00000 1.00000 1.00000 1.00000 1.00000 1.00000 1.00000 1.00000 1.00000
```

```
n = 19
K \ p = .1      .2      .3      .4      .5      .6      .7      .8      .9
----------------------------------------------------------------------------
 0 | 0.13509 0.01441 0.00114 0.00006 0.00000 0.00000 0.00000 0.00000 0.00000
 1 | 0.42026 0.08287 0.01042 0.00083 0.00004 0.00000 0.00000 0.00000 0.00000
 2 | 0.70544 0.23689 0.04622 0.00546 0.00036 0.00001 0.00000 0.00000 0.00000
 3 | 0.88500 0.45509 0.13317 0.02296 0.00221 0.00010 0.00000 0.00000 0.00000
 4 | 0.96481 0.67329 0.28222 0.06961 0.00961 0.00064 0.00001 0.00000 0.00000
 5 | 0.99141 0.83694 0.47386 0.16292 0.03178 0.00307 0.00011 0.00000 0.00000
 6 | 0.99830 0.93240 0.66550 0.30807 0.08353 0.01156 0.00062 0.00001 0.00000
 7 | 0.99973 0.97672 0.81803 0.48778 0.17964 0.03523 0.00282 0.00005 0.00000
 8 | 0.99996 0.99334 0.91608 0.66748 0.32380 0.08847 0.01054 0.00031 0.00000
 9 | 1.00000 0.99842 0.96745 0.81391 0.50000 0.18609 0.03255 0.00158 0.00000
10 | 1.00000 0.99969 0.98946 0.91153 0.67620 0.33252 0.08392 0.00666 0.00004
11 | 1.00000 0.99995 0.99718 0.96477 0.82036 0.51222 0.18197 0.02328 0.00027
12 | 1.00000 0.99999 0.99938 0.98844 0.91647 0.69193 0.33450 0.06760 0.00170
13 | 1.00000 1.00000 0.99989 0.99693 0.96822 0.83708 0.52614 0.16306 0.00859
14 | 1.00000 1.00000 0.99999 0.99936 0.99039 0.93039 0.71778 0.32671 0.03519
```

(continued)

Table A.1 (continued)

15	1.00000	1.00000	1.00000	0.99990	0.99779	0.97704	0.86683	0.54491	0.11500
16	1.00000	1.00000	1.00000	0.99999	0.99964	0.99454	0.95378	0.76311	0.29456
17	1.00000	1.00000	1.00000	1.00000	0.99996	0.99917	0.98958	0.91713	0.57974
18	1.00000	1.00000	1.00000	1.00000	1.00000	0.99994	0.99886	0.98559	0.86491
19	1.00000	1.00000	1.00000	1.00000	1.00000	1.00000	1.00000	1.00000	1.00000

n = 20

K \ p =	.1	.2	.3	.4	.5	.6	.7	.8	.9
0	0.12158	0.01153	0.00080	0.00004	0.00000	0.00000	0.00000	0.00000	0.00000
1	0.39175	0.06918	0.00764	0.00052	0.00002	0.00000	0.00000	0.00000	0.00000
2	0.67693	0.20608	0.03548	0.00361	0.00020	0.00001	0.00000	0.00000	0.00000
3	0.86705	0.41145	0.10709	0.01596	0.00129	0.00005	0.00000	0.00000	0.00000
4	0.95683	0.62965	0.23751	0.05095	0.00591	0.00032	0.00001	0.00000	0.00000
5	0.98875	0.80421	0.41637	0.12560	0.02069	0.00161	0.00004	0.00000	0.00000
6	0.99761	0.91331	0.60801	0.25001	0.05766	0.00647	0.00026	0.00000	0.00000
7	0.99958	0.96786	0.77227	0.41589	0.13159	0.02103	0.00128	0.00002	0.00000
8	0.99994	0.99002	0.88667	0.59560	0.25172	0.05653	0.00514	0.00010	0.00000
9	0.99999	0.99741	0.95204	0.75534	0.41190	0.12752	0.01714	0.00056	0.00000
10	1.00000	0.99944	0.98286	0.87248	0.58810	0.24466	0.04796	0.00259	0.00001
11	1.00000	0.99990	0.99486	0.94347	0.74828	0.40440	0.11333	0.00998	0.00006
12	1.00000	0.99998	0.99872	0.97897	0.86841	0.58411	0.22773	0.03214	0.00042
13	1.00000	1.00000	0.99974	0.99353	0.94234	0.74999	0.39199	0.08669	0.00239
14	1.00000	1.00000	0.99996	0.99839	0.97931	0.87440	0.58363	0.19579	0.01125
15	1.00000	1.00000	0.99999	0.99968	0.99409	0.94905	0.76249	0.37035	0.04317
16	1.00000	1.00000	1.00000	0.99995	0.99871	0.98404	0.89291	0.58855	0.13295
17	1.00000	1.00000	1.00000	0.99999	0.99980	0.99639	0.96452	0.79392	0.32307
18	1.00000	1.00000	1.00000	1.00000	0.99998	0.99948	0.99236	0.93082	0.60825
19	1.00000	1.00000	1.00000	1.00000	1.00000	0.99996	0.99920	0.98847	0.87842
20	1.00000	1.00000	1.00000	1.00000	1.00000	1.00000	1.00000	1.00000	1.00000

Table A.2 Cumulative Poisson distribution tables

						$\lambda = mean$						
X	0.01	0.05	0.1	0.2	0.3	0.4	0.5	0.6	0.7	0.8	0.9
0	0.990	0.951	0.905	0.819	0.741	0.670	0.607	0.549	0.497	0.449	0.407
1	1.000	0.999	0.995	0.982	0.963	0.938	0.910	0.878	0.844	0.809	0.772
2		1.000	1.000	0.999	0.996	0.992	0.986	0.977	0.966	0.953	0.937
3				1.000	1.000	0.999	0.998	0.997	0.994	0.991	0.987
4						1.000	1.000	1.000	0.999	0.999	0.998
5									1.000	1.000	1.000
X	1.0	1.1	1.2	1.3	1.4	1.5	1.6	1.7	1.8	1.9	2.0

(continued)

Table A.2 (continued)

0	0.368	0.333	0.301	0.273	0.247	0.223	0.202	0.183	0.165	0.150	0.135
1	0.736	0.699	0.663	0.627	0.592	0.558	0.525	0.493	0.463	0.434	0.406
2	0.920	0.900	0.879	0.857	0.833	0.809	0.783	0.757	0.731	0.704	0.677
3	0.981	0.974	0.966	0.957	0.946	0.934	0.921	0.907	0.891	0.875	0.857
4	0.996	0.995	0.992	0.989	0.986	0.981	0.976	0.970	0.964	0.956	0.947
5	0.999	0.999	0.998	0.998	0.997	0.996	0.994	0.992	0.990	0.987	0.983
6	1.000	1.000	1.000	1.000	0.999	0.999	0.999	0.998	0.997	0.997	0.995
7					1.000	1.000	1.000	1.000	0.999	0.999	0.999
8									1.000	1.000	1.000

X	2.2	2.4	2.6	2.8	3.0	3.5	4.0	4.5	5.0	5.5	6.0
0	0.111	0.091	0.074	0.061	0.050	0.030	0.018	0.011	0.007	0.004	0.002
1	0.355	0.308	0.267	0.231	0.199	0.136	0.092	0.061	0.040	0.027	0.017
2	0.623	0.570	0.518	0.469	0.423	0.321	0.238	0.174	0.125	0.088	0.062
3	0.819	0.779	0.736	0.692	0.647	0.537	0.433	0.342	0.265	0.202	0.151
4	0.928	0.904	0.877	0.848	0.815	0.725	0.629	0.532	0.440	0.358	0.285
5	0.975	0.964	0.951	0.935	0.916	0.858	0.785	0.703	0.616	0.529	0.446
6	0.993	0.988	0.983	0.976	0.966	0.935	0.889	0.831	0.762	0.686	0.606
7	0.998	0.997	0.995	0.992	0.988	0.973	0.949	0.913	0.867	0.809	0.744
8	1.000	0.999	0.999	0.998	0.996	0.990	0.979	0.960	0.932	0.894	0.847
9		1.000	1.000	0.999	0.999	0.997	0.992	0.983	0.968	0.946	0.916
10				1.000	1.000	0.999	0.997	0.993	0.986	0.975	0.957
11						1.000	0.999	0.998	0.995	0.989	0.980
12							1.000	0.999	0.998	0.996	0.991
13								1.000	0.999	0.998	0.996

(continued)

Table A.2 (continued)

X	6.5	7.0	7.5	8.0	9.0	10.0	12.0	14.0	16.0	18.0	20.0
14									1.000	0.999	0.999
15									1.000	0.999
16										1.000
0	0.002	0.001	0.001	0.000							
1	0.011	0.007	0.005	0.003	0.001						
2	0.043	0.030	0.020	0.014	0.006	0.003	0.001				
3	0.112	0.082	0.059	0.042	0.021	0.010	0.002				
4	0.224	0.173	0.132	0.100	0.055	0.029	0.008	0.002			
5	0.369	0.301	0.241	0.191	0.116	0.067	0.020	0.006	0.001		
6	0.527	0.450	0.378	0.313	0.207	0.130	0.046	0.014	0.004	0.001	
7	0.673	0.599	0.525	0.453	0.324	0.220	0.090	0.032	0.010	0.003	0.001
8	0.792	0.729	0.662	0.593	0.456	0.333	0.155	0.062	0.022	0.007	0.002
9	0.877	0.830	0.776	0.717	0.587	0.458	0.242	0.109	0.043	0.015	0.005
10	0.933	0.901	0.862	0.816	0.706	0.583	0.347	0.176	0.077	0.030	0.011
11	0.966	0.947	0.921	0.888	0.803	0.697	0.462	0.260	0.127	0.055	0.021
12	0.984	0.973	0.957	0.936	0.876	0.792	0.576	0.358	0.193	0.092	0.039
13	0.993	0.987	0.978	0.966	0.926	0.864	0.682	0.464	0.275	0.143	0.066
14	0.997	0.994	0.990	0.983	0.959	0.917	0.772	0.570	0.368	0.208	0.105
15	0.999	0.998	0.995	0.992	0.978	0.951	0.844	0.669	0.467	0.287	0.157
16	1.000	0.999	0.998	0.996	0.989	0.973	0.899	0.756	0.566	0.375	0.221
17		1.000	0.999	0.998	0.995	0.986	0.937	0.827	0.659	0.469	0.297
18			1.000	0.999	0.998	0.993	0.963	0.883	0.742	0.562	0.381
19				1.000	0.999	0.997	0.979	0.923	0.812	0.651	0.470

(continued)

Table A.2 (continued)

20		1.000	0.998	0.988	0.952	0.868	0.731	0.559
21			0.999	0.994	0.971	0.911	0.799	0.644
22			1.000	0.997	0.983	0.942	0.855	0.721
23				0.999	0.991	0.963	0.899	0.787
24				0.999	0.995	0.978	0.932	0.843
25				1.000	0.997	0.987	0.955	0.888
26					0.999	0.993	0.972	0.922
27					0.999	0.996	0.983	0.948
28					1.000	0.998	0.990	0.966
29						0.999	0.994	0.978
30						0.999	0.997	0.987
31						1.000	0.998	0.992
32							0.999	0.995
33							1.000	0.997
34								0.999
35								0.999
36								1.000

Table A.3 Cumulative standard normal distribution

Z	0	0.01	0.02	0.03	0.04	0.05	0.06	0.07	0.08	0.09
−3.40	0.0003	0.0003	0.0003	0.0003	0.0003	0.0003	0.0003	0.0003	0.0003	0.0002
−3.30	0.0005	0.0005	0.0005	0.0004	0.0004	0.0004	0.0004	0.0004	0.0004	0.0003
−3.20	0.0007	0.0007	0.0006	0.0006	0.0006	0.0006	0.0006	0.0005	0.0005	0.0005
−3.10	0.0010	0.0009	0.0009	0.0009	0.0008	0.0008	0.0008	0.0008	0.0007	0.0007
−3.00	0.0013	0.0013	0.0013	0.0012	0.0012	0.0011	0.0011	0.0011	0.0010	0.0010
−2.90	0.0019	0.0018	0.0018	0.0017	0.0016	0.0016	0.0015	0.0015	0.0014	0.0014
−2.80	0.0026	0.0025	0.0024	0.0023	0.0023	0.0022	0.0021	0.0021	0.0020	0.0019
−2.70	0.0035	0.0034	0.0033	0.0032	0.0031	0.0030	0.0029	0.0028	0.0027	0.0026
−2.60	0.0047	0.0045	0.0044	0.0043	0.0041	0.0040	0.0039	0.0038	0.0037	0.0036
−2.50	0.0062	0.0060	0.0059	0.0057	0.0055	0.0054	0.0052	0.0051	0.0049	0.0048
−2.40	0.0082	0.0080	0.0078	0.0075	0.0073	0.0071	0.0069	0.0068	0.0066	0.0064
−2.30	0.0107	0.0104	0.0102	0.0099	0.0096	0.0094	0.0091	0.0089	0.0087	0.0084
−2.20	0.0139	0.0136	0.0132	0.0129	0.0125	0.0122	0.0119	0.0116	0.0113	0.0110
−2.10	0.0179	0.0174	0.0170	0.0166	0.0162	0.0158	0.0154	0.0150	0.0146	0.0143
−2.00	0.0228	0.0222	0.0217	0.0212	0.0207	0.0202	0.0197	0.0192	0.0188	0.0183
−1.90	0.0287	0.0281	0.0274	0.0268	0.0262	0.0256	0.0250	0.0244	0.0239	0.0233
−1.80	0.0359	0.0351	0.0344	0.0336	0.0329	0.0322	0.0314	0.0307	0.0301	0.0294
−1.70	0.0446	0.0436	0.0427	0.0418	0.0409	0.0401	0.0392	0.0384	0.0375	0.0367
−1.60	0.0548	0.0537	0.0526	0.0516	0.0505	0.0495	0.0485	0.0475	0.0465	0.0455
−1.50	0.0668	0.0655	0.0643	0.0630	0.0618	0.0606	0.0594	0.0582	0.0571	0.0559
−1.40	0.0808	0.0793	0.0778	0.0764	0.0749	0.0735	0.0721	0.0708	0.0694	0.0681
−1.30	0.0968	0.0951	0.0934	0.0918	0.0901	0.0885	0.0869	0.0853	0.0838	0.0823

(continued)

Table A.3 (continued)

	0	0.01	0.02	0.03	0.04	0.05	0.06	0.07	0.08	0.09
-1.20	0.1151	0.1131	0.1112	0.1093	0.1075	0.1056	0.1038	0.1020	0.1003	0.0985
-1.10	0.1357	0.1335	0.1314	0.1292	0.1271	0.1251	0.1230	0.1210	0.1190	0.1170
-1.00	0.1587	0.1562	0.1539	0.1515	0.1492	0.1469	0.1446	0.1423	0.1401	0.1379
-0.90	0.1841	0.1814	0.1788	0.1762	0.1736	0.1711	0.1685	0.1660	0.1635	0.1611
-0.80	0.2119	0.2090	0.2061	0.2033	0.2005	0.1977	0.1949	0.1922	0.1894	0.1867
-0.70	0.2420	0.2389	0.2358	0.2327	0.2296	0.2266	0.2236	0.2206	0.2177	0.2148
-0.60	0.2743	0.2709	0.2676	0.2643	0.2611	0.2578	0.2546	0.2514	0.2483	0.2451
-0.50	0.3085	0.3050	0.3015	0.2981	0.2946	0.2912	0.2877	0.2843	0.2810	0.2776
-0.40	0.3446	0.3409	0.3372	0.3336	0.3300	0.3264	0.3228	0.3192	0.3156	0.3121
-0.30	0.3821	0.3783	0.3745	0.3707	0.3669	0.3632	0.3594	0.3557	0.3520	0.3483
-0.20	0.4207	0.4168	0.4129	0.4090	0.4052	0.4013	0.3974	0.3936	0.3897	0.3859
-0.10	0.4602	0.4562	0.4522	0.4483	0.4443	0.4404	0.4364	0.4325	0.4286	0.4247
0.00	0.5000	0.4960	0.4920	0.4880	0.4840	0.4801	0.4761	0.4721	0.4681	0.4641
Z	**0**	**0.01**	**0.02**	**0.03**	**0.04**	**0.05**	**0.06**	**0.07**	**0.08**	**0.09**
0.00	0.5000	0.5040	0.5080	0.5120	0.5160	0.5199	0.5239	0.5279	0.5319	0.5359
0.10	0.5398	0.5438	0.5478	0.5517	0.5557	0.5596	0.5636	0.5675	0.5714	0.5753
0.20	0.5793	0.5832	0.5871	0.5910	0.5948	0.5987	0.6026	0.6064	0.6103	0.6141
0.30	0.6179	0.6217	0.6255	0.6293	0.6331	0.6368	0.6406	0.6443	0.6480	0.6517
0.40	0.6554	0.6591	0.6628	0.6664	0.6700	0.6736	0.6772	0.6808	0.6844	0.6879
0.50	0.6915	0.6950	0.6985	0.7019	0.7054	0.7088	0.7123	0.7157	0.7190	0.7224
0.60	0.7257	0.7291	0.7324	0.7357	0.7389	0.7422	0.7454	0.7486	0.7517	0.7549
0.70	0.7580	0.7611	0.7642	0.7673	0.7704	0.7734	0.7764	0.7794	0.7823	0.7852
0.80	0.7881	0.7910	0.7939	0.7967	0.7995	0.8023	0.8051	0.8078	0.8106	0.8133
0.90	0.8159	0.8186	0.8212	0.8238	0.8264	0.8289	0.8315	0.8340	0.8365	0.8389
1.00	0.8413	0.8438	0.8461	0.8485	0.8508	0.8531	0.8554	0.8577	0.8599	0.8621
1.10	0.8643	0.8665	0.8686	0.8708	0.8729	0.8749	0.8770	0.8790	0.8810	0.8830
1.20	0.8849	0.8869	0.8888	0.8907	0.8925	0.8944	0.8962	0.8980	0.8997	0.9015
1.30	0.9032	0.9049	0.9066	0.9082	0.9099	0.9115	0.9131	0.9147	0.9162	0.9177
1.40	0.9192	0.9207	0.9222	0.9236	0.9251	0.9265	0.9279	0.9292	0.9306	0.9319
1.50	0.9332	0.9345	0.9357	0.9370	0.9382	0.9394	0.9406	0.9418	0.9429	0.9441
1.60	0.9452	0.9463	0.9474	0.9484	0.9495	0.9505	0.9515	0.9525	0.9535	0.9545
1.70	0.9554	0.9564	0.9573	0.9582	0.9591	0.9599	0.9608	0.9616	0.9625	0.9633
1.80	0.9641	0.9649	0.9656	0.9664	0.9671	0.9678	0.9686	0.9693	0.9699	0.9706
1.90	0.9713	0.9719	0.9726	0.9732	0.9738	0.9744	0.9750	0.9756	0.9761	0.9767
2.00	0.9772	0.9778	0.9783	0.9788	0.9793	0.9798	0.9803	0.9808	0.9812	0.9817
2.10	0.9821	0.9826	0.9830	0.9834	0.9838	0.9842	0.9846	0.9850	0.9854	0.9857
2.20	0.9861	0.9864	0.9868	0.9871	0.9875	0.9878	0.9881	0.9884	0.9887	0.9890
2.30	0.9893	0.9896	0.9898	0.9901	0.9904	0.9906	0.9909	0.9911	0.9913	0.9916
2.40	0.9918	0.9920	0.9922	0.9925	0.9927	0.9929	0.9931	0.9932	0.9934	0.9936
2.50	0.9938	0.9940	0.9941	0.9943	0.9945	0.9946	0.9948	0.9949	0.9951	0.9952

(continued)

Table A.3 (continued)

2.60	0.9953	0.9955	0.9956	0.9957	0.9959	0.9960	0.9961	0.9962	0.9963	0.9964
2.70	0.9965	0.9966	0.9967	0.9968	0.9969	0.9970	0.9971	0.9972	0.9973	0.9974
2.80	0.9974	0.9975	0.9976	0.9977	0.9977	0.9978	0.9979	0.9979	0.9980	0.9981
2.90	0.9981	0.9982	0.9982	0.9983	0.9984	0.9984	0.9985	0.9985	0.9986	0.9986
3.00	0.9987	0.9987	0.9987	0.9988	0.9988	0.9989	0.9989	0.9989	0.9990	0.9990
3.10	0.9990	0.9991	0.9991	0.9991	0.9992	0.9992	0.9992	0.9992	0.9993	0.9993
3.20	0.9993	0.9993	0.9994	0.9994	0.9994	0.9994	0.9994	0.9995	0.9995	0.9995
3.30	0.9995	0.9995	0.9995	0.9996	0.9996	0.9996	0.9996	0.9996	0.9996	0.9997
3.40	0.9997	0.9997	0.9997	0.9997	0.9997	0.9997	0.9997	0.9997	0.9997	0.9998

Table A.4 Chi-squared values for a specified right tail area

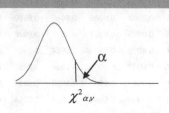

ν \ α	0.999	0.995	0.99	0.975	0.95	0.9	0.1	0.05	0.025	0.01	0.005	0.001
1	0.00	0.00	0.00	0.00	0.00	0.02	2.71	3.84	5.02	6.63	7.88	10.83
2	0.00	0.01	0.02	0.05	0.10	0.21	4.61	5.99	7.38	9.21	10.60	13.82
3	0.02	0.07	0.11	0.22	0.35	0.58	6.25	7.81	9.35	11.34	12.84	16.27
4	0.09	0.21	0.30	0.48	0.71	1.06	7.78	9.49	11.14	13.28	14.86	18.47
5	0.21	0.41	0.55	0.83	1.15	1.61	9.24	11.07	12.83	15.09	16.75	20.51
6	0.38	0.68	0.87	1.24	1.64	2.20	10.64	12.59	14.45	16.81	18.55	22.46

(continued)

Table A.4 (continued)

7	0.60	0.99	1.24	1.69	2.17	2.83	12.02	14.07	16.01	18.48	20.28	24.32
8	0.86	1.34	1.65	2.18	2.73	3.49	13.36	15.51	17.53	20.09	21.95	26.12
9	1.15	1.73	2.09	2.70	3.33	4.17	14.68	16.92	19.02	21.67	23.59	27.88
10	1.48	2.16	2.56	3.25	3.94	4.87	15.99	18.31	20.48	23.21	25.19	29.59
11	1.83	2.60	3.05	3.82	4.57	5.58	17.28	19.68	21.92	24.73	26.76	31.26
12	2.21	3.07	3.57	4.40	5.23	6.30	18.55	21.03	23.34	26.22	28.30	32.91
13	2.62	3.57	4.11	5.01	5.89	7.04	19.81	22.36	24.74	27.69	29.82	34.53
14	3.04	4.07	4.66	5.63	6.57	7.79	21.06	23.68	26.12	29.14	31.32	36.12
15	3.48	4.60	5.23	6.26	7.26	8.55	22.31	25.00	27.49	30.58	32.80	37.70
16	3.94	5.14	5.81	6.91	7.96	9.31	23.54	26.30	28.85	32.00	34.27	39.25
17	4.42	5.70	6.41	7.56	8.67	10.09	24.77	27.59	30.19	33.41	35.72	40.79
18	4.90	6.26	7.01	8.23	9.39	10.86	25.99	28.87	31.53	34.81	37.16	42.31
19	5.41	6.84	7.63	8.91	10.12	11.65	27.20	30.14	32.85	36.19	38.58	43.82
20	5.92	7.43	8.26	9.59	10.85	12.44	28.41	31.41	34.17	37.57	40.00	45.31
21	6.45	8.03	8.90	10.28	11.59	13.24	29.62	32.67	35.48	38.93	41.40	46.80
22	6.98	8.64	9.54	10.98	12.34	14.04	30.81	33.92	36.78	40.29	42.80	48.27
23	7.53	9.26	10.20	11.69	13.09	14.85	32.01	35.17	38.08	41.64	44.18	49.73
24	8.08	9.89	10.86	12.40	13.85	15.66	33.20	36.42	39.36	42.98	45.56	51.18
25	8.65	10.52	11.52	13.12	14.61	16.47	34.38	37.65	40.65	44.31	46.93	52.62
26	9.22	11.16	12.20	13.84	15.38	17.29	35.56	38.89	41.92	45.64	48.29	54.05
27	9.80	11.81	12.88	14.57	16.15	18.11	36.74	40.11	43.19	46.96	49.65	55.48

(continued)

Table A.4 (continued)

28	10.39	12.46	13.56	15.31	16.93	18.94	37.92	41.34	44.46	48.28	50.99	56.89
29	10.99	13.12	14.26	16.05	17.71	19.77	39.09	42.56	45.72	49.59	52.34	58.30
30	11.59	13.79	14.95	16.79	18.49	20.60	40.26	43.77	46.98	50.89	53.67	59.70
32	12.81	15.13	16.36	18.29	20.07	22.27	42.58	46.19	49.48	53.49	56.33	62.49
34	14.06	16.50	17.79	19.81	21.66	23.95	44.90	48.60	51.97	56.06	58.96	65.25
36	15.32	17.89	19.23	21.34	23.27	25.64	47.21	51.00	54.44	58.62	61.58	67.98
38	16.61	19.29	20.69	22.88	24.88	27.34	49.51	53.38	56.90	61.16	64.18	70.70
40	17.92	20.71	22.16	24.43	26.51	29.05	51.81	55.76	59.34	63.69	66.77	73.40
42	19.24	22.14	23.65	26.00	28.14	30.77	54.09	58.12	61.78	66.21	69.34	76.08
44	20.58	23.58	25.15	27.57	29.79	32.49	56.37	60.48	64.20	68.71	71.89	78.75
46	21.93	25.04	26.66	29.16	31.44	34.22	58.64	62.83	66.62	71.20	74.44	81.40
48	23.29	26.51	28.18	30.75	33.10	35.95	60.91	65.17	69.02	73.68	76.97	84.04
50	24.67	27.99	29.71	32.36	34.76	37.69	63.17	67.50	71.42	76.15	79.49	86.66
55	28.17	31.73	33.57	36.40	38.96	42.06	68.80	73.31	77.38	82.29	85.75	93.17
60	31.74	35.53	37.48	40.48	43.19	46.46	74.40	79.08	83.30	88.38	91.95	99.61
65	35.36	39.38	41.44	44.60	47.45	50.88	79.97	84.82	89.18	94.42	98.10	105.99
70	39.04	43.28	45.44	48.76	51.74	55.33	85.53	90.53	95.02	100.43	104.21	112.32
75	42.76	47.21	49.48	52.94	56.05	59.79	91.06	96.22	100.84	106.39	110.29	118.60
80	46.52	51.17	53.54	57.15	60.39	64.28	96.58	101.88	106.63	112.33	116.32	124.84
85	50.32	55.17	57.63	61.39	64.75	68.78	102.08	107.52	112.39	118.24	122.32	131.04
90	54.16	59.20	61.75	65.65	69.13	73.29	107.57	113.15	118.14	124.12	128.30	137.21

(continued)

Table A.4 (continued)

| 95 | 58.02 | 63.25 | 65.90 | 69.92 | 73.52 | 77.82 | 113.04 | 118.75 | 123.86 | 129.97 | 134.25 | 143.34 |
| 100 | 61.92 | 67.33 | 70.06 | 74.22 | 77.93 | 82.36 | 118.50 | 124.34 | 129.56 | 135.81 | 140.17 | 149.45 |

Table A.5 Values of T for a specified right tail area

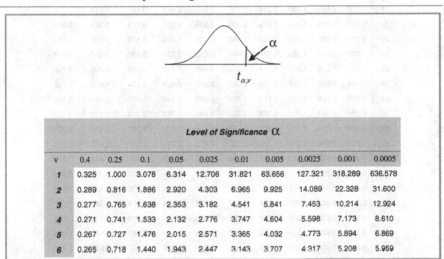

ν	0.4	0.25	0.1	0.05	0.025	0.01	0.005	0.0025	0.001	0.0005
1	0.325	1.000	3.078	6.314	12.706	31.821	63.656	127.321	318.289	636.578
2	0.289	0.816	1.886	2.920	4.303	6.965	9.925	14.089	22.328	31.600
3	0.277	0.765	1.638	2.353	3.182	4.541	5.841	7.453	10.214	12.924
4	0.271	0.741	1.533	2.132	2.776	3.747	4.604	5.598	7.173	8.610
5	0.267	0.727	1.476	2.015	2.571	3.365	4.032	4.773	5.894	6.869
6	0.265	0.718	1.440	1.943	2.447	3.143	3.707	4.317	5.208	5.959

Level of Significance α

(continued)

Table A.5 (continued)

7	0.263	0.711	1.415	1.895	2.365	2.998	3.499	4.029	4.785	5.408
8	0.262	0.706	1.397	1.860	2.306	2.896	3.355	3.833	4.501	5.041
9	0.261	0.703	1.383	1.833	2.262	2.821	3.250	3.690	4.297	4.781
10	0.260	0.700	1.372	1.812	2.228	2.764	3.169	3.581	4.144	4.587
11	0.260	0.697	1.363	1.796	2.201	2.718	3.106	3.497	4.025	4.437
12	0.259	0.695	1.356	1.782	2.179	2.681	3.055	3.428	3.930	4.318
13	0.259	0.694	1.350	1.771	2.160	2.650	3.012	3.372	3.852	4.221
14	0.258	0.692	1.345	1.761	2.145	2.624	2.977	3.326	3.787	4.140
15	0.258	0.691	1.341	1.753	2.131	2.602	2.947	3.286	3.733	4.073
16	0.258	0.690	1.337	1.746	2.120	2.583	2.921	3.252	3.686	4.015
17	0.257	0.689	1.333	1.740	2.110	2.567	2.898	3.222	3.646	3.965
18	0.257	0.688	1.330	1.734	2.101	2.552	2.878	3.197	3.610	3.922
19	0.257	0.688	1.328	1.729	2.093	2.539	2.861	3.174	3.579	3.883
20	0.257	0.687	1.325	1.725	2.086	2.528	2.845	3.153	3.552	3.850
21	0.257	0.686	1.323	1.721	2.080	2.518	2.831	3.135	3.527	3.819
22	0.256	0.686	1.321	1.717	2.074	2.508	2.819	3.119	3.505	3.792
23	0.256	0.685	1.319	1.714	2.069	2.500	2.807	3.104	3.485	3.768
24	0.256	0.685	1.318	1.711	2.064	2.492	2.797	3.091	3.467	3.745
25	0.256	0.684	1.316	1.708	2.060	2.485	2.787	3.078	3.450	3.725
26	0.256	0.684	1.315	1.706	2.056	2.479	2.779	3.067	3.435	3.707
27	0.256	0.684	1.314	1.703	2.052	2.473	2.771	3.057	3.421	3.689
28	0.256	0.683	1.313	1.701	2.048	2.467	2.763	3.047	3.408	3.674
29	0.256	0.683	1.311	1.699	2.045	2.462	2.756	3.038	3.396	3.660
30	0.256	0.683	1.310	1.697	2.042	2.457	2.750	3.030	3.385	3.646
40	0.255	0.681	1.303	1.684	2.021	2.423	2.704	2.971	3.307	3.551
60	0.254	0.679	1.296	1.671	2.000	2.390	2.660	2.915	3.232	3.460
120	0.254	0.677	1.289	1.658	1.980	2.358	2.617	2.860	3.160	3.373
∞	0.253	0.674	1.282	1.645	1.960	2.326	2.576	2.807	3.090	3.291

Table A.6 Values of F for a specified right tail area $F_{0.01(v_1,v_2)}$

X	0.01	0.05	0.1	0.2	0.3	0.4	0.5	0.6	0.7	0.8	0.9
						$\lambda = mean$					
0	0.990	0.951	0.905	0.819	0.741	0.670	0.607	0.549	0.497	0.449	0.407
1	1.000	0.999	0.995	0.982	0.963	0.938	0.910	0.878	0.844	0.809	0.772
2		1.000	1.000	0.999	0.996	0.992	0.986	0.977	0.966	0.953	0.937
3				1.000	1.000	0.999	0.998	0.997	0.994	0.991	0.987
4						1.000	1.000	1.000	0.999	0.999	0.998
5									1.000	1.000	1.000
X	1.0	1.1	1.2	1.3	1.4	1.5	1.6	1.7	1.8	1.9	2.0

(continued)

Table A.6 (continued)

df v_2	v_1																		
	1	2	3	4	5	6	7	8	9	10	12	15	20	24	30	40	60	120	∞
1	4052	4999	5404	5624	5764	5859	5928	5981	6022	6056	6107	6157	6209	6234	6260	6286	6313	6340	6366
2	98.5	99.0	99.2	99.3	99.3	99.3	99.4	99.4	99.4	99.4	99.4	99.4	99.4	99.5	99.5	99.5	99.5	99.5	99.5
3	34.1	30.8	29.5	28.7	28.2	27.9	27.7	27.5	27.3	27.2	27.1	26.9	26.7	26.6	26.5	26.4	26.3	26.2	26.1
4	21.2	18.0	16.7	16.0	15.5	15.2	15.0	14.8	14.7	14.5	14.4	14.2	14.0	13.9	13.8	13.7	13.7	13.6	13.5
5	16.3	13.3	12.1	11.4	11.0	10.7	10.5	10.3	10.2	10.1	9.89	9.72	9.55	9.47	9.38	9.29	9.20	9.11	9.02
6	13.7	10.9	9.78	9.15	8.75	8.47	8.26	8.10	7.98	7.87	7.72	7.56	7.40	7.31	7.23	7.14	7.06	6.97	6.88
7	12.2	9.55	8.45	7.85	7.46	7.19	6.99	6.84	6.72	6.62	6.47	6.31	6.16	6.07	5.99	5.91	5.82	5.74	5.65
8	11.3	8.65	7.59	7.01	6.63	6.37	6.18	6.03	5.91	5.81	5.67	5.52	5.36	5.28	5.20	5.12	5.03	4.95	4.86
9	10.6	8.02	6.99	6.42	6.06	5.80	5.61	5.47	5.35	5.26	5.11	4.96	4.81	4.73	4.65	4.57	4.48	4.40	4.31
10	10.0	7.56	6.55	5.99	5.64	5.39	5.20	5.06	4.94	4.85	4.71	4.56	4.41	4.33	4.25	4.17	4.08	4.00	3.91
11	9.65	7.21	6.22	5.67	5.32	5.07	4.89	4.74	4.63	4.54	4.40	4.25	4.10	4.02	3.94	3.86	3.78	3.69	3.60
12	9.33	6.93	5.95	5.41	5.06	4.82	4.64	4.50	4.39	4.30	4.16	4.01	3.86	3.78	3.70	3.62	3.54	3.45	3.36
13	9.07	6.70	5.74	5.21	4.86	4.62	4.44	4.30	4.19	4.10	3.96	3.82	3.66	3.59	3.51	3.43	3.34	3.25	3.17
14	8.86	6.51	5.56	5.04	4.69	4.46	4.28	4.14	4.03	3.94	3.80	3.66	3.51	3.43	3.35	3.27	3.18	3.09	3.00
15	8.68	6.36	5.42	4.89	4.56	4.32	4.14	4.00	3.89	3.80	3.67	3.52	3.37	3.29	3.21	3.13	3.05	2.96	2.87
16	8.53	6.23	5.29	4.77	4.44	4.20	4.03	3.89	3.78	3.69	3.55	3.41	3.26	3.18	3.10	3.02	2.93	2.84	2.75
17	8.40	6.11	5.19	4.67	4.34	4.10	3.93	3.79	3.68	3.59	3.46	3.31	3.16	3.08	3.00	2.92	2.83	2.75	2.65
18	8.29	6.01	5.09	4.58	4.25	4.01	3.84	3.71	3.60	3.51	3.37	3.23	3.08	3.00	2.92	2.84	2.75	2.66	2.57
19	8.18	5.93	5.01	4.50	4.17	3.94	3.77	3.63	3.52	3.43	3.30	3.15	3.00	2.92	2.84	2.76	2.67	2.58	2.49
20	8.10	5.85	4.94	4.43	4.10	3.87	3.70	3.56	3.46	3.37	3.23	3.09	2.94	2.86	2.78	2.69	2.61	2.52	2.42
21	8.02	5.78	4.87	4.37	4.04	3.81	3.64	3.51	3.40	3.31	3.17	3.03	2.88	2.80	2.72	2.64	2.55	2.46	2.36
22	7.95	5.72	4.82	4.31	3.99	3.76	3.59	3.45	3.35	3.26	3.12	2.98	2.83	2.75	2.67	2.58	2.50	2.40	2.31
23	7.88	5.66	4.76	4.26	3.94	3.71	3.54	3.41	3.30	3.21	3.07	2.93	2.78	2.70	2.62	2.54	2.45	2.35	2.26
24	7.82	5.61	4.72	4.22	3.90	3.67	3.50	3.36	3.26	3.17	3.03	2.89	2.74	2.66	2.58	2.49	2.40	2.31	2.21
25	7.77	5.57	4.68	4.18	3.85	3.63	3.46	3.32	3.22	3.13	2.99	2.85	2.70	2.62	2.54	2.45	2.36	2.27	2.17
26	7.72	5.53	4.64	4.14	3.82	3.59	3.42	3.29	3.18	3.09	2.96	2.81	2.66	2.58	2.50	2.42	2.33	2.23	2.13
27	7.68	5.49	4.60	4.11	3.78	3.56	3.39	3.26	3.15	3.06	2.93	2.78	2.63	2.55	2.47	2.38	2.29	2.20	2.10
28	7.64	5.45	4.57	4.07	3.75	3.53	3.36	3.23	3.12	3.03	2.90	2.75	2.60	2.52	2.44	2.35	2.26	2.17	2.06
29	7.60	5.42	4.54	4.04	3.73	3.50	3.33	3.20	3.09	3.00	2.87	2.73	2.57	2.49	2.41	2.33	2.23	2.14	2.03
30	7.56	5.39	4.51	4.02	3.70	3.47	3.30	3.17	3.07	2.98	2.84	2.70	2.55	2.47	2.39	2.30	2.21	2.11	2.01
40	7.31	5.18	4.31	3.83	3.51	3.29	3.12	2.99	2.89	2.80	2.66	2.52	2.37	2.29	2.20	2.11	2.02	1.92	1.80
60	7.08	4.98	4.13	3.65	3.34	3.12	2.95	2.82	2.72	2.63	2.50	2.35	2.20	2.12	2.03	1.94	1.84	1.73	1.60
120	6.85	4.79	3.95	3.48	3.17	2.96	2.79	2.66	2.56	2.47	2.34	2.19	2.03	1.95	1.86	1.76	1.66	1.53	1.38
∞	6.64	4.61	3.78	3.32	3.02	2.80	2.64	2.51	2.41	2.32	2.18	2.04	1.88	1.79	1.70	1.59	1.47	1.32	1.00

Table A.7 Values of F for a specified right tail area $F_{0.025(v_1, v_2)}$

df v_2	v_1																		
	1	2	3	4	5	6	7	8	9	10	12	15	20	24	30	40	60	120	∞
1	648	799	864	900	922	937	948	957	963	969	977	985	993	997	1001	1006	1010	1014	1018
2	38.5	39.0	39.2	39.2	39.3	39.3	39.4	39.4	39.4	39.4	39.4	39.4	39.4	39.5	39.5	39.5	39.5	39.5	39.5
3	17.4	16.0	15.4	15.1	14.9	14.7	14.6	14.5	14.5	14.4	14.3	14.3	14.2	14.1	14.1	14.0	14.0	13.9	13.9
4	12.2	10.6	9.98	9.60	9.36	9.20	9.07	8.98	8.90	8.84	8.75	8.66	8.56	8.51	8.46	8.41	8.36	8.31	8.26
5	10.0	8.43	7.76	7.39	7.15	6.98	6.85	6.76	6.68	6.62	6.52	6.43	6.33	6.28	6.23	6.18	6.12	6.07	6.02
6	8.81	7.26	6.60	6.23	5.99	5.82	5.70	5.60	5.52	5.46	5.37	5.27	5.17	5.12	5.07	5.01	4.96	4.90	4.85
7	8.07	6.54	5.89	5.52	5.29	5.12	4.99	4.90	4.82	4.76	4.67	4.57	4.47	4.41	4.36	4.31	4.25	4.20	4.14
8	7.57	6.06	5.42	5.05	4.82	4.65	4.53	4.43	4.36	4.30	4.20	4.10	4.00	3.95	3.89	3.84	3.78	3.73	3.67
9	7.21	5.71	5.08	4.72	4.48	4.32	4.20	4.10	4.03	3.96	3.87	3.77	3.67	3.61	3.56	3.51	3.45	3.39	3.33
10	6.94	5.46	4.83	4.47	4.24	4.07	3.95	3.85	3.78	3.72	3.62	3.52	3.42	3.37	3.31	3.26	3.20	3.14	3.08
11	6.72	5.26	4.63	4.28	4.04	3.88	3.76	3.66	3.59	3.53	3.43	3.33	3.23	3.17	3.12	3.06	3.00	2.94	2.88
12	6.55	5.10	4.47	4.12	3.89	3.73	3.61	3.51	3.44	3.37	3.28	3.18	3.07	3.02	2.96	2.91	2.85	2.79	2.73
13	6.41	4.97	4.35	4.00	3.77	3.60	3.48	3.39	3.31	3.25	3.15	3.05	2.95	2.89	2.84	2.78	2.72	2.66	2.60
14	6.30	4.86	4.24	3.89	3.66	3.50	3.38	3.29	3.21	3.15	3.05	2.95	2.84	2.79	2.73	2.67	2.61	2.55	2.49
15	6.20	4.77	4.15	3.80	3.58	3.41	3.29	3.20	3.12	3.06	2.96	2.86	2.76	2.70	2.64	2.59	2.52	2.46	2.40
16	6.12	4.69	4.08	3.73	3.50	3.34	3.22	3.12	3.05	2.99	2.89	2.79	2.68	2.63	2.57	2.51	2.45	2.38	2.32
17	6.04	4.62	4.01	3.66	3.44	3.28	3.16	3.06	2.98	2.92	2.82	2.72	2.62	2.56	2.50	2.44	2.38	2.32	2.25
18	5.98	4.56	3.95	3.61	3.38	3.22	3.10	3.01	2.93	2.87	2.77	2.67	2.56	2.50	2.44	2.38	2.32	2.26	2.19
19	5.92	4.51	3.90	3.56	3.33	3.17	3.05	2.96	2.88	2.82	2.72	2.62	2.51	2.45	2.39	2.33	2.27	2.20	2.13
20	5.87	4.46	3.86	3.51	3.29	3.13	3.01	2.91	2.84	2.77	2.68	2.57	2.46	2.41	2.35	2.29	2.22	2.16	2.09
21	5.83	4.42	3.82	3.48	3.25	3.09	2.97	2.87	2.80	2.73	2.64	2.53	2.42	2.37	2.31	2.25	2.18	2.11	2.04
22	5.79	4.38	3.78	3.44	3.22	3.05	2.93	2.84	2.76	2.70	2.60	2.50	2.39	2.33	2.27	2.21	2.14	2.08	2.00
23	5.75	4.35	3.75	3.41	3.18	3.02	2.90	2.81	2.73	2.67	2.57	2.47	2.36	2.30	2.24	2.18	2.11	2.04	1.97
24	5.72	4.32	3.72	3.38	3.15	2.99	2.87	2.78	2.70	2.64	2.54	2.44	2.33	2.27	2.21	2.15	2.08	2.01	1.94
25	5.69	4.29	3.69	3.35	3.13	2.97	2.85	2.75	2.68	2.61	2.51	2.41	2.30	2.24	2.18	2.12	2.05	1.98	1.91
26	5.66	4.27	3.67	3.33	3.10	2.94	2.82	2.73	2.65	2.59	2.49	2.39	2.28	2.22	2.16	2.09	2.03	1.95	1.88
27	5.63	4.24	3.65	3.31	3.08	2.92	2.80	2.71	2.63	2.57	2.47	2.36	2.25	2.19	2.13	2.07	2.00	1.93	1.85
28	5.61	4.22	3.63	3.29	3.06	2.90	2.78	2.69	2.61	2.55	2.45	2.34	2.23	2.17	2.11	2.05	1.98	1.91	1.83
29	5.59	4.20	3.61	3.27	3.04	2.88	2.76	2.67	2.59	2.53	2.43	2.32	2.21	2.15	2.09	2.03	1.96	1.89	1.81
30	5.57	4.18	3.59	3.25	3.03	2.87	2.75	2.65	2.57	2.51	2.41	2.31	2.20	2.14	2.07	2.01	1.94	1.87	1.79
40	5.42	4.05	3.46	3.13	2.90	2.74	2.62	2.53	2.45	2.39	2.29	2.18	2.07	2.01	1.94	1.88	1.80	1.72	1.64
60	5.29	3.93	3.34	3.01	2.79	2.63	2.51	2.41	2.33	2.27	2.17	2.06	1.94	1.88	1.82	1.74	1.67	1.58	1.48
120	5.15	3.80	3.23	2.89	2.67	2.52	2.39	2.30	2.22	2.16	2.05	1.94	1.82	1.76	1.69	1.61	1.53	1.43	1.31
∞	5.02	3.69	3.12	2.79	2.57	2.41	2.29	2.19	2.11	2.05	1.94	1.83	1.71	1.64	1.57	1.48	1.39	1.27	1.00

Table A.8 Values of F for a specified right tail area $F_{0.05(v_1, v_2)}$

df v_2	v_1 1	2	3	4	5	6	7	8	9	10	12	15	20	24	30	40	60	120	∞
1	161	199	216	225	230	234	237	239	241	242	244	246	248	249	250	251	252	253	254
2	18.5	19.0	19.2	19.2	19.3	19.3	19.4	19.4	19.4	19.4	19.4	19.4	19.4	19.5	19.5	19.5	19.5	19.5	19.5
3	10.1	9.55	9.28	9.12	9.01	8.94	8.89	8.85	8.81	8.79	8.74	8.70	8.66	8.64	8.62	8.59	8.57	8.55	8.53
4	7.71	6.94	6.59	6.39	6.26	6.16	6.09	6.04	6.00	5.96	5.91	5.86	5.80	5.77	5.75	5.72	5.69	5.66	5.63
5	6.61	5.79	5.41	5.19	5.05	4.95	4.88	4.82	4.77	4.74	4.68	4.62	4.56	4.53	4.50	4.46	4.43	4.40	4.37
6	5.99	5.14	4.76	4.53	4.39	4.28	4.21	4.15	4.10	4.06	4.00	3.94	3.87	3.84	3.81	3.77	3.74	3.70	3.67
7	5.59	4.74	4.35	4.12	3.97	3.87	3.79	3.73	3.68	3.64	3.57	3.51	3.44	3.41	3.38	3.34	3.30	3.27	3.23
8	5.32	4.46	4.07	3.84	3.69	3.58	3.50	3.44	3.39	3.35	3.28	3.22	3.15	3.12	3.08	3.04	3.01	2.97	2.93
9	5.12	4.26	3.86	3.63	3.48	3.37	3.29	3.23	3.18	3.14	3.07	3.01	2.94	2.90	2.86	2.83	2.79	2.75	2.71
10	4.96	4.10	3.71	3.48	3.33	3.22	3.14	3.07	3.02	2.98	2.91	2.85	2.77	2.74	2.70	2.66	2.62	2.58	2.54
11	4.84	3.98	3.59	3.36	3.20	3.09	3.01	2.95	2.90	2.85	2.79	2.72	2.65	2.61	2.57	2.53	2.49	2.45	2.40
12	4.75	3.89	3.49	3.26	3.11	3.00	2.91	2.85	2.80	2.75	2.69	2.62	2.54	2.51	2.47	2.43	2.38	2.34	2.30
13	4.67	3.81	3.41	3.18	3.03	2.92	2.83	2.77	2.71	2.67	2.60	2.53	2.46	2.42	2.38	2.34	2.30	2.25	2.21
14	4.60	3.74	3.34	3.11	2.96	2.85	2.76	2.70	2.65	2.60	2.53	2.46	2.39	2.35	2.31	2.27	2.22	2.18	2.13
15	4.54	3.68	3.29	3.06	2.90	2.79	2.71	2.64	2.59	2.54	2.48	2.40	2.33	2.29	2.25	2.20	2.16	2.11	2.07
16	4.49	3.63	3.24	3.01	2.85	2.74	2.66	2.59	2.54	2.49	2.42	2.35	2.28	2.24	2.19	2.15	2.11	2.06	2.01
17	4.45	3.59	3.20	2.96	2.81	2.70	2.61	2.55	2.49	2.45	2.38	2.31	2.23	2.19	2.15	2.10	2.06	2.01	1.96
18	4.41	3.55	3.16	2.93	2.77	2.66	2.58	2.51	2.46	2.41	2.34	2.27	2.19	2.15	2.11	2.06	2.02	1.97	1.92
19	4.38	3.52	3.13	2.90	2.74	2.63	2.54	2.48	2.42	2.38	2.31	2.23	2.16	2.11	2.07	2.03	1.98	1.93	1.88
20	4.35	3.49	3.10	2.87	2.71	2.60	2.51	2.45	2.39	2.35	2.28	2.20	2.12	2.08	2.04	1.99	1.95	1.90	1.84
21	4.32	3.47	3.07	2.84	2.68	2.57	2.49	2.42	2.37	2.32	2.25	2.18	2.10	2.05	2.01	1.96	1.92	1.87	1.81
22	4.30	3.44	3.05	2.82	2.66	2.55	2.46	2.40	2.34	2.30	2.23	2.15	2.07	2.03	1.98	1.94	1.89	1.84	1.78
23	4.28	3.42	3.03	2.80	2.64	2.53	2.44	2.37	2.32	2.27	2.20	2.13	2.05	2.01	1.96	1.91	1.86	1.81	1.76
24	4.26	3.40	3.01	2.78	2.62	2.51	2.42	2.36	2.30	2.25	2.18	2.11	2.03	1.98	1.94	1.89	1.84	1.79	1.73
25	4.24	3.39	2.99	2.76	2.60	2.49	2.40	2.34	2.28	2.24	2.16	2.09	2.01	1.96	1.92	1.87	1.82	1.77	1.71
26	4.23	3.37	2.98	2.74	2.59	2.47	2.39	2.32	2.27	2.22	2.15	2.07	1.99	1.95	1.90	1.85	1.80	1.75	1.69
27	4.21	3.35	2.96	2.73	2.57	2.46	2.37	2.31	2.25	2.20	2.13	2.06	1.97	1.93	1.88	1.84	1.79	1.73	1.67
28	4.20	3.34	2.95	2.71	2.56	2.45	2.36	2.29	2.24	2.19	2.12	2.04	1.96	1.91	1.87	1.82	1.77	1.71	1.65
29	4.18	3.33	2.93	2.70	2.55	2.43	2.35	2.28	2.22	2.18	2.10	2.03	1.94	1.90	1.85	1.81	1.75	1.70	1.64
30	4.17	3.32	2.92	2.69	2.53	2.42	2.33	2.27	2.21	2.16	2.09	2.01	1.93	1.89	1.84	1.79	1.74	1.68	1.62
40	4.08	3.23	2.84	2.61	2.45	2.34	2.25	2.18	2.12	2.08	2.00	1.92	1.84	1.79	1.74	1.69	1.64	1.58	1.51
60	4.00	3.15	2.76	2.53	2.37	2.25	2.17	2.10	2.04	1.99	1.92	1.84	1.75	1.70	1.65	1.59	1.53	1.47	1.39
120	3.92	3.07	2.6 8	2.45	2.29	2.18	2.09	2.02	1.96	1.91	1.83	1.75	1.66	1.61	1.55	1.50	1.43	1.35	1.25
∞	3.84	3.00	2.60	2.37	2.21	2.10	2.01	1.94	1.88	1.83	1.75	1.67	1.57	1.52	1.46	1.39	1.32	1.22	1.00

Table A.9 Values of F for a specified right tail area $F_{0.10(v_1,v_2)}$

df	v_1																		
v_2	1	2	3	4	5	6	7	8	9	10	12	15	20	24	30	40	60	120	∞
1	39.9	49.5	53.6	55.8	57.2	58.2	58.9	59.4	59.9	60.2	60.7	61.2	61.7	62.0	62.3	62.5	62.8	63.1	63.3
2	8.53	9.00	9.16	9.24	9.29	9.33	9.35	9.37	9.38	9.39	9.41	9.42	9.44	9.45	9.46	9.47	9.47	9.48	9.49
3	5.54	5.46	5.39	5.34	5.31	5.28	5.27	5.25	5.24	5.23	5.22	5.20	5.18	5.18	5.17	5.16	5.15	5.14	5.13
4	4.54	4.32	4.19	4.11	4.05	4.01	3.98	3.95	3.94	3.92	3.90	3.87	3.84	3.83	3.82	3.80	3.79	3.78	3.76
5	4.06	3.78	3.62	3.52	3.45	3.40	3.37	3.34	3.32	3.30	3.27	3.24	3.21	3.19	3.17	3.16	3.14	3.12	3.11
6	3.78	3.46	3.29	3.18	3.11	3.05	3.01	2.98	2.96	2.94	2.90	2.87	2.84	2.82	2.80	2.78	2.76	2.74	2.72
7	3.59	3.26	3.07	2.96	2.88	2.83	2.78	2.75	2.72	2.70	2.67	2.63	2.59	2.58	2.56	2.54	2.51	2.49	2.47
8	3.46	3.11	2.92	2.81	2.73	2.67	2.62	2.59	2.56	2.54	2.50	2.46	2.42	2.40	2.38	2.36	2.34	2.32	2.29
9	3.36	3.01	2.81	2.69	2.61	2.55	2.51	2.47	2.44	2.42	2.38	2.34	2.30	2.28	2.25	2.23	2.21	2.18	2.16
10	3.29	2.92	2.73	2.61	2.52	2.46	2.41	2.38	2.35	2.32	2.28	2.24	2.20	2.18	2.16	2.13	2.11	2.08	2.06
11	3.23	2.86	2.66	2.54	2.45	2.39	2.34	2.30	2.27	2.25	2.21	2.17	2.12	2.10	2.08	2.05	2.03	2.00	1.97
12	3.18	2.81	2.61	2.48	2.39	2.33	2.28	2.24	2.21	2.19	2.15	2.10	2.06	2.04	2.01	1.99	1.96	1.93	1.90
13	3.14	2.76	2.56	2.43	2.35	2.28	2.23	2.20	2.16	2.14	2.10	2.05	2.01	1.98	1.96	1.93	1.90	1.88	1.85
14	3.10	2.73	2.52	2.39	2.31	2.24	2.19	2.15	2.12	2.10	2.05	2.01	1.96	1.94	1.91	1.89	1.86	1.83	1.80
15	3.07	2.70	2.49	2.36	2.27	2.21	2.16	2.12	2.09	2.06	2.02	1.97	1.92	1.90	1.87	1.85	1.82	1.79	1.76
16	3.05	2.67	2.46	2.33	2.24	2.18	2.13	2.09	2.06	2.03	1.99	1.94	1.89	1.87	1.84	1.81	1.78	1.75	1.72
17	3.03	2.64	2.44	2.31	2.22	2.15	2.10	2.06	2.03	2.00	1.96	1.91	1.86	1.84	1.81	1.78	1.75	1.72	1.69
18	3.01	2.62	2.42	2.29	2.20	2.13	2.08	2.04	2.00	1.98	1.93	1.89	1.84	1.81	1.78	1.75	1.72	1.69	1.66
19	2.99	2.61	2.40	2.27	2.18	2.11	2.06	2.02	1.98	1.96	1.91	1.86	1.81	1.79	1.76	1.73	1.70	1.67	1.63
20	2.97	2.59	2.38	2.25	2.16	2.09	2.04	2.00	1.96	1.94	1.89	1.84	1.79	1.77	1.74	1.71	1.68	1.64	1.61
21	2.96	2.57	2.36	2.23	2.14	2.08	2.02	1.98	1.95	1.92	1.87	1.83	1.78	1.75	1.72	1.69	1.66	1.62	1.59
22	2.95	2.56	2.35	2.22	2.13	2.06	2.01	1.97	1.93	1.90	1.86	1.81	1.76	1.73	1.70	1.67	1.64	1.60	1.57
23	2.94	2.55	2.34	2.21	2.11	2.05	1.99	1.95	1.92	1.89	1.84	1.80	1.74	1.72	1.69	1.66	1.62	1.59	1.55
24	2.93	2.54	2.33	2.19	2.10	2.04	1.98	1.94	1.91	1.88	1.83	1.78	1.73	1.70	1.67	1.64	1.61	1.57	1.53
25	2.92	2.53	2.32	2.18	2.09	2.02	1.97	1.93	1.89	1.87	1.82	1.77	1.72	1.69	1.66	1.63	1.59	1.56	1.52
26	2.91	2.52	2.31	2.17	2.08	2.01	1.96	1.92	1.88	1.86	1.81	1.76	1.71	1.68	1.65	1.61	1.58	1.54	1.50
27	2.90	2.51	2.30	2.17	2.07	2.00	1.95	1.91	1.87	1.85	1.80	1.75	1.70	1.67	1.64	1.60	1.57	1.53	1.49
28	2.89	2.50	2.29	2.16	2.06	2.00	1.94	1.90	1.87	1.84	1.79	1.74	1.69	1.66	1.63	1.59	1.56	1.52	1.48
29	2.89	2.50	2.28	2.15	2.06	1.99	1.93	1.89	1.86	1.83	1.78	1.73	1.68	1.65	1.62	1.58	1.55	1.51	1.47
30	2.88	2.49	2.28	2.14	2.05	1.98	1.93	1.88	1.85	1.82	1.77	1.72	1.67	1.64	1.61	1.57	1.54	1.50	1.46
40	2.84	2.44	2.23	2.09	2.00	1.93	1.87	1.83	1.79	1.76	1.71	1.66	1.61	1.57	1.54	1.51	1.47	1.42	1.38
60	2.79	2.39	2.18	2.04	1.95	1.87	1.82	1.77	1.74	1.71	1.66	1.60	1.54	1.51	1.48	1.44	1.40	1.35	1.29
120	2.75	2.35	2.13	1.99	1.90	1.82	1.77	1.72	1.68	1.65	1.60	1.55	1.48	1.45	1.41	1.37	1.32	1.26	1.19
∞	2.71	2.30	2.08	1.94	1.85	1.77	1.72	1.67	1.63	1.60	1.55	1.49	1.42	1.38	1.34	1.30	1.24	1.17	1.00

Table A.10 Values of F for a specified right tail area $F_{0..25(v_1,v_2)}$

df v_2	v_1																		
	1	2	3	4	5	6	7	8	9	10	12	15	20	24	30	40	60	120	∞
1	5.83	7.50	8.20	8.58	8.82	8.98	9.10	9.19	9.26	9.32	9.41	9.49	9.58	9.63	9.67	9.71	9.76	9.80	9.85
2	2.57	3.00	3.15	3.23	3.28	3.31	3.34	3.35	3.37	3.38	3.39	3.41	3.43	3.43	3.44	3.45	3.46	3.47	3.48
3	2.02	2.28	2.36	2.39	2.41	2.42	2.43	2.44	2.44	2.44	2.45	2.46	2.46	2.46	2.47	2.47	2.47	2.47	2.47
4	1.81	2.00	2.05	2.06	2.07	2.08	2.08	2.08	2.08	2.08	2.08	2.08	2.08	2.08	2.08	2.08	2.08	2.08	2.08
5	1.69	1.85	1.88	1.89	1.89	1.89	1.89	1.89	1.89	1.89	1.89	1.89	1.88	1.88	1.88	1.88	1.87	1.87	1.87
6	1.62	1.76	1.78	1.79	1.79	1.78	1.78	1.78	1.77	1.77	1.77	1.76	1.76	1.75	1.75	1.75	1.74	1.74	1.74
7	1.57	1.70	1.72	1.72	1.71	1.71	1.70	1.70	1.69	1.69	1.68	1.68	1.67	1.67	1.66	1.66	1.65	1.65	1.65
8	1.54	1.66	1.67	1.66	1.66	1.65	1.64	1.64	1.63	1.63	1.62	1.62	1.61	1.60	1.60	1.59	1.59	1.58	1.58
9	1.51	1.62	1.63	1.63	1.62	1.61	1.60	1.60	1.59	1.59	1.58	1.57	1.56	1.56	1.55	1.54	1.54	1.53	1.53
10	1.49	1.60	1.60	1.59	1.59	1.58	1.57	1.56	1.56	1.55	1.54	1.53	1.52	1.52	1.51	1.51	1.50	1.49	1.48
11	1.47	1.58	1.58	1.57	1.56	1.55	1.54	1.53	1.53	1.52	1.51	1.50	1.49	1.49	1.48	1.47	1.47	1.46	1.45
12	1.46	1.56	1.56	1.55	1.54	1.53	1.52	1.51	1.51	1.50	1.49	1.48	1.47	1.46	1.45	1.45	1.44	1.43	1.42
13	1.45	1.55	1.55	1.53	1.52	1.51	1.50	1.49	1.49	1.48	1.47	1.46	1.45	1.44	1.43	1.42	1.42	1.41	1.40
14	1.44	1.53	1.53	1.52	1.51	1.50	1.49	1.48	1.47	1.46	1.45	1.44	1.43	1.42	1.41	1.41	1.40	1.39	1.38
15	1.43	1.52	1.52	1.51	1.49	1.48	1.47	1.46	1.46	1.45	1.44	1.43	1.41	1.41	1.40	1.39	1.38	1.37	1.36
16	1.42	1.51	1.51	1.50	1.48	1.47	1.46	1.45	1.44	1.44	1.43	1.41	1.40	1.39	1.38	1.37	1.36	1.35	1.34
17	1.42	1.51	1.50	1.49	1.47	1.46	1.45	1.44	1.43	1.43	1.41	1.40	1.39	1.38	1.37	1.36	1.35	1.34	1.33
18	1.41	1.50	1.49	1.48	1.46	1.45	1.44	1.43	1.42	1.42	1.40	1.39	1.38	1.37	1.36	1.35	1.34	1.33	1.32
19	1.41	1.49	1.49	1.47	1.46	1.44	1.43	1.42	1.41	1.41	1.40	1.38	1.37	1.36	1.35	1.34	1.33	1.32	1.30
20	1.40	1.49	1.48	1.47	1.45	1.44	1.43	1.42	1.41	1.40	1.39	1.37	1.36	1.35	1.34	1.33	1.32	1.31	1.29
21	1.40	1.48	1.48	1.46	1.44	1.43	1.42	1.41	1.40	1.39	1.38	1.37	1.35	1.34	1.33	1.32	1.31	1.30	1.28
22	1.40	1.48	1.47	1.45	1.44	1.42	1.41	1.40	1.39	1.39	1.37	1.36	1.34	1.33	1.32	1.31	1.30	1.29	1.28
23	1.39	1.47	1.47	1.45	1.43	1.42	1.41	1.40	1.39	1.38	1.37	1.35	1.34	1.33	1.32	1.31	1.30	1.28	1.27
24	1.39	1.47	1.46	1.44	1.43	1.41	1.40	1.39	1.38	1.38	1.36	1.35	1.33	1.32	1.31	1.30	1.29	1.28	1.26
25	1.39	1.47	1.46	1.44	1.42	1.41	1.40	1.39	1.38	1.37	1.36	1.34	1.33	1.32	1.31	1.29	1.28	1.27	1.25
26	1.38	1.46	1.45	1.44	1.42	1.41	1.39	1.38	1.37	1.37	1.35	1.34	1.32	1.31	1.30	1.29	1.28	1.26	1.25
27	1.38	1.46	1.45	1.43	1.42	1.40	1.39	1.38	1.37	1.36	1.35	1.33	1.32	1.31	1.30	1.28	1.27	1.26	1.24
28	1.38	1.46	1.45	1.43	1.41	1.40	1.39	1.38	1.37	1.36	1.34	1.33	1.31	1.30	1.29	1.28	1.27	1.25	1.24
29	1.38	1.45	1.45	1.43	1.41	1.40	1.38	1.37	1.36	1.35	1.34	1.32	1.31	1.30	1.29	1.27	1.26	1.25	1.23
30	1.38	1.45	1.44	1.42	1.41	1.39	1.38	1.37	1.36	1.35	1.34	1.32	1.30	1.29	1.28	1.27	1.26	1.24	1.23
40	1.36	1.44	1.42	1.40	1.39	1.37	1.36	1.35	1.34	1.33	1.31	1.30	1.28	1.26	1.25	1.24	1.22	1.21	1.19
60	1.35	1.42	1.41	1.38	1.37	1.35	1.33	1.32	1.31	1.30	1.29	1.27	1.25	1.24	1.22	1.21	1.19	1.17	1.15
120	1.34	1.40	1.39	1.37	1.35	1.33	1.31	1.30	1.29	1.28	1.26	1.24	1.22	1.21	1.19	1.18	1.16	1.13	1.10
∞	1.32	1.39	1.37	1.35	1.33	1.31	1.29	1.28	1.27	1.25	1.24	1.22	1.19	1.18	1.16	1.14	1.12	1.08	1.00

Table A.11 Control chart coefficients for variables

Subgroup size n	Control limits for $\overline{X}$-chart			Control limits for σ-chart					d_2	d_3	Control limits for $\overline{R}$-chart			
	A	A_2	A_3	c_4	B_3	B_4	B_5	B_6			D_1	D_2	D_3	D_4
2	2.121	1.880	2.659	0.7979	0	3.267	0	2.606	1.128	0.853	0	3.686	0	3.267
3	1.732	1.023	1.954	0.8862	0	2.568	0	2.276	1.693	0.888	0	4.358	0	2.575
4	1.500	0.729	1.628	0.9213	0	2.266	0	2.088	2.059	0.880	0	4.698	0	2.282
5	1.342	0.577	1.427	0.9400	0	2.089	0	1.964	2.326	0.864	0	4.918	0	2.115
6	1.225	0.483	1.287	0.9515	0.030	1.970	0.029	1.874	2.534	0.848	0	5.078	0	2.004
7	1.134	0.419	1.182	0.9594	0.118	1.882	0.113	1.806	2.704	0.833	0.204	5.204	0.076	1.924
8	1.061	0.373	1.099	0.9650	0.185	1.815	0.179	1.751	2.847	0.820	0.388	5.306	0.136	1.864
9	1.000	0.337	1.032	0.9693	0.239	1.761	0.232	1.707	2.970	0.808	0.547	5.393	0.184	1.816
10	0.949	0.308	0.975	0.9727	0.284	1.716	0.276	1.669	3.078	0.797	0.687	5.469	0.223	1.777
11	0.905	0.285	0.927	0.9754	0.321	1.679	0.313	1.637	3.173	0.787	0.811	5.355	0.256	1.744
12	0.866	0.266	0.886	0.9776	0.354	1.646	0.346	1.610	3.258	0.778	0.922	5.594	0.283	1.717
13	0.832	0.249	0.850	0.9794	0.382	1.618	0.374	1.585	3.336	0.770	1.025	5.647	0.307	1.693
14	0.802	0.235	0.817	0.9810	0.406	1.594	0.399	1.563	3.407	0.763	1.118	5.696	0.328	1.672
15	0.775	0.223	0.789	0.9823	0.428	1.572	0.421	1.544	3.472	0.756	1.203	5.741	0.347	1.653
20	0.671	0.18	0.680	0.9869	0.510	1.490	0.504	1.470	3.735	0.729	1.549	5.921	0.415	1.585
25	0.600	0.153	0.606	0.9896	0.565	1.435	0.559	1.420	3.931	0.708	1.806	6.056	0.459	1.541

For $n > 25$: The quantities are approximated as

$$A = \frac{3}{\sqrt{n}}, A_3 = \frac{3}{c_4\sqrt{n}}, c_4 = \frac{4(n-1)}{4n-3}; B_3 = 1 - \frac{3}{c_4\sqrt{2(n-1)}}, B_4 = 1 + \frac{3}{c_4\sqrt{2(n-1)}}; B_5 = c_4 - \frac{3}{\sqrt{2(n-1)}}, B_4 = c_4 + \frac{3}{\sqrt{2(n-1)}}$$

Table A.12 Process capability versus sigma level

Capability index (C_{pk})	Process Sigma		DPMO	Capability index (C_{pk})	Process Sigma		DPMO
	short term	long term			short term	long term	
2.00	6.0	4.5	3.4	1.13	3.4	1.9	28716
1.97	5.9	4.4	5.4	1.10	3.3	1.8	35930
1.93	5.8	4.3	8.5	1.07	3.2	1.7	44565
1.90	5.7	4.2	13.4	1.03	3.1	1.6	54799
1.87	5.6	4.1	21	1.00	3.0	1.5	66807
1.83	5.5	4.0	32	0.97	2.9	1.4	80757
1.80	5.4	3.9	48	0.93	2.8	1.3	96801
1.77	5.3	3.8	72	0.90	2.7	1.2	115070
1.73	5.2	3.7	108	0.87	2.6	1.1	135666
1.70	5.1	3.6	159	0.83	2.5	1.0	158655
1.67	5.0	3.5	233	0.80	2.4	0.9	184060
1.63	4.9	3.4	337	0.77	2.3	0.8	211855
1.60	4.8	3.3	483	0.73	2.2	0.7	241964
1.57	4.7	3.2	687	0.70	2.1	0.6	274253
1.53	4.6	3.1	968	0.67	2.0	0.5	308538
1.50	4.5	3.0	1350	0.63	1.9	0.4	344578
1.47	4.4	2.9	1866	0.60	1.8	0.3	382089
1.43	4.3	2.8	2555	0.57	1.7	0.2	420740
1.40	4.2	2.7	3467	0.53	1.6	0.1	460172
1.37	4.1	2.6	4661	0.50	1.5	0	500000
1.33	4.0	2.5	6210	0.47	1.4	−0.1	539828
1.30	3.9	2.4	8198	0.43	1.3	−0.2	579260
1.27	3.8	2.3	10724	0.40	1.2	−0.3	617911
1.23	3.7	2.2	13903	0.37	1.1	−0.4	655422
1.20	3.6	2.1	17864	0.33	1.0	−0.5	691462
1.17	3.5	2.0	22750	0.30			

Table A.13 Sigma value versus DPMO

Sigma value	DPMO	Sigma value	DPMO	Sigma value	DPMO
0.00	933193	4.05	5386	5.05	193
0.50	841345	4.10	4661	5.10	159
0.75	773373	4.15	4024	5.15	131
1.00	691462	4.20	3467	5.20	108
1.25	598706	4.25	2980	5.25	89
1.50	500000	4.30	2555	5.30	72
1.75	401294	4.35	2186	5.35	59
2.00	308537	4.40	1866	5.40	48
2.25	226627	4.45	1589	5.45	39
2.50	158655	4.50	1350	5.50	32
2.75	105650	4.55	1144	5.55	25.6
3.00	66807	4.60	968	5.60	20.7
3.25	40059	4.65	816	5.65	16.6
3.50	22750	4.70	687	5.70	13.4
3.60	17865	4.75	577	5.75	10.7
3.70	13904	4.80	483	5.80	8.5
3.75	12225	4.85	404	5.85	6.8
3.80	10724	4.90	337	5.90	5.4
3.90	8198	4.95	280	5.95	4.3
4.00	6210	5.00	233	6.00	3.4

Note This Table includes a 1.5 Sigma shift for all listed values of Z.

Table A.14 Critical values for the sign test

Test	α		
Two-tail	0.100	0.050	0.010
One-tail	0.050	0.025	0.005
n			
5	0		
6	0	0	
7	0	0	
8	1	0	0
9	1	1	0
10	1	1	0
11	2	1	0
12	2	2	1
13	3	2	1
14	3	2	1
15	3	3	2
16	4	3	2
17	4	4	2
18	5	4	3
19	5	4	3
20	5	5	3
21	6	5	4
22	6	5	4
23	7	6	4
24	7	6	5
25	7	7	5
26	8	7	6
27	8	7	6
28	9	8	6
29	9	8	7
30	10	9	7

The values of r_α, where $P(r \leq r_\alpha) = \alpha$, under H_0

Table A.15 Critical values for the Wilcoxon signed-rank test

Test	α			
Two-tail	0.100	0.050	0.020	0.010
One-tail	0.050	0.025	0.010	0.005
n				
4	0			
5	1	0		
6	2	1		
7	4	2	0	
8	6	4	1	
9	8	5	2	0
10	11	8	4	2
11	14	10	6	4
12	18	14	9	6
13	21	17	12	8
14	26	21	15	11
15	31	25	19	14
16	36	30	23	18
17	41	35	27	22
18	47	40	32	26
19	54	46	37	30
20	60	52	42	35
21	68	59	48	41
22	75	66	54	47
23	83	73	61	53
24	92	81	68	59
25	101	89	76	66

The values of w_α, where $P(w \leq w_\alpha) = \alpha$, under H_0

Table A.16 Critical values for the Mann-Whitney-Wilcoxon test

m	n α	2	3	4	5	6	7	8	9	10
2	0.01	4	6	8	10	12	14	16	18	20
	0.025	4	6	8	10	12	14	15	17	19
	0.05	4	6	8	9	11	13	14	16	18
	0.10	4	5	7	8	10	12	13	15	16
3	0.01		9	12	15	18	20	20	25	28
	0.025		9	12	14	16	19	21	24	26
	0.05		8	11	13	15	18	20	22	25
	0.10		7	10	12	14	16	18	21	23
4	0.01			16	19	22	26	29	32	36
	0.025			15	18	21	24	27	31	34
	0.05			14	17	20	23	26	29	32
	0.10			12	15	18	21	24	26	29
5	0.01				23	27	31	35	39	43
	0.025				22	26	29	33	37	41
	0.05				20	24	28	31	35	38
	0.10				19	22	26	29	32	36
6	0.01					32	37	41	46	51
	0.025					30	35	39	43	48
	0.05					28	33	37	41	45
	0.10					26	30	34	38	42
7	0.01						42	48	53	58
	0.025						40	45	50	55
	0.05						37	42	47	52
	0.10						35	39	44	48
8	0.01							54	60	66
	0.025							50	56	62
	0.05							48	53	59
	0.10							44	49	55
9	0.01								66	73
	0.025								63	69
	0.05								59	65
	0.10								55	61
10	0.01									80
	0.025									76
	0.05									72
	0.10									67

Note The critical values for the lower tail may be obtained by symmetry from the equation $u_{1-\alpha} = mn - u_\alpha$

Glossary

Accuracy The measured value that has little deviation from the actual value

Affinity diagrams The tools used to organize information and help achieve order out of the chaos that can develop in a brainstorming session

Analysis of variance A statistical method of identifying variations associated with various components

Arithmetic mean The most commonly used accuracy measure in Six Sigma project. It is defined as the ratio of sum of all observations to the total number of observations

Arrow diagram A tool to analyze the sequence of tasks necessary to complete a project and determine the critical tasks to monitor to execute the project efficiently. Also called activities network diagram

Balanced Scorecard The measurement tool used to assess degree of accomplishment of strategic results and the effectiveness of projects

Black Belt A Six Sigma role associated with an individual who is typically assigned full time to train and mentor Green Belts as well as lead improvement projects using specified methodologies such as DMAIC; DMADV and defined for Six Sigma (DFSS) tools, etc.

Brainstorming A useful and effective technique for idea generation, by the company's think tank, generally done in a group. These ideas are then scrutinized for their viability of execution and implementation thereafter

Capability maturity model integration A framework that organizes the project components used in generating models, training materials, and appraisal methods. This concept was first introduced and popularized by the software Engineering institute (SEI) of Carnegie Mellon University, Pittsburg

Cause and effect diagram Also called *Ishikava diagram* or *Fishbone diagram*, is to generate in a structured manner, maximum number of ideas regarding possible causes for a problem by using brainstorming technique

© Springer India 2015
K. Muralidharan, *Six Sigma for Organizational Excellence*,
DOI 10.1007/978-81-322-2325-2

Champions Champions are project sponsors, whose role is to ensure that the right projects are being identified and worked on, that teams are making good progress, and that the resources required for successful project completion are in place. Champions are generally guided by project managers in the organization

Coefficient of variation The ratio of standard deviation to the mean, used as a measure of consistency

Common cause variations The variation as the sum of the multitude of effects of a complex interaction of random or common causes. It is common to all occasions and places and always present in some degree

Competitive value matrix A method to depict the gaps between competitive products, brands, and services on the basis of two key drivers of value, quality, and price, and how the market evaluates each competitive offering

Competitive vulnerability matrix A powerful customer acquisition tool that identifies the low-hanging fruit—those customers who are not getting the value necessary to cement their loyalty to a competitor—and the basis of this poor value

Concordance studies Nominal measurement studies, which address repeatability and reproducibility

Confounding or aliasing A design technique for arranging a complete factorial experiment in blocks, where the block size is smaller than the number of treatment combinations in one replicate

Control charts A quality tool to monitor the stability of the process, determine when special cause is present, and when to take appropriate action

Correlation The extent of association between any two variables

Cost of quality The cost incurred in a process to assure that the quality standard is met. These costs do not add value to the customer of the process, but assure that high quality is achieved. Types of COQ include inspection and prevention costs. COQ should be kept as low as possible, while maintaining a high level of quality

Cost of poor quality The cost incurred when processes fail. These costs include the cost of repeating the process to get it right, costs to explain to the patient, MD, family what happened and how you will correct the problem, excessive malpractice costs, complaint management, risk management, legal costs, etc. These costs are categorized as Internal Failures and External Failures

CPM Acronym for critical path method enables one to find the shortest possible time in which to complete the project. It allows continuous monitoring of the schedule, which allows the project manager to track the critical activities, and alerts the project manager to the possibility that noncritical activities may be delayed beyond their total float, thus creating a new critical path and delaying project completion

Critical-to-quality They describe the requirements of quality in general, but lack the specificity to be measurable

Critical-to-cost They are similar to Critical to qualities but deal exclusively with the impact of cost on the customer

Critical-to-process They are typically the key process input variables (or independent variables)

Critical-to-safety They are stated customer needs regarding the safety of the product or process. Though identical to the CTQ and CTC, it is identified by the customer preference to quality

Critical-to-Delivery They represent those customers with stated needs regarding delivery

Customer The stakeholder who wants to use, influence, or is affected by the outcomes

Cycle time The time required to complete one cycle of an operation in a process, which includes the actual work time and wait time

Dashboard An easy to read, often single page, real-time user interface, showing a graphical presentation of the current status (snapshot) and historical trends of an organization's key performance indicators to enable instantaneous and informed decisions to be made at a glance

Defect Any characteristic of the product that fails to meet customer requirements

Defectives The total number of units containing different types of defects

Designed for Six Sigma A Six Sigma application of tools to product development and process design efforts with the goal of 'designing in' Six Sigma performance capability. DFSS is a methodology by which new products and services can be designed and implemented

DMADV The five-step approach (D-define, M-measure, A-analyze, D-design, V-validate) for redesign to improve a process

DMAIC The five-step (D-define, M-measure, A-analyze, I-improve, C-control), Six Sigma methodology to improvement a process by reducing variation. The hallmarks of this Six Sigma methodology are statistical methods and a strong commitment of executives and project managers

DMEDI A five-step (D-define, M-measure, E-explore, D-develop, I-implement) Lean Six Sigma approach to redesign a process to ensure efficiency or speed

DPU Defect per unit (see also defects)

DPMO Defects per million opportunities (see also defects)

Eco-Management and Audit Scheme A voluntary European-wide standard introduced by the European Union and applied to all European countries. The aim of EMAS is 'to recognize and reward those organizations that go beyond minimum legal compliance and continuously improve their environmental performance'

Estimate The observed value of the estimator. (See also estimator)

Estimator Any statistic used to estimate the value of an unknown parameter

Exhaustive events If two or more events together define the total sample space, the events are said to be collectively exhaustive events

Factor An independent variable or assignable cause that may affect the responses and of which different levels are included in the experiment. Factors are also known as explanatory variables, predictor variables, or input variables

Failure modes and effects analysis A set of guidelines, a process, and a form of identifying and prioritizing potential failures and problems in order to facilitate process improvement

Fault tree analysis A graphical means of evaluating the relationships between different parts of the system

Gantt chart A type of bar chart used in process/project planning and control to display planned work and finished work in relation to time

Goal A desired result to be achieved in a specified time

Goodness-of-fit-test A test to assess the adequacy of a particular distribution model

Green belt A Six Sigma role associated with an individual who retains his or her regular position within the firm but is trained in the tools, methods, and skills necessary to conduct Six Sigma improvement projects either individually or as part of larger teams

Green Six Sigma The qualitative and quantitative assessment of the direct and eventual environmental effects of all processes and products of an organization. The activities involve the systematic usage of infrastructure and manpower; optimum use of technology and accountability of sustainable business practices

GRPI model A diagnostic management tool to manage a project. Abbreviated for *Goals and responsibilities-Processes and procedures-Interpersonal relationships*

Histogram A visual display of the frequency distribution

House of quality A tool for depicting the relationship between quality metrics. The matrix includes all aspects of quality parameters extending beyond the technological solutions to management requirements housed on a particular performance barometer

Interrelationship diagram A diagram to identify the key factors for inclusion in a tree diagram

ISO An organization stand for the promotion of standards in the world with a view to facilitate international exchange of goods and services and to develop cooperation in the sphere of intellectual, scientific, technological, and economic activity including transfer of technology to developing countries

Kappa Useful for measuring both intra-rater and inter-rater rater agreement for nominal data

Kaizen A Japanese management concept of continuous improvement

Kanban A visual signal for overproduction. It eliminates wastes in a controlled way

Lean A quality improvement concept designed to reduce waste and process cycle time. Lean rules include focusing on the customer "aim", avoiding batching if customers will be delayed, waste elimination, and continuous flow / "pull". Lean is generally not as effective as a stand-alone approach, but rather incorporated into Six Sigma, due to Six Sigma's more robust execution method, DMAIC

Lean Six Sigma A quality improvement technique to reduce costs by eliminating product and process waste through a focused approach on eliminating non-value-added activities. See Lean above

Lower specification limit The smallest allowable value for quality characteristics

Master Black Belt (MBB) A Six Sigma role associated with an individual typically assigned full time to train and mentor Black Belts as well as lead the strategy to ensure improvement projects chartered are the right strategic projects for the organization. They often write and develop training materials, are heavily involved in project definition and selection, and work closely with business leaders called Champions

Matrix diagram A diagram to show the presence or absence of relationships among collected pairs of elements in a problem situation

Median An accuracy measure, which divides the distribution into two equal parts

Mean deviation Also called average deviation of a set of numbers is defined by the ratio of absolute difference of observations from mean to the number of observations

Mode An accuracy measure that is most frequently occurring value in a data set

Modular function deployment The method to establish customer requirements and to identify important design requirements with a special emphasis on modularity

Noise factor An independent variable that is difficult or too expensive to control as part of standard experimental conditions. Hence, they are also called random factors

Nominal scale It is a system of assigning number symbols to events in order to label them

Non-value-added activities The activities which generally do not contribute any value to the process activities and output

Non-value-added cost Same as COPQ. See COPQ definition above

Opportunity A value-added feature of a unit that should meet specifications proposed by the customer

Pareto chart A quality tool used for understanding the pattern of variations in attribute or categorical type data sets

PDCA Plan, Do, Check, Act cycle of improvement created by Walter Shewart in the 1930s, and enhanced by quality leaders like Edwards Deming and Joseph Juran over the years. The impact of effective use of PDCA concepts in an organization is that the culture evolves to one of high creativity, experimentation, and defeat of the status quo

PERT Acronym for project (or program) evaluation and review technique. It is one of the most commonly used techniques for project management control, facilitating critical path planning, and scheduling the project. It is based on representing the activities in a project by boxes (or nodes) that contain essential information calculated about the project

Poke Yoke A powerful method of eliminating waste, and controlling the inventories needed for a process. The philosophy allows building quality into the process and simplifies the identification of root causes of a particular problem

Population The entire set of units under study

PRESS Prediction error sum of squares, and is a measure of how well the model for the experiment is likely to predict in a new experiment

Prioritization matrix A matrix to evaluate options through a systematic approach of identifying weighing, and applying criteria to the options

Probability distribution A description of the possible values of a random variable and of their corresponding probabilities of occurrence

Project charter A document stating the purposes of the project

Process capability The determination of whether a process, with normal variation, is capable of meeting customer requirements or measure of the degree a process is/is not meeting customer requirements, compared to the distribution of the process

Precision The abridged range of an estimate of a characteristic

Process control The inspection of the items for analyzing quality problems and improving the performance of the production process

Process improvement A strategy of developing focused solutions so as to eliminate the root causes of business performance problems

Process management The totality of defined and documented processes, monitored on an ongoing basis, which ensure that measures are providing feedback on the flow and function of a process

Process performance A measure of actual results achieved by ongoing process

Primary data The information that is collected afresh and for the first time, and thus happens to be original in character

Project A chartered activity intended to achieve a stated result in a stated time. A project may vary in size and duration, involving a small group of people or large numbers in different parts of the organization. A project involves many processes and each such process progress with specific objectives

Project management A dynamic process that utilizes the appropriate resources of the organization in a controlled and structured manner to achieve some clearly defined objectives identifies as strategic needs conducted within a defined set of constraints

Project Charter A written declaration of the purpose and expected result of the project

Pugh Analysis A decision matrix, when a single option must be selected from several and multiple decisions

Quality function deployment The technique for documenting overall design logic. QFD is a method for translating customer requirements into an appropriate company program and technical requirements at each phase of the product realization cycle

Quartile deviation Half of the difference between the third quartile and the first quartile. Also called inter quartile range

Random sampling The procedure of drawing a sample randomly from a population

Random variable A real valued function associated with each outcome of a random experiment

Randomization A technique to assign treatments to experimental units so that each unit has an equal chance of being assigned a particular treatment, thus minimizing the effect of variation from uncontrolled noise factors

Range It is the difference between the largest and the smallest numbers in the set of observations

Redundancies Activities that are repeated at two points in the process, and can also be parallel activities that duplicate the same result

Reliability function (or survival function) The probability that a device will perform its intended function during a specified period of time under stated conditions

Repeatability The amount of variation due to measurement gauges and equipment

Reproducibility The amount of variation *due* to human (or appraisers) who use the gauges and equipment

Resolution In the context of experimental design, resolution refers to the level of confounding in a fractional factorial design

Response variable The output variable that shows the observed results or value of an experimental treatment. It is sometimes known as the dependent variable

Replication The repetition of the same treatment in a number of plots, purpose being to get an estimate of the error-variance of the experiment. Each repetition of the experiment is called a replicate. Replication increases the precision of the estimates of the effects in an experiment

Reverse logistics The process of planning, implementing, and controlling the efficient, cost-effective flow of raw materials, in-process inventory, finished goods, and related information from the point of consumption to the point of origin, for the purpose of recapturing value or proper disposal

Risk management A method of managing a project that focuses on identifying and controlling the areas or events that have the potential of creating and causing unwanted changes leading to unwanted results. Because of the complexity of risks it is impossible to derive a universal process for managing all risks in a project

Robustness The situation where the design of a product/process produces consistent, high-level performance, despite being subjected to a wide range of changing customer and manufacturing conditions

Secondary data The information that has already been collected and processed by some other agency

Significance level The probability of rejecting a true hypothesis

SMART A goal statement characteristic. Acronym for S-specific; M-measurable and measured; A-agreed upon by the manager and his/her supervisor; R-Realistic; yet stretch; T-time specific; results to be achieved by a defined date

Simulation methods A method of solving decision-making problems by designing, constructing, and manipulating a model of the real system

SIPOC diagram An extended version of the IPO (Input-Process-Output) diagram where Supplier-Customer interface is brought in for entire improvement activities of the organization

Six Sigma A disciplined, project-oriented, and methodology for eliminating defects in any process—from manufacturing to transactional and from product to service. Technically speaking, Six Sigma is described as a data-driven approach to reduce defects in a process or cut costs in a process or product, as measured by, "six standard deviations" between the mean and the nearest specification limits

Six Sigma marketing A fact-based data-driven disciplined approach to growing market share by providing targeted product/markets with superior value

Special cause variation The variations which are relatively large in magnitude and viable to identify, also called assignable causes of variation

Sponsor Someone who is accountable for the Six Sigma project and therefore is the appointed guardian of the project on behalf of the organization

Standard deviation The square root of the ratio of the sum of squares of observations taken from the mean to the number of observations and is denoted by σ

Standard error The standard deviation of the sampling distribution

Standard operating procedures A standard work instructions used anytime during the execution of a project

Statistical inference The process of drawing conclusions about the nature of some system on the basis of data subject to random variation

Statistical process control Methods for understanding, monitoring, and improving process performance over time

Statistical quality control Methods for understanding, monitoring, and identifying process variation (common cause and special cause) over time, and thereby, improving the quality of a process

SWOT analysis Acronym for strength-weakness-opportunities-threats analysis. The purpose of this analysis is to be able to generate ideas exploring the emerging opportunities, guarding against the threats, while keeping the organization's strengths and weaknesses in mind

Takt time The ratio of available production time to the rate of customer demand

Total quality management A strategy for implementing and managing quality improvement activities across organizations

Treatment A specific setting or combination of factor levels for an experimental unit

Tree diagram Also called systematic diagram and helps dissect a problem into subproblems and causes

RIZ Acronym for theory of inventive problem solving. A tactical planning of the processes

Upper specification limit The largest allowable value for quality characteristics

Value-added activities The activities that add value to the customers

Value stream All activities required to bring a product from conception to commercialization

Value stream mapping The mapping of all activities and process steps according to its time of occurrence required bringing a product, service, or capability to the client

Variance The square of the standard deviation and is denoted by σ^2

Voice of customer or voice of client A continuous process of collecting customer views on quality and can include customer needs, expectations, satisfaction, and perception

Voice of process The essential channel to establish a communication link between the customer requirements and the process parameters

Yield The total number of units handled correctly through the process steps

Printed in the United States
By Bookmasters